# LINUX
## with Operating
## System Concepts

# LINUX
## with Operating
## System Concepts

# Richard Fox

Northern Kentucky University
Highland Heights, Kentucky, USA

CRC Press
Taylor & Francis Group
Boca Raton   London   New York

CRC Press is an imprint of the
Taylor & Francis Group, an **informa** business

A CHAPMAN & HALL BOOK

CRC Press
Taylor & Francis Group
6000 Broken Sound Parkway NW, Suite 300
Boca Raton, FL 33487-2742

© 2015 by Taylor & Francis Group, LLC
CRC Press is an imprint of Taylor & Francis Group, an Informa business

No claim to original U.S. Government works

Printed on acid-free paper
Version Date: 20140623

International Standard Book Number-13: 978-1-4822-3589-0 (Paperback)

**Visit the Taylor & Francis Web site at**
**http://www.taylorandfrancis.com**

**and the CRC Press Web site at**
**http://www.crcpress.com**

*With all my love to*
*Cheri Klink, Sherre Kozloff, and Laura Smith*

# Contents

# Preface

Look around at the Unix/Linux textbook market and you find nearly all of the books target people who are looking to acquire hands-on knowledge of Unix/Linux, whether as a user or a system administrator. There are almost no books that serve as textbooks for an academic class. Why not? There are plenty of college courses that cover or include Unix/Linux. We tend to see conceptual operating system (OS) texts which include perhaps a chapter or two on Unix/Linux or Unix/Linux books that cover almost no OS concepts. This book has been written in an attempt to merge the two concepts into a textbook that can specifically serve a college course about Linux.

The topics could probably have been expanded to include more OS concepts, but at some point we need to draw the line. Otherwise, this text could have exceeded 1000 pages! Hopefully we have covered enough of the background OS concepts to satisfy all students, teachers, and Linux users alike.

Another difference between this text and the typical Unix/Linux text is the breadth of topics. The typical Unix/Linux book either presents an introduction to the OS for users or an advanced look at the OS for system administrators. This book covers Linux from both the user and the system administrator position. While it attempts to offer thorough coverage of topics about Linux, it does not cover some of the more advanced administration topics such as authentication servers (e.g., LDAP, Kerberos) or network security beyond a firewall, nor does it cover advanced programming topics such as hacking open source software.

Finally, this book differs from most because it is a textbook rather than a hands-on, how-to book. The book is complete with review sections and problems, definitions, concepts and when relevant, foundational material (such as an introduction to binary and Boolean logic, an examination of OS kernels, the role of the CPU and memory hierarchy, etc.).

Additional material is available from the CRC Website: http://crcpress.com/product/isbn/9781482235890.

# Acknowledgments and Contributions

Fɪʀsᴛ, I ᴀᴍ ɪɴᴅᴇʙᴛᴇᴅ to Randi Cohen and Stan Wakefield for their encouragement and patience in my writing and completing this text. I would also like to thank Gary Newell (who was going to be my coauthor) for his lack of involvement. If I was working with Gary, we would still be writing this! I would like to thank my graduate student officemate from Ohio State, Peter Angeline, for getting me into Unix way back when and giving me equal time on our Sun workstation.

I am indebted to the following people for feedback on earlier drafts of this textbook.

- Peter Bartoli, San Diego State University

- Michael Costanzo, University of California Santa Barbara

- Dr. Aleksander Malinowski, Bradley University

- Dr. Xiannong Meng, Bucknell University

- Two anonymous reviewers

I would like to thank the NKU (Northern Kentucky University) students from CIT 370 and CIT 371 who have given me feedback on my course notes, early drafts of this textbook, and the labs that I have developed, and for asking me questions that I had to research answers to so that I could learn more about Linux. I would like to specifically thank the following students for feedback: Sohaib Albarade, Abdullah Saad Aljohani, Nasir Al Nasir, Sean Butts, Kimberly Campbell, Travis Carney, Jacob Case, John Dailey, Joseph Driscoll, Brenton Edwards, Ronald Elkins, Christopher Finke, David Fitzer, James Forbes, Adam Foster, Joshua Frost, Andrew Garnett, Jason Guilkey, Brian Hicks, Ali Jaouhari, Matthew Klaybor, Brendan Koopman, Richard Kwong, Mousa Lakshami, David Lewicki, Dustin Mack, Mariah Mains, Adam Martin, Chris McMillan, Sean Mullins, Kyle Murphy, Laura Nelson, Jared Reinecke, Jonathan Richardson, Joshua Ross, Travis Roth, Samuel Scheuer, Todd Skaggs, Tyler Stewart, Matthew Stitch, Ayla Swieda, Dennis Vickers, Christopher Witt, William Young, John Zalla, and Efeoghene Ziregbe.

I would like to thank Charlie Bowen, Chuck Frank, Wei Hao, Yi Hu, and Stuart Jaskowiak for feedback and assistance. I would like to thank Scot Cunningham for teaching me about network hardware.

I would also like to thank everyone in the open source community who contribute their time and expertise to better all of our computing lives.

On a personal note, I would like to thank Cheri Klink for all of her love and support, Vicki Uti for a great friendship, Russ Proctor for his ear whenever I need it, and Jim Hughes and Ben Martz for many hours of engaging argumentation.

# How to Use This Textbook

THIS TEXTBOOK IS ENVISIONED for a 1- or 2-semester course on Linux (or Unix). For a 1-semester course, the book can be used either as an introduction to Linux or as a system administration course. If the text is used for a 1-semester junior or senior level course, it could potentially cover both introduction and system administration. In this case, the instructor should select the most important topics rather than attempting complete coverage. If using a 2-semester course, the text should be as for both introduction and system administration.

The material in the text builds up so that each successive chapter relies, in part, on previously covered material. Because of this, it is important that the instructor follows the chapters in order, in most cases. Below are flow charts to indicate the order that chapters should or may be covered. There are three flow charts; one each for using this text as a 1-semester introductory course, a 1-semester system administration course, and a 1-semester course that covers both introduction and system administration. For a 2-semester sequence, use the first two flow charts; one per semester. In the flow charts, chapters are abbreviated as "ch." Chapters appearing in brackets indicate optional material.

1-semester introductory course:

$$\text{ch 1} \rightarrow \text{ch 2} \rightarrow [\text{appendix}] \rightarrow \text{ch 3} \rightarrow \begin{bmatrix} \text{ch 4} \\ \\ \text{ch 5} \end{bmatrix} \rightarrow \text{ch 6} \rightarrow \text{ch 7} \rightarrow \text{ch 8}$$

1-semester system administration course:

$$\begin{bmatrix} [\text{ch 5}] \\ [\text{ch 6}] \\ [\text{ch 7}] \end{bmatrix} \rightarrow \text{ch 8} \rightarrow \text{ch 9} \rightarrow \text{ch 10} \rightarrow \text{ch 11} \rightarrow [\text{appendix}]$$
$$\rightarrow \text{ch 12} \rightarrow \text{ch 13} \rightarrow \text{ch 14} \rightarrow \text{ch 15}$$

1-semester combined course:

$$[\text{ch 1}] \rightarrow \text{appendix}^* \rightarrow \text{ch 2} \rightarrow \text{ch 3} \rightarrow \text{ch 4} \rightarrow \text{ch 5}^* \rightarrow \text{ch 6}^* \rightarrow \text{ch 7}$$
$$\rightarrow \text{ch 8}^* \rightarrow \text{ch 9} \rightarrow \text{ch 10} \rightarrow \text{ch 11} \rightarrow \text{ch 12} \rightarrow \text{ch 13}^* \rightarrow \text{ch 14}$$

As the 1-semester combined course will be highly compressed, it is advisable that chapters denoted with an asterisk (*) receive only partial coverage. For Chapter 5, it is recommended that only vi be covered. For Chapter 6, it is recommended that the instructor skips the more advanced material on *sed* and *awk*. In Chapter 8, only cover operating system installation. In Chapter 13, the instructor may wish to skip material on open source installation and/or gcc.

The following topics from the chapters are considered optional. Omitting them will not reduce the readability of the remainder of the text.

- Chapter 1: the entire chapter (although a brief review of the Gnome GUI would be useful)

- Chapter 2: interpreters

- Chapter 3: Linux file system structure (if you are also going to cover Chapter 10); secondary storage devices; compression algorithms

- Chapter 5: all material excluding vi

- Chapter 6: sed and awk

- Chapter 8: the Linux kernel; virtual memory, SELinux

- Chapter 12: writing your own network scripts

- Chapter 13: the gcc compiler; the Open Source Movement

- Chapter 14: disaster planning and recovery

- Chapter 15: the entire chapter

This textbook primarily describes Red Hat version 6 and CentOS version 2.6. As this textbook is being prepared for publication, new versions of Linux are being prepared by both Red Hat (version 7) and CentOS (version 2.7). Although most of these changes do not impact the material in this textbook, there are a few instances where version 7 (and 2.7) will differ from what is described here. See the author's website at www.nku.edu/~foxr/linux/ which describes the significant changes and provides additional text to replace any outdated text in this book.

# Author

**R**ICHARD FOX IS A PROFESSOR of computer science at Northern Kentucky University (NKU). He regularly teaches courses in both computer science (artificial intelligence, computer systems, data structures, computer architecture, concepts of programming languages) and computer information technology (IT fundamentals, Unix/Linux, web server administration). Dr. Fox, who has been at NKU since 2001, is the current chair of NKU's University Curriculum Committee. Prior to NKU, Dr. Fox taught for nine years at the University of Texas—Pan American. He has received two Teaching Excellence awards, from the University of Texas—Pan American in 2000 and from NKU in 2012.

Dr. Fox received a PhD in computer and information sciences from Ohio State University in 1992. He also has an MS in computer and information sciences from Ohio State (1988) and a BS in computer science from University of Missouri Rolla (now Missouri University of Science and Technology) from 1986.

Dr. Fox published an introductory IT textbook in 2013 (also from CRC Press/Taylor & Francis). He is also author or coauthor of over 45 peer-reviewed research articles primarily in the area of artificial intelligence. He is currently working on another textbook with a colleague at NKU on Internet infrastructure.

Richard Fox grew up in St. Louis, Missouri and now lives in Cincinnati, Ohio. He is a big science fiction fan and progressive rock fan. As you will see in reading this text, his favorite composer is Frank Zappa.

# CHAPTER 1

# Introduction to Linux

THIS CHAPTER'S LEARNING OBJECTS are

- To understand the appeal behind the Linux operating system

- To understand what operating systems are and the types of things they do for the user

- To know how to navigate using the two popular Linux GUIs (graphical user interfaces)

- To understand what an interpreter is and does

- To understand virtual machines (VMs)

- To understand the history of Unix and Linux

- To understand the basic components of a computer system

## 1.1 WHY LINUX?

If you have this textbook, then you are interested in Linux (or your teacher is making you use this book). And if you have this textbook, then you are interested in computers and have had some experience in using computers in the past. So why should you be interested in Linux? If you look around our society, you primarily see people using Windows, Macintosh OS X, Linux, and Unix in one form or another. Below, we see an estimated breakdown of the popularity of these operating systems as used on desktop computers.*

- Microsoft Windows: 90.8%

  - Primarily Windows 7 but also Vista, NT, XP, and Windows 8

- Macintosh: 7.6%

  - Mac OS X

---

* Estimates vary by source. These come from New Applications as of November 2013. However, they can be inaccurate because there is little reporting on actual operating system usage other than what is tracked by sales of desktop units and the OS initially installed there.

- Linux or Unix: 1.6%

   - There are numerous distributions or "flavors" of Linux and Unix

If we expand beyond desktop computers, we can find other operating systems which run on mainframe computers, minicomputers, servers, and supercomputers. Some of the operating systems for mainframe/minicomputer/server/supercomputer are hardware-specific such as IBM System i (initially AS/200) for IBM Power Systems. This harkens back to earlier days of operating systems when each operating system was specific to one platform of hardware (e.g., IBM 360/370 mainframes). For mobile devices, there are still a wide variety of systems but the most popular two are iOS (a subset of Mac OS X) and Android (Linux-based).

With such a small market share for Linux/Unix, why should we bother to learn it? Certainly for your career you will want to learn about the more popular operating system(s). Well, there are several reasons to learn Linux/Unix.

First, the Macintosh operating system (Mac OS X) runs on top of a Unix operating system (based on the Mach Unix kernel). Thus, the real market share is larger than it first appears. Additionally, many of the handheld computing devices run a version of Linux (e.g., Google Android). In fact, estimates are that Android-based devices make up at least 20% of the handheld device market. There are other hardware devices that also run Linux or Unix such as firewalls, routers, and WiFi access points. So Linux, Unix, Mac OS X, and the operating systems of many handheld devices comes to more than 15% of the total operating system usage.*

Second, and more significant, is the fact that a majority of the servers that run on the Internet run Linux or Unix. In 2011, a survey of the top 1 million web servers indicated that nearly 2/3s ran Unix/Linux while slightly more than 1/3 ran Microsoft Windows. The usage of Unix/Linux is not limited to web servers as the platform is popular for file servers, mail servers, database servers, and domain name system servers.

That still might not explain why you should study Linux. So consider the following points.

- Open source (free)—The intention of open source software is to make the source code available. This gives programmers the ability to add to, enhance or alter the code and make the new code available. It is not just the Linux operating system that is open source but most of its application software. Enhancements can help secure software so that it has fewer security holes. Additions to the software provide new features. Alterations to software give users choices as to which version they want to use. Linux, and all open source software, continues to grow at an impressive rate as more and more individuals and groups of programmers contribute. As a side effect, most open source software is also freely available. This is not true of all versions of Linux (e.g., Red Hat Enterprise Linux is a commercial product) but it is true of most versions.

---

* Other sources of operating system statistics aside from Net Applications include NetMarketShare, StatCounter, ComputerWorld, AT Internet, canalys, Gartner, zdnet, and Linux Journal. See, for instance, marketshare.hitslink.com.

- Greater control—Although Windows permits a DOS shell so that the user can enter commands, the DOS commands are limited in scope. In Linux/Unix, the command line is where the power comes from. As a user, you can specify a wide variety of options in the commands entered via the command line and thus control the operating system with greater precision. Many users become frustrated with the Windows approach where most control is removed from the users' hands. The Windows approach is one that helps the naive user who is blissfully unaware of how the operating system works. In Linux, the user has the option to use the GUI but for those who want to be able to more directly control the operating system rather than be controlled by the operating system, the command line gives that level of access.

- Learning about operating systems—As stated in the last bullet point, Windows helps shelter users from the details of what the operating system is doing. This is fine for most users who are using a computer to accomplish some set of tasks (or perhaps are being entertained). But some users will desire to learn more about the computer. With the command line interface (CLI) in Unix/Linux, you have no choice but to learn because that is the only way to know how to use the command line. And learning is fairly easy with the types of support available (e.g., man pages).

- It's cool—well, that depends upon who you are.

In this textbook, we primarily concentrate on the CentOS 6 version of Red Hat Linux.* As will be discussed later in this chapter, there are hundreds of Linux distributions available. One reason to emphasize CentOS is that it is a typical version of Red Hat, which itself is one of the most popular versions of Linux. Additionally, CentOS is free, easy to install, and requires less resources to run efficiently and effectively than other operating systems.

## 1.2 OPERATING SYSTEMS

The earliest computers had no operating system. Instead, the programmer had to define the code that would launch their program, input the program from storage or punch cards and load it in memory, execute the program and in doing so, retrieve input data from storage or punch cards, and send output to storage. That is, the programmer was required to implement all aspects of running the program. At that time, only one program would run at a time and it would run through to completion at which point another programmer could use the computer.

By the late 1950s, the process of running programs had become more complicated. Programs were often being written in *high-level programming languages* like FORTRAN, requiring that the program to be run first be translated into machine language before it could be executed. The translation process required running another program, the *compiler*. So the programmer would have to specify that to run their program, first they would have to load and run the FORTRAN compiler which would input their program. The FORTRAN

---

* At the time of writing, the current version of CentOS is 6.4. This is the version primarily covered in this chapter.

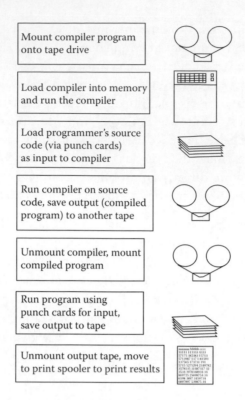

FIGURE 1.1   Compiling and running a program.

compiler would execute, storing the result, a compiled or machine language program, onto tape. The programmer would then load their program from tape and run it. See the process as illustrated in Figure 1.1. In order to help facilitate this complex process, programmers developed the *resident monitor*. This program would allow a programmer to more easily start programs and handle simple input and output. As time went on, programmers added more and more utilities to the resident monitor. These became operating systems.

An operating system is a program. Typically it comprises many parts, some of which are required to be resident in memory, others are loaded upon demand by the operating system or other applications software. Some components are invoked by the user. The result is a layered series of software sitting on top of the hardware.

The most significant component of the operating system is the *kernel*. This program contains the components of the operating system that will stay resident in memory. The kernel handles such tasks as process management, resource management, memory management, and input/output operations. Other components include services which are executed on-demand, device drivers to communicate with hardware devices, tailored user interfaces such as shells, personalized desktops, and so on, and operating system utility programs such as antiviral software, disk defragmentation software, and file backup programs.

The operating system kernel is loaded into memory during the computer's booting operation. Once loaded, the kernel remains in memory and handles most of the tasks issued to the operating system. In essence, think of the kernel as the interface between the users

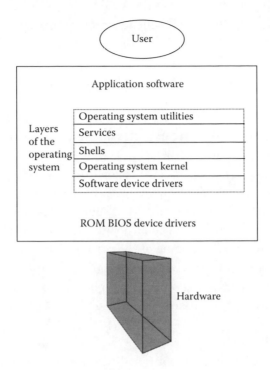

FIGURE 1.2    Layers of the computer system.

and application software, and the hardware of the computer system. Figure 1.2 illustrates the various levels of the computer system from the user at the top to the hardware at the bottom. The idea is that the user controls the computer through layers of software and hardware components. First, the user interfaces with the running application software. The user might for instance select Open and then select a file to open in the software. The application software then communicates with the OS kernel through system calls. The kernel communicates with the hardware via device drivers. Further, the user might communicate directly with the operating system through services and an OS shell. We will explore operating system concepts in more detail as we move throughout the chapter.

## 1.3 THE LINUX OPERATING SYSTEM: GUIs

As with most operating system today, Linux can be operated through a GUI. Many users only have experience with a GUI and may not even be aware that there is a text-based (command-line) interface available. It is assumed, because you are using this book, that you want to understand the computer's operating system at more than a cursory level. We will want to explore beyond the GUI. In Linux, it is very common to use the CLI as often, or even more often, than the GUI. This is especially true of a system administrator because many tasks can only be performed through the CLI. But for now, we will briefly examine the Linux GUI.

The two common GUIs in Red Hat Linux are Gnome and KDE. Both of these GUIs are built on top of an older graphical interface system called X-Windows. X-Windows, or the X Window System, was first developed at MIT in 1984. It is free and open source.

The windowing system is architecture-independent meaning that it can be utilized on any number of different platforms. It is also based on a client–server model where the X Server responds to requests from X Clients to generate and manipulate various graphical items (namely windows). The clients are applications that use windows to display their contents and interact with the users. The clients typically run on the same computer as the server but can also run remotely allowing a server to send its windowing information to computers over a network. This allows the user to access a remote computer graphically. The current version of X Windows is known as X11.

With X Windows implemented, users have a choice between using X or a more advanced GUI that runs on top of (or is built on top of) X Windows. In our case, we will examine two such windowing systems: Gnome and KDE. Both GUIs offer very similar features with only their appearances differing. And although each comes with its own basic set of application software, you can access some of the software written for one interface in the other interface (for instance, many KDE applications will run from within Gnome). We will briefly examine both windowing systems here but it is assumed that the reader will have little trouble learning either windowing system.

### 1.3.1 User Account and Logging In

To use Linux, you must of course have it installed on your computer. Many people will choose to install Linux as a dual booting operating system. This means that when you first boot your computer, you have a choice of which operating system to boot to. In most cases, the dual booting computers have a version of Windows and (at least) one version of Linux. You can also install Linux as the only operating system on your computer. Alternatively, you can install Linux inside of a VM. VMs are described in Section 1.5 of this chapter.

To use Linux, you must also have a user account on the machine you are seeking to access. User accounts are installed by the system administrator. Each user account comes with its own resources. These include a home directory and disk space, a login shell (we cover shells in Chapter 2), and access rights (we cover permissions in Chapter 3). Every account comes with a username and an associated password that only the user should know. We explore the creation of user accounts and other aspects of user accounts in Chapter 9.

After your computer has booted to Linux, you face a login screen. The login mechanism supports operating system security to ensure that the user of the computer (1) is an authorized user, and (2) is provided certain access rights. Authentication is performed through the input of the user's username and password. These are then tested against the stored passwords to ensure that the username and password match.

In CentOS 6, the login screen provides the user with a list of the system's users. If your username appears in this list, you can click on your name to begin the login process. Alternatively, if your name does not appear, you must click on the "Other…" selection which then asks you to enter your user name in a textbox. In either case, you are then asked to enter your password in a textbox. Figure 1.3 illustrates this first login screen.

If you select "Other…," the login window changes to that shown in Figure 1.4. Now you are presented with a textbox to enter your user name. At this point, you may either click on "Log In" or press the enter key.

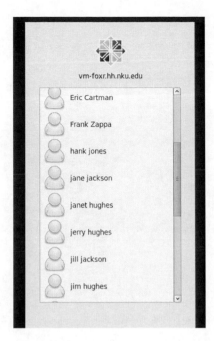

FIGURE 1.3    Login screen.

The login process is complete if the user name and password match the stored password information. If not, you will be asked to try to log in again.

The default is to log you into the GUI. However, you can also set up the login program to log you into a simple text-based, single window for command line interaction. We will concentrate on the two most common GUIs in Red Hat Linux, Gnome, and KDE here.

### 1.3.2  Gnome

The Gnome desktop is shown in Figure 1.5. Along the top of the desktop are menus: Applications, Places, System. These menus provide access to most of the applications software, the file system, and operating system settings, respectively. Each menu contains

FIGURE 1.4    Logging in by user name.

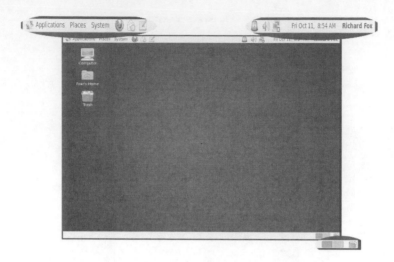

FIGURE 1.5    Gnome desktop.

submenus. The Applications menu consists of Accessories, Games, Graphics, Internet, Office, Programming, Sound & Video, and System Tools.

The Accessories submenu will differ depending on whether both Gnome and KDE are installed, or just Gnome. This is because the KDE GUI programs are available in both windowing systems if they are installed.

Two accessories found under this submenu are the Archive Manager and Ark, both of which allow you to place files into and extract from archives, similar to Winzip in Windows. Other accessories include two calculators, several text editors (gedit, Gnote, KNotes, KWrite), two alarm clocks, a dictionary, and a screen capture tool. Figure 1.6 shows the full Accessories menu, showing both Gnome-oriented GUI software and KDE-oriented GUI software (KDE items' names typically start with a "K"). The KDE GUI software would not be listed if KDE accessories have not been installed.

The Games, Graphics, and Internet menus are self-explanatory. Under Internet, you will find the Firefox web browser as well as applications in support of email, ftp, and voice over IP telephony and video conferencing. Office contains OpenOffice, the open source version of a productivity software suite. OpenOffice includes Writer (a word processor), Calc (a spreadsheet), Draw (drawing software), and Impress (presentation graphics software). The items under the Programming menu will also vary but include such development environments as Java and C++. Sound & Video contain multimedia software.

Finally, System Tools is a menu that contains operating system utilities that are run by the user to perform maintenance on areas accessible or controllable by the user. These include, for instance, a file browser, a CD/DVD creator, a disk utility management program, and a selection to open a terminal window, as shown in Figure 1.7. To open a terminal window, you can select Terminal from the System Tools menu. If KDE is available, you can also select Konsole, the KDE version of a terminal emulator.

The next menu, Places, provides access to the various storage locations available to the user. These include the user's home directory and subdirectories such as Documents,

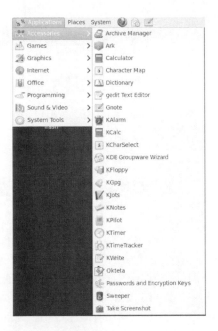

FIGURE 1.6    Accessories menu.

Pictures, Downloads; Computer to provide access to the entire file system; Network, to provide access to file servers available remotely; and Search for Files… which opens a GUI tool to search for files. The Places menu is shown in Figure 1.8. Also shown in the figure is the Computer window, opened by selecting Computer from this menu. You will notice two items under Computer, CD/DVD Drive, and Filesystem. Other mounted devices would also appear here should you mount any (e.g., flash drive).

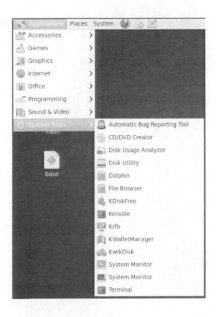

FIGURE 1.7    System tools submenu.

FIGURE 1.8   Places menu.

Selecting any of the items under the Places menu brings up a window that contains the items stored in that particular place (directory). These will include devices, (sub)directories, and files. There will also be links to files, which in Linux, are treated like files. Double clicking on any of the items in the open window "opens" that item as another window. For instance, double clicking on Filesystem from Figure 1.8 opens a window showing the contents of Filesystem. Our filesystem has a top-level name called "/" (for root), so you see "/" in the title bar of the window rather than "Filesystem." The window for the root file system is shown in Figure 1.9. From here, we see the top-level Linux directories (we will discuss these in detail in Chapters 3 and 10). The subdirectories of the root filesystem are predetermined in Linux (although some Linux distributions give you different top-level directories). Notice that two folders have an X by them: lost+found and root. The X denotes that these folders are not accessible to the current user.

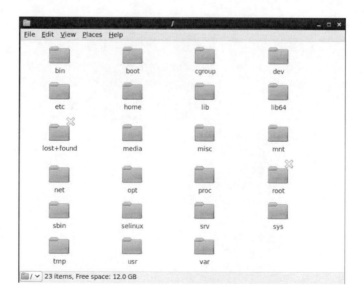

FIGURE 1.9   Filesystem folder.

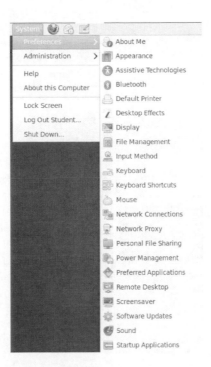

FIGURE 1.10   Preferences submenu under system menu.

The final menu in the Gnome menu bar is called System. The entries in this menu allow the user to interact with the operating system. The two submenus available are Preferences and Administration. Preferences include user-specific operating preferences such as the desktop background, display resolution, keyboard shortcuts, mouse sensitivity, screen saver, and so forth. Figure 1.10 illustrates the Preferences submenu.

Many of the items found under the Administration submenu require system administrator privilege to access. These selections allow the system administrator to control some aspect of the operating system such as changing the firewall settings, adding or deleting software, adding or changing users and groups, and starting or stopping services. These settings can also be controlled through the command line. In some cases in fact, the command line would be easier such as when creating dozens or hundreds of new user accounts. The Administration submenu is shown in Figure 1.11.

The System menu includes commands to lock the screen, log out of the current user, or shut the system down. Locking the screen requires that the user input their password to resume use.

There are a number of other features in the desktop of Gnome. Desktop icons, similar to Windows shortcut icons, are available for the Places entries of Computer, the user's home directory, and a trash can for deleted files. Items deleted via the trash can are still stored on disk until the trash is emptied. You can add desktop icons as you wish. From the program's menu (e.g., Applications), right click on your selection. You can then select "Add this launcher to desktop." Right clicking in the desktop brings up a small pop-up

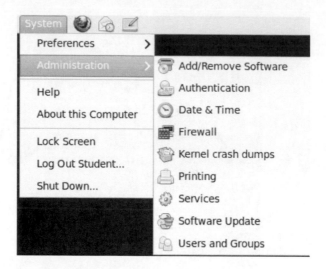

FIGURE 1.11   Administration submenu.

menu that contains a few useful choices, primarily the ability to open a terminal window. Figure 1.12 shows this menu.

Running along the top of the desktop are a number of icons. These are similar to desktop icons. Single clicking on one will launch that item. In the case of the already existing icons, these pertain to application software. You can add to the available icons by selecting an application from its menu, right clicking on it and selecting "Add this launcher to panel."

As seen in Figure 1.5 (or Figure 1.11), the preexisting icons are (from left to right) the Firefox web browser, Evolution email browser, and Gnote text editor. Toward the right of the top bar are a number of other icons to update the operating system, control speaker volume, network connectivity, the current climate conditions, date and time, and the name of the user. Clicking on the user's name opens up a User menu. From this menu, you can obtain user account information, lock the screen, switch user or quit, which allows the user to shut down the computer, restart the computer, or log out.

FIGURE 1.12   Desktop menu.

At the bottom of the desktop is a panel that can contain additional icons. At the right of the panel are two squares, one blue and one gray. These represent two different desktops, or "workspaces." The idea is that Gnome gives you the ability to have two (or more) separate work environments. Clicking on the gray square moves you to the other workspace. It will have the same desktop icons, the same icons in the menu bar, and the same background design, but any windows open in the first workspace will not appear in the second. Thus, you have twice the work space. To increase the number of workspaces, right click on one of the workspace squares and select Properties. This brings up the Workspace Switcher Preferences window from which you can increase (or decrease) the number of workspaces.

### 1.3.3 KDE Desktop Environment

From the login screen you can switch desktops. Figure 1.13 shows that at the bottom of the login window is a panel that allows you to change the language and the desktop. In this case, we can see that the user is selecting the KDE desktop (Gnome is the default). There is also a selection to change the accessibility preferences (the symbol with the person with his or her arms outspread) and examine boot messages (the triangle with the exclamation point). The power button on the right of the panel brings up shutdown options.

Although the Gnome environment is the default GUI for Red Hat, we will briefly also look at the KDE environment, which is much the same but organized differently. Figure 1.14 shows the desktop arrangement. One of the biggest differences is in the placement of the menus. Whereas Gnome's menus are located at the top of the screen, KDE has access to menus along the bottom of the screen.

There are three areas in the KDE Desktop that control the GUI. First, in the lower left-hand corner are three icons. The first, when clicked on, opens up the menus. So, unlike Gnome where the menus are permanently part of the top bar, KDE uses an approach more like Windows "Start Button menu." Next to the start button is a computer icon to open up the Computer folder, much like the Computer folder under the Places menu in Gnome. The next icon is a set of four rectangles. These are the controls to switch to different workspaces. The default in KDE is to have four of them.

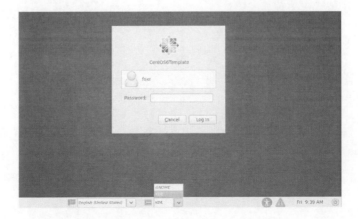

FIGURE 1.13  Selecting the KDE desktop.

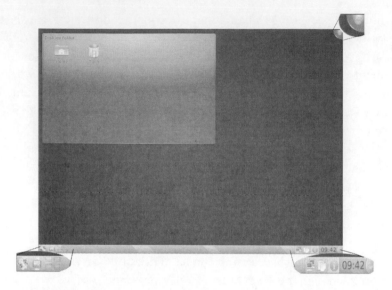

FIGURE 1.14    KDE desktop.

On the bottom right of the screen, are status icons including the current time of day and icons that allow you to manipulate the network, the speaker, and so on. In the upper right-hand corner is a selection that brings up a menu that allows you to tailor the desktop by adding widgets and changing desktop and shortcut settings.

Figure 1.15 illustrates the start button menu. First, you are told your user name, your computer's host name, and your "view" (KDE, Desktop). A search bar is available to search

FIGURE 1.15    Start button applications menu.

for files. Next are the submenus. Much like Gnome, these include Graphics, Internet, Multimedia, and Office. Settings are similar to Preferences from Gnome while System and Utilities make up a majority of the programs found under Administration in Gnome. Find Files/Folders is another form of search. Finally, running along the bottom of this menu are buttons which change the items in the menu. Favorites would list only those programs and files that the user has indicated should be under Favorites while Computer is similar to the Computer selection from the Places menu in Gnome. Recently Used lists those files that have been recently accessed. Finally, Leave presents a pop-up window providing the user with options to log out, switch user, shut down, or restart the computer.

Finally, as with Gnome, right clicking in the desktop brings up a pop-up menu. The options are to open a terminal window using the Konsole program, to open a pop-up window to enter a single command, to alter the desktop (similar to the menu available from the upper right-hand corner), or to lock or leave the system.

The decision of which GUI to use will be a personal preference. It should also be noted that both Gnome and KDE will appear somewhat differently in a non-Red Hat version of Linux. For instance, the Ubuntu version of Gnome offers a series of buttons running along the left side of the window. To learn the GUI, the best approach is to experiment until you are familiar with it.

## 1.4 THE LINUX COMMAND LINE

As most of the rest of the text examines the CLI, we only present a brief introduction here. We will see far greater detail in later chapters.

### 1.4.1 The Interpreter

The CLI is part of a *shell*. The Linux shell itself contains the CLI, an *interpreter*, and an environment of previously defined entities like functions and variables. The interpreter is a program which accepts user input, interprets the command entered, and executes it.

The interpreter was first provided as a mechanism for programming. The programmer would enter one program instruction (command) at a time. The interpreter takes each instruction, translates it into an executable statement, and executes it. Thus, the programmer writes the program in a piecemeal fashion. The advantage to this approach is that the programmer can experiment *while* coding. The more traditional view of programming is to write an entire program and then run a compiler to translate the program into an executable program. The executable program can then be executed by a user at a later time.

The main disadvantage of using an interpreted programming language is that the translation task can be time consuming. By writing the program all at once, the time to compile (translate) the program is paid for by the programmer all in advance of the user using the program. Now, all the user has to do is run the executable program; there is no translation time for the user. If the program is interpreted instead, then the translation task occurs at run-time every time the user wants to run the program. This makes the interpreted approach far less efficient.

While this disadvantage is significant when it comes to running applications software, it is of little importance when issuing operating system commands from the command line. Since the user is issuing one command at a time (rather than thousands to millions of program instructions), the translation time is almost irrelevant (translation can take place fairly quickly compared to the time it takes the user to type in the command). So the price of interpretation is immaterial. Also consider that a program is written wholly while your interaction with the operating system will most likely vary depending upon the results of previous instructions. That is, while you use your computer to accomplish some task, you may not know in advance every instruction that you will want to execute but instead will base your next instruction on feedback from your previous instruction(s). Therefore, a compiled approach is not appropriate for operating system interaction.

## 1.4.2 The Shell

The interpreter runs in an environment consisting of previously defined terms that have been entered by the user during the current session. These definitions include instructions, command shortcuts (called aliases), and values stored in variables. Thus, as commands are entered, the user can call upon previous items still available in the interpreter's environment. This simplifies the user's interactions with the operating system.

The combination of the interpreter, command line, and environment make up what is known as a shell. In effect, a shell is an interface between the user and the core components of the operating system, known as the kernel (refer back to Figure 1.2). The word shell indicates that this interface is a layer outside of, or above, the kernel. We might also view the term shell to mean that the environment is initially empty (or partially empty) and the user is able to fill it in with their own components. As we will explore, the components that a user can define can be made by the user at the command line but can also be defined in separate files known as scripts.

The Linux environment contains many different definitions that a user might find useful. First, commands are stored in a history list so that the user can recall previous commands to execute again or examine previous commands to help write new commands. Second, users may define aliases which are shortcuts that permit the user to specify a complicated instruction in a shortened way. Third, both software and the user can define variables to store the results of previous instructions. Fourth, functions can be defined and called upon. Not only can commands, aliases, variables and functions be entered from the command line, they can also be placed inside scripts. A script can then be executed from the command line as desired.

The Linux shell is not limited to the interpreter and environment. In addition, most Linux shells offer a variety of shortcuts that support user interaction. For instance, some shells offer command line editing features. These range from keystrokes that move the cursor and edit characters of the current command to tab completion so that a partially specified filename or directory can be completed. There are shortcut notations for such things as the user's home directory or the parent directory. There are also wild card characters that can be used to express "all files" or "files whose names contain these letters," and so on. See Figure 1.16 which illustrates the entities that make up a shell. We will examine the Bash shell in detail in Chapter 2. We will also compare it to other popular shells such as csh (C-shell) and tcsh (T-shell).

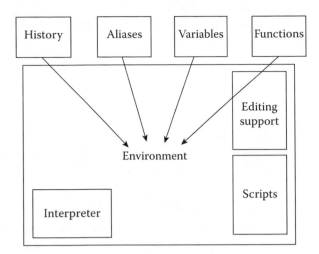

FIGURE 1.16  Components of a Linux shell.

A shell is provided whenever a user opens a terminal window in the Linux GUI. The shell is also provided when a user logs into a Linux system which only offers a text-based interface. This is the case if the Linux system is not running a GUI, or the user chooses the text-based environment, or the user remotely logs into a Linux system using a program like telnet or ssh. Whichever is the case, the user is placed inside of a shell. The shell initializes itself which includes some predefined components as specified by initialization scripts. Once initialized, the command line prompt is provided to the user and the user can now begin interacting with the shell. The user enters commands and views the responses. The user has the ability to add to the shell's components through aliases, variables, commands, functions, and so forth.

### 1.4.3  The CLI over the GUI

As a new user of Linux you might wonder why you should ever want to use the CLI over the GUI. The GUI is simpler, requires less typing, and is far more familiar to most users than the command line. The primary difference between using the GUI and the CLI is a matter of convenience (ease) and control.

In favor of the GUI:

- *Less to learn*: Controlling the GUI is far easier as the user can easily memorize movements; commands entered via the command line are often very challenging to remember as their names are not necessarily descriptive of what they do, and many commands permit complex combinations of options and parameters.

- *Intuitiveness*: The idea of moving things by dragging, selecting items by pointing and clicking, and scrolling across open windows becomes second nature once the user has learned the motions; commands can have archaic syntax and cryptic names and options.

- *Visualization*: As humans are far more comfortable with pictures and spatial reasoning than we are at reading text, the GUI offers a format that is easier for us to interpret and interact with.

- *Fun*: Controlling the computer through a mouse makes the interaction more like a game and takes away from the tedium of typing commands.

- *Multitasking*: Users tend to open multiple windows and work between them. Because our computers are fast enough and our operating systems support this, multitasking gives the user a great deal of power to accomplish numerous tasks concurrently (i.e., moving between tasks without necessarily finishing any one task before going on to another). This also permits operations like copying and pasting from one window to another. Although multitasking is possible from the command line, it is not as easy or convenient.

In favor of the command line:

- *Control*: Controlling the *full* capabilities of the operating system or applications software from the GUI can be a challenge. Often, commands are deeply hidden underneath menus and pop-up windows. Most Linux commands have a number of options that can be specified via the command line to tailor the command to the specific needs of the user. Some of the Linux options are not available through the GUI.

- *Speed*: Unless you are a poor typist, entering commands from the command line can be done rapidly, especially in a shell like Bash which offers numerous shortcuts. And in many instances, the GUI can become cumbersome because of many repetitive clicking and dragging operations. On the other hand, launching a GUI program will usually take far more time than launching a text-based program. If you are an impatient user, you might prefer the command line for these reasons.

- *Resources*: Most GUI programs are resource hogs. They tend to be far larger in size than similar text-based programs, taking up more space on hard disk and in virtual memory. Further, they are far more computationally expensive because the graphics routines take more processing time and power than text-based commands.

- *Wrist strain*: Extensive misuse of the keyboard can lead to carpal tunnel syndrome and other health-related issues. However, the mouse is also a leading cause of carpal tunnel syndrome. Worse still is constant movement between the two devices. By leaving your hands on the keyboard, you can position them in a way that will reduce wrist strain. Keyboards today are often ergonomically arranged and a wrist rest on a keyboard stand can also reduce the strain.

- *Learning*: As stated in Section 1.1, you can learn a lot about operating systems by exploring Linux. However, to explore Linux, you mostly do so from the command line. By using the command line, it provides you with knowledge that you would not gain from GUI interaction.

The biggest argument in favor of the command line is control. As you learn to control Linux through the command line, hopefully you will learn to love it. That is not to say that you would forego the use of the GUI. Most software is more pleasing when it is controlled through the GUI. But when it comes to systems-level work, you should turn to the command line first and often.

## 1.5 VIRTUAL MACHINES

We take a bit of a departure by examining VMs (virtual machines). A VM is an extension to an older idea known as *software emulation*. Through emulation, a computer could emulate another type of computer. More specifically, the emulator would translate the instructions of some piece of incompatible software into instructions native to the computer. This would allow a user to run programs compiled for another computer with the right emulator.

The VM is a related idea to the emulator. The VM, as the name implies, creates an illusionary computer in your physical computer. The physical computer is set up to run a specific operating system and specific software. However, through emulation, the VM then can provide the user with a different operating system running different software.

One form of VM that you might be familiar with is the Java Virtual Machine (JVM), which is built into web browsers. Through the JVM, most web browsers can execute Java Applets. The JVM takes each Java Applet instruction, stored in an intermediate form called *byte code*, decodes the instruction into the machine language of the host computer, and executes it. Thus, the JVM is an interpreter rather than a compiler. The JVM became so successful that other forms of interpreters are now commonly available in a variety of software so that you can run, for instance, Java or Ruby code. Today, just about all web browsers contain a JVM.

Today's VMs are a combination of software and data. The software is a program that can perform emulation. The data consist of the operating system, applications software, and data files that the user uses in the virtual environment. With VM software, you install an operating system. This creates a new VM. You run your VM software and boot to a specific VM from within. This gives you access to a nonnative operating system, and any software you wish to install inside of it. Interacting with the VM is like interacting with a computer running that particular operating system. In this way, a Windows-based machine could run the Mac OS X or a Macintosh could run Windows 7. Commonly, VMs are set up to run some version of Linux. Therefore, as a Linux user, you can access both your physical machine's operating system (e.g., Windows) and also Linux without having to reboot the computer.

The cost of a VM is as follows:

- The VM software itself—although some are free, VM software is typically commercially marketed and can be expensive.

- The operating system(s)—if you want to place Windows 7 in a VM, you will have to purchase a Windows 7 installation CD to have a license to use it. Fortunately, most versions of Linux are free and easily installed in a VM.

- The load on the computer—a VM requires a great deal of computational and memory overhead, however modern multicore processors are more than capable of handling the load.

- The size of the VM on hard disk—the image of the VM must be stored on hard disk and the size of the VM will be similar in size to that of the real operating system, so for instance, 8 GBytes is reasonable for a Linux image and 30 GBytes for a Windows 7 image.

You could create a Linux VM, a Windows VM, even a mainframe's operating system, all accessible from your computer. Your computer could literally be several or dozens of different computers. Each VM could have its own operating system, its own software, and its own file system space. Or, you could run several VMs where each VM is the same operating system and the same software, but each VM has different data files giving you a means of experimentation. Table 1.1 provides numerous reasons for using virtualization.

VM software is now available to run on Windows computers, Mac OS X, and Linux/Unix computers. VM software titles include vSphere Client and Server, VMware Workstation,

TABLE 1.1    Reasons to Use Virtualization

| Reason | Rationale |
| --- | --- |
| Multiplatform experimentation | For organizations thinking of switching or branching out to different operating systems, virtualization gives you the ability to test out other operating systems without purchasing new equipment |
| Cost savings | Software with limited number of licenses can be easily shared via virtualization as opposed to placing copies of the software on specific machines. An alternative is to use a shared file server. Additionally, as one computer could potentially run many different operating systems through virtualization, you reduce the number of different physical computers that an organization might need |
| Scalability | Similar to cost savings, as an organization grows there is typically a need for more and more computing resources but virtualization can offset this demand and thus help support scalability |
| Power consumption reduction | As with cost savings and scalability, having fewer resources can lead to a reduction of overall power consumption |
| Cross-platform software support | Software that only runs on a specific platform can be made accessible to all users no matter what platform they are using |
| Security | The VM is an isolated environment that cannot directly impact the true (outer) computer. Thus, malware impacts only the VM |
| Fault tolerance | Using two or more VMs, you can achieve fault tolerance by shifting the work load from one of another in case one VM fails |
| Administrative experience | Virtualization allows you to set up VMs for people who are learning to be system administrators so that any mistakes made will not harm the outer (true) computer |
| Controlling users' environments | An administrator can take remote control of a user's VM to solve problems |
| Collaboration | Virtualization allows people to share the same desktop environment in support of collaboration |
| Remote access | Through a VM server, you can access VMs remotely and allow more than one user to access a VM at any time |

VMware Player, Virtual Box, CoLinux, Windows Virtual PC, Parallels Desktop, VM from IBM, Virtual Iron, QEMU, and Xen. The latter two titles are open source and many of these titles have free versions available.

With VM software readily available, we can expand on its capabilities by implementing virtualization. With virtualization, an organization hosts a number of VMs through one or more VM servers. The servers operate in a client–server networking model where a client runs a VM program on the user machine and requests access to one of the stored VMs. The VM servers typically store the VMs on a storage area network (SAN). The collection of VM servers, the SAN, and the data that make up individual VMs can be called a VM Farm.

Now, accessibility to your VM is one of starting your VM client software and logging into the VM server. You select your VM from a list of choices, log into it, and the VM server then runs (emulates) the VM for you. Your client operates as an input/output device while the processing and storage take place on the VM server. As you make modifications to your VM, they are saved in the SAN.

There are numerous advantages to virtualization. First is accessibility. You are able to access your VM from any computer that has Internet access and runs the proper VM client software. Second is cost savings. If your company is small and you cannot afford all of the hardware needed for your employees, you can lease or purchase time and space in a VM farm where the VM servers and SAN become the hardware that you use.

You can also rely on the company hosting your VMs to handle security and backups alleviating some of the IT needs from your own organization. Today, more and more organizations are taking advantage of virtualization to improve their efficiency and lower costs.

## 1.6 UNIX AND LINUX

Unix is an old operating system, dating back to 1969. Its earliest incarnation, known as MULTICS, was developed for a single platform. It was developed by AT&T Bell Labs. Two of the employees, Dennis Ritchie and Ken Thompson, wanted to revise MULTICS to run as a platform-independent operating system. They called their new system Unics, with its first version being written in the assembly language of the DEC PDP-11 computer so that it was not platform-independent. They rewrote the operating system in the C programming language (which Ritchie developed in part for Unix) to make it platform independent. This version they named Unix.

Numerous versions of Unix were released between 1972 and the early 1980s including a version that would run on Intel 8086-based computers such as the early IBM PC and PC-compatible computers. Unix was not a free operating system. In spite of it being implemented as a platform-independent operating system, it was not available for all hardware platforms.

In 1983, Richard Stallman of MIT began the GNU Project, an effort to complete a Unix-like operating system that was both free and open source. GNU stands for GNU's Not Unix, an indication that GNU would be a Unix-like operating system but separate from Unix. His goal was to have anyone and everyone contribute to the project. He received help from programmers around the world who freely contributed to the GNU operating system, which they wrote from scratch. Although a completed version of GNU was never released, the approach taken was to lead to what we now call the open source community. Stallman formed the Free Software Foundation (FSF) and defined the GNUs General Public License (GPL). We explore open source and its significance in Chapter 13.

At around the same time as the initiation of the GNU Project, researchers at the University of California Berkeley developed their own version of Unix, which was given the name BSD (Berkeley Standard Distribution) Unix. This version includes networking code to support TCP/IP so that these Unix computers could easily access the growing Internet. In time, BSD 4.2 would become one of the most widely distributed versions of Unix.

The result of several competing forms of Unix led to what some have called the "Unix Wars." The war itself was not restricted to fighting over greater distribution. In 1992, Unix System Laboratories (USL) filed a lawsuit against Berkeley Software Design, Inc and the Regents of the University of California. The lawsuit claimed that BSD Unix was built, at least partially, on source code from AT&T's Unix, in violation of a software license that UC Berkeley had been given when they acquired the software from AT&T. The case was settled out of court in 1993.

By 1990, the open software foundation, members of the open source community had developed standardized versions of Unix based on BSD Unix. Today, there are still many different distributions of Unix available which run on mainframe, minicomputers, and servers.

In 1991, a student from Finland, Linus Torvalds, was dissatisfied with an experimental operating system that was made available through an operating systems textbook of Andrew Tanenbaum. The operating system was called Minix. It was a scaled down Unix-like operating system that was used for educational purposes. Torvalds decided to build his own operating system kernel and provide it as source code for others to play with and build upon.* Early on, his intention was just to explore operating systems. Surprisingly, many programmers were intrigued with the beginnings of this operating system, and through the open source community, the operating system grew and grew.

The development of Linux in many ways accomplished what Stallman set out to do with the GNU project. Stallman and many in the FSF refer to Linux as GNU/Linux as they claim that much of Linux was built on top of the GNU project code that had been developed years earlier.

According to some surveys, roughly 75% of Linux has been developed by programmers who work for companies that are investing in Linux. The GPL causes many of these programmers to publish their code rather than keeping the code proprietary for the companies they work for. Additionally, 18% of the code is developed strictly by volunteers who are eager to see Linux grow.

Today, Linux stands on its own as a different operating system from Unix. In many cases, Linux is freely available in source code and the open source community continues to contribute to it. And like Unix, there are many distributions. Unlike Unix, however, Linux'

---

* Torvalds has claimed that had the GNU project kernel been available, he would not have written his own; cited from http://web.archive.org/web/20121003060514/http://www.dina.dk/~abraham/Linus_vs_Tanenbaum.html.

popularity is far greater because, while Unix can run on personal computers, Linux is geared to run on any platform and is very effective on personal computers.

Although there are dozens of dialects of Unix, there are hundreds of different Linux distributions. Navigating between the available dialects can be challenging. Nearly all of the dialects can be categorized into one of four ancestor paths.

- *Debian*: This branch includes the very popular Ubuntu which itself has spawned dozens of subdialects. Another popular spin-off of Debian is Knoppix.

- *Red Hat*: There are as many or more subdialects of Red Hat as Debian. The most popular subdialect is Fedora. Another popular descendant is Mandrake and another is CentOS. Another distribution that is increasing in popularity is Scientific Linux, produced by Fermi National Accelerator Laboratory and CERN.

- *SLS/Slackware*: This branch was first produced by a German company which led to SuSE Linux. Although there are dozens of spin-offs of SLS/Slackware, it is far less popular than either Debian or Red Hat.

- *Miscellany*: There are dozens of dialects that either led nowhere or have few successors.

Linux and Unix operating systems are partially or completely POSIX conforming. POSIX is the portable operating system interface, a set of standards that operating system developers might attempt to target when they implement their systems. POSIX defines an application programming interface (API) so that programmers know what functions, data structures, and variables they should define or utilize to implement the code they are developing for the operating system.

In the development of Linux, the POSIX API has been used to generate a standard called the Linux Standard Base (LSB). Anyone implementing a dialect of Linux who wishes to include this standard knows what is expected by reading the LSB. The LSB, among other things, defines the top-level directory structure of Linux and the location of significant Linux files such as libraries, executables, and configuration files, a base set of Linux commands and utilities to be implemented, and implementations for such programs as gcc, the C compiler. Thus, underlying most dialects of Linux, you will find commonalities. In this way, learning one version of Linux is made easier once you have learned any other version of Linux.

## 1.7 TYPES OF USERS

The Linux operating system provides for two classes of users that we might term normal users and superusers. The term superuser is commonly used in many operating systems although in Linux, we call such a user *root*. The root user (or users) has access to all system commands and so can access all system resources through those commands. Normal users have greatly restricted access in that they can execute public programs, access public files and access their own file space. We can add a third category of user account to this list: software accounts. We look at the types of users in detail in Chapter 9 when we introduce user account creation.

The role of root is to properly administer the computer system. In this section, we will discuss the types of duties of a system administrator. Operating systems divide accessibility into two or more categories, the ordinary user and the administrator (sometimes called privileged mode). Some operating systems have intermediate categories between the two extremes where a user is given more privileges but not full administrator privileges. We will see in Chapter 9 that we can do this in Linux through the sudo command.

The reason for the division between normal user and administrator modes is to ensure that normal users can not impact other users. Consider a home computer shared by a family. If all users had administrator privileges, then they could view and manipulate each other's files. If a parent had important financial data, one of the children could delete or alter the data without adequate protection. In a work environment, keeping data secure becomes even more important. Different users would have access to different types of data (financial, personnel, management, research), based on their identified role within the organization. Performing administrator duties (creating accounts, scheduling backups, installing software, etc.), if performed by the wrong person, can cause disastrous results if the person does not do things correctly. Imagine that when installing new software, the person unknowingly wipes out the boot sector of the disk. Upon reboot, the computer no longer functions correctly.

And so we have a unique account in all operating systems that is capable of full system access. It is through this account that all (or most) administrative functions will be performed. What does a system administrator do? The role of the administrator will vary based on the number of users of the computer system(s), the complexity of the computer system(s), the types of software made available, and more significantly, the size of the organization.

A small organization with just a few employees might employ a single system administrator who is also in charge of network administration, computer security, and user training. In a large organization, there may be several system administrators, several network administrators, a few people specifically in charge of all aspects of security, and another group in charge of training.

The following list is common to many system administrators:

- Install the operating system

- Update the operating system when needed

- Configure the operating system to fit the needs of the users in the organization

- Secure the operating system

- Configure and maintain network communication

- Install, configure, and maintain application software

- Create and manage user accounts and ensure the use of strong passwords

- Install and troubleshoot hardware connected to computers directly or through a network

- Manage the file system including partitioning the disk drives and performing backups

- Schedule operations as needed such as backing up file systems, mounting and unmounting file systems, updating the operating system and other application software, examining log files for troubleshooting and suspicious activity

- Define (for your organization's management) computer usage policies and disaster recovery plans

- Create documentation and training materials for users

- Make recommendations for system upgrades to management

System administrators may not be responsible for all of the above duties. Other forms of administration (e.g., network administration, webserver administration, database administration, DNS administration, and computer security specialist) may take on some of the duties or have overlapping duties with the system administrator(s). For instance, a network administrator would be in charge of installing, configuring, and securing the network but the system administrator may also be involved by configuring each individual workstation to the network. A webserver administrator would be in charge of configuring, maintaining, and troubleshooting the webserver but the system administrator may be in charge of installing it and setting up a special account for the webserver administrator so that he/she can access some system files.

For the first half of this book (through Chapter 7), we will concentrate on Linux from the normal user's perspective, introducing computer concepts that should be known of all users and the Linux commands that support the normal user. Starting in Chapter 8, we will focus on Linux from the system administrator's perspective. From that point forward, we will focus on more advanced concepts and the Linux commands that usually require root access.

## 1.8 WHAT IS A COMPUTER?

We should probably more formally discuss a computer before we wrap up this introductory chapter. A computer is generally taken to be an electronic device that performs the IPOS (input-process-output-storage) cycle. It should be noted that computers do not have to be electronic (the earliest computing devices were mechanical in nature, using rotating gears or other movable parts). However, by making the computer electronic, it is capable of performing at truly amazing speeds because of how fast electrical current can flow through digital circuitry.

### 1.8.1 The IPOS Cycle

The IPOS cycle represents the four basic tasks of any computer: input, processing, output, and storage. Input is obtaining data from the outside world and delivering it to one or more components in the computer. Output is taking information stored in the computer and delivering it to the outside world, usually for humans to view. You might think of input as the raw data for a process and output as the results of the process.

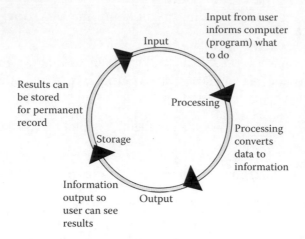

FIGURE 1.17    The IPOS cycle. (From Richard Fox, 2013, *Information Technology: An Introduction for Today's Digital World.*)

Processing is the execution of one or more computer programs on the data. Processes can vary greatly in size, complexity, and function. Processes will often perform mathematical operations, string operations, computer graphics, storage operations, and/or operating system activities.

Storage is applied today to both data and program code. Storage usually refers to secondary storage, that is, the permanent placement of information on a device like a disk, but it can also refer to temporary storage in some form of memory such as main memory (RAM) or short-term storage such as registers and cache.

Figure 1.17 illustrates the IPOS cycle. Although a computer will do all four of these, it is not necessarily the case that all four of them are done in this sequence. For instance, a program may require some input followed by processing followed by additional input followed by more processing. Also, results from a process can either be output or stored in a file, thus omitting either output or storage.

### 1.8.2 Computer Hardware

All computers have one or more processors. The processor, or central processing unit (CPU), is the device which controls the components within the computer. It performs an operation called the *fetch-execute cycle*. This cycle requires that the next program instruction be fetched from where it is stored (some form of memory), decoded from machine language into operational code (often called microcode) and executed.

During the course of the fetch–execute cycle, the CPU signals various devices to perform operations. These include memory to perform a read operation (to obtain the instruction), movement between storage devices in the CPU called registers, and movement between arithmetic and logic circuits to perform computations. The CPU contains a control unit which is responsible for handling the fetch–execute cycle and sending out control signals.

The CPU also contains an arithmetic-logic unit. This component contains digital circuits, each of which is responsible for handling one form of arithmetic or logic operation such as an adder, a multiplier, a shifter, and a comparator. The CPU also contains data registers. These are temporary storage locations used in the CPU to store partial computations as the CPU executes program code.

As a simple example, if the current program requires performing the operation A = B * (C + D), then C and D are loaded from memory into registers. They are then added together and the result stored in a register. B is loaded from memory and stored in a register. The values of B and C + D are then multiplied and the result is stored in a register. Finally, the result is moved from register to memory. This instruction requires the use of the adder, multiplier, and some registers.

Aside from the data registers, the CPU has control registers. These are used by the control unit portion of the CPU to handle the fetch-execute cycle. There are several important control registers to note. The program counter register (PC) stores the location in memory of the next instruction to fetch. The instruction register (IR) stores the instruction currently being executed. The status flags (SF) are individual bits stored in a single register to store information about the last operation such as whether it resulted in a positive, negative, or zero value, whether it produced a carry, and whether it caused an interrupt. The stack pointer (SP) points to the location in memory corresponding to the top of the run-time stack. Different processors may have other control registers.

Every computer also has one or more forms of memory. Types of memory include registers, cache, main memory, and secondary storage. Registers are built into the CPU. Cache is fast memory that allows the CPU to access data and program code without delay. However, cache is expensive and so a limited amount is made available. Often today, computers have several caches, some on the CPU and some separated off the CPU (known as on-chip and off-chip cache, respectively). The on-chip caches provide faster access than off-chip caches because of their location, but must be smaller in size because of the limited space available on the CPU.

Main memory, often called DRAM, is a good deal slower than cache but is far cheaper. While a computer might have 8 GB of DRAM, it will probably only have 1–2 MB of cache (if you are unfamiliar with storage capacity terminology, see the appendix).

You might ask why we need main memory when cache is so much faster. Think of DRAM as an extension to cache. That is, we prefer to keep as much of our running programs and data in cache memories, but because of their limited size, we extend memory by placing the rest in DRAM. Alternatively, you can think of cache as being a special area to place the items that are most useful so that the CPU can look there first in hopes of avoiding access to DRAM.

Even with the large DRAM sizes of our modern computers, DRAM cannot usually store all of the program code that users are executing. For instance, some GUI programs are dozens of GBytes in size by themselves whereas DRAM might be limited to just 8 GByte. So we need a back-up to DRAM. For this, we use the hard disk in two forms. First, temporary hard disk storage that backs up DRAM is called *virtual memory* and is stored in a special area called the *swap space*. Second, for permanence, we store everything on hard disk in user storage space. We differentiate between main memory (cache, DRAM) and secondary storage even though both are used to store our program code and data. We will discuss virtual memory in more detail in Chapters 8 and 10 and the file system in Chapters 3 and 10.

This organization of memory is known as the *memory hierarchy*. The idea is that the CPU, when it wants something, will try the top level of the hierarchy first (on-chip cache) and only if it fails to find what it wants will it resort to moving down the hierarchy. Moving down the hierarchy results in slower access but a greater likelihood that what it is looking for will be found because each lower level consists of a storage type that has a greater capacity.

If, for instance, the item sought is found in DRAM, then it and its neighboring memory locations are copied up the hierarchy. The reason to bring its neighbors is the hopes that what the CPU is looking for now is next to memory locations storing items that the CPU will look for in the near future. However, in copying items up the hierarchy, other items will have to be discarded. For instance, copying some new items into on-chip cache will require discarding items from that cache because of its limited size. So the process of moving items around in memory is something of a guessing game: can we remove items that we do not need in favor of items we will need soon?

As mentioned above, computers have some form of memory but not necessarily all of these forms. It depends on the type of computer. Handheld devices will often forego hard disk storage in favor of using DRAM and in some cases, handheld devices also forego DRAM and use only flash memory.

Another component in our computer is the I/O system. We refer to this as a single system although in fact it is made up of numerous devices and interfaces (connections). Collectively, the devices in the I/O system give the user the ability to interact with the computer. The devices will include input devices like a keyboard, mouse or other pointing device, touch screen and microphone and output devices like a monitor (possibly a touch screen), printer, and speakers. Additionally, our secondary storage devices are part of the I/O system as, aside from storing files, they are used for input (when we open a file) and output (when we save a file).

So we can think of the computer as being divided into three general categories: the processor, memory, and I/O/storage. We can also differentiate the levels of memory and storage into a hierarchy of devices. At the top of the hierarchy are the fastest but also most expensive forms of memory and so we limit the amount because of their expense. As you descend the hierarchy, your computer generally has more and more because the cost per storage unit is greatly reduced. This hierarchy consists of from top to bottom: registers, on-chip cache, off-chip cache, DRAM, swap space, hard disk user space, network storage devices, removable storage devices (e.g., flash drive, optical disk drive, and magnetic tape drive).

### 1.8.3 Software and Users

There are two additional components to any computer system, although these are not hardware. These are software and people (users). The computer itself is a general purpose processing machine. It is set up to execute any type of program. But without the program, the computer has nothing to do. The hardware by itself has no motivation or goal. So instead, humans run programs.

```
#include <stdio.h>

int main(){
      int a, b, c;
      printf("Enter three numbers: ");
      scanf("%d %d %d", &a, &b, &c);
      if(a>b&&b>c)
            printf("%d is the greatest\n", a);
      else if(b>a&&b>c)
            printf("%d is the greatest\n", b);
      else
            printf("%d is the greatest\n", c);
      return 0;
}
```

FIGURE 1.18   Simple example C program.

A program is a step-by-step description of how to solve a problem. The description must be in a language that the computer can understand. The only language that a computer is capable of executing is that of its native machine language. However, programmers often write their program code in other languages and use program translation software to convert their program into machine language. An example program written in C is shown in Figure 1.18.

Without a user, there is no one or nothing to tell the computer which program to run and at what time. Nor, is there anything to provide input or view the output. It is the user that controls the computer. More correctly, it is the user who uses the computer. This is because in many cases, the user makes requests of the computer. The computer does not fulfill every request as the user must have proper access rights for the request to be carried out.

The computer system then consists of a series of layers. At the top layer is the user who uses the computer to accomplish some task(s). The user's interactions cause applications software to execute. Both the user's interactions and the applications software must access computer hardware. The operating system acts as an intermediary between the user/application and hardware. The operating system is itself layered with a shell for interaction, utilities and services, the kernel, device drivers and ROM BIOS (or firmware), and the hardware at the bottom layer (see Figure 1.2).

## 1.8.4 Types of Computers

Computers can vary in size, power, and expense. At the large end, we have supercomputers which consist of dozens, hundreds, or thousands of processors combined by some internal network. These computers, which can cost millions of dollars, are the fastest computers on the planet. They are often used to handle computationally intensive tasks like weather forecasting, modeling earthquakes, operations related to genetics and the human genome, and simulating the spread of viruses to name just a few of their applications.

At the other end of the spectrum of computers are handheld devices like smart phones. A smart phone is primarily designed to be a telephone but can run a number of other applications (apps). Typically, the smart phone stores the apps in a small internal memory. The processor of the smart phone is one that contains technology of perhaps 10–15 years earlier. Because of these limitations, smart phones often can only handle one or perhaps two applications at a time. Additionally, to permanently save information recorded on a smart phone, you will have to transfer it to another computer. Slightly more capable than the smart phone are tablet devices.

In between the two extremes of handheld devices and supercomputers are the personal computers: laptops and desktops. This type of computer will contain several storage devices including an optical disk drive and an internal hard disk. You can also connect external disk drives to the device and for desktop units, additional internal devices such as a second hard disk or a floppy disk drive.

More sophisticated than the desktop computer is the server. The idea behind the server is that it is a networked computer that services requests of other networked computers. The server might provide file storage for a few dozen computers or it might handle HTTP requests over the Internet (a web server) or it may send and receive email messages. Depending upon the number of expected users, a server may have similar computing power and main memory as a desktop computer, or it may require faster processing and more memory.

For instance, a file server which is utilized by no more than 10 users may not require a superior processor. On the other hand, some types of servers are expected to perform a good deal of processing. Web servers for large websites for instance must execute server side code. A web site like Google would use high-end servers which contain much more powerful processors than what you will find in a desktop computer. What is certain though is that a server will utilize a greater amount of disk storage than a desktop computer and it is likely that a server might have dozens of hard disk drives.

There are also computers that fall between the supercomputer and the server: mainframe and minicomputers. These computers are more historically interesting as they were the only platform of computer available in the first few decades of computer usage. The mainframe and minicomputer category consists of computers expected to handle the processing duties of dozens or hundreds of users at a time. With the inexpensive availability of personal computers though, mainframe and minicomputer usage is far less common.

Table 1.2 illustrates some of the characteristics of each of these types of computers. It should be noted that these values tend to change rapidly in the computing field. The values shown in Table 1.2 are estimates from 2013. The breakdown of computer types as shown in Table 1.2 are not precise. You might notice for instance that high-end mainframe computers and low-end supercomputers are the same. We might think of this grouping as one class of computer which ranges at one extreme from computers worth $100,000 to those that are worth tens of millions of dollars. We tend to see the greater expense as we add more processors (from dozens to thousands) and as we increase the amount of DRAM and storage. Similarly, the laptop and desktop computers run together. Laptops can have just as great an amount of DRAM, processor cores, and storage space as a desktop computer. About the only two aspects that are clearly different between the two types of computers

TABLE 1.2  Characteristics of Computers (as of October 2013)

| Type | Number of CPUs/Cores | DRAM | Storage Devices | Typical Cost | Operations Per Second[a] |
|---|---|---|---|---|---|
| Super computer | Thousands to hundreds of thousands | 100s–1000s GBytes | Storage area networks | $ Tens of millions | TFLOP to PFLOP range |
| Mainframe | Hundreds to thousands (or more) | Dozens to 100s of GBytes | Server storage or storage area network | $100,000–$1 million | GFLOP to TFLOP range |
| Server | 1–8 multicore (up to 8 cores) | Up to 32 GBytes | Many hard disk drives (up to dozens) | $5,000–$15,000 | GFLOP range |
| Desktop | 1–2 multicore (up to 8 cores) | 4–16 GBytes | 1–2 Hard disk drives, optical disk drive, numerous USB ports | Under $2,400, often around $1,200 | GFLOP range |
| Laptop | 1 multicore (up to 8 cores) | 4–16 GBytes | 1 Hard disk drive, optical disk drive, a few USB ports | Under $1,200, often around $900 | GFLOP range |
| Tablet | 1 | 4–8 GBytes | USB port, small capacity hard disk drive | $300–$700 | Up to GFLOP range |
| Smart phone | 1 | Up to 2 GBytes | USB port, internal storage (flash or disk) up to 64 GBytes | Under $500, often under $200 | Hundreds of MFLOPs up to 1 GFLOP |

[a] The word FLOP means "floating point operations per second" and is a common way of gaging processor speed. The letter preceding FLOP indicates the magnitude. P for peta, or quadrillion, T for tera, or trillion, G for giga, or billion, M for mega, or million.

is the laptop is built to be portable and so is lightweight and compact, and the desktop, because it is larger, can have a greater number of devices inside (additional disk drive units) or connected to it (e.g., external keyboard and speakers).

## 1.9  THIS TEXTBOOK

Linux (and Unix) textbooks generally fall into two categories. There are users guides that are less textbook oriented and more "how to" texts. These are fine if you already know something about Linux and want to become more of an expert, or alternatively want to learn enough to use the operating system. What they fail to do is teach you underlying concepts; the "why" instead of the "how." The other category of book is true operating system concept books that might spotlight Linux or Unix with one or a few chapters dedicated to these operating systems.

The book you hold in your hands is somewhat unique in that it combines both styles of textbooks. This book does teach you the "how" of Linux but also emphasizes the "what" and "why" of the operating system.

Additionally, Linux (and Unix) textbooks tend to focus on either teaching you how to use the operating system or teaching you how to administer the operating system. This textbook again differs because it offers both perspectives. Roughly the first half of the textbook covers the "user" perspective while the second half covers the "administration" perspective.

What you will see from here on out are topics on the following from the user perspective.

- *CLI*: How to issue commands and the many shortcuts available.

- *The Linux file system*: How to navigate and use it.

- *Process management*: How to issue, monitor, and terminate your processes.

- *Regular expressions*: What they are, how to use them, and why you might use them.

- *Scripting*: How to write basic Bash and Csh scripts.

From the system administrator perspective, we view these topics.

- *Installation*: How to install CentOS Linux.

- *Account creation*: How to create user and group accounts and provide them with privileges.

- *The Linux file system*: How to administer the Linux file system.

- *The boot process*: Understanding and managing the Linux boot and initialization process.

- *Services*: Understanding the role of the operating system service, how to start and stop them, and how to configure them.

- *The network*: Configuring and understanding the Linux services and software related to network communication.

- *Software installation*: Installing and configuring open source software.

- *Troubleshooting*: Dealing with services, log files, and other areas of Linux.

- *Servers*: Understanding how to install and configure some of the more commonly used servers in Linux.

Throughout the textbook, operating system concepts are introduced in an attempt to provide the reader with an understanding of what is going on beneath the command line. This book should not be confused with concepts of operating systems textbook which would be far less detailed in any specific operating system.

## 1.10 CHAPTER REVIEW

Concepts introduced in this chapter:

- Command line interface (CLI)—an interface used in Linux in which the user enters commands by typing instructions and parameters from a prompt. The command line requires the use of an interpreter which will determine how to carry out the command entered. Linux interpreters permit the user to enter complex commands.

- Computer system—the collection of computer hardware, software, and users. The hardware would do nothing without software. Software are the programs that we run on the computer. The user is required to tell the computer what to do and when.

- Gnome—one of two common GUIs that are used in Linux, Gnome is the default GUI for Red Hat Linux.

- Graphical user interface (GUI)—allows the user to interact with the computer through pointing, dragging, and clicking motions of the mouse. The GUI is the more common approach used today whereas the CLI is the older of the approaches.

- Interpreter—a program which takes instructions, converts them into executable code, and executes them. Linux shells have built-in interpreters to execute command line instructions and run shell scripts.

- KDE—an alternate GUI from Gnome available in most Linux platforms.

- Kernel—the core component of the operating system responsible for most of the OS duties. Once the computer is booted, the kernel is loaded into memory where it remains until the computer is shut down or rebooted.

- Linux—an operating system developed starting in the early 1990s, based on the Unix operating system in part. Linux has become a popular operating system for servers and supercomputers and computer hobbyists although it only has a small presence in the overall marketplace.

- Linux market share—in spite of Linux' popularity on the Internet as the principle operating system used for servers and supercomputers, the Linux market share trails behind Windows dramatically and is also behind Mac OS. Including mobile device forms of Linux and Unix, the market share is under 15%.

- Linux standard base—a standard for Linux developers to implement commonalities in the underlying mechanisms of Linux so that all Linux dialects retain similar features.

- Open source—software that is developed by programmers in the open source community. The software is made available for free in either executable form or source code. The idea is that other programmers can freely obtain the software and modify it as desired, contributing back to the community with new or improved software.

- Operating system—a suite of programs that manage our computer system for us so that we as users do not have to worry about the details. The operating system is broken into several components, the most important of which is the kernel. Other components include device drivers so that the operating system can communicate with specific hardware devices, shells which offer the user an interface that they can personalize services which handle communication and requests coming from different agents, and utilities which are add-on programs to further help manage and manipulate the computer system.

- Operating system platforms—the different types of operating systems used on different devices. Aside from Linux, other popular operating systems are Windows (which has the biggest marketshare), Unix, iOS (the operating system found on Apple handheld devices), and Mac OS X.

- POSIX—Portable Operating System Interface which describes an application programmer interface that most Unix and Linux versions attempt to implement (at least partially) so that programmers can more easily contribute.

- Root—a special account in Linux for the system administrator(s).

- Shell—a user-tailored environment to permit easier usage of the OS. In a GUI, the user can tailor the desktop and menus. From the command line, the user can specify his or her own definitions for functions, aliases, variables, and more. In Linux, a shell includes an interpreter to handle commands entered in the shell.

- System administrator—a special type of user who is in charge of administering a computer system who has access to all system software and resources.

- Terminal window—in Linux, this opens a shell and provides the user with a command line prompt from which the user can enter commands.

- Unix—an older operating system which looks similar to Linux in many ways. Although the Linux kernel was not modeled on Unix itself, the interface (commands, directory structure) and much of the functionality is the same or similar.

- User account—the means by which a user is able to log in and access a computer. User accounts come with a username, a password, and user resources such as a home directory space.

- Virtual machine—using software to create the illusion of a separate computer. The VM allows us to run other computer platforms on a given computer for experimentation or safety.

## REVIEW QUESTIONS

1. On what type(s) of devices can Linux run?

2. Do handheld devices running a Linux OS, such as Android, count toward Linux' market share?

3. Rank these operating system platforms in order of most to least commonly used in today's computing market: Linux, Windows, IBM System i, Mac OS.

4. Define open source software.

5. How does Linux offer the user greater control over the Windows OS?

6. List two reasons why you might want to learn Linux.

7. All computers have and have always had operating systems. True or false. Explain.

8. What is a resident monitor?

9. What is an operating system kernel? At what point is the kernel loaded into memory? How long does it stay in memory?

10. Order the following layers of a computer system from top (user) to bottom (hardware): OS kernel, device drivers, application software, OS services, shells, ROM BIOS, OS utilities.

11. In order to log into a Linux system, you must know your username and your _____.

12. A user's account will often come with access to what resource(s)?

13. What is a login screen?

14. What is Gnome? What is KDE?

15. Both Gnome and KDE are based on what older GUI?

16. From the Gnome desktop, list two ways to open a terminal window.

17. From the Gnome desktop, which menu would you look under to open a window listing your user home directory?

18. From the Gnome desktop, which menu would you look under to find commands to access system administration programs like the Linux firewall and adding users and groups?

19. What is a workspace?

20. What does the Linux interpreter do?

21. What are the components that make up a Linux shell?

22. What types of definitions can you place inside an interpreted environment in Linux?

23. Which interface, the GUI or command-line requires more memorization and skill to use?

24. Which interface, the GUI or command-line requires more system resources to use?

25. Provide three reasons for why a user might want to use the CLI rather than a GUI in Linux.

26. How does emulation differ from virtualization?

27. What is a VM?

28. In what way could using a VM make your computer safer?

29. Provide three reasons (aside from what was asked in 28) for why you might want to create a VM.

30. What operating system looks similar to Linux?

31. What was the original name for the Unix operating system?

32. What is BSD Unix?

33. How did the development of Linux differ from that of Unix?

34. Which dialect of Linux does CentOS derive from?

35. Why do operating systems have (at least) two types of accounts, normal users and administrators?

36. As a user of a Linux computer system, what documentation do you feel a system administrator should produce for you, if any?

37. Why would an organization want separate network administrators from system administrators?

38. It is important for a system administrator to understand general computer concepts such as how the operating system functions, virtual memory, process management, computer networks, and so forth. Why do you suppose this is true?

39. Research the role of a computer security specialist and enumerate some of the duties involved.

40. What is the IPOS cycle?

41. Define a computer.

42. How does hardware differ from software?

43. What is the memory hierarchy? List these types of memories in order from the fastest to slowest: on-chip cache, DRAM, hard disk storage, off-chip cache, registers.

44. What are the components of a computer system?

45. Rank these computing devices in terms of speed from fastest to slowest: mainframe, server, smart phone, supercomputer, desktop computer.

46. What does the term PetaFLOPS mean? What type(s) of computer achieves this?

47. You can find Linux or Unix running on which of the following types of computers? Mainframe, server, smart phone, supercomputer, desktop computer.

# CHAPTER 2

# The Bash Shell

THIS CHAPTER'S LEARNING OBJECTS are

- To understand the role of the shell in Linux

- To be able to enter commands via the command line prompt in Linux

- To understand what options and parameters are in Linux commands

- To be able to apply aliases, history, command-line editing, brace expansion, tilde expansion, and wildcards in Bash

- To be able to use variables from the command line prompt and understand the roles of the common environment variables

- To understand the Linux commands ls and man

- To be able to apply redirection in Linux commands

- To be able to tailor a shell using .bashrc, .bash_profile

## 2.1  INTRODUCTION

A shell is a user interface. Recall from Chapter 1, there are different types of user interfaces. In Linux, we use the graphical user interface (GUI) and the command line interface (CLI). The CLI is older, and in many users minds, better. See Figure 2.1 for reasons why a Linux user might want to use the command line rather than the graphical interface. Of course, there are disadvantages as well with ease of use and the learning curve to understand how to issue commands from the command line being the top two.

A shell runs an interpreter. The interpreter is a program which accepts user input and determines how to execute that input. The original Unix shell was the Bourne shell with a built-in Bourne interpreter. This shell was named after the program's author, Stephen Bourne. The intention of a shell was to provide the user with a command-line interpreter (a program which would interpret commands and execute them), but to also provide the user with an environment to work in. Today, the Bourne shell is not used, but in its place

1. Control—the command line permits users to include options that provide both flexibility and power over the defaults when issuing commands by GUI.

2. Speed—many programs issued at the command line are text-based and therefore will start faster and run faster than GUI-based programs.

3. Resources—as with speed, a text-based program uses fewer system resources.

4. Remote access—if you are using a program like ssh to reach a computer, you may be limited to a text–based interface, so learning it is important and once learned, you gain the other benefits.

5. Reduced health risk—yes, this one sounds odd, but mouse usage is one of the most harmful aspects of using a computer as it strains numerous muscles in your hand and wrist, while typing is less of a strain (although poor hand placement can also damage your wrist).

FIGURE 2.1    Advantages of the Command Line over a GUI.

is the popular Bourne-again Shell, or BASH (or Bash). The Bash shell* is the default shell for most Linux users. There are other shells, and at the end of this chapter, we will compare Bash to a few of those.

The shell environment can be tailored. The idea is one of the user "filling in" the environment through definitions of different kinds. The different types of definitions include short cuts to existing commands, defining new commands, defining variables, and entering commands. The power of a Linux shell goes beyond the ability to fill in your own pieces though. A shell also offers very powerful command-line editing features and a variety of typing short cuts.

In this chapter, we examine the Bash shell. To see how to use some of these features, we also introduce some of the more basic Linux commands. Topics include methods of obtaining help, Bash-specific features, comparison to other shells, and details on how the Bash interpreter works. Although we introduce some basic Linux commands, keep in mind that we will continue to introduce Linux commands throughout the textbook. Those covered in this chapter will include the ls command to view contents of the Linux file system, but we concentrate on the Linux file system in later chapters.

## 2.2  ENTERING LINUX COMMANDS

So you have logged into Linux and opened your first shell (see Section 2.4.2). You see a prompt awaiting your command. Perhaps it looks like this:

```
[foxr@localhost ~]$
```

---

* Note that "Bash shell" is redundant because the "sh" portion of Bash means shell. However, most people will still refer to it as the Bash shell.

The prompt is telling you who you are, where you are, who you are, and where you are. The first "who you are" is your user name (foxr in this case). The first "where you are" is the machine you are on. Localhost means that the shell you have opened is on *this* computer. Had you logged in remotely to another computer, you would receive a different value. The second "where you are" is provided by the ~ (tilde) character. The ~ means "user's home directory." This tells you that your current working directory is ~, which for this user is probably /home/foxr. The second "who you are" is provided by the $. The $ represents your prompt. This will precede any input you type.

Generally in Linux, we see two types of prompts. The $ indicates that you are a normal user whereas a # would indicate that you have logged in to this shell as root (the system administrator).

So we see by the prompt several pieces of information. As you will probably know who you are and which machine you are on, the most useful piece of information in the prompt is the current working directory. We will discuss later how you can modify your prompt.

The Bash interpreter awaits your command. To use the interpreter, you enter a command and press the <enter> key. Whatever was typed is now interpreted.

You can also type multiple commands at the prompt. You would do this by separating each individual Linux command with the semicolon character. You press <enter> at the end of the last command.

## 2.2.1 Simple Linux Commands

Although the prompt tells you a lot of information, there are several simple Linux commands that can tell you who you are, where you are, and what can be found at your current location. These commands are:

- whoami—output the user's username

- pwd—output the current working directory

- hostname—output the name of the host (this will either be localhost.localdomain or the IP address or IP alias of the machine logged in to)

- ls—list the contents of the current working directory

If you are following along in a Bash shell, try each of them. You will see that whoami, pwd, and hostname respond with information that you will find in your prompt. If you are logged in as root, whoami returns root. Note that if the user prompt tells you that you are currently at ~, pwd would still return the full directory path, /home/foxr in this case. Most users' directories are typically found under the /home directory and their home directory name will be the same as their user name. This may not always be the case. For instance, you might find that /home has subdirectories to divide up the user space into groupings, so for instance pwd might return /home/a/foxr or /home/b/zappaf.

Some other useful, simple Linux commands are:

- passwd—used to change your password. You are first prompted to input your current password and then your new password (twice). If the password you enter is not sufficiently strong, you are warned.

- uname—output basic information about your operating system.

- arch—output basic information about your computer's hardware (architecture).

- who—list the users currently logged in to the computer you are operating on, including their terminal (which terminal window they have opened or whether they are on console).

- bash—start a new Bash shell (if in a shell, this starts a new session so that the outer shell session is hidden). A variation is sh, to start a new shell. The sh program may not start a Bash shell however depending on which version of Linux you are using.

- exit—leave the current Bash shell and if this is the outermost Bash shell, close the window.

As stated earlier, you can enter multiple commands from the prompt by separating each with a semicolon. What follows is what you would expect to see in response to entering the three commands uname, arch and who individually and then combined. These responses would vary computer to computer and session to session based on who is logged in.

```
[foxr@localhost ~]$ uname
Linux
[foxr@localhost ~]$ arch
x86_64
[foxr@localhost ~]$ who
foxr        tty7        2013-10-11 09:42 (:0)
foxr        pts/0       2013-10-11 15:14 (:0)

[foxr@localhost ~]$ uname; arch; who
Linux
x86_64
foxr        tty7        2013-10-11 09:42 (:0)
foxr        pts/0       2013-10-11 15:14 (:0)
```

## 2.2.2 Commands with Options and Parameters

Most Linux commands expect arguments. Arguments come in two basic forms, options and parameters. Options are user-specified variations to the command so that the command performs differently. As a simple example, the ls command has numerous options that provide the user with different information than just a list of the entities in the current directory. One option is −1 (hyphen lower-case L). This provides a "long" listing of the entities. That is, it provides more details than just the name of each item.

Parameters are commonly either file names or directories (or both). A filename or directory name specifies to the command what entity to operate on. The format of a Linux command is

```
command [option(s)] [parameter(s)]
```

where the [] indicate that the item is optional.

For instance, if you wish to see the long listing of a single file called file1.txt you would issue the command

```
ls -l file1.txt
```

Many Linux commands permit *any number* of parameters. The above instruction could just as easily have operated on several files. This would be the case if the command were

```
ls -l file1.txt file2.txt file3.txt file4.txt
```

We will explore an easier way to specify a list of files later in this chapter by using wildcards.

Through options, you are able to take greater control of Linux. In fact, it is through options that you will find the real power behind the command line. One of the problems with using options is remembering them. Fortunately, we can view all of the options of a command easily through the command's man page (discussed later) or by issuing the command followed by the option --help (not all Linux commands have this option but many do).

Let us explore the ls command in more detail. The ls command lists the current directory's contents. Typing ls will display for you the *visible* files and subdirectories in the current directory. The term visible indicates entries whose names do not start with a period (.). For instance, file1.txt is a visible file while .file1 is not visible. To view all entries through ls, the option −a is used. This option specifies that ls should "list all" entries. The −a option will not only list the hidden, or dot, files (those that start with a dot), it will also list . and .. which are the current directory and the parent directory, respectively. We explore . and .. in more detail in the next chapter.

As already mentioned, the option −l (a lower case "L") provides a *long* listing. The long listing provides several properties of each item:

- The type of object (file, directory, symbolic link, etc. where a hyphen means a regular file)

- The permissions of the object (who can read, write, and execute it, which appear using r, w, and x, with a hyphen indicating no permission, we examine permissions in detail in Chapter 3)

- The number of names linked to the object

```
-rw-r--r--. 1 zappaf zappaf 24735 Aug 13  2007 joes.garage.act.i
-rw-r--r--. 1 zappaf zappaf 23996 Aug 13  2007 just.another.band.from.l.a
-rw-r--r--. 1 zappaf zappaf   891 Aug 13  2007 make_a_jazz_noise_here
-rw-r--r--. 1 zappaf zappaf  2424 Aug 13  2007 nando
-rw-r--r--. 1 zappaf zappaf 11001 Aug 13  2007 overnite.sensation
-rw-r--r--. 1 zappaf zappaf  2772 Aug 13  2007 sofa
-rw-r--r--. 1 zappaf zappaf  1807 Aug 13  2007 sofa-lyrics
-rw-r--r--. 1 zappaf zappaf   473 Aug 13  2007 sofa_no_2
-rw-r--r--. 1 zappaf zappaf  6727 Aug 13  2007 some-lyrics.txt
-rw-r--r--. 1 zappaf zappaf  9535 Aug 13  2007 studio_tan
-rw-r--r--. 1 zappaf zappaf 14475 Aug 13  2007 the.man.from.utopia
-rw-r--r--. 1 zappaf zappaf 11688 Aug 13  2007 uncle_meat
-rw-r--r--. 1 zappaf zappaf   736 Aug 13  2007 weasles_ripped_my_flesh
-rw-r--r--. 1 zappaf zappaf   557 Aug 13  2007 ycdtosa6
-rw-r--r--. 1 zappaf zappaf 14761 Aug 13  2007 yellow-snow.txt
-rw-r--r--. 1 zappaf zappaf 27771 Aug 13  2007 you.are.what.you.is
-rw-r--r--. 1 zappaf zappaf  2361 Aug 13  2007 your-mouth
-rw-r--r--. 1 zappaf zappaf 10970 Aug 13  2007 zappa.in.new.york
```

FIGURE 2.2   A long listing.

- The user owner and group owner of the object

- The size of the object

- The last modification (or creation) date and time of the object

- The name of the object (and if a symbolic link, the item it is linked to)

Figure 2.2 shows an example of an ls  –l operation applied to a portion of a user's home space.

An instruction can receive any combination of options. If you wish to provide a long listing of all files, you would use ls  –al (or ls  –la). You could separate these to be ls –a  –l or ls  –l  –a but it is more common to combine all options as desired to save typing.

You might notice the period that appears after the first series of characters (the hyphens, r and w entries). The period indicates that the given item (file, directory) has an SELinux context. SELinux is security-enhanced Linux and provides a greater degree of protection over those available through ordinary Linux permissions. We cover Linux permissions in Chapter 3 and briefly overview SELinux in Chapter 8.

While ls has a number of options, there are only a few options that you would typically use. Table 2.1 describes these (omitting –a and –l from earlier).

As mentioned at the beginning of this section, commands receive options and parameters. For ls, the parameters are a listing of files and directories. These names are the entities that are listed. If no such argument(s) is provided, the ls command lists all entries in the current directory. If parameters are provided, they can consist of one or several files and/or directories.

Consider the following instruction:

```
ls file.txt
```

TABLE 2.1    Common ls Options

| Option | Meaning |
| --- | --- |
| -A | Same as -a except that . and .. are not shown |
| -B | Ignore items whose names end with ~ (~is used for backup files) |
| -C | List entries in columns (fits more items on the screen) |
| -d | List directories by name, do not list their content |
| -F | Append listings with item classifications (ends directory names with/, ends executable files with *) |
| -g | Same as –l except that owner is not shown |
| -G | Same as –l except that group owner is not shown |
| -h | When used with –l, modifies file sizes to be "human readable" |
| -i | Include inode number of items (inodes are discussed in Chapter 10) |
| -L | Dereference links—that is, display information about item being linked to, not the link itself (links are discussed in Chapters 3 and 10) |
| -r | List items in reverse alphabetical order |
| -R | Recursive listing (list contents of all subdirectories) |
| -s | When used with –l, outputs sizes in blocks rather than bytes |
| -S | Sort files by size rather than name |
| -t | Sort files by modification time rather than name |
| -X | Sort files by extension name rather than name |
| -1 | (numeric 1) List files one per line (do not use columns) |

If this file exists, the ls command responds by listing it. Otherwise, ls returns the response

```
ls: file.txt: No such file or directory
```

This instruction is not a very informative instruction as it only tells you if the file exists or not.

A user more commonly will want to see what is in the directory. Therefore, there seems little value to supply such an argument. However, when used with options such as –l, it becomes far more meaningful. For instance, when issuing ls –l file.txt, we might see

```
-rwxr--r--. 1    foxr  foxr  183   Jul 23 2011 file.txt
```

If we use ls in conjunction with a directory name, then the command displays all of the contents of the given directory (rather than the current working directory). We can use wildcards to indicate possible variations in name. The * wildcard is used to match anything. Thus, ls *.txt will list all files in the current working directory whose names end with .txt no matter what appears on the left-hand side of the period. We explore wildcards in Chapter 3.

## 2.3 MAN PAGES

How do you learn about the options available for a command? Another command available in Linux is called man, which is short for manual page. This command displays the manual contents for a given Linux command. These contents explain the intention of the command and how to use it. The information can be cryptic for introductory users however. So, it is worthwhile reviewing man pages as you learn commands to gain a better understanding of the command as well as to learn how to interpret man pages.

The man command expects the name of a Linux command as its argument. The result of man is displayed in the current terminal window, filling the screen. In almost every case, the man contents are longer than one page, so at the end of each page, you receive a prompt of : which waits for one of several keystrokes described below. The items displayed in a man page are described in Table 2.2.

What follows are the entries listed under the man page for the `ls` command.

- NAME

    - `ls`—list directory contents

- SYNOPSIS

    - `ls` [OPTION].... [FILE]...

OPTION is the list of optional options and FILE is the optional list of files/directories. The [] indicate that these entries are optional.

This is followed by approximately nine screens' worth of options and their descriptions.

- AUTHOR

    - Written by Richard Stallman and David MacKenzie.

- REPORTING BUGS

    - Report bugs to <bug-coreutils@gnu.org>.

TABLE 2.2   Man Page Contents

| Entry | Meaning |
| --- | --- |
| Name | Command name with one line description |
| Synopsis | One or more entries that show the syntax of the command's usage |
| Description | Detailed description of what the command does |
| Options | Description of every available option, including additional parameters expected by options, and options that options can have (options may be listed under description) |
| Arguments/Parameters | Additional parameters expected or allowed |
| Environment | Environment variables that this command might use |
| Bugs | Known errors or fixes |
| Notes | Additional information that might be useful |
| Files | Files that this command might access |
| Exit or return values | Error or exit codes |
| Examples | In rare cases, some man pages include examples of how to use the command |
| Functions | Names of functions that can be called by the command |
| Author(s) | Person/people who wrote the command's programming code |
| See also | Other Linux instructions related to this one or worth exploring |

- COPYRIGHT

  - The copyright is listed.

- SEE ALSO

  - The full documentation for ls is maintained as a Texinfo manual.

If the info and ls programs are properly installed at your site, the command

```
info ls
```

should give you access to the complete manual.

After each screen, the man page pauses with the ":" prompt. At this prompt, man awaits your keystroke to command how man should continue. You have the following options:

- Forward one screen—'f,' 'z,' space bar

- Forward half a screen—'d'

- Forward one line—'e,' 'j,' enter key, down arrow

- Back one screen—'b,' 'w'

- Back one half screen—'u'

- Back one line—'b,' up arrow

- Return to top of man page—'1G'

- Go to bottom of man page—'G'

- Go to line #—#G (# is a number like 10G for line 10)

- Obtain help—'h,' 'H'

- To move forward # lines—# (# is a number)

- To search forward for a string—/string <enter>

- To search backward for a string—?string <enter>

- Exit man page—'q,' 'Q'

Nearly all Linux commands have options available. As we explore other Linux commands, we will go over the most useful of the options. To learn more about any command, access its man page.

Another form of help is an instruction called apropos. This instruction is most useful if you are not sure of the name of a command but have an idea of what the instruction should do. For instance, if you want to identify the instruction that will respond with a summary of virtual memory usage, you might issue the command apropos virtual memory or apropos "virtual memory." Note that these two instructions result in different listings. The former provides a list of over 100 instructions while the latter matches three as shown below.

```
mremap      (2)     - re-map a virtual memory address
vfork       (3p)    - create a new process; share virtual memory
vmstat      (8)     - Report virtual memory statistics
```

In this case, vmstat would be the instruction sought. The number within parentheses indicates the section within the Linux man pages that contains the given instruction's entry.

Following apropos with a string results in a listing of all instructions which contained that string as part of its description (as found in the man page). If your string contains multiple words, apropos will respond differently depending on whether you enclose the string in single or double quote marks or not. Without the quote marks, apropos finds all instructions whose descriptions contain any of the words found in the string. With double quote marks, apropos matches the string exactly. Remember apropos when you cannot remember the instruction's name!

We will continue to explore Linux commands in nearly every chapter of this text. For now, we move forward with more features of the Bash shell.

## 2.4 BASH FEATURES

### 2.4.1 Recalling Commands through History

Every time you enter a command, that command is stored in a history list. To recall the history list, type history. This will provide you a list of every command that has been entered. The number of entries may be limited, but you should see dozens, perhaps hundreds of commands. As an example, see Figure 2.3 which provides a short history list.

Each entry in the history list is preceded by a number. You can re-execute any command in the history list by typing !# where # is the number as shown in the history list. For instance, to repeat the vi command from Figure 2.3, enter !9.

```
1    ls
2    cd ~
3    pwd
4    man cd
5    ls -l
6    cd /bin
7    ls -l | grep root
8    cd ~
9    vi script1
10   cat script1
11   rm script1
12   history
```

FIGURE 2.3   Sample History.

There are several different ways to recall commands from the history list using the ! (sometimes called a "bang"). These are:

- !#—Recall the instruction from line # in the history list

- !!—Recall the last instruction from the history list (in the case of Figure 2.3, it would re-execute history)

- !*string*—Where *string* is a string of character to recall the most recent instruction that started with the given string

If you typed !c, the cat command would execute whereas if you entered !cd, the cd ~ command (number 8) would execute. Notice that !c would yield an error message because command #11 deleted the file and so using cat to display the file would not work. We explore cat, rm, cd, and other commands listed above in Chapter 3.

Another way to recall instructions is to step through the history list from the command line. This can be done by either pressing the up arrow key or typing cntrl+p (i.e., pressing the control key and while holding it down, typing p). By repeatedly pressing the up arrow key or cntrl+p, you are moving instruction by instruction through the history list from the most recent instruction backward. When you reach the instruction you want, press the enter key. Using cntrl+p or the up arrow key is just one form of command line editing. You can also step your way forward through the history list once you have moved back some number of commands by using cntrl+n or the down arrow key.

As with ls, history has a number of options available. The command history by itself lists the entire history list up to a certain limit (see Section 2.5). Useful options are as follows.

- -c—Clear the history list

- -d #—Where # is a number, delete that numbered entry from the history list

- -w—Write the current history list to the history file

- -r—Input the history file and use it instead of the current history

- #—Where # is a number, display only the last # items in the list

## 2.4.2  Shell Variables

A shell is not just an environment to submit commands. As with the history list, the shell also remembers other things, primarily variables and aliases. Variables and aliases can be established in files and input to the interpreter or entered via the command line.

A variable is merely a name that references something stored in memory. Variables are regularly used in nearly every programming language. Variables, aside from referencing memory, have associated with them a type. The type dictates what can be stored in that memory location and how that memory location should be treated.

In many programming languages, variables must be declared before they can be used. This is not the case in Bash. Instead, when you want to first use a variable, you assign it a value. The assignment statement takes the form

```
var = value
```

where *var* is a variable name and *value* is a the initial value stored in the variable. Spaces around the equal sign are not permissible. The variable name can consist of letters, digits, and underscores (_) as long as it begins with a letter or underscore, and can include names that had already been used. However, it is not recommended that you use names of environment variables as the environment variables are set up for specific uses and overriding their values can be dangerous.

A variable in Bash is *typeless*. This means that you do not have to state what type of item will be stored in a variable, and the type can change over time. The only types available in Bash are strings, integers, and arrays. If the number contains a decimal point, the number is treated as a string. Only integers can be used in arithmetic operations. In fact, all variables store strings only but as we will learn in Chapter 7, we can force Bash to treat a string as an integer when performing arithmetic operations.

Once a variable has a value, it can be used in other assignment statements, output statements, selection statements, and iteration statements. We will save the discussion of selection and iteration statements until Chapter 7. To retrieve a value stored in a variable precede the variable name with a $ as in $X or $NAME.

The command to output the value of a variable is *echo*. This is the general purpose output statement for Linux. The format is

```
echo string
```

where *string* can be any combination of literal characters and variable names. The variable names must be preceded by a $ in order to output the values stored in the variables. For instance, assume name = Richard. Then

```
echo name
```

outputs

```
name
```

whereas

```
echo $name
```

outputs

```
Richard
```

The echo command has a few basic options:

- -n—Do not output a newline character at the end. If you use –n, then the next echo statement will output to the same line as this echo statement. By default, echo outputs a newline character at the end of the output.

- -e—Enable interpretation of backslash escapes (explained below).

- -E—Disable interpretation of backslash escapes.

The echo statement accepts virtually anything as a string to be output. Some examples are shown below. Assume that we have variables first and last which are storing Frank and Zappa, respectively.

```
echo $first $last
     Frank Zappa
echo hello $first $last
     hello Frank Zappa
echo hello first last
     hello first last
echo $hello $first $last
     Frank Zappa
```

Notice in the last case, the variable hello has no value. In Linux, we refer to this as the value null. Trying to output its contents using $hello results in no output rather than an error.

If the string is to include a character that has a special meaning, such as the $, you must force the Bash interpreter to treat the character literally. You do this by preceding the character with a \ (backslash). For instance, if amount stores 183 and you want to output this as a dollar amount, you would use

```
echo \$ $amount
```

which would output

```
$ 183
```

The space between \$ and $ was inserted for readability. The instruction echo \$$amount would output

```
$183
```

Other escape characters are shown in Table 2.3.

It is not very common to create your own variables from the command line prompt. However, there are multiple variables already defined that you might find useful. These are

TABLE 2.3    Escape Characters for Linux echo

| Escape Character | Meaning |
| --- | --- |
| \\ | Output a \ |
| \b | Backspace (back cursor up 1 position) |
| \n | Newline—start a new line |
| \t | Tab |
| \! \$ \& \; \' \" | !, $, &, ;, ' and " respectively |

known as environment variables. To see the environment variables defined, use the command env. Some of the more useful variables defined in the Bash environment are shown in Table 2.4.

The PS1 variable can be altered by the user which will then alter the appearance of the user's prompt. If you output PS1 using echo, you might find its contents to be particularly cryptic. For instance, it might appear as [\u@\h \W]$. Each character stored in the prompt variable is output literally unless the character is preceded by a \, in which case it is treated as an escape character. But unlike the escape characters of Table 2.3, these characters inform the Bash interpreter to fill them in with special information from various environment variables.

Table 2.5 indicates the more useful escape characters available for PS1. For instance, we might modify PS1 to store [\t - \!]$ which would output as a prompt [*time - number*]$ where *time* is the current time and *number* is the history list number as in [11:49 - 162]$.

There are also options for changing the color of the prompt and formatting the time and date.

TABLE 2.4    Useful Environment Variables in Bash

| Variable | Meaning (Value) |
| --- | --- |
| DESKTOP_ SESSION | GUI being used (Gnome, KDE) |
| DISPLAY | The terminal window being used, denoted as :0.0 if you are logged in through a GUI |
| HISTSIZE | Number of entries retained in the history list |
| HOME | User's home directory name |
| HOSTNAME | Name of computer |
| LOGNAME | User's user name |
| MAIL | Location of user's email directory |
| OLDPWD | Last directory visited before current directory (see PWD) |
| PATH | The list of directories that the interpreter checks with each command |
| PS1 | Specifies the user's prompt (explained in more detail in Table 2.5) |
| PWD | The current directory |
| SHELL | Name of the program being used as the user's shell (for Bash, this is usually /bin/bash) |
| TERM | The terminal type being used |
| USER | The user's username |
| USERNAME | The user's name (if provided) |

TABLE 2.5    Escape Characters Available for PS1

| Character | Meaning |
| --- | --- |
| \u | User name |
| \h | Host name |
| \w | Current working directory |
| \! | Number of current command (as it will appear in the history list) |
| \t | Current time |
| \d | Date (as in Mon Aug 13) |
| \$? | Status of last command |
| \a | Bell character (makes a tone) |
| \j | Number of jobs in the current shell (we explore jobs in Chapter 4) |
| \n | New line |
| ['*command*'] | Output of the Linux *command* |

### 2.4.3 Aliases

The history list provides one type of command line shortcut. Another very useful shortcut is the *alias*. As with a variable, an alias is an assignment of a name to a value, but in this case the value is a Linux command. The command itself can but does not have to include options and arguments. To define an alias, use an assignment statement preceded by the word alias. The form is

```
alias name = command
```

where *name* is the alias name and *command* is the Linux command.

If the command includes any blank spaces at all, the command must be enclosed in single quote marks (' '). It is a good habit to always use the quote marks whether there are spaces in the command or not.

Aliases are used by users for several reasons:

- To reduce the amount of typing

- So that complicated instructions do not have to be remembered

- So that common typos made by the user can be interpreted correctly

- So that dangerous commands are made safer

To illustrate each of these ideas, we look at some examples

To reduce the amount of typing, if there is a common instruction that a user issues often, an alias can be defined. For instance, the command cd .. changes the current working directory to the current directory's parent directory (we talk about this in Chapter 3 if that does not make sense). To reduce the amount of typing, the user may define the following alias:

```
alias ..='cd ..'
```

Now, the user can type `..` at the command line to perform `cd ..` and thus save typing a few characters.

The mount command mounts a file partition into the file space so that it can be accessed. However, the mount command can be a challenge to apply correctly or to remember. The following alias defines a simple way to mount the CD-ROM drive with its options.

```
alias mountcd = 'mount/dev/cdrom/cd iso9660 ro,user,noauto'
```

At this point, you may not be very comfortable with command line input. As you become more familiar, chances are you will start typing commands at a faster pace which will invariably lead to typos. Simple typos might include misspelling a command such as by accidentally typing sl for ls or lss for less. Imagine that a user has experienced mistyping less as lss and ls as sl. The following two aliases might then be useful:

```
alias lss = less
alias sl = ls
```

Here, we are adding "lss" as a definition to the Linux environment so that, if typed, it is replaced by less. Notice that, if the user were to enter the command `sl ~`, the interpreter handles this in three separate steps. First, the interpreter substitutes the alias as defined, so `sl` becomes `ls`. Next, the parameter is handled. In this case, ~ indicates the user's home directory, which is stored in the environment variable $HOME. Finally, the command is executed, in this case, `ls $HOME`.

Dangerous commands are those that manipulate the file system without first asking the user for confirmation. As an example, the command `rm *` will remove all files in the current directory. The safer mode of rm (the remove, or delete, command), is to use the option –i. This option is known as "interactive mode." What this means is that the command will pause and ask the user for confirmation before any deletion is performed. If there are 25 entries in the current directory, the interpreter will ask the user for confirmation 25 times. This can be annoying, but it is far safer than deleting items that you may actually want to keep. So, the user may define the alias

```
alias rm = 'rm –i'
```

Now, `rm *` becomes `rm –i *`.

To view the aliases already defined, issue the `alias` command with no assignment statement. You can also remove an alias by using `unalias` as in

```
unalias sl
```

to remove the alias for sl.

## 2.4.4 Command Line Editing

To support the user in entering commands, the Bash interpreter accepts a number of special keystrokes that, when entered, move the cursor, copy, cut, or paste characters. These

TABLE 2.6    Command Line Editing Keystrokes

| Keystroke | Meaning |
|---|---|
| c+a | Move cursor to beginning of line |
| c+e | Move cursor to end of line |
| c+n (also up arrow) | Move to next instruction in history list |
| c+p (also down arrow) | Move to previous instruction in history list |
| c+f  (also right arrow) | Move cursor one character to the right |
| c+b (also left arrow) | Move cursor one character to the left |
| c+d (also delete key) | Delete character at cursor |
| c+k | Delete all characters from cursor to end of line |
| c+u | Delete everything from the command line |
| c+w | Delete all characters from front of word to cursor |
| c+y | Yank, return all deleted characters (aside from c+d) to cursor position |
| c+_ | Undo last keystroke |
| m+f | Move cursor to space after current word |
| m+b | Move cursor to beginning of current word |
| m+d | Delete the remainder of the word |

keystrokes are based on keystrokes from the emacs text editor. Learning these keystrokes helps you learn emacs (or vice versa, learning emacs will help you perform command line editing). The keystrokes are combinations of the control key or the escape key and another key on the keyboard. We have already seen two such keystrokes cntrl+p and cntrl+n used to move through the history list.

Table 2.6 shows many of the keystrokes available. In the table, control+*key* is indicated as c+*key* and escape+*key* is indicated as m+*key*. The use of 'm' for escape is because Unix users often referred to the escape key as "meta." The control and escape keys are used differently in that the control key is held down and then the key is pressed while the escape key is pressed first (and released) followed by the key. So for instance, c+p means "hold control and while holding control, hit p" while m+p would be "press escape and then press p."

Let us consider an example. Your Linux system has five users named dukeg, marst, underwoodi, underwoodr, and zappaf. Each of these users has a home directory under /home. You wish to view the contents of each directory. The command you would issue would be ls /home/*username* as in ls /home/zappaf. Rather than entering five separate commands from scratch, after entering the first command you can use it and the command line editing features to reduce the effort in entering the other four commands. We start with the first command.

```
ls /home/dukeg <enter>
```

This displays the contents of dukeg's directory.

Now you wish to see the next user's home directory, marst. Here are the keystrokes you could enter to accomplish this.

- c+p—Move back one instruction in the history list, now the command line has the previous command, placing ls  /home/dukeg on the command line.

- m+b—Move back to the beginning of the current word, the cursor is now at the 'd' in dukeg in ls  /home/<u>d</u>ukeg.

- c+k—Kills the rest of the line, that is, it deletes all of dukeg leaving ls  /home/  on the command line.

Now type marst <enter> which causes ls  /home/marst to execute. You are shown marst's directory. Next, you repeat the same three keystrokes and enter underwoodi <enter>. Now you are shown underwoodi's directory. Next, you can enter the following:

- c+p—Move back one instruction in the history list.

- <backspace>.

And then type r <enter>. This deletes the 'i' in underwoodi and replaces it with 'r'. Finally, you repeat the three keystrokes above (c+p, m+b, c+k) and type zappaf <enter>.

We will visit another example in Chapter 3 which will hopefully further convince you of the value of command line editing.

## 2.4.5 Redirection

In Linux, there are three predefined locations for interaction with the Linux commands (programs). These are standard input (abbreviated as STDIN), standard output (STDOUT), and standard error (STDERR). Typically, a Linux command will receive its input from STDIN and send its output to STDOUT. By default, STDIN is the keyboard and STDOUT is the terminal window in which the instruction was entered.

These three locations are actually file descriptors. A file descriptor is a means for representing a file to be used by a process. In Linux, file descriptors are implemented as integer numbers with 0, 1, and 2 reserved for STDIN, STDOUT, and STDERR, respectively. A redirection is a command that alters the defaults of STDIN, STDOUT, and STDERR.

There are two forms of output redirection, two forms of input redirection, and a fifth form of redirection called a *pipe*. Output redirection sends output to a file rather than to the terminal window from which the program was run. One form writes to the file while the other appends to the file. The difference is that if the file previously existed, the first form overwrites the file. One form of input redirection redirects input from a text file to the program while the other redirects input from keyboard to the program. The Linux pipe is used to send the output of one program to serve as the input to another. This allows the user to chain commands together without having to use temporary files as an intermediate.

The forms of redirection are as follows:

- \> -Send output to the file whose name is specified after the >; if the file already exists, delete its contents before writing.

- \>> -Append output to the file whose name is specified after the >>; if the file does not currently exist, create it.

- < -Use the file whose name is specified after the < as input to the command.

- << -Use STDIN (keyboard) as input. As the <enter> key can be part of any input entered, we need to denote the end of input. There are two mechanisms for this. The user can either use control+d to indicate the end of input or place a string after the << characters to be used as an "end of input" string.

- |-Redirect the output of one command to serve as the input to the next command.

Let us examine the pipe. You want to count the number of items in a directory. The ls command only lists the items found in a directory. The wc command is a word counter. It counts the number of items found in an input and outputs three pieces of information: the number of lines in the input, the number of words in the input, and the number of characters in the input. If we use wc  -l (a lower case "L"), it only outputs the number of lines. The ls  -1 (the number 1) outputs the listing of items in a directory all in one column.

We can combine these two instructions with a pipe. A pipe is denoted with the character | (vertical bar) between the two instructions. In order to pipe from the long listing of ls to word count, we issue the command ls  -1  |  wc  -l. This instruction consists of two separate commands. The first outputs the items in the current directory, all in one column. This output is not sent to the terminal window but instead used as input for the wc command. The wc command expects input to come from a file, but here it comes from ls  -1. The wc command counts the number of lines, words and characters, and outputs the number of lines.

The '<' redirection is probably the least useful because most Linux commands expect input to come from file. We will see, later in the text, that the < form of redirection is used in some instances when a user wants to use a file as input to a user-defined script. We will also find << to be of limited use. Here, we examine all four, starting with > and >>.

If you wish to store the results of a Linux program to a file, in most cases, you will redirect the output. If you want to collect data from several different commands (or the same command entered on different parameters or at different times), you would probably want to append the output to a single file.

Consider that you want to create a file that stores all of the items in various users' directories. You want to put them all into a single file called user_entries.txt. Let us use the same five users from the example in Section 2.4.4. The five commands would be

```
ls/home/dukeg > user_entries.txt
ls/home/marst >> user_entries.txt
```

```
ls /home/underwoodi >> user_entries.txt
ls /home/underwoodr >> user_entries.txt
ls /home/zappaf >> user_entries.txt
```

Notice that the first redirection symbol is '>' to create the file. If the file already existed, this overwrites the file. Had we used '>>' instead and the file already existed, we would be appending not only the latter four outputs but all five of them to the file. NOTE: using command line editing features would simplify the five instructions. Can you figure out how to do that?

The cat command performs a concatenate operation. This means that it joins files together. We can use cat in several different ways. For instance,

```
cat filename
```

will output the result of filename to the terminal window. This lets you quickly view the contents of a text file. Another use of cat is to take multiple files and join them together. The command

```
cat file1.txt file2.txt file3.txt
```

will output the three files to the terminal window, one after another. Instead of dumping these files to the window though, we might be interested in placing them into a new file. We would use redirection for that as in

```
cat file1.txt file2.txt file3.txt > joined_files.txt
```

Yet another use of cat is to create a file from input entered by the keyboard. This requires two forms of redirection. First, we alter cat's behavior to receive input from the keyboard rather than from one or more files. The < redirection redirects input to come from a file rather than a keyboard. We want to use << instead. To establish the end of the input, we will use a string rather than control+d. This means that as the user enters strings and presses <enter>, each string is sent to cat. However, if the string matches our end-of-input string, then cat will terminate. We will use the string "done." The command is so far

```
cat << done
```

This tells the interpreter to receive anything that the user enters via the keyboard, including <enter> keystrokes, terminating only when the user enters "done  <enter>."

Because cat defaults to outputting its results to the terminal window, whatever you entered via cat << done is output back to the window (except for "done" itself). Consider for instance the following interaction. Notice that typical $ is the user prompt while > is the prompt used while the user is entering text in the cat program.

```
$ cat << done
> pear
> banana
> cherry
> strawberry
> done
pear
banana
cherry
strawberry
$
```

As can be seen, the four lines that the user entered are echoed to the window. The string "done" is not. This seems like a useless pursuit; why bother to repeat what was entered? What we would like is for cat to output to a file. So we have to add another redirection clause that says "send the output to the specified file." The full command then might look like this.

cat << done > fruit

Thus, the list of entered items is saved to the file name fruit.

All five forms of redirection can be useful, but the pipe is probably the one you will use the most. We will hold off on further details of the pipe until we reach Chapter 3.

### 2.4.6 Other Useful Bash Features

There are numerous other useful features in Bash. We have already alluded to two of them: tilde expansion and wildcards. In tilde expansion, the ~ is used to denote a user's home directory. If provided by itself, it means the current user's home directory. For instance, if you are foxr, then ~ is foxr's home directory. If used immediately before a username, then the tilde indicates that particular user's home directory. For instance, ~zappaf would mean zappaf's home directory. In most cases, user home directories are located under /home, so ~zappaf would be /home/zappaf.

Wildcards are a shortcut approach to specify a number of similarly named files without having to enumerate all of them. The easiest wildcard to understand, and the most commonly used wildcard, is the *. When the * is used, the Linux interpreter replaces it with all entities in the given directory. For instance, rm * would delete everything in the current directory. The * can be used as a portion of a name. All of the items that start with the name "file" and end with ".txt" but can have any characters in between would appear as file*.txt. This would, for instance, match file1.txt, file21.txt, file_a.txt, and also file.txt. There are other wildcards available. The wildcards are covered in more detail in Chapter 3.

Another highly useful feature is called tab completion. When entering a filename (or directory name), you do not have to type the full name. Instead, once you have typed a unique portion of the name, pressing <tab> will cause Bash to complete it for you. If there are multiple matches, the shell beeps at you. Pressing <tab> <tab> will cause Bash to respond with all possible matches.

For instance, imagine that your current directory stores the following three files that start with the letter 'f.'

```
foolish.txt forgotten.txt funny.txt
```

Typing the command `ls for <tab>` causes Bash to complete this for you as `ls for-gotten.txt`. Typing `ls fo <tab>` causes Bash to beep at you because there are two possible matches. Typing `ls fo<tab><tab>` will cause Bash to list both `foolish.txt` and `forgotten.txt` and repeat the command line prompt with `ls fo` on it, waiting for you to continue typing. Tab completion is extremely useful in that it can reduce the amount of your typing but it can also help you identify a filename or directory name if you cannot remember it.

Finally, and perhaps most confusingly is brace expansion. The user can specify a list of directories inside of braces (curly brackets, or {}). Each item is separated by a comma. This is perhaps not as useful as other types of shortcuts because most Linux commands can accept a list of files or directories to begin with. For instance, `ls /home/foxr /home/zappaf /home/marst /home/dukeg` would list all four of these users' directories. With the braces though, the user has less to type as this can become

```
ls /home/{foxr,zappaf,marst,dukeg}
```

There are more complex uses of the braces, and like the wildcards, this will be covered in more detail in Chapter 3.

Recall that a shell contains an environment. This environment consists of all of the entities defined during this session. What happens if you want some definitions to persist across sessions? There are two mechanisms to support this. First, you can define scripts that run when a new shell starts. The scripts contain the definitions that you want to persist. To support this, Linux has pre-set script files available. As a user, all you have to do is edit the files and add your own definitions. Second, you can export variables that already have values using the export instruction, as in

```
export varname
```

to export the variable *varname*.

The following script files are made available for users to edit. By editing one of these files, you can tailor your shell so that the definitions are available whenever you start a new session.

- `.profile`—Executes whenever the user logs in (no matter which shell is used).

- `.bashrc`—Executes whenever a new bash shell is started.

- `.bash_profile`—Executes whenever the user logs in to a Bash shell.

If you only use Bash, you can put your definitions in .bashrc. This ensures that the definitions take effect whenever you start any new shell. If you start a shell of a different type, then the two bash files (.bashrc and .bash_profile) do not execute. However, there are other, similar, files for other shells. For instance, as .bashrc runs whenever a new Bash shell is started, .cshrc runs whenever a new C-shell (csh) is started.

Note that making changes to one of these files will not immediately impact your current session(s). These files are only read when you log in or start a new bash session. In order for you to force the changes to take effect, you can use the instruction `source` on the given file, as in `source ~/.bashrc`. This causes the interpreter to read the file, executing the statements therein.

## 2.5 OTHER SHELLS

Bash is only one shell available in Linux and Unix operating systems. Linux and Unix usually come with other shells that the user can select either from within a bash shell or the system administrator can specify as their default shell. Let us briefly compare the Bash shell to other shells.

The earliest shell was called the Thompson shell, introduced in 1971 in Unix. This was replaced in 1977 by the Bourne shell, again in Unix. Both shells were denoted by sh, the name of the program that would provide the shell environment and interpreter.

The primary features of the Bourne shell were that shell scripts (programs) could be run by name as if they were any other program in Unix. Programs run in a Bourne shell could be interactive or run in a batch processing mode. Redirection and pipes were made available. Variables were typeless, meaning that any variable could change the type of entity it stored at any time. Quotation marks were introduced for use in commands. Variables could be exported.

In addition, the Thompson shell defined a number of programming instructions separate from the Unix kernel for use in shell scripts. The Bourne shell defined additional instructions such as nested if–then–else constructs, a case (switch) statement, and a while loop. The syntax for the Bourne shell programming constructs was loosely based on Algol.

In the late 1970s, Bill Joy wrote a new Unix shell that he called C shell (csh). The main differences between the Bourne shell and the C shell are that the syntax is largely based on the C programming language and the C shell contains a number of useful editing and shortcut features. In fact, because of the utility of these features, they have been incorporated into many shells since C shell. Among the editing and shortcut features introduced in C shell were:

- History

- Command line editing

- Aliases

- Directory stack (we visit this in Chapter 3)

TABLE 2.7    TC Shell History Commands

| Keystroke | Meaning |
|---|---|
| !! | Execute previous command |
| !n | Execute command n on the history list |
| !string | Execute the most recent command that starts with the characters *string* |
| !?string | Execute the most recent command that contains the characters *string* |
| command !* | Execute *command* but use all arguments from the previous command |
| command !$ | Execute *command* but use the last argument from the previous command |
| command !^ | Execute *command* but use the first argument from the previous command |
| command !:n | Execute *command* but use the nth argument from the previous command |
| command !:m-n | Execute *command* but use arguments m-n from the previous command |

- Tilde completion

- Escape completion (in Bash, it is tab completion)

- Job control (covered in Chapter 4)

A variation of the C shell was created between 1975 and 1983, known as TC shell (tcsh) where the 'T' indicates the TENEX operating system. The TC shell, or the T shell for short, added greater operability to accessing commands in the history. Table 2.7 shows the variations made available in TC shell.

The KornShell (ksh) was developed by David Korn of Bell Labs and released in 1983. It shares many of the same features as C Shell but includes WYSIWYG style command line editing features, unlike the emacs-style command line features of Bash, C shell, and TC shell. It also includes floating point variables and operators.

The next shell released was the Almquist shell, known as ash. This was released in 1989. It is a variation of the Bourne shell written for computers with modest resources because it was smaller and faster than the Bourne shell. Later versions began to incorporate features found in C shell such as command line editing and history. The ash shell has largely been replaced by dash, a newer version available in Debian Linux, or discarded altogether.

In 1989, Brian Fox released the Bourne Again Shell (Bash) for the GNU project. As we have already explored the Bash shell through this chapter, we will not go over the details here. However, as you have seen, it shares many features with C Shell and TC Shell.

Finally, in 1990, the Z shell (zsh) was released as an extension of both the Bourne shell and the Korn shell. In addition to the features obtained from the previous shells, it adds spelling correction, customizable prompts, and modules that directly support both networking and mathematical computations. In addition, some shells have adopted previous shells and added features to support specific programming languages, whether they be Java (Rhino, BeanShell), Perl (psh), SQL (sqsh), or Scheme (scsh).

Most users today will either use a variation of Bourne/Bash or C Shell (often TC shell). So here, we conclude with a comparison between TC Shell and Bash. The differences, relevant to the discussion of this chapter, are illustrated in Table 2.8. If features are not listed

TABLE 2.8    Differences between Bash and TC Shell

| Feature | Bash | TC Shell |
|---|---|---|
| User prompt | $ | % |
| Assignment statement format | var=value | set var=value |
| Export statement | export var=value | setenv var value |
| Alias definition format | alias term=command | alias term command |
| Default user definition file | .bashrc | .cshrc |
| Automated spelling correction | no | yes |

in the table, assume they are the same between the shells. For instance, <, <<, >, >>, |, *, and ~ all work the same between the two shells.

## 2.6  INTERPRETERS

We wrap up this chapter by considering the interpreter. As described in Chapter 1, an interpreter is a program whose task is to execute instructions.

### 2.6.1  Interpreters in Programming Languages

We generally categorize programming languages as interpreted and compiled languages. A compiled language, which includes languages like C, C++, Ada, and Pascal, requires that the program written in that language be translated all at once from its source code into the executable code of machine language. In most of these languages, the entire program must be completely written before translation can take place. C and C++ are exceptions in that pieces of code known as functions can be compiled without having to compile pieces of code that these functions might call upon.

The compiler is a program which performs this translation process, converting an entire program into an executable program. The compiled approach to programming is commonly used for large-scale applications as will be described below.

The interpreted approach to programming utilizes an environment in which the programmer enters instructions one at a time. Each instruction, upon being entered, is interpreted. That is, the interpreter accepts the input, parses it, converts it into an appropriate executable instruction(s), and executes it. This differs greatly from the compiled approach because it allows the programmer to experiment with instructions while writing a program. Thus, a program can be written one instruction at a time. This certainly has advantages for the programmer who might wish to test out specific instructions without having to write an entire program around those instructions.

Another disadvantage with compiled programs is that they are compiled to run on a specific platform. If for instance we compile a C program for a Windows computer, then that compiled program would not run in a Linux environment nor could it run on an HP mainframe. The programmer(s) must compile the source program in every format that the program is destined for.

While the translation time is roughly the same between compiling any single instruction and interpreting any single instruction, we desire to have applications compiled before we make them available to users. The reason for this is that the entire translation process can be time consuming (seconds, minutes, even hours for enormous programs). The typical user will not want to wait while the program is converted, instruction by instruction, before it can be executed. So we compile the program and release the executable code. In addition, many users may not have the capability of running the compiler/interpreter or understanding errors generated during the translation process.

There are many situations where interpreting comes in handy. These primarily occur when the amount of code to be executed is small enough that the time to interpret the code is not noticeable. This happens, for instance, with scripts that are executed on a web server (server-side scripting) to produce a web page dynamically or in a web browser to provide some type of user interaction. As the scripts only contain a few to a few dozen instructions, the time to interpret and execute the code is not a concern.

There are numerous interpreted programming languages available although historically, most programming languages have been compiled. One of the earliest interpreted languages was LISP, a language developed for artificial intelligence research. It was decided early on that a LISP compiler could be developed allowing large programs to be compiled and thus executed efficiently after compilation. This makes LISP both an interpreted language (the default) and compiled (when the programmer uses the compiler) giving the programmer the ability to both experiment with the language while developing a program and compile the program to produce efficient code. More recent languages like Ruby and Python are interpreted while also having their own compilers.

In the 1990s, a new scheme was developed whereby a program could be partially compiled into a format called *byte code*. The byte code would then be interpreted by an interpreter. The advantage of this approach is that the byte code is an intermediate version which is platform independent. Thus, a program compiled into byte code could potentially run on any platform that has the appropriate interpreter. In addition, the byte code is close enough to executable code that the interpreting step is not time consuming so that run-time efficiency is at least partially retained from compilation. Java was the first programming language to use this approach where the interpreter is built into software known as the Java Virtual Machine (JVM). Nearly all web browsers contain a JVM so that Java code could run inside your web browser. Notice that JVM contains the term "virtual machine." The idea behind the JVM is that it is a VM capable of running the Java environment, unlike the more generic usage of VM introduced in Chapter 1.

## 2.6.2 Interpreters in Shells

When it comes to operating system usage, using an interpreter does not accrue the same penalty of slowness as with programming languages. There are a couple of reasons for this. First, the user is using the operating system in an interactive way so that the user needs to see the result of one operation before moving on to the next. The result of one operation might influence the user's next instruction. Thus, having a precompiled list of steps would make little or no sense. Writing instructions to be executed in order will only work if the user is absolutely sure of the instructions to be executed.

Additionally, the interpreter is available in a shell and so the operating system is running text-based commands. The time it takes to interpret and execute such a command will almost certainly be less than the time it takes the user to interact with the GUI, which is much more resource intensive.

The other primary advantage to having an interpreter available in the operating system is that it gives the user the ability to write their own interpreted programs. These are called scripts. A script, at its most primitive, is merely a list of operating system commands that the interpreter is to execute one at a time. In the case of Linux, the shell has its own programming instructions (e.g., assignment statements, for loops, if–then statements) giving the user a lot more flexibility in writing scripts. In addition, instructions entered via the command line can include programming instructions.

The other advantage to interpreted code is that the interpreter runs within an environment. Within that environment, the user can define entities that continue to persist during the session that the environment is active. In Linux, these definitions can be functions, variables, and aliases. Once defined, the user can call upon any of them again. Only if the item is redefined or the session ends will the definition(s) be lost.

### 2.6.3 The Bash Interpreter

The Bash interpreter, in executing a command, goes through a series of steps. The first step is of course for the user to type a command on the command line. Upon pressing the <enter> key, the interpreter takes over. The interpreter reads the input.

Now, step-by-step, the interpreter performs the following operations:

- The input is broken into individual tokens.
  - A token is a known symbol (e.g., << , |, ~, *, !, etc) or
  - A word separated by spaces where the word is a command or
  - A word separated by spaces where the word is a filename, directory name, or path or
  - An option combining a hyphen (usually) with one or more characters (e.g., -l, -al)
- If the instruction has any quotes, those quotes are handled.
- If an alias is found, that alias is replaced by the right-hand side of the definition.
- The various words and operators are now broken up into individual commands (if there are multiple instructions separated by semicolons).
- Brace expansion unfolds into individually listed items.
- ~ is replaced by the appropriate home directory.
- Variables are replaced by their values.
- If any of the commands appears either in ` ` or $(), execute the command (we explore this in Chapter 7).

- Arithmetic operations (if any) are executed.

- Redirections (including pipes) are performed.

- Wildcards are replaced by a list of matching file/directory names.

- The command is executed and, upon completion, the exit status of the command (if necessary) is displayed to the terminal window (if no output redirection was called for).

## 2.7 CHAPTER REVIEW

Concepts and terms introduced in this chapter:

- Alias—A substitute string for a Linux instruction; aliases can be entered at the command line prompt as in `alias rm = 'rm –i'`, or through a file like .bashrc.

- Bash—A popular shell in Linux, it is an acronym for the Bourne Again Shell. Bash comes with its own interpreter and an active session with variables, aliases, history, and features of command line editing, tab completion, and so forth.

- Bash session—The shell provides an environment for the user so that the user's commands are retained, creating a session. The session contains a history of past commands as well as any definitions entered such as aliases, variables, or functions.

- Byte code—An approach taken in Java whereby Java programs are compiled into an intermediate form so that a JVM can interpret the byte code. This allows compiled yet platform-independent code to be made available on the Internet yet does not require run-time interpreting that a purely interpreted language would require, thus permitting more efficient execution.

- Command line editing—Keystrokes that allow the user to edit/modify a command on the command line prompt in a Bash session. The keystrokes mirror those found in the Emacs text editor.

- Compiler—A program that converts a high-level language program (source code) into an executable program (machine language). Many but not all high-level languages use a compiler (e.g., C, C++, Java, FORTRAN, COBOL, Ada, PL/I). Prior to the 1990s, few high-level languages were interpreted (Lisp being among the first) but today numerous languages are interpreted (e.g., PHP, Perl, JavaScript, Ruby, Python) but still may permit compilation for more efficient execution.

- C-Shell—A popular Unix and Linux shell that predates Bash.

- Environment variable—A variable defined by a program (usually the operating system or an operating system service) that can be used by other programs. These include HOME (the user's home directory), PWD (the current working directory), OLDPWD (the previous working directory), PS1 (the user's command line prompt definition), and USER (the user's username) to name a few.

- History list—Bash maintains the list of the most recent instructions entered by the user during the current session. This list is usually 1000 instructions long. You can recall any of these instructions by using the ! character. !! recalls the last instruction, !*n* recalls instruction *n* from the list, and !*string* recalls the last instruction that matches *string*.

- Home directory (~)—The user's home directory, typically stored at /home/*username*, can be denoted using the ~ (tilde) character. You can also reference another user's home directory through ~*username*.

- Interpreter—A program that interprets code, that is, receives an instruction from the user, translates it into an executable form, and executes it. Linux shells contain an interpreter so that the user can enter commands in the shell via a command line.

- Java virtual machine—A program which contains a Java byte-code interpreter. Nearly all web browsers contain a JVM so that you can run compiled Java code allowing programmers to place compiled code on the Internet without having to worry about compiling the code for every available platform. This also protects the programmer's intellectual property in that the code is not available in its source format so that people cannot see how the program works.

- Long listing—The ls command in Linux lists the contents of the current (or specified) directory. The long listing displays additional information about the contents including permissions, sizes, and owners.

- Man pages—Help information available for nearly every Linux instruction.

- Options—Linux instructions permit a series of options that allow the user to more concretely specify how the instruction should operate. For instance, ls permits –l to provide a long listing, -a to print hidden files, and so forth. To view the options, use the man page.

- Parameters—In addition to options, many Linux commands expect or permit parameters such as files to operate on.

- Pipe—A form of redirection in which the output of one Linux command is used as the input to another. The pipe operator is | which is placed between two commands such as `ls  -l  |  wc  -l`.

- Prompt—The shell presents a prompt when waiting for the user's next command. The PS1 environment variable allows you to tailor the prompt. By default, it outputs the user's name, the computer's hostname, the current working directory, and either $ or # to indicate that you are a normal user or root.

- PS1 variable—This variable is predefined for users but the user can redefine it as desired. By default it is [\u@\h \W]\$ which means to output the user's username, an @ symbol, the hostname, a space, the current working directory, all in [], followed by \$, the prompt character (usually either $ or #). This could be redefined to output the current date, the current time, the command number (i.e., the number in the history list that this command will be stored as), and/or the result of the last command, among others.

- Redirection—The ability to change the direction of input or output of the command being specified. By default, most Linux commands accept input from a file, but this could be redirected to come from keyboard, and send output to the terminal window, but this could be redirected to go to a file. Redirection characters are <, << for redirecting input, >, >> for redirecting output, and | for a pipe between two instructions.

- STDERR—Default output location for error messages.

- STDIN—Default input source for running programs.

- STDOUT—Default output for running programs.

- TC-Shell—A variation of C-Shell with updated features, many of which are found in Bash such as command line editing and filename completion.

- Wildcard—Characters that are used to represent other characters to match against in order to simplify file and directory specification. For instance, * means to match anything so that ls *.txt will list all files that end with .txt.

Linux commands covered in this chapter:

- alias—To define an alias, the form is `alias name = instruction`. The string *name* can then be used in place of the instruction from the command line. If the instruction contains spaces, you must enclose it in single quote marks. The alias exists for the duration of the current Bash session so you might move the definition into your .bashrc file.

- apropos—Find instructions whose descriptions match the given string.

- arch—Output the architecture (processor type) of the computer.

- bash—Start a new bash session. Once in this session, any definitions of your previous bash session are "hidden" unless they were exported. To leave this bash session, use exit.

- cat—Concatenate the file(s). By default, output is sent to the terminal window but this is often used with redirection to create new files. Using `cat << string > filename` allows you to input text from the keyboard and store the result in the new file *filename*.

- echo—Output instruction to output literal strings and the values of variables. Echo will be of more use in scripts (see Chapter 7).

- exit—Used to exit a bash session. If there was no previous bash session, this will close the terminal window.

- export—Used to export a variable into the environment. This can be used from within a script or so that a variable can be used in a new (or older) bash session.

- history—Output the last part of the history list.

- hostname—Output the computer's host name.

- ls—Provide a listing of the contents of the specified directory, or if no directory is specified, the current working directory.

- man—Display the given instruction's man(ual) page.

- passwd—Allow the user to change passwords, or if specified with a username, allow root to change the specified user's password.

- pwd—Print the current working directory.

- source—Execute the given file by the current interpreter.

- unalias—Remove an alias from the environment.

- uname—Print the operating system name.

- wc—Perform a word count for the given textfile(s) which outputs the number of bytes (characters), words and lines.

- who—List all logged in users and their login point (e.g., on console, remotely from an IP address).

- whoami—Output the current user's username.

## REVIEW QUESTIONS

1. Why do we refer to the user's environment as a shell?

2. What is the history of the Bash shell?

3. Given the following prompt, answer the questions below.

   `[zappaf@frenchie /etc]#`

   a. Who is logged in?

   b. Is the user logged in as themself or root?

   c. What is the current working directory?

   d. What is the hostname?

4. What command would you use to determine your username?

5. What command would you use to find out your current working directory?

6. What command would you use to change your password?

7. What is the difference between the arch command and the uname command?

8. Assume that foxr is logged in to a Linux computer on console and user zappaf is logged into the computer remotely from IP address 1.2.3.4. What would the command who show?

9. From your Bash shell, you type `bash`. After working for a while, you type `exit`. What happens?

10. What is the difference between `ls  -l` and `ls  -L`?

11. What is the difference between `ls  -a` and `ls  -A`?

12. You want to perform a recursive listing using ls. What option(s) would you specify?

13. You type ls and you receive 3 columns worth of file names. You want to see the output as a single column listing. What can you do?

14. What does the synopsis portion of a man page tell you? If you are not sure, try `man ls`, `man  mount`, and `man  ip` as examples and see what you get.

15. You want to reexecute the last instruction you entered in your Bash session. Describe three ways to do this.

16. Given the history list from Figure 2.3, what command is recalled with each of the following?

    a.  `!!`

    b.  `!5`

    c.  `!ls`

    d.  `!cd`

    e.  `!c`

17. What does `history  -c` do?

18. Define a variable to store your first initial, your middle initial, and your last name. Include proper spaces and punctuation.

19. Assume the variable AGE stores 21. What is the output for each of the following?

    a.  `echo AGE`

    b.  `echo $AGE`

    c.  `echo $age`

    d.  `echo \$AGE`

    e.  `echo $AGE\!`

20. Provide an instruction to alter your prompt so that it displays the following information:

(time : instruction number) [username] prompt character as in
(10:04 am : 31) [foxr] $

21. Write the command to define as an alias ... to move you up two directories.

22. Write the command to define as an alias stuff to perform a long listing of your home directory.

23. Match the following command line editing function with the keystroke that provides it

    a.   Move the cursor to the end of the line          i. c+k

    b.   Move back one character          ii. c+ _

    c.   Move forward one word          iii. c+e

    d.   Delete one character          iv. m+f

    e.   Retrieve the last instruction entered          v. c+b

    f.   Undo the last keystroke          vi. c+d

    g.   Delete all characters from here to the end of the line    vii. c+p

24. Provide an instruction to redirect the output of ls -l to the file ~/listing.txt.

25. How does the operator >> differ from the operator >?

26. Explain why wc < somefile does not require the < operator.

27. What is the difference between cat file1.txt file2.txt file3.txt and cat file1.txt file2.txt > file3.txt?

28. Provide an example of why you might use << as a form of redirection.

29. The following instruction does not do what we expect in that we want to output the word count for the items listed in the current directory. What does it do and how can you fix it?

ls > wc

30. Name three compiled languages.

31. Name three interpreted languages.

32. Why is Bash an interpreted environment rather than a compiled one?

33. Explain the role of the following characters as used in Bash.

    a.   ~

    b.   !

    c.   –

    d.   *

# CHAPTER 3

# Navigating the Linux File System

THIS CHAPTER'S LEARNING OBJECTIVES are

- To be able to utilize Linux file system commands cat, cd, cmp, comm, cp, diff, head, join, less, ln, ls, more, mkdir, mv, paste, pwd, rm, rmdir, sort, tail, and wc

- To be able to navigate through the Linux file system using both absolute and relative paths

- To understand the hierarchical layout of the Linux file system including the roles of the top-level directories

- To be able to apply wildcards to simplify file commands

- To be able to use find to locate files

- To understand hard and soft links

- To understand the Linux permissions scheme, how to alter permissions using chmod and how to convert permissions to their 3-digit value

- To understand the different types of storage devices and the advantages that the hard disk has over the other forms

- To understand lossy versus lossless file compression

## 3.1 INTRODUCTION

The operating system is there to manage our computers for us. This leaves us free to run applications without concern for how those applications are run. Aside from starting processes and managing resources (e.g., connecting to a printer or network, adding hard disks, etc.), the primary interaction that a user has with an operating system is in navigating through the file system. It should be noted that the term file system has multiple meanings. In this chapter, we generally mean the entire collection of storage devices that make up the file space. In Chapter 10, we will see that the term file system also means the type of storage used for a partition.

The file system is a logical way to describe the collection of storage devices available to the computer. The primary storage device for any computer system is the internal hard disk. There may be additional hard disks (internal or external) as well as mounted optical disk, flash drive(s), networked drives (either optical or hard disk), and mounted magnetic tape.

The word *logical* is used to denote the user's view of the file system, which is a collection of partitions, directories, and files. It is an abstraction or generalization over the physical implementation of the file system, which consists of blocks, disk surfaces (tracks, sectors), file pointers, and the mechanisms of a storage unit (read/write arm, spindle, etc.). It is far easier for us to navigate a file system by using the abstract concepts of files, directories, partitions, and paths.

After a short introduction, we will explore how to get around the Linux file system. We view this strictly from the logical point of view as the user is not required (nor even permitted) to interact with the physical file system. It is the operating system which translates the user's commands into physical operations. First, we will look at paths to express source and destination locations. We then examine the common Linux commands for navigating the file system. We then look in detail at the default structure of a Linux file system so that we can learn where programs and files are placed. We also introduce Linux permissions and the concept of file compression. Later in the text, we will return to the file system from a system administrator's perspective to see how to manage the file system.

### 3.1.1 File System Terminology

File system: the collection of storage devices which, combined, make up the system's file space. We can view the file system at two different levels. The logical, or abstract, level expresses the collection of files that we have stored, organized hierarchically through directories and subdirectories. The physical levels are the devices and how they physically store the information such as through magnetic information. At the physical level, files are broken into blocks. We will primarily consider the logical level in this chapter, concentrating on files and directories. Most users will group common files together into a directory. Directories and files may be placed into other directories. This provides a hierarchical structure to the file system. If organized well, a hierarchy may be several levels deep so that the user can easily find a file without resorting to searching for it. Top-level directories are placed onto various partitions.

Partition: a disk partition is a *logical* division of the collection of storage devices into independent units. This organizes storage so that different sections are dedicated to different portions of the file system. In Linux, many of the top-level directories are placed in separate partitions, such as /home and /var. How does a partition differ from a directory? The operating system is able to perform a maintenance operation on one partition without impacting other partitions. For instance, one partition might be removed (unmounted) while other partitions remain accessible. One partition might be specified as read-only while others are readable and writable. Performing a backup on a partition renders it unavailable while the backup is being performed, but other partitions can be available if they are not part of the backup. Although partitioning is a logical division, it also sets aside physical sections of disk surfaces to different partitions. Partitions are created by the system administrator. Directories can be created by any user.

In Linux, common partitions are for the user directory portion (/home), the operating system kernel and other system files (/, /bin, /sbin, /etc) and the swap space (virtual memory). Finer grained partitioning can be made, for instance by providing /usr, /var, and /proc their own partitions. Another partition that might be useful is to give the boot system (/boot) its own partition. This especially becomes useful if you plan to make your computer dual-booting.

Directory: a user-specified location that allows a user to organize files. In Linux, files and directories are treated similarly. For instance, both are represented using an inode and both have many properties in common (name, owner, group, creation date, permissions). Directories will store files, subdirectories, and links. A link is a pointer to a file in another directory. There are two forms of links, hard links and symbolic links.

File: the basic unit of storage in the logical file system, the file will either store data or an executable program. Data files come in a variety of types based on which software produced them. Typically, a file is given an extension to its name to indicate its type, for instance, .docx, .mpg, .mp3, .txt, .c. In Linux, many data files are stored as text files meaning that they can be viewed from the command line using commands such as cat, more, and less, and can be edited using a text editor (vi, emacs, gedit). Files, like directories, have a number of properties. The file's size may range from 0 bytes (an empty file) to as large as necessary (megabytes or gigabytes, for instance).

Path: a description of how to reach a particular file or directory. The path can be absolute, starting at the Linux file system root (/) or relative, starting at the current working directory.

inode: a data structure stored in the file system used to represent a file. The inode itself contains information about the file (type, owner, group, permissions, last access/create/ modification date, etc.) and pointers to the various disk blocks that the file contains. We examine inodes and disk blocks in detail in Chapter 10.

Link: a combination of a file's name and the inode number that represents the file. A directory will store links to the files stored in the directory. Through the link, one can gain access to the file's inode and from the inode, to the file's blocks. There are two types of links, hard and soft (symbolic).

### 3.1.2 A Hierarchical File System

Linux is a hierarchical operating system. This means that files are not all located in one place but distributed throughout the file system based on a (hopefully) useful organizing principle. To organize files, we use directories. A hierarchical file system is one that contains directories and subdirectories. That is, any given directory not only stores files but can also store directories, called subdirectories. Since a subdirectory is a directory, we can apply this

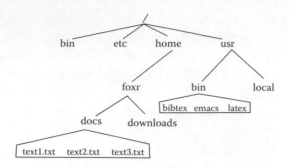

FIGURE 3.1   Partial example of a hierarchical file space.

definition so that this directory can have its own directories, which we might call subsubdirectories. The pathway from the topmost level to the bottommost level of the file system represents the depth of the file system. In Linux, there is no limit to the maximum depth.

Most Linux dialects are based on a preestablished standard (the LSB, see Chapter 1). This standard includes much of the initial structure of the Linux file system. Thus, you will find in just about every Linux operating system top-level directories named /bin, /boot, /dev, /etc, /home, /lib, /tmp, /usr, /var, and possibly others. It is this organization that helps not only system administrators but all Linux users easily learn where to find useful files.

As a user, you are given your own directory under /home. Within this directory, you are allowed to store files and create subdirectories as desired (to within the limit dictated by any disk quota, if any). Subdirectories can themselves contain subsubdirectories to any depth.

Figure 3.1 illustrates the concepts of the Linux hierarchy file system. In the figure, we see some of the top-level Linux directories, many of which contain subdirectories and subsubdirectories. One user, foxr, is shown, and only two subdirectories of /usr are shown.

When issuing a command, a path is required if the command is not stored in the current directory. This is also true if the command is to operate on a specified file which is not stored in the current directory. For instance, imagine that the user foxr wants to run the program emacs, located in `/usr/bin`, and open the file text1.txt located in foxr's docs subdirectory. The user is currently in his home directory (foxr). A path is required to reach both emacs and text1.txt. There are multiple ways to specify paths, as covered in the next section. One command is

```
/usr/bin/emacs /home/foxr/docs/text1.txt
```

where the first part is the command (emacs) and the second is a file to open within emacs.

## 3.2  FILENAME SPECIFICATION

### 3.2.1  The Path

In this and the next subsection, refer to Figure 3.2 as needed. In this figure, we see two top-level directories, etc and home with home having two subdirectories, foxr and zappaf. The items in italics are files.

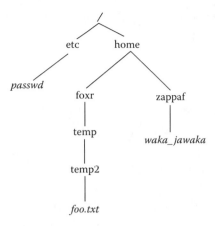

FIGURE 3.2  Example file space.

A path is a description of how to reach a particular location in the file system. A path is required if the Linux command operates on a file or directory that is not in the current working directory. Paths can be expressed in one of two ways: absolute and relative.

An absolute path starts at the root level of the file system and works its way down the file system's hierarchy, directory by directory. The length of an absolute path is determined by the number of directories/subdirectories that exist between the root level and the target file or directory. Each new directory listed in a path is denoted by a slash (/) followed by the directory name. Each directory/subdirectory listed in the path is separated by a /. As an example, the passwd file is stored in the top-level directory etc. From root, this is denoted as /etc/passwd.

Consider a user, foxr, who has a directory in the top-level directory home, and has a sub-directory, temp, which has a subdirectory, temp2, which contains the file foo.txt (see Figure 3.2). The absolute path to this file is /home/foxr/temp/temp2/foo.txt indicating a greater depth than /etc/passwd.

Some commands operate on directories instead of files. In such a case, the / that follows the directory name is not needed. For instance, /home/foxr/temp is acceptable although /home/foxr/temp/ is also acceptable. If the command is to operate on the items stored in the directory, then the specification may look like this: /home/foxr/temp/*. The use of * is a wildcard, as introduced in Chapter 2.

A relative path starts at the current working directory. The working directory is the directory that the user last changed into (using the cd command). The initial working directory upon logging in and opening a terminal window is either the user's home directory (e.g., /home/foxr for user foxr) or the user's Desktop directory (depending on how the user opened the terminal window). Recalling foxr's directory structure from Figure 3.2, if foxr's current directory is temp, then the file foo.txt is found through the relative path temp2/foo.txt. Notice that the subdirectory, temp2, is specified without a leading slash (/). If you were to state /temp2/foo.txt, this would lead to an error because /temp2 is an absolute reference meaning that temp2 is expected to be found among the top-level directories. If you were in the directory /etc, then the file passwd is indicated by relative path as just passwd.

You can identify the current working directory in three ways: through the command pwd, through the variable PWD (for instance, by entering the command echo $PWD), or by looking at the command line prompt (in most cases, the current working directory is the last item in the prompt before the $).

### 3.2.2 Specifying Paths above and below the Current Directory

To specify a directory or file beneath the current directory, you specify a downward motion. This is done by listing subdirectories with slashes between the directories. If you are using a relative path, you do not include a leading /. For instance, if at /home/foxr, to access foo.txt, the path is temp/temp2/foo.txt.

To specify a directory above the current directory, you need to indicate an upward motion. To indicate a directory above the current directory, use .. as in ../foo.txt which means the file foo.txt is found one directory up. You can combine the upward and downward specifications in a single path. Let us assume you are currently in /home/foxr/temp/temp2 and wish to access the file waka_jawaka in zappaf's home directory. The relative path is ../../../zappaf/waka_jawaka. The first .. moves you up from temp2 to temp. The second .. moves you from temp to foxr. The third .. moves you from foxr to home. Now from /home, the rest of the path moves you downward into zappaf and finally the reference to the file. In a relative path, you can use as many .. as there are between your current directory and the root level of the file system. The absolute reference for the file waka_jawaka is far easier to state: /home/zappaf/waka_jawaka.

As another example, if you are currently at /home/foxr/temp/temp2, to reach the file /etc/passwd, you could specify absolute or relative paths. The absolute path is simply /etc/passwd. The relative path is ../../../etc/passwd.

To denote the current directory, use a single period (.). This is sometimes required when a command expects a source or destination directory and you want to specify the current working directory. We will see this when we use the move (mv) and copy (cp) commands. If you wish to execute a script in the current working directory, you precede the script's name with a period as well.

As explained in Chapter 2, ~ is used to denote the current user's home directory. Therefore, if you are foxr, ~ means /home/foxr. You can also precede any user name with ~ to indicate that user's home directory so that ~zappaf denotes /home/zappaf.

### 3.2.3 Filename Arguments with Paths

Many Linux file commands can operate on multiple files at a time. Such commands will accept a list of files as its arguments. For instance, the file concatenation command, cat, will display the contents of all of the files specified to the terminal window. The command

```
cat file1.txt file2.txt file3.txt
```

will concatenate the three files, displaying all of their content.

Each file listed in a command must be specified with a path to that file. If the file is in the current working directory, the path is just the filename itself. Otherwise, the path may be specified using the absolute pathname or the relative pathname.

Assume that file1.txt is under /home/foxr, file2.txt is under /home/foxr/temp, and file3. txt is under /home/zappaf, and the current working directory is /home/foxr. To display all three files using cat, you could use any of the following (along with other possibilities):

- `cat file1.txt temp/file2.txt ../zappaf/file3.txt`

- `cat file1.txt /home/foxr/temp/file2.txt`

  `/home/zappaf/file3.txt`

- `cat ./file1.txt temp/file2.txt ../zappaf/file3.txt`

Given a filename or a full path to a file, the instructions `basename` and `dirname` return the file's name and the path, respectively. That is, basename will return just the name of the file and dirname will return just the path to the file. If, for instance, we issue the following two commands:

```
basename /etc/sysconfig/network-scripts/ifcfg-eth0
dirname /etc/sysconfig/network-scripts/ifcfg-eth0
```

we receive the values `ifcfg-eth0` and `/etc/sysconfig/network-scripts`, respectively. For the commands `basename ~` and `dirname ~`, we receive `username` and `/home`, respectively where *username* is the current user's username.

### 3.2.4 Specifying Filenames with Wildcards

A wildcard is a way to express a number of items (files, directories, links) without having to enumerate all of them. The most common wildcard is *, which represents "all items." By itself, the * will match any and every item in the current directory. For instance, `ls *` lists all items in the current directory. You can combine the * with a partial description of item names. When appended with .txt, the * specifies "match all items whose name ends with .txt." For instance, `ls *.txt` lists all items whose names end with the .txt extension in the current working directory.

The Linux interpreter performs an operation called *filename expansion*, or *globbing*, prior to executing an instruction. The interpreter converts any wildcard(s) given in the instruction into a list of all items that match the combination of wildcard(s) and literal characters. In the command `ls *.txt`, the interpreter first expands *.txt into the list of filenames ending with .txt. Another example would be `ls a*` which would match all items whose names start with the lower case 'a'. The interpreter compiles the list of all matching items, and then replaces *.txt or a* with the list generated. Now, the ls command can be performed.

TABLE 3.1    Wildcards in Linux

| Wildcard | Explanation |
| --- | --- |
| * | Match anything (0 characters or more) |
| ? | Match any one character |
| [chars] | Match any one of the characters in the list |
| [char1-char2] | Match any one of the characters in the range from char1 to char 2 (e.g., [0–9], [a-e], [A-Z]) |
| {word1,word2,word3} | Match any one of the words |
| [!chars] | Match any one character not in the list |

Table 3.1 lists the available wildcards in Bash. There are additional wildcards that can be used in Bash, but they require specific installation.

Assume the current working directory has the following files:

```
file.txt file1.txt file2.txt file2a.txt file21.txt file_a.txt
   file_21.txt file.dat file1.dat file2.dat abc bcd aba bb
```

Table 3.2 provides several example listings using ls and different wildcard combinations. Wildcards can be confusing. The best way to learn them is through practice. Later in the text, we will look at regular expressions, which also use *, ?, and []. Unfortunately, the interpretation of * and ? differ in regular expressions, which will further complicate your ability to use them.

One set of wildcards, as shown in Figure 3.2, is the braces, or {} characters. Here, you can enumerate a list of items separated by commas. The Bash interpreter uses brace expansion to unfold the list. Recall that the ls command can receive a list of items as in the following instruction.

TABLE 3.2    Examples Using ls with Wildcards

| ls Command | Result |
| --- | --- |
| `ls *.txt` | `file1.txt file21.txt file2a.txt` `file2.txt file_a.txt file.txt` |
| `ls *.*` | `file1.txt file21.txt file2a.txt` `file2.txt file_a.txt file.txt` `file.dat file1.dat file2.dat` |
| `ls *` | `All files` |
| `ls file?.*` | `file1.txt file2.txt file1.dat` `file2.dat` |
| `ls file??.txt` | `file21.txt file2a.txt file_a.txt` |
| `ls file?.{dat,txt}` | `file1.dat file1.txt file2.dat` `file2.txt` |
| `ls [abc][abc][abc]` | `aba abc` |
| `ls [!a]*` | All files except for aba and abc |

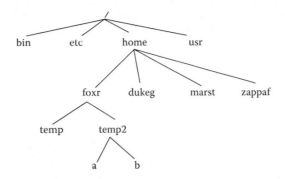

FIGURE 3.3  Example directory structure.

```
ls file1.txt file2.txt file1.dat
```

Thus, enumerating the list in braces does not seem worthwhile. However, the power of brace expansion comes when the items in braces are directories and subdirectories. If you use this correctly, it saves you from having to type full paths to each directory.

Consider the directory structure shown in Figure 3.3. If you wanted to perform some operation, say ls, on all of the subdirectories of foxr, you could do so with the following listing:

```
ls /home/foxr/temp /home/foxr/temp2 /home/foxr/temp2/a
    /home/foxr/temp2/b
```

But this listing can be greatly simplified (reduced in size) using braces as shown here:

```
ls /home/foxr/{temp,temp2/{a,b}}
```

The above instruction is unfolded into the full list /home/foxr/temp, /home/foxr/temp2, /home/foxr/temp2/a, /home/foxr/temp2/b, without the user having to do so. Through brace expansion, you do not have to repeat all of the directory paths.

Through the use of .., ., ~, *, ?, [], and {}, the user is provided a number of very useful shortcuts to simplify paths. Recall from Section 2.6.3 that the Bash interpreter performs several steps before executing an instruction. With respect to these symbols, Bash first expands items found in braces. This is followed by replacing ~ with the appropriate home directory name. Next, globbing, or filename expansion, takes place replacing wildcards with lists of matching files and directories. Finally, the instruction can be executed on the resulting list(s) generated.

## 3.3  FILE SYSTEM COMMANDS

In this section, we examine many common Linux file system commands. This section does not include commands reserved for system administrators. Keep in mind that if you are attempting to apply an instruction on an item owned by another user, the command may

TABLE 3.3    Common Linux File Commands

| Command | Description | Common Options |
|---------|-------------|----------------|
| pwd | Display current directory | |
| cd | Change directory | |
| ls | List contents of directory | -a (all), -l (long listing), -r (reverse), -R (recursive), -S (sort by file size), -1 (display in one column) |
| mv | Move or rename file(s)/directory(ies) | -f (force), -i (interactive), -n (do not overwrite existing file) |
| cp | Copy file(s) | same as mv, also -r (recursive) |
| rm | Delete file(s) | -f (force), -i (interactive), -r (recursive) |
| mkdir | Create a directory | |
| rmdir | Delete a directory | |
| cat | Concatenate files (display to window) | -n (add line numbers), -T (show tabs) |
| less | Display file screen-by-screen | -c (clear screen first), -f (open nonregular files) |
| more | Display file screen-by-screen | -num # (specify screen size in # rows), +# (start at row number #) |
| find | Locate file(s) | Covered in Table 3.6 |

not work depending on that item's permissions. We explore permissions in Section 3.5. We also visit some more special purpose file system commands at the end of this chapter.

One aspect of Linux which is most convenient is that many file system commands will operate on different types of file system constructs. These include files, directories, symbolic links, pipes, I/O devices, and storage devices. We discuss the difference between these types in Chapter 10 although we also look at symbolic links in this chapter and pipes in other chapters.

The `file` command will describe the type of entity passed to the command. For instance, `file /etc/passwd` will tell us that this is ASCII text (a text file) while `file /` will tell us that / is a directory and `file /dev/sda1` will tell us that sda1 is a block special device (a special file for a block device). Other types that might be output from the file command include empty (an empty text file), a particular type of text file (e.g., C program, bash shell script, Java program), symbolic link, and character special (character-type device). Using the -i option causes file to output the MIME type of the file as in `text/plain`, `application/x-directory`, `text/x-shellscript`, and others.

Table 3.3 displays the commands we will cover in Sections 3.3.1 through 3.3.5, along with a description of each command and the more common options. To gain a full understanding of each command, it is best to study the command's main page.

### 3.3.1 Directory Commands

You will always want to know your current working directory. Without this knowledge, you will not be able to use relative paths. The command `pwd` is used to print the current working directory. It is one of the simplest commands in Linux as it has no options. The command responds with the current working directory as output. For instance:

```
$ pwd
/home/foxr
$
```

Of course, you can also identify the current working directory by examining your user prompt. Unless you have changed it, it should contain the working directory, for instance:

```
[foxr@localhost ~] $
```

Unfortunately, the prompt only displays the current directory's name, not its full path. If you were in a subdirectory of foxr's called TEMP, the prompt would look like this:

```
[foxr@localhost TEMP] $
```

To navigate around the file system, you need to change the directory. The command to do that is `cd`. The command expects the path to the destination directory. This path can be relative or absolute. You would only specify a directory, not a filename, in the command. Referring back to Figure 3.3, and assuming you are currently in /home/zappaf, the following cd commands can be used to navigate about this file space.

- To reach /bin either cd /bin or cd ../../bin

- To reach marst either cd /home/marst or cd ../marst

- To reach temp either cd /home/foxr/temp or cd ../foxr/temp

- To reach a either cd /home/foxr/temp2/a or cd ../foxr/temp2/a

If you were in b, you would issue these commands.

- To reach a either /home/foxr/temp2/a or ../a

- To reach temp either /home/foxr/temp or ../../temp

- To reach dukeg either /home/dukeg or ../../../dukeg

If you have a lot of different directories that you wish to visit often, you might place them onto the directory stack. The `dirs` command displays all directories on the stack. To add a directory, use `pushd` *dirname*. The push operation places the new directory on the "top" of the stack, so you are actually building your list backward. To remove the top directory from the stack, use `popd`.

Once we have moved to a new location in the file system, we may want to view the contents of the current location. This is accomplished through the ls command. As we already explored the ls command in some detail in Chapter 2, we will skip it here.

### 3.3.2 File Movement and Copy Commands

Now that we can navigate the file system and see what is there, we want to be able to manipulate the file system. We will do this by moving, renaming, copying, deleting files and directories, and creating files. The commands mv, cp, and rm (move/rename, copy, delete) can work on both files and directories. For each of these, however, the user issuing the command must have proper permissions. We will assume that the commands are being issued on items whose permissions will not raise errors.

The mv (move) command is used to both move and rename files/directories. The standard format for mv is

```
mv [options] source destination
```

If you wish to rename a file, then source and destination are the old name and the new name, respectively. Otherwise, source and destination *must* be in different directories. If destination is a directory, then the file's name is not changed. For instance,

```
mv ~foxr/temp/foo2.txt .
```

would move foo1.txt to the user's subdirectory temp without changing the name. However,

```
mv foo1.txt ~/temp/foo2.txt
```

would move the file and rename it from foo1.txt to foo2.txt.

To indicate that a file from a directory other than the current directory should be moved to the current directory, you would specify the directory and filename as source and use . as destination. For instance,

```
mv ~foxr/temp/foo2.txt .
```

would move foxr's file foo2.txt, which is in his subdirectory temp, to the current working directory. Notice the use of ~ here, for the specification of the directory, you can use ~ and .. as needed.

If you are moving a file to the same directory, the command acts as a rename command. Otherwise, the file is physically moved from one directory to another, thus the file in the source directory is deleted from that directory. This physical "movement" is only of a link. The file's contents, stored in disk blocks, is usually not altered.[*]

The mv command can operate on multiple files if you are moving them all to a new directory and only if you use some form of filename expansion. For instance, you would not be allowed to specify

```
mv file1.txt file2.txt file3.txt /home/zappaf
```

as there are too many parameters. You are not allowed to rename multiple files in a single mv command as there is only one "newname" specified in the instruction and renaming multiple files into the same destination would not work.

---

[*] The disk file may have to be moved only if the file is being moved to a different disk partition.

If the mv operation would cause the file to be moved to the destination location where a file of the same name already exists, the user is prompted to see if the mv should take place and thus overwrite the existing file in the destination directory. You can force mv to operate without prompting the user by supplying the -f option. The -i option will prompt the user before overwriting the file (-i is the default so it does not need to be specified).

The copy command is cp. It's format is much the same as mv

```
cp [options] source destination
```

Unlike mv, the file is physically copied so that two versions will now exist. There are three different combinations of source and destination that can be used. First, the destination specifier can be another directory in which case the file is copied into the new directory and the new file is given the same name as the original file. Second, if destination is both a directory and file name, then the file is copied into the new directory and given a different name. Third, the destination is a filename in which case the file is copied into the current directory under the new filename.

Let us look at some examples.

```
cp foo.txt ~zappaf/foo1.txt
```

will copy the file foo.txt from the current working directory to zappaf's home directory, but name the new file foo1.txt rather than foo.txt.

Source can consist of multiple files by using wildcards. The following command provides an example:

```
cp *.txt ~
```

In this command, all files in the current directory that end with a .txt extension are copied to the user's home directory.

Note that if the source and destination contain directories and they are the same directory, then cp will only work if the destination filename is different from the source filename. For instance,

```
cp foo1.txt .
```

will not work because you cannot copy foo1.txt onto itself.

The copy command has a number of different options available. The most common options are shown in Table 3.4.

Without the -p option, all copied items are given the current user's ownership. For instance, if the files foo1.txt, foo2.txt, and foo3.txt in zappaf are all owned by zappaf with the group cit371, the operation

```
cp ~zappaf/foo*.txt .
```

TABLE 3.4    Common Options for the cp Command

| | |
|---|---|
| -b | Create a backup of every destination file |
| -f | Force copy, as with mv's -f |
| -i | Prompt the user, as with mv's -i |
| -I, -s | Create hard/symbolic link rather than physical copy |
| -L | Follow symbolic links |
| -p | Preserve ownership, permissions, time stamp, etc. |
| -r | Recursive copy (recursion is discussed below) |
| -u | Copy only if source newer than destination or destination missing |
| -v | Verbose (output each step as it happens) |

will copy each of these three files to your current working directory, and change the owner and group to be yours. Similarly, creation date and time are given the current date and time. But if you use -p, the copied items have the same owner, group, creation date and time, and permissions as the original files. This may be useful when you are copying files as root.

As noted in Table 3.4, cp has -i and -f options that are essentially the same as those found with the mv command. However, cp has another option worth discussing, -r. What does a *recursive* copy do? The term recursive means that the copy is performed recursively. A recursive operation is one that operates on the files in the current directory but also works on the subdirectories. By recursive, we mean that the files (and subdirectories) in the sub-directories are also copied. Consider the filespace illustrated in Figure 3.4. Assume t1, t2, t3, f1, f2, f3, b1, and b2 are files and all other items are directories.

The current working directory is ~dukeg. The command

```
cp -r ~foxr/temp .
```

would copy directory temp2 with the files f1, f2, and f3, but also subdirectories a and b, and the files b1 and b2 from directory b. These would be copied into ~dukeg so that dukeg

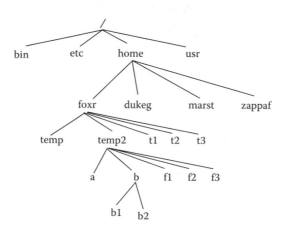

FIGURE 3.4    Example file space.

would now have directory temp2, files f1, f2, f3, and temp2 would have subdirectories a and b, files f1, f2, f3, and subdirectory b would have b1 and b2. The recursive copy is a very convenient way to copy an entire portion of the file system to another location. Of course, to copy the files and subdirectories, you would need to have adequate permissions.

### 3.3.3 File Deletion Commands

The next command to examine is `rm`, for remove or delete. The rm command has the syntax

```
rm [options] file(s)
```

As with mv and cp, the rm command can work on multiple files either by listing each item (separated by spaces) or using wildcards, or both. If a directory contained files file1.txt, file2.txt, and file3.txt (and no other files ending in .txt), and you wanted to delete all three, either of the following instructions would work

```
rm file*.txt
rm file1.txt file2.txt file3.txt
```

Obviously, the first instruction is simpler. However, if the directory did contain other files whose names were similar, for instance, file4.txt and file10.txt, and you did not want those deleted, you would have to use the latter instruction.

As rm permanently deletes files, it is safest to use rm with the option -i. This option causes rm to pause before each file is deleted and ask the user if he or she is sure about deleting that file. The user responds with a 'y' or 'n' answer. Users may not think to use `rm` `-i`, so it is common for a system administrator to create an alias of rm to rm -i so that, by default, rm is always run with the -i option.

If you are applying rm to a lot of files, for instance by doing `rm *.txt`, you may not want to be bothered with having to respond to all of the prompts (one per item being deleted) and so you might use `rm -f *.txt` to override the prompting messages. This can be dangerous though, so only use -f when you are sure that you want to delete the file(s) specified.

The -r option is the same as with cp, but in this case, the files and subdirectories are deleted. If you (or the system administrator) have aliased rm to be rm -i, then with -r you are asked before each deletion to confirm. This can be time consuming and undesirable. In this case, you might combine -f and -r, as in

```
rm -fr *
```

This will delete everything in the current directory, any subdirectories and their contents as well. Of course you should explore the current directory before using such a command to ensure that you do not delete content mistakenly.

Notice that `rm -r` will delete directories. Generally, rm cannot delete a directory. You would have to use rmdir (explained below). But in the case of a recursive deletion, directories can be removed once their contents are deleted. Referring back to Figure 3.4, the command

```
rm -fr /home/foxr/temp2
```

would delete the files and directories in this order: f1, f2, f3, b1, b2, a, b, temp2.

### 3.3.4 Creating and Deleting Directories

To create a directory, use `mkdir` followed by the directory name as in

```
mkdir TEMP
```

The directory is created in the current working directory, unless a path is specified. The directory is initially empty. The permissions for the directory default to permissions that you have established previously through the `umask` command. If you have not used umask, then they default to the umask value established by the system administrator. We look at umask in Chapter 9.

The mkdir command only allows for a few basic options. With -m or --mode=*MODE*, you can specify the initial permissions of the directory, otherwise they default to the permissions set up with umask. Alternatively, -Z or --context=*CTX* can be used to specify SELinux (security enhanced) rather than normal permissions to control access to the directory. The context, *CTX*, provides for rules that describe access rights. SELinux is discussed in Chapter 8.

To remove a directory, it must be empty. Therefore, you would have to delete its contents first. Once a directory is empty, to delete it, use the instruction `rmdir`. Attempting to delete a nonempty directory yields the error message:

```
rmdir: failed to remove `dirname': Directory not empty.
```

To delete the contents of the directory, you would want to use rm. By using the * wildcard, it simplifies the rm operation. Now you will perform these three operations:

```
rm  * (in the directory will delete its contents)
cd  .. (to move to the parent directory)
rmdir directory
```

It is a lot easier to delete the directory and its contents using the recursive deletion, `rm -r directory`. This instruction will delete the contents stored in *directory* first and then delete the directory itself. If rm has been aliased to rm -i, then you will be asked for approval before each item is deleted. To avoid this, you could use `rm -rf directory` although this could be risky.

The rmdir command also has an option, -p, to recursively delete parent directories as well as the current directory. In such a case, you specify the path of the directories to delete from top to bottom. Referring back to Figure 3.3, if all of the directories foxr, temp2 and b were empty, and you wanted to delete them all, then from /home you could issue the command `rmdir -p foxr/temp2/b`.

### 3.3.5 Textfile Viewing Commands

There are a number of commands that allow you to obtain information about files. Three commands to display the contents of one or more files to the terminal window are `cat`, `less`, and `more`. The cat command (short for concatenate) displays the contents of all of the files specified, without pausing. The cat command is more commonly used with redirection so that the contents of multiple files can be combined and placed into a new file. For instance,

```
cat *.txt > all_text_files.txt
```

would take all .txt files of the current directory and combine them into one file, also in the current directory. The command

```
cat foo.txt > foo2.txt
```

is the same as `cp foo.txt foo2.txt`.

If you wish to quickly see the contents of one or more files, you can also use cat, but because the contents will not pause screen-by-screen, you might want to use `more` or `less`. Both of these commands will display the contents of the file(s), but will pause after each screen to let the user control what happens next.

In more, the user either hits the enter key to move one line ahead, the space bar to move one screen ahead, or q to quit. The more command has a number of options to help more precisely control the program's behavior, for instance by using -n *number* to specify the number of lines that appear in each screen or *+linenumber* to start the display at the given line number. The less command, which is newer and more useful, gives the user more control as the user can scroll both forward and backward through the file, for instance by using the arrow keys.

You can also view the first part or last part of a file easily using the commands `head` and `tail`, respectively. These two commands display the first and last 10 lines of a file, respectively. You can control the amount of the file displayed with options -c and -n. Both options expect an integer number to follow the option to indicate the amount of the file to display. The option -c is used to specify the number of bytes (characters) to output and -n is used to specify the number of lines to output.

For head, you can also precede the integer with a minus sign to indicate that the program should skip some number of bytes or lines. Similarly, for tail, precede the integer with a plus sign to indicate the starting point within the file.

For instance, `head -c -20` would start at the 21st byte of the file. The instruction `tail -n +12` will display the file starting at line 12. Some additional examples are shown in Table 3.5. Assume that file.txt is a text file that consists of 14 lines and 168 bytes.

TABLE 3.5    Examples Using head and tail

| Command | Result |
| --- | --- |
| head file.txt | Output first 10 lines of file.txt |
| tail file.txt | Output last 10 lines of file.txt |
| head -c 100 file.txt | Output first 100 bytes of file.txt |
| head -n -3 file.txt | Output first 11 lines of file.txt (all but the last three lines) |
| tail -n 5 file.txt | Output last 5 lines of file.txt |
| tail -c +100 file.txt | Output all characters of file.txt starting at the 100th byte |

Another file operation is sort, to sort a given file line-by-line in increasing order. The option -r causes sort to work in decreasing order. Sort can work on multiple files in which case the lines are mixed, as if the files were first combined using cat. Another useful option for sort is -f which causes sort to ignore case so that 'a' and 'A' are treated equally (otherwise 'A' comes before 'a').

### 3.3.6 File Comparison Commands

The instructions cmp, comm, and diff are all available to compare the contents of text files. The diff instruction can operate on any number of files or whole directories. Both cmp and comm expect only two files and comm expects the files to be sorted.

In diff, for each line that is in the first file but not the second, it is output preceded by a '<' and for every line that is in the second file but not in the first, it is output preceded by a '>'. For each line or group of lines that differ, a summary is provided indicating how the two differ. For instance, if file 1 contains a line that is not found in file 2, you will be told that the line had to be deleted to match file 2. If file 2 contains a line that is not found in file 1, you will be told that the line had to be added. If two corresponding lines between the two files do not match, then you will be told that the line had to be changed. These are indicated using the letters 'a' for added, 'd' for deleted, and 'c' for changed. This notation might look like 3a5,6 to indicate that at line 3 of the first file, we had to add lines 5 to 6 of the second file.

If the first filename is a directory, diff will find the file in the directory whose name matches that of the second file. The option -i will cause diff to ignore case so that upper and lower case letters are treated as one. Other options will cause diff to ignore white space (blanks, blank lines, tabs). If a file does not exist, diff responds with an error. Additionally, if diff is provided with two directories, it will compare all pairs of files who share the same name in both directories. To operate on more than two files, you must supply diff with the option --from-file=. For instance, if you wanted to compare file1 to all of file2, file3, file4, and file5, you would use the following instruction:

```
diff --from-file=file1 file2 file3 file4 file5
```

In such a case, the output of the comparisons is separated by --- to indicate that the next file is being compared.

The cmp instruction compares exactly two files. Similar to diff, it compares these files byte-by-byte and line-by-line but it stops once it finds a mismatch, returning the byte and line of the difference. You can force cmp to skip over a specified number of bytes for each file or stop after reaching a specified limit. The following instruction

```
cmp file1 file2 -i 100:150 -n 1024
```

will compare file1 and file2, starting at byte 101 of file1 and 151 of file2, comparing 1024 bytes.

The comm instruction expects to receive two files that contain sorted content. If the content is not sorted, while comm will still work, you may not get the results you hope for. The comm instruction works through the two files, comparing them line-by-line and returns for each line whether it appeared in the first file, the second file, or both files. This output is organized into columns where column 1 features lines only appearing in the first file, column 2 features lines common to both files, and column 3 features lines only appearing in the second file.

A related program to diff, cmp, and comm is uniq. Unlike these previous commands, uniq operates on a single file, searching for consecutive duplicate lines. Based on the parameters supplied, you can use uniq to remove any such duplicates although uniq does not overwrite the file but instead outputs the file without the duplicates. You would then redirect the output to a new file, as in

```
uniq file.txt > filewithoutduplicates.txt
```

As uniq only compares adjacent lines, it would not find duplicate lines that do not appear one after the other. Options allow you to output the found duplicates or count the number of duplicates rather than outputting the file without the duplicates, and you can ignore case or skip over characters.

### 3.3.7 File Manipulation Commands

The join command allows you to join two files together when the files contain a common value (field) per line. If this is the case, when the two files each have a row that contains that same value, then those two lines are joined together. As an example, consider two files, first.txt and last.txt, each of which contain a line number followed by a person's first or last name. Joining the two files would append the value in last.txt to the value found in first.txt for the given line number. If we had for instance in the two files:

1. Frank          1. Zappa

2. Tommy          2. Mars

then the join would yield

1. Frank Zappa

2. Tommy Mars

For lines that do not share a common value, those lines are not joined. The options -1 *NUM* and -2 *NUM* allow you to specify which fields should be used on the first and second files, respectively, where *NUM* is the field number desired. The option -i would cause join to ignore case if the common values included letters. The option -e *STRING* would cause *STRING* to be used in case a field was empty, and -a 1 or -a 2 would output lines from the first or second file which did not contain a match to the other file.

Another means of merging files together is through the paste command. Here, the contents of the files are combined, line by line. For instance, paste *file1 file2* would append file2's first line to file1's first line, file2's second line to file1's second line, and so forth. The option -s will serialize the paste so that all of file2 would appear after all of file1. However, unlike cat, paste would place the contents of file1 onto one line and the contents of file2 onto a second line.

The split instruction allows you to split a file into numerous smaller files. You specify the file to split and a prefix. The *prefix* is used to name the new files which are created out of the original file. For each file created, its name will be prefix followed by a pattern of letters such as aa, ab, ac, through az, followed by ba, bb, bc, and so forth. For instance,

```
split file1.txt file1
```

would result in new files named file1aa, file1ab, file1ac, file1ad, etc where each of these files comprises equal-sized portions of file1.txt.

You control the split point of the file through either -b *size* or -C *lines* where *size* indicates the size of the division in bytes (although the value can include K, KB, M, MB, G, GB, etc) and *lines* indicates the number of lines to include in each file. If, for instance, file1. txt has 40 lines, then the command

```
split file1.txt -C 10 file1
```

would place the first 10 lines in file1aa, the next 10 in file1ab, the next 10 in file1ac and the last 10 in file1ad.

The option -d will use numeric suffixes (00, 01, 02, etc) rather than the two-letter suffixes. The split command does not alter the original file. It should be noted that while the previously discussed file commands described in this section are intended for text files, split can operate on binary files, thus dividing a binary file into multiple subfiles.

A related command is cut. The cut command will remove portions of each line of a file. This is a useful way of obtaining just a part of a file. The cut command is supplied options that indicate which parts of a line to retain, based on a number of bytes (-b *first-last*), specific characters (-c *charlist*), a delimiter other than tab (-d '*delimiter*'), or field numbers (-f *list*) if the line contains individual fields. Alternatively, --comple-ment will reverse the option so that only the cut portion is returned.

Let us consider a file of user information which contains for each row, the user's first name, last name, user name, shell, home directory and UID, all delineated by tabs. We want to output for each user, only their user name, shell, and UID. We would issue the command cut -f 3,4,6 *filename*. If each row's contents were separated by spaces, we would add -d ' ' to indicate that the delimiter is a space.

The cut command can be useful in other ways than viewing partial contents of a file. Imagine that we want to view the permissions of the contents of a directory but we do not want to see other information like ownership or size. The ls -l command gives us more information than desired. We can pipe the result to cut to eliminate the content that we do not want. For instance, if we only want to see permissions and not names, we could simply use

```
ls -l | cut -c1-10
```

This would return only characters 1 through 10 of each line. If we do not want to see the initial character (the — for a file, d for a directory, etc), we could use -c2-10. This solution however only shows us permissions, not file names. We could instead use -f and select fields 1 and 9. The ls -l command uses spaces as delimiters, so we have to add -d ' ', or

```
ls -l | cut -f 1,9 -d ' '
```

Unfortunately, our solution may not work correctly as the size of the listed items may vary so that some lines contain additional spaces. For instance, if one file is 100 bytes and another is 1000 bytes, the file of 100 bytes has an added space between the group owner and the file size in its long listing. This would cause field 9 to become the date instead of the filename. In Chapter 6, we will look at the awk instruction which offers better control than cut for selecting fields.

One last instruction to mention in this section is called strings. This instruction works on files that are not necessarily text files, outputting printable characters found in the file. The instruction outputs any sequence of 4 or more printable characters found between unprintable characters. The option -n *number* overrides 4 as the default sequence length, printing sequences of at least *number* printable characters instead.

### 3.3.8 Other File Commands of Note

The wc command performs a count of the characters, words, and lines in a text file. The number of characters is the same as the file's byte count (characters stored in ASCII are stored in 1 byte apiece). The number of lines is the number of new line (\n) characters found. You can control the output through three options, -c or -m for characters only (-m counts characters while -c counts bytes), -l for lines only and -w for words only. Additionally, the option -L prints the length of the longest line.

You can use wc in a number of different ways. For instance, if you want to quickly count the number of entries in a directory, you might use ls | wc -l. If you want to count the number of users logged in, you can use who | wc -l.

There are numerous options to create files. Many applications software will create files. In Linux, from the command line, you might use a text editor like vi or emacs. You can launch either of these with the filename as an argument, as in vi newfile.txt, or you can use the "save" command to save the file once you have entered text into the editor. The first time you save the file, you will be prompted for a name. We visit vi and emacs later in Chapter 5.

The touch command creates an empty text file. The command's true purpose is to modify the access and modification time stamps of a file. However, if the file does not currently exist, an empty text file is created. This can be useful if you wish to ensure that a file exists which will eventually contain output from some other program such as a shell script. Or, touch can be used to start a text file that you will later edit via vi. The options -a and -m modify the last access time and date and last modification time and date to the current time.

The touch command also has the option -t to specify a new access and/or modification date and time. The -t option is followed by the date and time using the format [[CC]YY]MMDDhhmm[.ss], which means the current year, month, and date as a 6- or 8-digit number (if CC is omitted, the year is implied to be from the current century, and if YY is omitted, the year is implied to be this year). The .ss notation permits the seconds if included. For instance, -t 1112131415.16 would be 2:15.16 p.m. on December 13, 2011. Finally, you can use -r filename so that the given filename's time/date is used in place of the current time/date.

Another way to create a file is to call upon a Linux command and redirect the output from terminal window to a file. For instance,

```
cat file1.txt file2.txt file3.txt | sort -r > newfile.txt
```

will combine the three text files, sort the entries in reverse order and store the result to newfile.txt. If newfile.txt already exists, it is overwritten. The use of >> will append the results to an already existing file, or create the file if it does not already exist.

### 3.3.9 Hard and Symbolic Links

A directory stores items such as files and subdirectories. In reality, the directory stores links to these items. The links are formally known as *hard links*. Each hard link consists of the item's name and the inode number. The inode number is then used to index into or reach the inode itself. Thus, the link provides a pointer to the file. You generally have a single link per file. However, you are allowed to create duplicate links to be stored in other directories so that accessing the item can be handled more conveniently.

There are two types of links. The hard link is a duplicate of the original link. The *soft*, or *symbolic*, link stores a pointer to the original link.

You might want to create a link to an existing file for convenience to provide a means of accessing the file from multiple locations. To create a link, you will use the ln command. With the command, you specify whether the link should be a hard link (the default) or a soft link (by using the option -s).

The format of the instruction is ln [-s] existingfile newfile where the existing file must permit links (there are permissions that can be established that forbid links to a file). In this example, *newfile* is a pointer, not a new file or a copy of *existingfile*.

The symbolic link is very useful in that you can place the link in a directory where you expect that users might try to access a file. For instance, if users think a particular program is stored in /bin but in fact is in /usr/local/bin, a symbolic link from /bin/someprogram to /usr/local/bin/someprogram would solve this problem. This would be better than adding /usr/local/bin to all users' path variables. Additionally, if a system administrator (or user) were to move a file that other users reference, then a symbolic link can ensure that the users can still access the file in spite of it being moved.

Another application is for a user to place a symbolic link in his or her home directory, pointing to another location. In this way, the user can reference items outside of their directory without needing to specify a full path. As an example, imagine that the user has written a game which pulls words out of the Linux dictionary file, /usr/share/dict/words. Let us assume the user issues the following command:

```
ln -s /usr/share/dict/words ~/words
```

This creates a symbolic link in the user's home directory called words that links to the dictionary. Now the user's program can reference words without needing the path.

You should note that the symbolic link is merely a pointer and the pointer is not aware of the file it points to. If someone were to delete the file being linked to after the symbolic link had been created, the symbolic link would still exist. Using that symbolic link will yield an error because the file that it points to no longer exists. Alternatively, deleting the symbolic link does not impact the file itself.

You can see if an item is a file or a symbolic link by using `ls  -l`. The symbolic link is denoted in two ways. First, the first character in the list of permissions is a 'l' for link. Second, the name at the end of the line will have your symbolic link name, an arrow (->), and the file being pointed to. There are several symbolic links in the /etc directory. For instance, /etc/grub.conf is a symbolic link pointing to /boot/grub/grub.conf, and the scripts located in /etc/init.d are actually symbolic links pointing to the files in /etc/rc.d/init.d.

You can also see if a file is pointed to by multiple hard links. The ls -l command displays in the second column an integer number. This is the number of hard links pointing at the file. Most files will contain only a single link, so the number should be 1. Directories often have multiple links pointing at them. If you were to delete a file, the hard link number is decremented by 1 so for instance if a file has 3 hard links and you delete one, it becomes 2. Only if you delete a file whose hard link number is 1 will the file actually be deleted from disk.

If you find the ln instruction confusing because of the -s option, you can also create links using `link` to create a hard link and `symlink` to create a symbolic link.

## 3.4 LOCATING FILES

In this section, we examine different mechanisms for searching through the Linux file system for particular files, directories, or items that match some criteria.

### 3.4.1 The GUI Search Tool

The easiest mechanism is to use the GUI Search for Files tool. In the Gnome desktop, this is found under the Places menu. This tool is shown in Figure 3.5. Under "Name contains," you enter a string. The tool will search starting at the directory indicated under "Look in folder." Any file found whose name contains the string specified in that portion of the file system will be listed under Search results. By selecting "Select more options," the user can also specify search criteria for files that contain certain strings, files that are of a specified size, modification date, ownership, and other details. The GUI program then calls upon find, locate (both of which are described below), and grep (which is covered in Chapter 6). While the Search for Files tool is easy to use, you might want to gain familiarity with both find and grep as you will often prefer to perform your searches from the command line.

### 3.4.2 The Find Command

The find program is a very powerful tool. Unfortunately, to master its usage requires a good deal of practice as it is very complicated. The find command's format is

```
find directory options
```

where directory is the starting position in the file system for the search to begin. There are several different forms of options discussed below but at least one form of the options (expression) is required.

The simplest form of find is to locate files whose name matches a given string. The string can include wildcard characters. As an example, if you want to find all .conf files in the /etc directory (and its subdirectories), you would issue the instruction

FIGURE 3.5   GUI search for files tool.

```
find /etc -name "*.conf"
```

Notice that the instruction

```
ls /etc/*.conf
```

is similar but in ls, you are only examining files in /etc, not its subdirectories. You could use -R in ls to recursively search all subdirectories so it performs similarly to find. A variation of the option -name is -iname, which does the same thing but is case insensitive.

The find command provides numerous options which increase its usefulness over the ls command. Options are divided into three categories. First are options similar to most Linux commands which cause the program to function differently. Second are options that the user specifies to tailor the search. These are *search criteria*. Table 3.6 illustrates some of the more useful expressions. Third are options that control what find will do when it has located an item that matches the expression(s). We will consider these later in this subsection.

The notation [+-]n from Table 3.6 needs a bit of explanation. These expressions can either be followed by an integer number, a +integer or a –integer. The integer by itself, n, means an exact match. The values +n and –n indicate "greater than n" and "less than n," respectively. For instance, $-size\ 1000c$ means "exactly 1000 bytes in size" while $-size$ $-1000c$ means "less than 1000 bytes in size."

The various expressions listed in Table 3.6 can be combined using Boolean connectors. To specify that two (or more) expressions must be true, combine them with –and or –a (or just list them). For instance, the following three are all equivalent, which will seek out all files under the user's home directory that are between 100 and 200 bytes.

- find ~ -size +100c -size -200c
- find ~ -size +100c -a -size -200c
- find ~ -size +100c -and -size -200c

To specify that any of the expressions should be true, combine them with –o or –or. The following expression looks for any item owned by user 500 or group 500.

- find ~ -uid 500 -o -gid 500

To specify that the given expression must not be true, use either ! expression or –not. For instance, the following two both search for noncharacter-based files.

- find ~ ! -type c
- find ~ -not -type c

If your expression contains numerous parts and you need to specify operator precedence, place them within parentheses. Without parentheses, not is applied first, followed by and, followed by or.

TABLE 3.6    Expressions Used in Find

| Expression | Meaning | Example |
|---|---|---|
| -amin [+—]*n* | Find all files accessed >, < or = *n* minutes ago | -amin +10 |
| -anewer *file* | Find all files accessed more recently than *file* | -anewer file1.txt |
| -atime [+—]n | Find all files accessed >, < or = *n* days ago | -atime −1 |
| -cmin [+—]n | Find all files created >, < or = *n* minutes ago | -cmin 60 |
| -cnewer *file* | Find all files created more recently than *file* | -cnewer abc.dat |
| -ctime [+—]n | Find all files created >, < or = *n* days ago | -ctime +3 |
| -mmin [+—]n | Find all files last modified >, <, = *n* minutes ago | -mmin −100 |
| -mtime [+—]n | Find all files last modified >, <, = *n* days ago | -mmin 1 |
| -newer *file* | Find all files modified more recently than *file* | -newer foo.txt |
| -empty | Find all empty files and directories | |
| -executable, -readable, -writable | Find all files that are executable, readable, or writeable | |
| -perm *test* | Find all files whose permissions match *test* | -perm 755 |
| -fstype *type* | Find all files on a file system of type *type* | -fstype nfs |
| -uid n -gid n | Find all files owned by user *n* or group *n* | -uid 501 |
| -user name -group name | Find all files owned by user name or group name | -user zappaf |
| -regex *pattern* | Find all files whose name matches the regular expression in *pattern* | -regex [abc]+\.txt |
| -size [+—]n | Find all files whose size is >, <, = *n*, *n* can be followed by b (512-byte blocks), c (bytes), w (2-word bytes), k (kilobytes), M (megabytes), and G (Gigabytes) | size +1024c size +1 k |
| -type *type* | Find all files of this type, *type* is one of b (block), c (character), d (directory), p (pipe), f (regular file), l (symbolic link), s (socket) | |

There are five options available for find. The first three deal with the treatment of symbolic links. The option -P specifies that symbolic links should never be followed. This is the default behavior and so can be omitted. The option -L forces symbolic links to be followed. What this means is that if an item in a directory is actually a symbolic link, then the expression is applied not to the link but to the item being pointed at. So for instance, if a symbolic link points to a character-type file and the expression is -type c, then the file will match. If instead, option -P (or the default) is used, the file does not match because the symbolic link is not a character-type file.

The option -H is a little trickier to understand. Like -P, it prevents find from following symbolic links when searching for files. However, as we will see below, find can also have an action to perform on any item that matches the expression. With -H, while find does not follow symbolic links to identify a file, symbolic links can be followed when performing an action. As this is a complex concept, we will not go into it in any more detail.

The final two options are to receive debugging information and ordering of expressions. Debugging information is useful if you suspect that your find operation has some errors in its logic or is not efficiently organized. To specify debugging use the option -D and follow it by one or more debugging options of help, tree, stat, opt, and rates.* Ordering of expressions allows find to attempt to optimize the expressions provided in order to reduce search time. Ordering is specified by -O followed by an optimization number with no space between them, as in -O3. The optimization number is between 0 and 3 with 3 being the most optimized. Note that while optimization will reduce search time, it takes longer to set up the search query. Therefore, optimization has a tradeoff and the higher the optimization level, the slower find will be to start. Depending upon the size of the file space being searched, optimization may or may not be useful.

Aside from these options, you can also specify options in the expression portion of the command. These include -daystart so that -amin, -atime, -cmin, -ctime, -mmin, and -mtime all start from the start of the day rather than 24 hours earlier. For instance, by saying "search for files created 1 day ago," if you add -daystart, then "1 day ago" literally means at any time of the day before midnight" rather than "exactly 24 hours ago."

The option -depth indicates that item should be sought (or acted upon) before the current directory itself is acted upon. This is useful when doing a recursive deletion. Next, -maxdepth, which is followed by a nonnegative integer, indicates how far down the file system find should search from the given directory. Typically, find will search all subdirectories from the directory specified to the leaf level of the file space. By indicating -maxdepth 1, aside from the current directory, find will only search subdirectories of the current directory. There is also a -mindepth. Finally, -mount indicates that find should not go down any directories that are mounted.

The default action for find is to list all matches found to the terminal window. Find can also perform other operations, as indicated in Table 3.7.

As the find program is far more complicated than the GUI Search for Files program, you might question its use. However, keep in mind that the GUI program takes more resources to run and would be unavailable if you are accessing a Linux machine through a text-only interface. Additionally, the find command offers a greater degree of control. Finally, find can be used in a shell script or in a pipe and obviously the GUI search program cannot. It is worthwhile learning find, but be patient in learning to use it correctly.

### 3.4.3 Other Means of Locating Files

There are several other options to locate items in the file system. To locate an executable program, use which *name* where *name* is the program's name. This works as long as the program is located in a directory that is part of the user's PATH variable and as long as the program is accessible by the user. Try for instance which useradd and you will receive an error because you, the user, do not have access to useradd. The program whereis is similar in that it will locate a program's location. But whereis does not rely on the PATH variable. Additionally, if available, whereis will also report on all locations relevant to the program: the executable program itself (binary), the source code (usually stored in a /lib subdirectory) and the manual page (usually stored under /usr/share/man).

---

* See find's man page to explore debugging options in more detail.

TABLE 3.7    Find Actions

| Option | Meaning | Usage |
|---|---|---|
| -delete | Delete all files that match, if a deletion fails (for instance because the current user does not have adequate permissions), output an error message | `find ~ -empty -delete`<br>Deletes all empty files |
| -exec *command* \; | Execute *command* on all files found. As most commands require a parameter, we specify the parameter (i.e., each found file) using {} | `find ~ -type f -exec`<br>`wc -l {} \;`<br>Executes wc -l on each file of type f (regular file) |
| -ok *command* \; | Same as exec except that find pauses before executing the command on each matching file to ask the user for permission | `find / -perm 777 -ok`<br>`rm {} \;`<br>Ask permission before deleting each file whose permissions are 777 |
| -ls, -print | Output found item using `ls -dils` or by full file name, respectively | `find ~ -name *core*`<br>`-print` |
| -printf *format* | Same as -print but using specified format which can include such information as file's last access time, modification date, size, depth in the directory tree, etc. | `find ~ -size +10000c`<br>`-printf%b`<br>Output size in disk blocks of all files greater than 10,000 bytes |
| -prune | If a found item is a directory, do not descend into it | `find ~ -perm 755`<br>`-prune` |
| -quit | If a match is found, exit immediately | `find ~ -name *core*`<br>`-quit` |

Two other available programs are `locate` and `slocate`. Both of these commands access a database storing file locations. To update (or create) the database for the locate command, use the command `updatedb`. The option -U allows you to specify a portion of the filespace to update. For instance, `updatedb -U /home/foxr` would only update the portion of the database corresponding to user foxr's home directory. Once the database is established, the locate instruction is simply

```
locate [option] string
```

which responds with the location within the file system of all files that match the given *string*. The common options are -c to count the number of matches found and output the count rather than the matched files, -d *file* to use *file* as a substitute database over the default database, -L to follow symbolic links, -i to ignore case, -n # to exit after # matches have been found, and -r *regex* to use the regular expression (*regex*) instead of the string.

The slocate instruction is newer and is used with SELinux. Unlike locate, slocate can operate without updatedb by using the -u or -U option. With -u, the database is updated by examining the entire file system. With -U, as with updatedb, you can specify a portion of the file system.

Keep in mind that updating the database can be time consuming. Users are not allowed to update the database because they will not have write access to the database. Therefore, updating is done by the system administrator. And to prevent the update from taking too much time, a system administrator should schedule the update to take place in some off hour time.

## 3.5 PERMISSIONS

So far, we have mentioned permissions but have not discussed them. If you have already been experimenting with some of the instructions described in Sections 2 and 3, you may have found that you did not have proper permissions for some operations. So here we examine what permissions are.

### 3.5.1 What Are Permissions?

Permissions are a mechanism to support operating system *protection*. Protection ensures that users do not misuse system resources. A system's resources include the CPU, memory, and network, but also include the partitions, directories, and files stored throughout the file system. Permissions protect the entities in the file system so that a user cannot misuse the entities owned by another user.

In Linux, permissions are established for all files and directories. Permissions specify who can access a file or directory, and the types of access. All files and directories are owned by a user. If we limited ownership by user only, then the only permissions that a file or directory could support are one set for the owner and one set for everyone else.

For instance, imagine that you have a file, abc.txt. You could make it so that you had read and write access and no one else could access the file at all. However, if there are a few other users that you want to give access to, your only recourse would be to give those users your password so that they could log in as you to access the file (bad idea), send them a copy of the file (but then, if you modified the file, they would have an out-of-date version), or alter the file's permissions so that everyone could read it.

This is where groups come in. By default, every Linux user account comes with a private group account whose name is the same as the user's username. Whenever you create a new file or directory, that item is owned by you and your private group. You can change the group ownership to be of another group. This is how you can share the item with other users.

Let us imagine that we have a group called *itstaff* and that you and I are both in that group. I create a file that I want all itstaff to be able to read, but no one else. I establish the file's permissions to be readable and writable by myself, readable by anyone in the itstaff group, and not accessible to anyone else. Next, I change the group ownership from my private group to itstaff. Now you can read it (but not write to it).

In Linux, permissions are controlled at three levels:

- Owner (called user, or 'u' for short)

- Group ('g' for short)

- The rest of the world (called other, or 'o' for short)

For every item, you can establish eight different types of access for each of owner, group, and world. These eight types consist of all of the combinations of read, write, execute, and no access. We will denote read access as 'r,' write access as 'w,' and execute access as 'x.' For instance, one file might be read-only (r), another might be readable and writable (rw), another might be readable, writable, and executable (rwx). Although there are eight combinations, some will probably never be used such as "write only." We will explore why in a little while.

The following list describes what each level of access provides:

- Read—for a file, it can be viewed or copied

- Write—for a file, it can be overwritten (e.g., using save as)

- Execute—for a file, it can be executed (this is necessary for executable programs and shell scripts)

- Read—for a directory, the directory's contents can be viewed by ls

- Write—for a directory, files can be written there

- Execute—for a directory, a user can cd into it

**Note:** to delete a file, you must have write access to the directory that contains it.

### 3.5.2 Altering Permissions from the Command Line

To change a file's permission, the command is chmod. The command's syntax is

```
chmod permissions file(s)
```

where *permissions* can be specified using one of three different approaches described below. The parameter *file(s)* is one or more files (or directories). As with the previous file system commands, if multiple files are to be altered, you may list them with spaces between their names or use wildcards if appropriate.

The first approach for specifying permissions is to describe the changes to be applied as a combination of u, g, o along with r, w, x. To add a permission, use + and to remove a permission, use –. Let us assume that we have the file file1.txt, which is currently readable and writable by u and g and readable by o. We want to remove writable by group and remove readable by other. The command would be

```
chmod g-w,o-r file1.txt
```

Notice in this command that there are no spaces within the permission changes itself. We see that g is losing w (group is losing writability) and o is losing r (other is losing readability). If instead we wanted to add execute to owner and group, it would be

```
chmod u+x,g+x file1.txt
```

If you want to change a permission on all three of u, g, and o, you can use 'a' instead for all. For instance, given the original permissions, we want to add execute to everyone. This could be done with

```
chmod a+x file1.txt
```

The second approach to altering permissions uses an = instead of + or −. In this case, you assign new permissions rather than a change to the permissions. For instance, if you want to make file1.txt readable, writable, and executable to the user, readable to the group, and nothing to the world, this could be done with

```
chmod u=rwx,g=r,o= file1.txt
```

In this case, since we want other to have no access, we place no letters after the o=. You can omit u=, g=, or o= if no change is taking place to that particular party.

Additionally, you can combine =, +, and − as in

```
chmod u = rwx,g-w,o-r file1.txt
```

You can make multiple changes by combining + and − to any of u, g, and o, as in

```
chmod u = rwx,g-w + x,o-r file1.txt
```

The final approach is similar to the = approach in that you are changing all of the permissions. In this approach, you replace the letters with a 3-digit number. Each digit is the summation of the access rights granted to that party (user, group, other) where readability is a 4, writability is a 2, and executability is a 1.

Readability, writability, and executability would be $4 + 2 + 1 = 7$. Readability and executability would be $4 + 1 = 5$. No access at all would be 0. Let us assume that we want file1.txt to have readable, writable, and executable access for the owner, readable, and executable access for the group and no access for the world. The command would be

```
chmod 750 file1.txt
```

Another way to think of each digit is as a binary number. Assume that a 1 bit means "access" and a 0 bit means "no access." The readability bit comes first followed by the writability bit followed by the executability bit, or rwx. To provide rwx, this would be 111 which is binary for 7 $(4 + 2 + 1)$. For rw- (i.e., readable and writable but not executable), this would be $110 = 6$ $(4 + 2)$. For r-x (readable and executable), this would be $101 = 5$ $(4 + 1)$. For r-- (read only), this would be $100 = 4$.

The combinations rwx, rw-, r-x, r-- might look familiar to you, as well they should. When you do a long listing (ls  -l), the first 10 characters of a line are a combination of letters and hyphens. The first letter denotes the type of file (as we discussed in the find

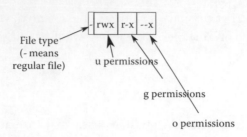

FIGURE 3.6    Reading permissions.

command). The remaining nine characters denote the permissions. For instance, `rwxr-xr--` means that the user has read, write, and execute access, the group has read and execute access and other has only read access. Figure 3.6 illustrates the grouping of these first 10 characters. The leading hyphen indicates the file is a regular file (whereas a 'd' would refer to a directory and an 'l' means a symbolic link; we discuss the other types in Chapter 10). Here, we see a file whose access rights are rwxr-x--x which makes it readable, writable, and executable by user, readable and executable by group and executable by world. The three-digit value for this item is 751.

While you are free to use any of the three combinations for chmod, the 3-digit approach is the least amount of typing. But it does require learning the possible values. See Table 3.8

TABLE 3.8    Example 3-Digit Permissions

| Permission | Explanation | Usefulness |
|---|---|---|
| 000 | No one has any type of access to the item | Not useful other than system files accessible only by root |
| 200 | Write only | Not useful, need to be able to read the file before writing back to it |
| 400 | Read only for owner | Useful if file contains confidential read-only information, not common |
| 644 | Access to owner, read-only for all others | Very common |
| 664 | Access to owner and group, read-only for world | Common for file sharing with group members |
| 660 | Access to owner and group, no access for world | Common if file should not be accessed outside of group |
| 646 | World can write but group members cannot | Not useful as we would expect group to have equal or greater access than world |
| 755 | Adding execute access to 644, used for directories, script files, and executable programs | Very common |
| 745 | Execute privilege added to owner and world | Often used instead of 755 for script files |
| 775 | Adds write access to group | Less common than 755 but useful if group needs write access to directory or script file, or members of group might compile the program |
| 711 | Group members and world can execute the file only | Often used for web server directories |
| 666/777 | Read and write access to everyone | Dangerous |

which discusses several different 3-digit combinations and why they might be useful or not useful.

### 3.5.3 Altering Permissions from the GUI

You can also modify a file's permissions through the GUI. From the Places menu (in Gnome), select a location. This opens a file browser. You can double click on any directory to open it up into a new window. Files appear in a variety of different icon shapes. Figure 3.7 illustrates some of the types you can find. Here, we have a text file, a text file storing a script, an executable file, an archive or package, and a directory, respectively from left to right.

Right clicking on a file brings up a pop-up menu. Select Properties and the file's (directory's) Properties window will appear (this is similar to a file's properties window found in Windows operating systems). There are several tabs including Basic to describe the file, Open With so that you can select software to open the file into, and Permissions. You will see the file's owner and group owner, and the access currently provided for owner (u), group (g), and world (o). Drop down menus allow you to change any of these along with the group's owner and the SELinux context. You also see the last changed date. Figure 3.8 demonstrates a file's Properties window, in this case, a file named foo.

For files, permissions available in the drop down menus are limited to "Read only" and "Read and write" for the owner and "None," "Read only," and "Read and write" for group

FIGURE 3.7   Different file icons.

FIGURE 3.8   Properties window for the file foo.

and others. Notice that "executable" is not an option even if the file is an executable program or script. For directories, permissions available are "List files only," "Access files," "Create and delete files," "---," and "None" for group and world. You also have the ability to alter file access for all files within the directory, selecting one of "Read-only," "Read and write," "---," and "None" for group and world. Given the limitations available through the GUI, the chmod command is a more expressive way to alter permissions and can be much easier if you want to change permissions on multiple files.

### 3.5.4 Advanced Permissions

There are three other types of permissions to cover. One of these, SELinux, utilizes a far more complex scheme than the ugo/rwx mechanism described here. We introduce SELinux in Chapter 8 although as a Linux user and even system administrator, you would probably not change the SELinux settings from their default. Another form of permission deals with the user ID bit and the group ID bit. This influences how a process will execute. This is reserved for executable files only and will be discussed in Chapter 4. The last remaining topic is known as the *sticky bit*.

Historically, the sticky bit was used to indicate whether a process, which was no longer executing, should remain in swap space (in case it was executed again in the near future). The use of the sticky bit was only provided in Unix and was largely abandoned. Linux implementations however have used the sticky bit for a different purpose. For Linux (and modern versions of Unix), the sticky bit applies only to directories and indicates that a writable directory has restrictions on the files therein.

Let us consider a scenario to motivate the use of the sticky bit. You have created a directory to serve as a repository for group work for users of the Linux system. You make the directory writable so that users can store files there. Thus, the directory's permissions would need to be 777 (so that the world can write to it). This can be a dangerous setting as it would allow the world to not only access the contents of the directory, but also rename or delete the directory and any of its contents. By setting the sticky bit, the files within the directory can be controlled by the file's owner or the directory's owner, but not the rest of the world.

To establish that a directory should have its sticky bit set, you can use chmod and set the permissions to 1777. The initial '1' bit sets the world's executable status to 't.' You can also set the sticky bit through the instruction chmod o+t *directoryname* or chmod a+t *directoryname*. To remove the sticky bit, reset the permissions to 777 (or a more reasonable 755) or use o-t (or a-t).

Assume the directory mentioned in our scenario is called pub (for public) and was created by the user foxr. He has set his home directory's permissions to 755 and pub's permissions to 777. Now, user zappaf has written (or copied) the file foo.txt to pub. Performing an ls -al on pub would show the following three lines (assuming that there are no other items in pub).

```
drwxrwxrwx. 2  foxr    foxr    4096  Jul  29  08:24  .
drwxr-xr-x. 32 foxr    foxr    4096  Jul  29  08:24  ..
-rw-rw-r--. 1  zappaf  zappaf  1851  Jul  29  08:25  foo.txt
```

Notice that the file is not writable by anyone other than zappaf, however the directory, indicated by . in the `ls  -al` listing, is writable. The directory indicated by .. is the parent directory. If the parent directory were not accessible then users other than foxr would not be able to cd or ls into this subdirectory.

Assume dukeg accesses the directory ~foxr/pub and performs `rm  foo.txt` or `mv foo.txt  foo2.txt`. Either command would work for him because the directory is world-writable. Normally, we would not want to make the directory world-writable because this is a security violation that permits anyone to change or delete a file.

But suppose foxr alters this directory's permissions by adding the sticky bit. He does this by issuing either `chmod  1777  ~/pub` or `chmod  o+t  ~/pub`. A long listing of this directory will now appear with one slight change, the directory itself has a different permission on world execute status:

```
dwrxrwxrwt. 2 foxr foxr 4096 Jul 29 08:26  .
```

As a result, while other users can still write files to the directory, they cannot affect the files already stored there. Now dukeg attempts to do the rm command and receives this message:

```
rm: remove write-protected regular file 'foo.txt'?
```

If dukeg responds with y (yes), the following error arises:

```
rm: cannot remove 'foo.txt': Operation not permitted
```

Similarly, an attempt to mv the file results in

```
mv: cannot move 'foo.txt' to 'foo2.txt': Operation not permitted
```

dukeg is allowed to copy the file, but then this would be the case if the directory and file were both readable no matter if the directory was writable or had the sticky bit set.

## 3.6 LINUX FILE SYSTEM STRUCTURE

Upon installing Linux, you will find the file system is already populated with thousands of files and dozens of directories. The structure of these directories and files has been pre-established so that system administrators and users alike can learn where to find important files. The structure is also established to support the PATH variable and initialization scripts. In this section, we take a look at this structure, concentrating on what is often called the "top-level directories." It is less critical for a user to understand this structure; it is essential for a system administrator to know it. So, while we introduce the structure here, we will revisit it in Chapter 10.

The top-level directories in Linux are given in Table 3.9. Some of these top-level directories are empty. Others contain subdirectories with their own structure.

TABLE 3.9    Linux Top-Level Directory Structure

| Directory | Use | Subdirectories (If Any) |
|---|---|---|
| /bin | User executable programs | None |
| /boot | Linux boot programs including boot loader program (grub in most cases) | Boot loaders in subdirectories |
| /dev | Physical devices are treated as files, stored in this directory | Various to organize specific types of devices, see Table 3.10 |
| /etc | Configuration files and scripts | Numerous |
| /home | User home directory space | One subdirectory per user |
| /lib | Libraries used by the kernel and programs stored in /bin, /sbin | modules, security, udev, etc |
| /lost+found | Recovered files from disk errors due to an unexpected shutdown of the system | None |
| /media | Initially empty, removable media can be mounted here | None |
| /mnt | Mount point for temporary file systems | None |
| /opt | Initially empty, this optional directory can be used to install some application software | Application software subdirectories (if any) |
| /proc | Kernel records of running process information | Subdirectories created dynamically of each running process |
| /root | The system administrator's home directory | None, or subdirectories for profile storage such as .config, .gconf, .mozilla |
| /sbin | Like /bin, system executable programs, but these are typically used by system administrators, not users | None |
| /tmp | Temporary storage for running processes | None initially |
| /usr | Application software files and executables | bin, etc, include, lib, local, sbin, share, src, tmp |
| /var | Data files that grow over time (printer spool files, email repositories, log files, etc.) | Varies by system, includes cache, log, mail, spool, www |

Some of these top-level directories are worth exploring in more detail. Let us start with /dev. In Linux, I/O and storage devices are treated as files. By treating devices as files, Linux programs can communicate with the various devices through Linux instructions and redirection. Table 3.10 provides a partial listing of the devices. The device name is sometimes very cryptic, for instance, sda and hda are device descriptions for hard disk drives (sda for SCSI and SATA, hda for IDE). Because there might be multiple instances of a device, most of the devices are numbered. You might find for instance hda0, hda1, hda2, sda1, sda2, sda3, and so forth.

As a Linux user, the most important directories are /home, /usr, and /bin. You should find that /usr/bin, /usr/local, /usr/sbin, and /bin are all part of your PATH variable (and if not, they should be added). This allows you to access the programs in these directories without having to denote a path to the programs. For instance, you need only type ls rather than /bin/ls, and env instead of /usr/bin/env. Later in the text, we will explore the /etc, /proc, and /var directories in detail.

TABLE 3.10  Common/dev Devices

| Name | Description |
| --- | --- |
| autofs | File system |
| cdrom, cdrw | CD drives |
| core | A link to the proc directory of a file that represents memory |
| cpu | CPU register information |
| dvd, dvdrw | DVD drives |
| fd0, fd1, ... | Floppy disk drives |
| input | A directory that contains event devices, events are generated by input devices |
| hda0, hda1, ... | IDE-style hard drives |
| loop0, loop1, ... | Mounted devices not accounted for by other device names (e.g., cdrom, dvd, usb) |
| lp0, lp1, ... | Printers |
| net | Network interface |
| null | Discards anything sent to it (used like a trash can) |
| ppp | Point-to-point network devices |
| ram0, ram1, ... | Ramdisks |
| random | Generates a random number |
| sda0, sda1, ... | SCSI and SATA-style hard drives |
| tty0, tty1, ... | Terminal windows |
| usb | For USB compatible devices |
| zero | Produces the value 0 for processes that need it |

## 3.7  SECONDARY STORAGE DEVICES

Let us focus on secondary storage devices to get a better idea of what they are and how they operate. Storage devices differ from main memory in three ways. First, we use storage devices for permanent storage whereas main memory is used to store the program code and data that the processor is using, has used recently or will use in the near future. Second, storage devices are usually nonvolatile forms of storage meaning that they can retain their contents even after the power has been shut off. This is not the case with main memory (DRAM), cache or registers, all of which are volatile. Third, storage devices have a far greater storage capacity than main memory. The amount per byte of storage for hard disk is extremely cheap when compared to that of DRAM. On the other hand, DRAM access time is far shorter than hard disk access time.

There are several different technologies used for storage devices. The primary form is the hard disk drive. There are also optical disk drives, magnetic tape, solid-state drives (SSD, including flash memory drives) and in the case of older computers, floppy disk drives. Each of these technologies stores information on a physical medium that is separate from the device that accesses the information. In the case of the floppy disk, optical disk and magnetic tape, the media is removable from the device. In the case of the hard disk drive, the disk platters are permanently sealed inside the disk drive. Finally, the USB drive is removable from the computer where it is inserted into a USB port. Table 3.11 illustrates differences between these devices.

TABLE 3.11   Characteristics of Storage Devices*

| Device | Maximum Transfer Rate | Common Size | Common Cost | Common Usage |
|---|---|---|---|---|
| Hard disk | 1 Gbit/s | 1 TB | $100–150 | Store the operating system, application software, and most user data |
| Optical disk (DVD) | 400 Mbit/s | 5 GB | $1 | Store multimedia data and software, available for portability |
| Magnetic tape | 140 Mbit/s | 5 TB | Varies based on size, up to $500 | Store backup and archival data |
| SSD | Up to 600 Mbit/s | 256 MB (up to 2 TB) | Varies based on size, up to $700 for 2 TB, $200 for 256 GB | Used in place of or in addition to hard disk, often supports storage area networks (SAN) |
| USB (3.0) | 400 Mbit/s | Up to 32 GB | $20–40 | Store data files for portability |
| Floppy disk | 1 Mbit/s | 1.5 MB | $.50 | Obsolete |

### 3.7.1 The Hard Disk Drive*

The hard disk drive is the most common form of secondary storage. Every laptop and desktop computer will come with an internal hard disk drive. In most cases, the operating system will be stored here (there are exceptions such as using a USB or optical disk to live boot Linux). Additionally, the user(s) of the computer will most likely install all software and store all data files onto this drive. Alternatively, some data files may be saved on other drives such as an external hard disk or optical disk.

The hard disk drive is a self-contained unit that combines the media storing the files with the technology to read and write data to the media. The media are several platters of "hard" disks. The disk drive unit also contains a spindle to spin the platters, two read/write heads for each platter (one above, one below) and the circuitry to move the read/write heads and perform the disk access. The surface of the hard disk is broken into logical segments called tracks and sectors. This segmentation is performed during disk formatting. See Figure 3.9 for an illustration of the hard disk drive.

Hard disk drives spin their disks very rapidly (up to 15,000 revolutions per minute) and have a high transfer rate (higher than all other storage devices) although far lower than the access time of DRAM. The hard disk drive is sealed so that the environment (dust, particles of food, hair, etc.) cannot interact with the disk surface. The hard disk drive uses a form of magnetic storage. Information is placed onto the surface of the disk as a sequence of magnetized spots, called magnetic domains or magnetic moments. These magnetized domains correspond to the magnetization of the atoms of the region. One magnetized spot is denoted with these atoms being aligned in one of two directions. We can refer to the two

---

* Transfer rate, common size, and common cost are based on values available in October 2013. These values tend to change frequently.

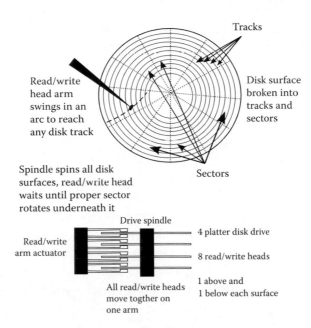

FIGURE 3.9 Hard disk technology: access, read/write heads, spindle. (From Richard Fox, 2013, *Information Technology: An Introduction for Today's Digital World.*)

directions as being positive or negative (representing the 1s and 0s of binary). Hard disk technology continues to improve in terms of both transfer rate and storage capacity.

Information stored on disk is broken into blocks. Each block is of a fixed size. This allows a file to be distributed in blocks across the surface of the disk and even across multiple surfaces. The advantages of this are twofold. First, because all blocks are the same size, deleting a file is handled by returning blocks back to be reused. We will see when we explore tape storage that it is unlikely that deleted file space can be reused. Second, as the disk drive rotates the disk very rapidly, placing disk file content in contiguous locations would not be efficient because transfer of data takes some time and so the disk will have rotated beyond the next disk block location when the next block is ready for reading/writing. The drive would have to wait for another revolution before access could resume.

Disk access time consists of three distinct parts: seek time, rotational delay (or rotational latency), and transfer time. Seek time is the time it takes for the read/write head to move across the surface of the disk from its current location to the intended disk track. Rotational delay is the time it takes for the desired sector to rotate underneath the waiting read/write head. Transfer time is the time it takes to perform the magnetization onto the disk's surface and move that information as binary data to memory, or the time it takes to receive the binary information from memory and write each bit as a new magnetic moment onto the disk. Transfer time can be as rapid as 1 Gb/s although most hard disk drives seldom reach this rate. Seek time is dependent on where the read/write head is now and where it has to move to and the speed of the read/write head's arm. Rotational latency is dependent on which sector is currently under/over the read/write head, which sector is being sought, and how rapidly the disk is spinning. Also, since sectors take up more space as you move outward from the center of the disk, it also depends on the desired track. In spite of these delays, hard disk access is the fastest of all storage devices by far.

There are two drawbacks to the hard disk: expense and lack of portability. For around $80, you can buy a 1 TB external hard disk drive, or you can buy a 4 TB hard disk for under $200. This gives you storage at the rate of 200 Mb per penny. The storage capacity per penny for a tape cartridge is even greater and the tape permits portability. However, tape, as we will see below, is not a very efficient form of storage. The storage capacity per penny of an optical disk and a USB drive are both greater than the hard disk making both forms impractical for large-scale storage (e.g., hard disk backups).

The hard disk and the floppy disk are similar in that they both use read/write heads, store magnetic information, and rotate to provide access. The floppy disk is removable from its disk drive. This causes the floppy disk's surface to be exposed to the environment. Because of this, the floppy disk has a much lower density of magnetic regions when compared to hard disks. The result is that the floppy disk storage capacity is much less. Because of first optical disks and second USB drives, floppy disks, which have slow access times, have become obsolete.

Hard disk drives are the primary form of storage today. We rely on them to store our application software and data files. Ensuring the integrity of this storage space becomes critical. Later in the text we consider approaches to ensure integrity through backups and through redundancy in the form of RAID.

### 3.7.2 Magnetic Tape

The magnetic tape is often thought of as obsolete, just like the floppy disk. Like the floppy disk, magnetic tape stores information as magnetized spots on the surface of the tape. And like the floppy disk, magnetic tape is removable from the tape drive. However, more like hard disk, magnetic tape can store a great deal of information on one unit (tape). Today, storage capacities of tape cartridge exceed 5 TB. The advantage then of tape is its ability to back up an entire hard disk on a single cartridge. Although tape also offers a form of portability, tape is never used in this way because few users use tape and the tape must be compatible with the user's tape drive.

The drawback of tape is in its performance. There are two problems with magnetic tape. First, files are stored in one contiguous block. If there are three files stored on tape and the middle file is deleted, can that space be reused? If you have used either audio cassettes or video tapes to record material, you probably know the issue here. You do not want to record something in between two or more items for fear of running over the latter item. Recording over a part of a data file corrupts that file, possibly to the point of the entire file being lost. Thus, reusing freed up space is often not possible on tape. See Figure 3.10 where the file currently under the read/write head, if deleted, may or may not allow sufficient space for another file.

The second performance issue is that tape stores files wholly. Access to a file will require fast forwarding or rewinding the tape to move the beginning of the desired file under the read/write head. As tapes can be lengthy in size, the amount of time to fast forward or rewind can be counted in minutes. Disk access often takes under a second.

Assume this file is deleted. Can we place a new file
in its location such that it does not flow into the next
file on the tape? If not, we cannot reuse this space!

FIGURE 3.10 Contiguous files on tape do not permit easy reuse of space.

Because of these drawbacks, tape tends to be used for only two purposes. The first is to back up the contents of a hard disk. As your hard disk usually stores all of your software and data files, losing any part of or the entire hard disk's contents can be devastating. Performing timely backups ensures that you will not lose much if your hard disk fails or is damaged. However, since your hard disk is quite large, you must back it up onto some media that can contain it all. This might be a second hard disk or it might be a magnetic tape. Large organizations often use magnetic tape for just this purpose because it is more cost effective than purchasing numerous external hard drives. The second use of magnetic tape is related, storing information as an archive. The archive is essentially data from a previous time period (e.g., last year) that the organization hopes to not need but stores anyway just in case it is needed. An audit or an inquiry or research into previous customer trends may require that the archive be accessed. For either the backup or archive, the items stored will be used infrequently enough that we can afford the poor access time in favor of reduced cost.

### 3.7.3 Optical Disk and USB Drive

Optical disk was first introduced to the personal computer market in the 1980s. The technology centers around the use of a laser to shine onto an optical disk. The surface of the disk was normally reflective but pits would be burned onto the surface. A pit would swallow up the light while the surface itself, without a pit, would cause the light to shine back. In this way, 1s and 0s could be written onto a disk.

Optical disk was most commonly used as a form of read-only memory because most of the disks of that day had information permanently burned onto the surface of the disk. As such, this technology was often called a CD-ROM (compact disk read-only memory). The CDs would either store music or application software and multimedia data files. Once the technology was made available, optical disk drives were being sold that could burn onto a blank disk. These were called WORMs (write-once, read-many).

A different form of optical drive was introduced in 1985 with the NeXT workstation. It used a form of magneto-optical drive (CD-MO). Here, a laser would be used to read the magnetic state on the surface of the disk to pick up the corresponding bit. To write to the disk, first the given disk location would have to be erased and then written requiring two separate accesses. In addition to the laser, an electromagnet would be applied where the laser would heat up the disk and the electromagnet would adjust the bit stored in the given location. Although this technology was superior to the CD-ROM/WORM approach, it was too cost prohibitive to find a large audience.

In the late 1990s, a different form of readable/writable optical technology was released. In this case, the surface is comprised of a crystalline structure which in its natural state is reflective. Instead of burning information onto the surface and permanently altering its contents, a heated laser is used to alter the crystalline structure so that it is no longer reflective. This allows you to erase the entire disk and reuse it, which lead to the more common CD-R and CD-RW formats.

Unlike magnetic disk or tape, the optical disk can only be erased wholly or not at all. The optical disk led to both portable storage and a lesser costing storage device over the hard disk drives of the era. However, optical disks have a vastly limited storage capacity compared to hard disks and this differential has grown over time as hard disk technology improves. Today, DVD and Blu-Ray formats allow greater storage capacity over CDs but they are still not very useful forms of storage to perform backups as it would take dozens to hundreds of disks to back up a single hard disk drive.

The USB drive is the most recent form of secondary storage device. Unlike the other types of storage which combine a moving medium with a drive unit that moves the media and accesses the content, the USB drive is a form of nonvolatile memory called flash memory. Storage capacities for flash memory have been improving particularly in the last decade to reach many gigabytes. Further, flash memory is portable and nearly all computers today have USB ports. The main drawbacks of USB storage are the relatively slow access speed compared to hard disk and a limitation in the number of times the drive can be written to. Flash memory tends to wear out over time, particularly with each erasure. Estimates indicate that flash memory will become unusable after perhaps 1000 to 100,000 erasures. The actual number of erasures before a failure occurs will vary by specific device but it does indicate a limited lifetime for the device.

As a historical note, magnetic tape is the oldest of these storage devices, dating back to the earliest computers. In those days, the media was reel-to-reel tape. Later came audio cassette and video cassette tape. The tape cartridges available today are far more cost effective because they can store as much or more than hard disk drives. In the 1960s, magnetic floppy disk was introduced, followed by magnetic hard disk. Optical disks were not introduced until the 1980s. Flash memory was available in a rudimentary form in the 1980s but the USB drive was not introduced until around 2000.

In order for the CPU to communicate with a storage device (whether storage or other), we need to install the proper device driver. A device driver is software that we add onto the operating system. Its role is to act as an intermediary between the CPU and the device itself. This is a requisite because the CPU will not know (nor should it) how to communicate with each type of device. Every device has its own functions, features, and peculiarities. Therefore, we add the device driver program to translate from generic CPU commands to the specific actions needed to control the given device.

It used to be the case that aside from some hardware device drivers installed in ROM, you would have to install a device driver for every new piece of hardware. Today, many popular drivers are pre-installed in the operating system. Other popular device drivers are easily installed over the Internet. For Linux, most device drivers are pre-installed and available through kernel modules. You can find these modules under /lib/modules/ *version*/kernel/drivers where *version* is the Linux version installed.

## 3.8 FILE COMPRESSION

We wrap up this chapter with a brief examination of file compression and the tools available in Linux. The reason to use file compression is to reduce the impact that a file's size has on the available disk storage and during Internet transfer.

### 3.8.1 Types of File Compression

There are two forms of compression that we use to reduce file size: *lossy* and *lossless* compression. In lossy compression, information is actually discarded from the file to reduce its size. This loss cannot be regained and so if the information is important, you have degraded the file's information. We typically do this for multimedia files where the loss is either not noticeable or where the loss is acceptable. Audio lossy compression, for instance, will often eliminate data of sounds that are outside or at the very range of human hearing as such audio frequencies are not usually missed. The JPEG image compression may result in a slight blurring of the image as pixels are grouped together. The JPEG format can reduce a file's size to as much as one-fifth of the original file at the expense of having an image that does not match the original's quality.

There are specialized formats for video, streaming video, and human speech recordings. For video, one idea is to encode the starting frame's image and then describe in a frame-by-frame manner the changes that are taking place. The idea is that we can describe the change between frame i and frame i + 1 much more succinctly than we can describe frame i + 1 in its entirety or in a compressed way. Certain frames are stored as true images because, starting with the given frame, a new image/view/perspective is being presented in the video. Lossy compression of video, image, and audio files can reduce file sizes as much as 10-fold for audio and images and 100-fold for video!

Lossless compression is file compression technique whereby the contents of the file are being changed but the original file can be restored at a later time. Thus, the compressed file does not result in a loss of data. Most lossless compression algorithms revolve around the idea of searching for and exploiting redundant or repetitive data found in the original file. While this works fine for text files, it can also be applied to binary files although in many cases with less success. As many of our Linux data files are text files, we might look to compress these files to save disk space. We will also bundle together files using an archive program like tar. As we will see in Chapter 13, we will almost always compress and bundle source code to make the transfer of the numerous files quicker and easier. We look at file bundling and the tar program in Chapters 10 and 13.

### 3.8.2 The LZ Algorithm for Lossless Compression

There are numerous lossless compression algorithms available. Most are based on the idea of searching for repeated strings in the original file and then replacing those strings with shorter strings. The first example of such an algorithm is known as the Lempel Ziv (LZ) algorithm. The LZ algorithm searches a text file for strings and creates a dictionary of all strings found. Two common forms of this algorithm are known as LZ77 and LZ78, based on papers published by the two authors (Lempel and Ziv) in 1977 and 1978, respectively. The two algorithms work very similarly by replacing each string in the given text file with the location in the dictionary of the string's entry. As not all strings are replaced, our final compressed file will consist of literal characters and replacement locations.

Let us assume for an example that we are going to only search for words in a text file. All blank spaces and punctuation marks will remain in the file. Starting from the beginning of the file, we take each new word found and add it to a growing dictionary. Once we have compiled our dictionary, we start again at the beginning of the file and this time we replace each word with its location in the dictionary. We will assume that the locations will be integer numbers.

Imagine that we have a file of 2000 words in which only 250 are unique. Our dictionary then will consist of 250 entries. We replace the 2000 words in the dictionary with a number between 1 and 250 (or 0–249 if you prefer). As each integer number in this range can be stored in 1 byte, we are replacing longer words with single bytes. Our new text file might look something like the following. Remember that words are stored in alphabetic order so the sequence of numbers in our example looks random.

205 51 67 147 98 104 16! 205 51 88 140 221. 205 14 171 34 51 67 12 199.

Notice the replaced version of the text still retains spaces and punctuation marks. How much savings does our compressed version provide for us?

Let us assume that the 250 words average 5 characters apiece. Stored in ASCII, the 2000 words then make up 2000 × 5 = 10,000 bytes. We add to this the spaces and punctuation marks, also stored in ASCII. With 2000 words, there will be approximately 2000 spaces and punctuation marks. Let us assume in fact 2250 total spaces and punctuation marks, requiring a further 2250 bytes of storage. So our total file is some 12,250 bytes in length. The compressed version will replace each of the 2000 words with a single byte integer. Thus, the 10,000 bytes is now reduced to 2000 bytes. We also have the 2250 bytes for spaces and punctuation marks, or 4250. We are not quite done though. We also have to store the dictionary itself, which consists of 250 words with an average of 5 bytes per word, or 1250 bytes. Our compressed file is 5500 bytes long. Our compressed file is then 5500/12,250 = 44.9%, or less than half the original's size.

Could we do better? Yes, there are many additional tricks we can play to help reduce the size. For instance, we did not include any blank spaces and yet nearly every word will have a blank or punctuation mark afterward. If we were to add these to our words so that, for instance, we are storing "The " rather than "The", we would reduce the compressed file's size even more although we would be increasing both the number of entries in the dictionary and the length of each string in the dictionary. We could also search for strings irrelevant of word boundaries. For instance, we might find the string "the" which is found in many words like "the," "they," "then," "them," "anthem," "lathe," and "blithe." More complex algorithms will look for the most common substrings and use those rather than full words. Punctuation marks and digits can also be incorporated into these strings.

The lossless compression algorithms trade off search time for compressed file size. The more time the algorithm can take to compress a file, the more likely the resulting file will be even smaller. A user, compressing or uncompressing a file, may not want to wait a lengthy period of time. And therefore, the algorithm will have to search fewer combinations resulting in a less compressed file.

### 3.8.3 Other Lossless Compression Algorithms

Aside from the LZ algorithm, another popular algorithm is Burrows–Wheeler. This algorithm first sorts the text in strings to reorder the characters so that repeated characters appear multiple times. This allows those repetitions to be encoded into shorter strings. To obtain the greatest amount of compression, the algorithm looks for the longest repeated substrings within a group of characters (one word, multiple words, phrase, sentence). Once rearranged, compression can take place. Compression uses a substitution code like Huffman coding. In Huffman coding, strings are replaced such that the more commonly occurring strings have smaller codes. For instance, if the string "the" appears more times than anything else, it might be given the code 110 while less commonly occurring strings might have four, five, six, seven, eight, or even nine bits like 1000, 11100, 101011, and so forth.

In Huffman coding, we have to make sure that codes can be uniquely identified when scanned from left to right. For instance, we could not use 00, 01, 10, 11, 000, 001, 011, 101 in our code because the sequence 000101101 would not have a unique interpretation. We might think that 0001011 is 00 followed by 01 followed by 011 followed by 01, or it could be interpreted as 000 followed by 101 followed by 101. A code must be interpretable without such a conflict. We might therefore pick a better code using sequences like the following:

- Most common string: 000

- Second most common string: 001

- Third most common string: 010

- Fourth most common string: 011

- All other strings start with a 1 bit and are at least 4 bits long

In this way, 00000101100101010100001010101011101010101 could clearly be identified as starting with 000 001 011 001 010 1010001010101011101010101. As we have not enumerated the rest of the code, after the first five encoded strings, it is unclear what we have, but the first five are clearly identifiable.

In Linux, there are several programs available for performing compression and decompression on text files. The most popular two are gzip and bzip2. The gzip program is based on the LZ77 and Huffman codes. The use of LZ77 creates a dictionary and the use of Huffman codes permits the most commonly occurring strings to be encoded in as few bits as possible. The operation is gz *filename* where the result produces *filename*.gz. The compressed file cannot be viewed until it is decompressed. The decompression algorithm is either gunzip or gzip -d and it will remove the .gz extension.

The bzip2 program uses the Burrows–Wheeler sorting algorithm combined with Huffman codes for compression. There is a bzip program although it is based on the LZ77 and LZ78 algorithms. The bzip2 program operates much the same as gzip/gunzip with the instruction being bzip2 *filename* which creates *filename*.bz2. To decompress, use either bunzip2 or bzip2 -d. As with gunzip, these programs will remove the .bz2 extension.

There is also a zip program which is used to both bundle and compress files. Its format is different because it produces a single bundled file in place of multiple input files.

```
zip destination.zip sourcefile(s)
```

To unzip, use unzip *filename*.zip. The zip program uses the DEFLATE compression algorithm which combines both LZ77 and Huffman codes.

## 3.9 CHAPTER REVIEW

Concepts and terms introduced in this chapter:

- Absolute path—path from root to directory/file.

- Compression—reducing files' sizes by either discarding data or replacing data with replacement strings.

- Decompression—or uncompression, restoring the file to its original format. Only lossless compressed files can be decompressed.

- Device driver—a program that permits the CPU to communicate with a hardware device. The driver is installed as part of the operating system.

- Directory—a unit of organization which can contain files, subdirectories, and links.

- File—the basic unit of storage in the logical file system.

- File system—the collection of devices storing the files.

- Home directory—the directory reserved for the given user.

- Link—a pointer to an object in the file system.

- Logical file system—the user's view of the file system consisting of files, directories, paths, and partitions.

- Mounting—the act of making a file system accessible.

- Parent directory—the directory of which the current directory is contained.

- Partition—a logical way to separate groups of directories so that each partition can be treated independently.

- PATH variable—an environment variable which stores the directories that the Bash interpreter searches for an executable program when a command is entered.

- Permissions—access control to determine who can read, write, or execute a file or directory.

- Physical file system—the physical mechanisms that implement the file system including disk drives (and other storage devices); a file is stored in the physical file system as a collection of blocks distributed across one or more devices, connected by pointers.

- Recursive copy/delete—an operation to copy or delete not just the contents of a directory, but all subdirectories and their contents.

- Relative path—path specified from the current working directory.

- Top-level Linux directories—a standard collection of directories that help Linux users and administrators know where to look.

- Wildcards—characters that can be used in Bash to specify multiple files such as * matching all files and ? matching any one character of a file name.

- Working directory—the directory that the user is currently interacting in, reached using cd commands.

Linux commands covered in this chapter:

- bzip2/bunzip2—compress/decompress files using Burrows–Wheeler and Huffman codes.

- cat—concatenate files; often used to display files' contents to the terminal window.

- cd—change the working directory.

- chmod—change the permissions of a file.

- comm—compare two sorted files, displaying all differences.

- cmp—compare two files, byte by byte, displaying the point where the two files differ.

- cp—copy one or more files or entire directories to another location.

- cut—remove portions of every line of a file.

- diff—compare two files and display their differences.

- file—output the type of entity passed as a parameter such as a text file, a directory, a block device.

- find—locate files in the file system based on some criteria such as name, permissions, owner, last access time.

- gzip/gunzip—compress/uncompress files using LZ77 and Huffman codes.

- head—display the first part of a file.

- join—join lines of files together which have a common value for one of its fields.

- less—display a file screen-by-screen; allows forward and backward movement.

- locate/slocate—locate files in the file system based on name; uses a preestablished database of file names.

- ln—create a hard link; creates a symbolic link if you supply it with the option -s.

- ls—list the contents of the given or current directory.

- mkdir—create a new directory.

- more—display a file screen-by-screen; allows only forward movement.

- mv—move one or more files to a new location or rename a file.

- paste—unite files together by fields.

- pwd—print the current working directory.

- rm—delete one or more files; can delete directories if the -r recursive option is used.

- rmdir—delete an empty directory.

- sort—sort the contents of a file.

- split—split a file's contents into multiple files.

- strings—output sequences of 4 or more printable characters found in non-text-based files.

- tail—display the last part of a file.

- touch—create an empty text file.

- uniq—examine a file for duplicated consecutive lines.

- updatedb—create or update the file database as used with locate/slocate.

- wc—perform a character, word, and line count on the given file(s).

## REVIEW QUESTIONS

1. Which of the following Linux file system commands can you issue

   a. On a directory?

   b. On a file?

   ```
   ls cd mv rm cp sort touch less cat wc chmod
   ```

2. Referring to Figure 3.4, assume you are currently in the directory /home/foxr/
   temp2/a.

   a. Write an absolute path to take you to zappaf.

   b. Write a relative path to take you to zappaf.

   c. Write an absolute path to take you to etc.

   d. Write a relative path to take you to etc.

3. You are user zappaf. You were in the directory /usr and then performed cd /home/
   foxr/temp. What values would you find in these environment variables? PWD
   OLDPWD HOME.

4. What is the difference between home/foxr and /home/foxr?

5. What does ~ represent? What does ~zappaf represent?

6. What is the difference between . and ..?

For questions 7 through 16, assume that the current directory contains the following items
(all of which are files):

   abc.txt   abc1.txt   abb.txt   bbb.txt   bbb.dat   bbb   bbbb   c123.txt   ccc.txt

7. List the files that the following command will display: ls *

8. List the files that the following command will display: ls *.txt

9. List the files that the following command will display: ls a*.*

10. List the files that the following command will display: ls ab?.txt

11. List the files that the following command will display: ls bbb*

12. List the files that the following command will display: ls bbb.*

13. List the files that the following command will display: ls c??.txt

14. List the files that the following command will display: ls [ab][ab][ab].*

15. List the files that the following command will display: ls [a-c]??.???

16. List the files that the following command will display: ls [!c]*

17. Refer back to Figure 3.4. Assume that you are currently in the subdirectory temp2 of
    foxr. Assume the subdirectory a contains five files and subdirectory b is empty.

    a. What is the result of doing rmdir a?

    b. What is the result of doing rmdir b?

    c. What is the result of doing rm -ir a*?

    d. How would rm -fr a* differ from rm -ir a* from part c?

For questions 18 through 21, assume the directory zappaf has the following partial long listing:

```
-rw-rw-r--     zappaf    zappaf    . . .    foo1.txt

-rw-rw-r--     zappaf    zappaf    . . .    foo2.txt

-rw-rw----     zappaf    zappaf    . . .    foo3.txt
```

18. Write a command, as foxr, to copy these files to foxr's temp directory.

19. Will your command from 18 successfully copy all three files? Why or why not?

20. For your copy command from #18, what would differ if you added the option -p?

21. How does the command mv ~zappaf/foo1.txt. differ from the command mv ~zappaf/foo1.txt foo4.txt?

22. The -f option in rm forces a file to be deleted without asking for permission. What then does the -f option do for cp?

23. What is the difference between `tail -n 10` and `tail -n +10` when issued on a file of 100 lines?

24. What is the difference between `tail -c 10` and `tail -n 10` when issued on a file of 100 lines and 1000 characters?

25. Assume you have two files that are nearly identical. To view the differences between them, which file comparison program should you use?

26. Assume you have two files that are sorted. To view the differences between them, which file comparison program should you use?

27. Which of the file comparison programs can compare one file to several other files?

28. How does uniq differ from diff and cmp?

Assume you have two files: names.txt and phone.txt which each contain enumerated (numbered) lists of names (names.txt) and phone numbers (phone.txt). We assume that the $i$th entry in file 1 corresponds to the $i$th entry in file 2 for all entries. Answer questions 29 and 30.

29. What command could you issue to create a file containing all names and phone numbers, properly organized?

30. Assume that the phone.txt file is incomplete so that some entries do not contain phone numbers. The numbers between the two files still correspond (for instance, if the 3rd person's phone number is missing, the phone.txt file skips a line with the entry 3). Revise your command from question 29 so that missing phone numbers are indicated with the word UNKNOWN.

31. Explain what the split instruction does.

32. Imagine that you have a large multimedia (1 GB) file (e.g., a movie stored as an mpg or avi file). You want to break up the file into smaller 100 MB files. Provide a command to accomplish this so that the file components are all numbered like 00, 01, 02, 03 after their name.

33. What is the difference between `wc foo1.txt` and `wc -l foo1.txt`?

34. Referring to Figure 3.3, write a command to create a symbolic link of the file in zappaf's directory foo1.txt from foxr's temp directory. How would your command differ if you wanted to create a hard link?

35. Assume that file alink is a symbolic link to the file afile located in another directory. What happens to afile if you delete alink? What happens to alink if you delete afile?

36. Repeat question 35 assuming that alink is a hard link instead of a symbolic link.

37. Write a find command to find and list all files in the current directory accessed within the last 3 days.

38. Write a find command to find all files in your home directory modified less than 60 min ago and delete them.

39. Write a find command to perform wc -l on all files in your home directory newer than the file myfile.txt.

40. Write a find command to find all files in the entire file system (starting at /) whose permissions are 666 and change them to 664.

41. Write a find command to find all files in the current directory whose size is between 1000 and 10,000 bytes.

Assume for questions 42 through 50, we have the following partial long listing of files and subdirectories in the current directory. Also assume that you are not user foxr but you are in the group cool. Answer each question independently (i.e., do not assume that a previous instruction changed the permissions as shown below).

```
drwxrwxr--      foxr    foxr    . . .    subdir1
drwxr-xr--      foxr    cool    . . .    subdir2
-rw-rw-r--      foxr    foxr    . . .    foo1
-r-xr-xr--      foxr    foxr    . . .    foo2
-rwxr-----      foxr    cool    . . .    foo3
-rw-r--r--      foxr    foxr    . . .    foo4
-rw-rw-rw-      foxr    foxr    . . .    foo5
```

42. Provide the 3-digit permissions for each of the items listed here.

43. Write a chmod command to permit execute access both group and world for foo3 using the ugo= approach.

44. Write a chmod command to permit execute access both group and world for foo3 using the ugo+/- approach.

45. Write a chmod command to permit execute access both group and world for foo3 using the 3-digit approach.

46. Write a chmod command to permit write access to foo4 for group but remove read access for world using the ugo+/- approach.

47. Write a chmod command to remove all group and world access to fool using the ugo+/- approach.

48. Write a chmod command to remove all group and world access to fool using the 3-digit approach.

49. As a member of cool, do you have the ability to cd into subdir2?

50. Of the items listed here, which have permissions that you think should be altered? Why?

51. Which of the following permissions you generally do not see in a Linux system? Why not?

    a. 765

    b. 646

    c. 404

    d. 200

    e. 111

52. What does the permission 1777 mean? How would the permissions of such a directory appear when you use `ls -al`?

53. Why might you make a directory world-writable? If you do so, what are the consequences on files stored there?

54. Why would you set a directory's sticky bit?

55. How does the /proc directory differ from all other top-level directories?

56. Briefly explain the difference between the types of programs you would find in /bin versus /sbin.

57. Why is /root located under / and not /home?

58. Of the following directories, which would you generally not anticipate changing for weeks or more at a time? /bin, /boot, /home, /sbin, /var

59. Under which top-level directory will the system administrator probably spend time editing files?

60. Explain why a user may wish to invest in a magnetic tape drive and magnetic tape cartridges.

61. Research the following compression formats and specify whether they are lossy or lossless: flac, gif, jpeg, mp3, mpeg-1, mpeg-2, mpeg-4, png, tiff, and wma.

62. A file stores the following text. Define a dictionary of words for this file and then substitute each word with a number as its replacement where the number is the location of the word, alphabetically, in the dictionary. Periods, spaces, and line breaks should be retained in the file. Include your dictionary in your replacement file. How much compression do you achieve with your replacement file?

The blue fox jumped over the red cow.

The red cow ate the green grass.

The green cow ate the blue grass.

The blue fox then jumped over the green cow.

The green cow ate more blue grass.

# CHAPTER 4

# Managing Processes

THIS CHAPTER'S LEARNING OBJECTIVES are

- To understand the parent–child relationship between processes in Linux

- To understand the different ways an operating system can manage processes

- To understand the role of the effective UID and GID for processes

- To understand the difference between foreground and background processes and be able to move Linux processes between the two

- To understand the tools available in Linux to monitor running processes

- To understand niceness and how to adjust process priorities

- To understand the ways that processes can terminate and how to use kill, killall

## 4.1 INTRODUCTION

A *process* is a running program. Processes can be started from the GUI or the command line. Processes can also start other processes. Some programs interact with the user and others run to completion and report their results, either to a window or a disk file. In this chapter, we look at how to manage processes, primarily from the command line interface. We will also look at process and resource management GUI programs.

Whenever a process runs, Linux keeps track of it through a process ID (PID). After booting, the first process is an initialization process called init. It is given a PID of 1. From that point on, each new process gets the next available PID. If your Linux system has been running for a while, you might find PIDs in the tens of thousands.

A process can only be created by another process. We refer to the creating process as the *parent* and the created process as the *child*. The parent process *spawns* one or more child processes. The spawning of a process can be accomplished in one of several ways. Each requires a system call (function call) to the Linux kernel. These function calls are fork(), vfork(), clone(), wait(), and exec().

The fork system call duplicates the parent to initialize the child. The child then obtains the same contents as is stored in memory for the parent including such values as the current working directory, the terminal window in which the parent (and child) was launched, umask values, and so on. The only difference between the two is the PIDs of the processes and the location in memory of the processes. For vfork, the two processes (parent and child) actually share the same memory. The clone system call is similar to vfork. The two processes share memory, but also share other resources.

With clone, you can create threads. The wait system call is like fork except that the parent goes into a wait (sleep) mode until the child terminates. At that point, the parent can resume execution.

The exec system call causes a process to start a program which replaces the current process. That is, the process invoking exec will be replaced by the new running process. The new running process is given the parent process' PID (in actuality, the new process merely replaces the old). With exec, we actually have several different system calls available. They vary in that the different forms launch the new process using a different collection of environments. For instance, execl uses the same environment for the child as the parent except that it uses the current process' PATH variable to locate the executable.

As the parent process is replaced with exec, it is common to instead pair up fork and exec. First, the parent process issues a fork to create a new process with a new PID. This new process is a duplicate of the parent. Now the child invokes exec to change the process to one of a different program.

As an example, imagine that you have opened a terminal window which is running a bash shell. From this shell, you invoke a program, say vi. The bash shell first issues a fork call to duplicate itself and then the child issues an exec to replace itself with the vi program. We will explore function calls in more detail in Chapter 8.

Let us consider a more concrete example of parents and children. The first process to run once the Linux kernel has initialized itself is init. After init initializes the operating system (including the user interface), we have a child process, the GUI environment itself (let us assume Gnome). To run some application, it is Gnome that spawns a child process through fork and exec. If you are operating in a terminal window, then the terminal window represents one process and spawns a shell process (for instance, bash). Commands that the user enters from the command line interface are spawned by bash. Usually, a process run from the command line executes and terminates before the command line becomes available to the user.

Entering a command which contains a pipe actually spawns multiple processes. For instance, from the command line, if you enter the command

```
ls -al | grep zappaf | wc -l
```

then bash spawns first an ls process whose output is sent to grep, a second process, and that output is redirected to wc  -l, a third process.

Consider as a user you have logged into a Linux computer using the GUI. You open a terminal window. Your default shell is bash, so the terminal window, upon starting, runs bash. From the bash shell, you launch a background process such as a script (we discuss background processes in Section 4.2). Then you launch a foreground process, perhaps a find operation. You have created a chain of processes. The init process spawns the GUI which spawns a terminal window which spawns bash which spawns both a background and a foreground process. The chain might look like this:

```
init\
     \--gnome\
             \--gnome_terminal\
                              \--bash\
                                     \--./somescbash
                                     --find
```

There are different types of processes that run in the system. As a user, we are mostly interested in our own processes (which include threads). From a system's point of view, there are also services. We will ignore services in this chapter but cover them in detail later in the text.

## 4.2 FORMS OF PROCESS MANAGEMENT

Before we specifically address processes in Linux, let us first consider what the process is and the different ways processes are run in operating systems. As defined in the previous section, a process is a running program. Consider that a program is a piece of software. It contains code and has memory space reserved for data storage. Most of this storage is dedicated to variables that are declared and used in the program. The program exists in one of two states, the original source code or the executable code. In either case, the program is a *static* entity.

The process, on the other hand, is an *active* entity. As the process executes, the data change—variables change values, portions of the program move between swap space and memory, and possibly also move within memory as well, and the location within the program that the processor is about to execute also changes, instruction by instruction. In the latter case, the memory location of the next instruction to be executed is stored in a register often called the program counter (PC). The PC along with other registers (instruction register or IR, stack pointer or SP, status flags or SF, data registers such as an accumulator, or AC, and so forth) constitute the state of the running program along with the current values of the data (variables). Thus, the state changes, often as each new instruction is executed.

The operating system must keep track of the program's state and does so through a data structure often called the *process status word* (PSW). The PSW will be a collection of the most important pieces of information about the running process such as the current values in the PC, IR, SF, and so on.

Aside from keeping track of a process' status, the operating system is in charge of scheduling when processes run, and of changing the process' status as needed. For instance, a process which requires time-consuming input will be moved from "ready for CPU" to "waiting for input." Handling the scheduling and movement of processes is called process management.

Different operating systems have implemented different forms of process management. These are

- Single tasking

- Batch

- Multiprogramming

- Multitasking

- Multithreading

- Multiprocessing

We briefly discuss these here although Linux primarily performs multitasking and multithreading.

### 4.2.1 Single Process Execution

In single tasking, the operating system starts a process and then the CPU's attention is entirely held by the one process. This means that the computer is limited to running only the one process until either that process terminates or the user suspends that process. In many operating systems that performed single tasking, suspending a process was only possible by terminating the process. Single tasking has many drawbacks. Among them, there is no chance of having the computer do more than one thing until that particular task is over. Additionally, if the process has to perform some time-consuming input or output task, the processor remains idle while the I/O takes place. Only the most rudimentary operating systems today would be limited to single tasking because of these drawbacks.

Batch processing is much like single tasking with three differences. First, in batch processing, there may be multiple users wishing to use the system at one time. Since the system is single tasking though, the system will only concentrate on one user's process at a time. Therefore, the second difference is that users submit their programs to run (jobs) using a separate, possibly offline system. This system is responsible for scheduling when the submitted job will be selected by the CPU for execution. Third, since the job may run any time from immediately to days later, the job cannot be expected to run interactively.

To support the lack of interactivity, a user must submit the input with the job as well as the destination for the output. Typically the input will be a file stored on some storage device or placed as a series of strings of data to be read by the program during execution. A common way to specify the data is through "cards." In earlier days of computing, cards were literally punch cards containing the data. This was eventually replaced by lines of code using an operating system language like JCL (Job Control Language). Today, if batch processing is used, the input is stored in a disk file whose name is provided when the process runs. Similarly, when submitting the program, the user would specify the output's destination.

The offline system not only receives user requests to run processes but also schedules them using some type of scheduling algorithm. During the era of batch processing, a number of different scheduling algorithms were developed. The simplest, and perhaps fairest, is first come first serve, often known as FIFO for first-in first-out.

In an organization where some users' processes are more critical than others, a priority scheme could be used. At a university, for instance, the highest priority might be given to administration processes with faculty processes given the next highest priority. Processes submitted by graduate students might come next followed by those of undergraduate students. Although not a fair scheme, it represents the needs or priorities of the university. Within each priority level, submitted jobs might be organized in a simple first come first serve order.

Another scheduling algorithm is called shortest job first. In this case, the scheduler estimates the execution time needed by the process and orders the waiting jobs in ascending order of this estimated time. It has been shown that this scheduling algorithm guarantees the minimum average waiting time. The opposite approach is called longest job first. As both of these scheduling algorithms estimate execution time, they may not actually yield the minimum (or maximum) average waiting time.

Whichever algorithm is applied for scheduling, it is the scheduler's task to order the jobs and place them in a *queue* (waiting line) reflecting the scheduled ordering. In the case of the priority scheme, we would have one queue per priority level.

### 4.2.2 Concurrent Processing

One of the great inefficiencies of batch processing is that of I/O. Imagine, for instance, that input is to be read from magnetic tape. During execution of the current process, if it needs to input from tape file, this will cause a pause in the execution because of the relative slowness of tape input. During this time, the CPU idles.

It would be nice to give the CPU something to do while waiting. This leads us to an improved form of process management called multiprogramming. In multiprogramming, the CPU continues to only execute a single process, but if that process needs to perform time consuming I/O, that process is removed from its current position of being executed by the CPU and moved by the operating system into an I/O queue. The operating system then selects the next waiting process for execution by the CPU. In doing so, to the user it appears as if there is more than one process currently executing. We will call this concurrent processing. Multiprogramming is one form, we look at others below.

The movement of processes requires a new operation known as a *context switch*. The context switch is the switching of processes from the one that the CPU was running to the next process. To perform a context switch, the operating system gets involved. First, the currently executing process' status must be captured and saved. This is typically stored somewhere in the operating system's area of main memory.

The process' status, as explained earlier, includes the values stored in the PC, IR, SF, AC, and SP registers among others. That is, we are saving what the processor was doing (IR), where it was doing it (PC), and what the results of the latest operation were (SF, possibly AC, SP). Now the operating system starts or resumes another process by loading its status (IR, PC, SF, SP, AC) values from memory into the corresponding registers. Once done, the CPU starts up again but is now executing on a different process in a different area of memory. To the CPU, it is merely performing its fetch–execute cycle but on a different set of instructions. But to us, and the operating system, the computer has switched contexts from one process to another.

As the context switch requires moving values between the CPU and memory, it causes the CPU to pause. This is not nearly as time-consuming as pausing the CPU to wait for I/O, but it is a pause. High-performance computers use multiple sets of registers so that the switching of status values can be done solely within the CPU. This comes with a greater expense since such CPUs require more registers.

The idea behind multiprogramming is that a process will voluntarily surrender its access to the CPU because it has to wait for some other event. Above, we referred solely to some input or output, but it could also occur because it is waiting on data from another running process. We refer to the idea that a process voluntarily giving up the CPU as *cooperative multitasking*. In essence, the computer is multitasking (doing several things at once) by switching off between the tasks whenever the current task requires some waiting period. Now the CPU switches to another task and works on it in the interim.

In multiprogramming, once the waiting process has completed its wait (whether for I/O or for communication with another process), a context switch occurs whereby this process resumes and the executing process again waits. The operating system maintains a ready queue to handle the switching of processes. This queue consists of those processes that the CPU is actively switching between.

The next step in our evolution of process management is to force a context switch so that no single process can monopolize the CPU. We add a piece of hardware called a timer, which counts clock cycles. The timer is initialized to some value, say 10,000. Then, after each clock cycle, the timer is decremented. When it reaches 0, it alerts the operating system which performs a context switch with the next process waiting in the ready queue. The amount of time awarded to a process is known as its *time slice*. This form of process management goes beyond multiprogramming (cooperative multitasking) and is known as *preemptive multitasking*, although we could also refer to it as multitasking as long as it is clear that this differs from multiprogramming. So we now see two reasons for a context switch, a process voluntarily gives up the CPU or the timer causes the switch.

Cooperative and preemptive multitasking both require scheduling. In this case, the selection of the next process to execute when the CPU becomes free (either because the current process terminates or is moved to a waiting situation) will be made by the operating system. For multiprogramming, scheduling may be accomplished via first come first serve, priority, or a round robin scheduling algorithm.

In round robin scheduling, processes are handled in the order that they were submitted (first come first serve) except that the OS loops around to repeat the earliest processes after the last processes in the ready queue have been moved into a waiting situation.

Preemptive multitasking usually uses round robin scheduling. However, we could add to this strategy a priority for each process. If process priorities differ, then the time slices allotted to the processes can also differ. Imagine each process has a priority value from 1 to 10. For those with the highest priority (10), they are given 10,000 cycles of CPU time. Those given the lowest priority (1) might be initialized with only 1000 cycles of CPU time. We can also age processes so that, as time goes on, their priority lessens so that the CPU can give more time to newer processes.

Figure 4.1 illustrates the concepts of the context switch in multiprogramming (cooperative multitasking) and (preemptive) multitasking. In the top portion of the figure, process P3 is executing and the CPU's registers reference information about this process (its current instruction in the IR, the next instruction's location in the PC, etc.). At some point, P3 needs to perform I/O so the operating system, saves its process status information to memory, moves P3 to an I/O queue (not shown in the figure), and starts or restores the next process, P7. In the bottom of the figure, P3 is executing and with each instruction, the timer decrements. Upon reaching 0, the CPU is interrupted and the operating system performs the same context switch except that P3 moves to the back of the ready queue.

The main detractor of cooperative and preemptive multitasking is the time it takes to perform the context switch. Threads are similar to processes but they represent different instances of the same running program. Because they are of the same program, they can share some of the information that processes would not share. For instance, two threads of the Mozilla Firefox web browser would use the same executable code and some of the same data. They would differ in that they would have different register values and their own unique sets of data as well. The advantage of using threads is that context switching

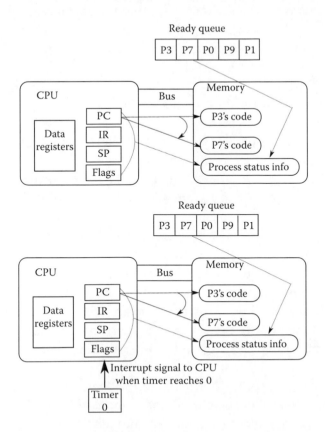

FIGURE 4.1    Context switch in multiprogramming (top) and multitasking (bottom). (From Richard Fox, 2013, *Information Technology: An Introduction for Today's Digital World*.)

between threads is often handled faster than context switching between processes. For one reason, switching between processes may involve loading program code from virtual memory into main memory. This should not happen when switching between threads. Threads are a newer idea in programming but are available in most operating systems. However, to take advantage of threads, the program must be written using threads and you must be running multiple threads of the same program. A multitasking system that is switching off between threads is known as a multithreaded system. Thus, a multitasking system can also be multithreading.

Another form of process management occurs when the computer has multiple processors. This is known as multiprocessing. In this case, the operating system selects the processor to run a given process. In this way, multiple processors can execute their own processes. Typically, once launched onto a processor, the process remains there. However, load balancing allows the operating system to move processes between processors if necessary. In addition, some programs are set up to themselves be distributed to multiple processors in which case one process might execute in parallel on more than one processor, each processor in charge of some portion of the overall process. Today's multicore processors permit multiprocessing.

Linux by default uses cooperative and preemptive multitasking and multithreading. Programmers who wish to take advantage of multiprocessing must include instructions in the program that specify how to distribute the process. Linux can also perform batch processing on request using the batch instruction (covered in Chapter 14).

### 4.2.3 Interrupt Handling

We mentioned above that the timer interrupts the CPU. What does this mean? Left to itself, the CPU performs the fetch–execute cycle repeatedly until it finishes executing the current program. But there are occasions where we want to interrupt the CPU so that it can focus its attention elsewhere. The timer reaching 0 is just one example.

We refer to the situation where something needs attention by the CPU as an *interrupt request* (IRQ). The IRQ may originate from hardware or software. For hardware, the IRQ is carried over a reserved line on the bus connecting the hardware device to the CPU (or to an interrupt controller device). For software, an IRQ is submitted as an interrupt signal (this is explained in Section 4.6.2 later in this chapter).

Upon receiving an interrupt, the CPU finishes its current fetch–execute cycle and then decides how to respond to the interrupt. The CPU must determine who raised the interrupt (which device or if it was raised by software). IRQs are prioritized so that if the CPU is handling a low-priority IRQ, it can be interrupted to handle a higher priority IRQ. The CPU acknowledges the IRQ to the interrupting device. Now, it must handle the interrupt.

To handle the interrupt, the CPU switches to the operating system. For every type of interrupt, the operating system contains an *interrupt handler*. Each interrupt handler is a piece of code written to handle a specific type of interrupting situation. The CPU performs a context switch from the current process to the proper interrupt handler. The CPU must save what it was doing with respect to the current process as explained in the last subsection.

In Linux, interrupt handlers are part of the kernel. As the interrupt handler executes, information about the interrupt is recorded under /proc/interrupts. This file stores the number of interrupts per I/O device as well as the IRQ number for that device. If you have multiple CPUs or multiple cores, you receive a listing for each of the CPUs/cores.

We see an example of the /proc/interrupts file below for a computer with two CPUs (or two cores). The first column lists the IRQ associated with the given device. The second column is the number of interrupts from the device for each of the two CPUs. The third column is the device's driver and the last column is the device's name. Notice that the timer has the most number of interrupts by far, which is reasonable as the timer interrupts the CPU many times per second.

```
         CPU0          CPU1
 0:      20342577      20342119      IO-APIC-edge      timer
12:      3139          381           IO-APIC-edge      i8042
59:      22159         6873          IO-APIC-edge      eth0
64:      3             0             IO-SAPIC-EDGE     ide0
80:      4             12            IO-SAPIC-edge     keyboard
```

The file /proc/stat also contains interrupt information. It stores for each IRQ the number of interrupts received. The first entry is the sum total of interrupts followed by each IRQ's number of interrupts. For instance, on the processor above, we would see an entry of intr where the second number would be 40684696 (the sum of the interrupts from the timer, IRQ 0). The entire intr listing will be a lengthy sequence of integer numbers. Most likely, many of these values will be zero indicating that no interrupt requests have been received from that particular device.

## 4.3 STARTING, PAUSING, AND RESUMING PROCESSES

Starting a Linux process from the GUI is accomplished through the menus. The GUI is launched at OS initialization time and is spawned by the init process. As the user runs programs, those are spawned by the GUI (e.g., Gnome). Some application processes will spawn additional processes as well. For instance, the user may be running the Firefox web browser. Opening a new tab requires spawning a child process (actually a thread). Sending a web page's content to the printer spawns another process/thread.

### 4.3.1 Ownership of Running Processes

Many running processes require access to files (as well as other resources). For instance, if you are running the vi text editor, you may want to open a file in your home directory or save a file to your home directory. How then does the vi process obtain access rights (especially since you probably do not permit others to write to your directory)? When a process is launched by the user, the process takes on the user's UID and the user's private group's

GID. That is, the process is running as an extension of you. Therefore, the process has the same access rights that you have. This is perfectly suitable in many cases.

However, some applications require access to their own file space and their own files. For instance, the lp command (print files) copies the output to a printer spool file, located in /var/spool. This space is not accessible to the user. So how does lp write the data there? When some software is installed, a special account is created for that software. Thus, lp has its own user account. You might examine the /etc/passwd file for other software accounts (you will find, among others, accounts for mail, halt, ftp, and the apache web server). For lp, it has a home directory of /var/spool/lpd and a login shell of /sbin/nologin. There is no login shell because we do not want people to attempt to log in using lp's account and gain access to system files. You will see that the various software accounts have home directories in the top level directories of /sbin, /proc, or /var. In running lp, the operating system switches from you the user to lp the software account. Now lp has access to its own file space.

In other situations, the process does not need its own account but it still has to gain access to system files. Consider the passwd command. This program is owned by root. When a user runs passwd, the program accesses the password file (which is stored in/ etc/shadow, not/etc/passwd) first to ensure that the user has entered the correct current password. Then the program accesses the file again to modify the password. When a user runs passwd, it should run under the user's account. But the shadow file is only accessible to root. The way Linux gets around this problem is to provide a special bit in the permissions that indicate that the process should be owned not by the user who launched the process, but by the owner of the file containing the program's executable code. In the case of passwd, this is root. Therefore, because this special bit is set, when running passwd, it takes on root's ownership. Now it can access the /etc/shadow file that is only accessible to root.

To set up a program to execute under the ownership of the file's owner, you need to alter the permissions (using chmod as described below). To determine if a process will run under the user's ownership or the file's ownership, inspect the file's permissions through a long listing. You will find that the owner's executable bit is not set to 'x' but to 's.' If you alter the owner's execution status, the permission changes from 'x' to 's.' Similarly, you can change the group execution. If the executable bit is set to s, this is known as setting user permission ID or setting group permission ID (SUID, GUID). By altering the executable bit, the program, when running, switches it's owner or group ID. The programs passwd, mount, and su, among others, are set up in this way.

Let us take passwd as a specific example. If you look at the program, stored in /usr/bin, you find that it has permissions of -r-s--x--x and its owner and group are both root. Thus, root is able to read it and for execute, it has the letter 's.' The world has execute access. So, any user can run passwd. But when running passwd, the program is granted not the user's permissions but the file owner's permissions. That is, passwd runs under root and not the user. If you look at the /etc/passwd file, which the passwd program may manipulate, it has permissions of -rw-r--r-- and also has root for its owner and group. So passwd, running under your user account would not be able to write to the /etc/passwd file. But passwd running under root can. Thus, the executable bit for passwd is 's' for the owner.

When the execution permission for owner is changed to 's,' this changes the effective user of the program. The real user is still the user who issued the command. The effective user is now the program's owner (root in this case). This is also known as the EUID (effective user ID). Similarly, setting the group ID on execution changes the effective group (EGID).

To change the execution bit use chmod. Recall from Chapter 3 that there were three ways to use chmod, ugo+/-, ugo=, and the three-digit permission. You can change the execution bit from x to s (or from nothing to s) and from s to x (or from s to nothing) using the +/- approach and the three-digit approach. The ugo= approach cannot be used. With the ugo +/- approach, merely substitute 's' for 'x' as in u+s or g+s. You can also reset the value back to 'x' using u-s+ x (or g-s+x), and you can remove execute access entirely with u-s or g-s.

The three-digit approach is different in that you use a four-digit number. If you want to establish the 's' for the user, prepend a 4 to the 3-digit number. For instance, if you want a file to have permission of -rwsr-xr-x, you would normally use 755 to indicate that everyone should have execute access. Now, however, you would prepend a 4, or 4755, to state that the file should be rwxr-xr-x but with an 's' for the user's execution bit. Similarly, you would prepend a 2 to change the group's execution from 'x' to 's.' You would use 6 to change both user and group execution status. Below are several examples.

```
4744 _ rwsr--r--      4754 _ rwsr-xr--
4755 _ rwsr-xr-x      6755 _ rwsr-sr-x
2750 _ rwxr-sr--      2755 _ rwxr-sr-x
```

To reset the execution bit from 's' to 'x,' prepend a 0, as in 0755 to reset rwsr-xr-x to rwxr-xr-x. Recall from Chapter 3 that prepending a 1 to the three-digit number, as in 1777, adds the sticky bit to the executable permission for world. With SUID and GUID, we used 4xxx and 2xxx to change the executable permission for owner and group, respectively. The sticky bit though is used on directories to control write access to already existing files within the directory.

### 4.3.2 Launching Processes from a Shell

Launching processes from within a shell is accomplished by specifying the process name. You have already seen this in Chapters 2 and 3. If the file storing the program is not in a directory stored in the PATH variable, then the command must include the directory path. This can be specified either with an absolute or relative path. As an example, /sbin may not be in the user's PATH variable because it contains system administration programs. To execute the ifconfig program (which displays information about network connectivity and IP addresses), you might have to enter /sbin/ifconfig.

Launching your own programs is slightly different from launching system programs. These programs will include any that you have written and compiled into executable programs as well as shell scripts. The notation to run such a program is ./filename if the program is in the current directory. If the file is in another directory, you do not need to use the ./ notation.

To this point, most of the programs you have executed from the command line have run quickly. The result of the program is some output sent to the terminal window and then the command line prompt is made available to you again. What happens if you have a lengthy process to run and yet you want to continue to work through the command line? If the program you are going to run can run in a batch mode, then you can launch it and regain the command line prompt while the program is still executing. Batch mode means that the process will run independently of the user, or without interaction. Consider, for instance, that you want to use find. Recall that find, when run on the entire file system, can take a lot of time. The command you want to execute is

```
find / —perm 000 > results.txt
```

Because this instruction needs no input from the user, and because its output is being sent to a file and so will not interfere with the user running other programs, we should be able to launch it as a batch process. In order to do so, when issuing the command, end the command with the character &. This denotes that the process should run in the *background*, as opposed to a process being in the *foreground* which is the default in Linux.

What is a background process? A background process has two different meanings depending on the context. One meaning is that the process runs when the CPU has time for it. This is the case when you want to run a time-consuming process that takes up a lot of system resources, but you do not want it to interfere with other tasks. So for instance, the process may run when you have taken a break and then move back to the background when you start working again. In our case, background means that the process will work alongside of other processes, but will do so without interaction with the user. In both Windows and Linux, GUI processes can be minimized or moved to the back of the desktop. In such cases, you are no longer interacting with them, so they are in the background.

The use of & tells the operating system that the process, launched from the command line, should run in the background so that the user can interact with other processes. The interpreter responds with a message to indicate that the process has been launched. This message has the form

```
[1]  3811
```

The value in the [] is the job number (discussed shortly) and the second value is the process ID. The process then executes until completion. Upon completion of the process, the terminal window displays a message to indicate that the process has completed, which looks like this

```
[1]+ Done                    command
```

where *command* is the full command entered by the user.

If you launch a program in the background and do not redirect its output, upon termination, the program's output will be displayed to the terminal window. This might be confusing to you the user as, while you are performing command line input, you might find suddenly a bunch of output appearing.

You may run as many processes as you wish in the background. A terminal window, however, can only have one process running in the foreground. The foreground process is the one that provides interaction with the user, whether that is to receive the user input data or commands from the user, or to output information to the window, or both. However, this does not mean that while a process is running in the foreground, you are not able to launch other processes or move background processes to the foreground.

Before discussing how to move processes between foreground and background, we need a mechanism to identify the foreground and background processes in the terminal window. We obtain this information with the command jobs. The jobs command responds with all of the active jobs in the terminal window. These are processes that have been started but not yet terminated. A process will have one of three different statuses. It may be in the foreground, it may be in the background, or it may be stopped (suspended). However, a process will not be in the foreground if you are executing jobs because jobs require that you have access to the command line. So what you will find as processes listed by jobs is that they are either stopped or in the background.

To stop a process that is in the foreground, type control+z in the terminal window. This presents you with information like this:

```
[2]+ Stopped           find ~ -name *.txt
```

This line tells us that job #2, find, was stopped. The plus sign is described below.

To resume this process, type fg. The command fg moves this process into the foreground. Alternatively, to resume this job in the background, type bg. Thus, we have two ways to get a job into the background. We can launch it in the background using & or we can launch the process, stop it, and then move it to the background using bg. We can similarly switch running processes between fg and bg. Let us take a closer look.

Imagine that we have started top. We suspend it with control+z. We then start vi and suspend it followed by a man jobs command and then a grep process. We resume man, stop it, then resume top, and stop it. Typing jobs gives us the following output:

```
[1]+ Stopped           top
[2]  Stopped           vi file1.txt
[3]- Stopped           man jobs
[4]  Stopped           grep files *.*
```

This tells us that there are four active processes in the current terminal window. All four are current stopped. The numbers in the listing indicate the order that the processes were started. The + and − indicate the most recent two processes that were in the foreground. In this case, top was the most recent foreground process and man jobs the second most recent.

Now, we want to move the grep instruction into the background. We do so typing bg 4. If we were to type bg without a number, it would move top into the background because it was the last process to be referenced (it has the +). Moving any of top, vi, or man into the background makes little sense because each of these processes require user interaction (grep does not). If we were to type fg 2, it would move vi into the foreground. Typing control+z followed by jobs would now yield the following output. Notice how the + and − have changed.

```
[1]   Stopped                     top
[2]+  Stopped                     vi file1.txt
[3]   Stopped                     man jobs
[4]−  Stopped                     grep files *.*
```

Note that the job number is not in any way related to the process ID. If there is only a single job, you can move it to the foreground or background using fg or bg respectively without a job number. Otherwise, fg and bg will move the job with the + to the foreground or background. A stopped process resumes executing if it is moved to the foreground or the background. However, the resumed process, if moved to the background, may not execute until the CPU has time for it.

## 4.4 MONITORING PROCESSES

You might find that, as a user, you seldom have to keep tabs on running processes. The operating system can run multiple processes efficiently with little or no user intervention. Yet there are times when you might want to see what a process is doing. For instance, a process may be using an inordinate amount of system resources or a process might have died on you.

### 4.4.1 GUI Monitoring Tools

There are several different tools available to monitor processes and system resources. The primary GUI tool is called the System Monitor. There are two different versions, one for Gnome and one for KDE. Figure 4.2 shows the Gnome version. There are four tabs along the top of the GUI, System, Processes, Resources, and File Systems.

In Figure 4.2, Processes is selected and the active processes are listed in the window. They are ordered, in this case, by CPU utilization. You might notice that the process requiring the most CPU attention is the system monitor itself! It is typical that GUI programs will require far greater CPU utilization than text-based programs. The next program in CPU utilization is a find command, running inside a gnome-terminal window. You might notice that the gnome-terminal process is "sleeping." In fact, it is running the find command, so while find is executing, the terminal window is not accessible (unless we suspend find as we discussed in Section 4.3). We will examine the meaning behind the values under Status in the next subsection when we review the ps command.

Other information in this window describes for each process the process ID, niceness, memory utilization, waiting channel, and session (not shown in the figure). We have already discussed the process ID (PID) and we will continue to reference it in this chapter. The niceness of a process describes its priority. We will talk about that in more detail in

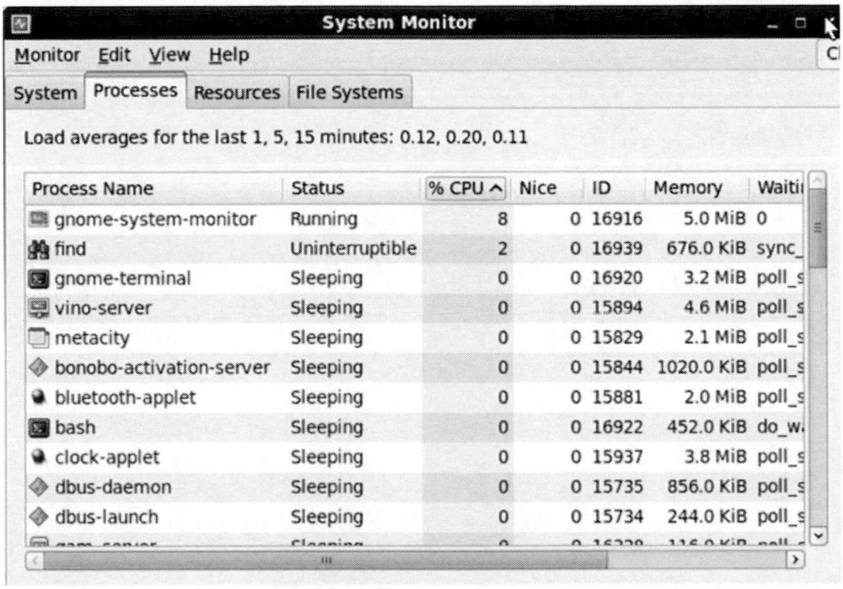

FIGURE 4.2  Process listing shown in system monitor.

the next section. Memory utilization is the amount of memory that the process is currently holding on to. We will ignore the waiting channel and session information.

Also notice that the window provides a summary of load averages. These averages describe the amount of CPU utilization, where 0.12 means 12%. We see the load averages for the previous 1, 5, and 15 min.

From the Processes tab, aside from viewing the processes, you are able to select any single process and stop it, resume it, kill it, end it, or change its priority. Ending a process terminates it as if you select exit from the process' GUI (e.g., "exit" from the "file" menu). Killing a process ends it immediately which may damage open files or cause processes spawned by this process to be abandoned. It is better to end a process than kill it unless the process has stopped responded.

The Resources tab provides a summary of CPU, memory and swap space usage, and network utilization over time. An example is shown in Figure 4.3. This tab is strictly output. You might notice in this case almost no swapping has been done. This is because the user has not run many large processes. Network utilization has increased, which was caused by the user starting Firefox.

The File Systems tab displays statistics about the file system. This information is similar to that provided by the program df. Finally, the System tab summarizes operating system specifications.

The KDE system monitor has two tabs, Process Table and System Load. While both are similar to the Processes and Resources tab from the Gnome system monitor, they both contain a useful summary along the bottom of the window. The summary provides the total number of processes, current CPU utilization, current memory use, and swap space use.

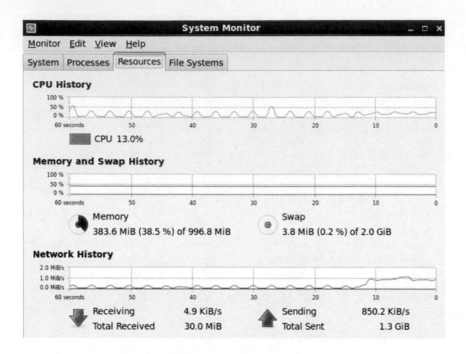

FIGURE 4.3  Resource utilization as shown through system monitor.

### 4.4.2  Command-Line Monitoring Tools

Anything that is available via the GUI is also available by the command line. The `top` command is an interactive program that outputs current process resource utilization. The top program refreshes itself every 3 s. You can alter the update period by adding `-d s.t` where *s* is the number of seconds and *t* is the number of tenths of seconds. For instance, `-d 1` would change the delay from 3 s to 1 while `-d 0.5` would change the delay to every half second. You could provide an even shorter interval if desired (for instance, `-d 0.001`) but that is probably unnecessary and also resource consuming. The value supplied must be positive or 0.

The top program's output is illustrated in Figure 4.4. The first thing you might notice when running top is that it fills the entire terminal window no matter what size the terminal window is. Along the top of the output are system statistics. They include the current number of users, the average CPU load over several intervals, the number of tasks (processes, services, threads) and summaries of their statuses (e.g., running, sleeping, stopped), CPU statistics (percentage of time in user versus system modes, percentage of low priority processes, idle processes, processes waiting on I/O, and others), memory utilization, and swap space (virtual memory) utilization.

The lower portion of the output consists of statistics on the running processes. The default is to order the processes by CPU time. The information shown for each process is the process' ID, the process' priority and niceness values (nice is discussed below), CPU statistics, memory statistics and swap space statistics, and finally the process' name. Other information, such as the parent process' ID, the user who launched the process given as

```
┌─────────────────────────────────────────────────────────────────────────┐
│ ▣            Student@CentOS6Template:~/Desktop            _ □ ✕          │
├─────────────────────────────────────────────────────────────────────────┤
│ File  Edit  View  Search  Terminal  Help                                 │
│ top - 08:53:03 up 12 days, 23:59,  3 users,  load average: 0.13, 0.03, 0.01 │
│ Tasks: 152 total,   2 running, 150 sleeping,   0 stopped,   0 zombie      │
│ Cpu(s):  1.0%us,  0.7%sy,  0.0%ni, 98.3%id,  0.0%wa,  0.0%hi,  0.0%si,  0.0%st │
│ Mem:   1020752k total,   809320k used,   211432k free,    12892k buffers  │
│ Swap:  2064376k total,      140k used,  2064236k free,   445224k cached   │
│                                                                           │
│   PID USER      PR  NI  VIRT  RES  SHR S %CPU %MEM   TIME+  COMMAND       │
│  1924 root      20   0  144m  35m 8508 S  1.0  3.6  0:32.11 Xorg          │
│ 10993 Student   20   0  292m  12m 9224 S  0.3  1.3  0:00.53 gnome-terminal│
│     1 root      20   0 19404 1460 1192 S  0.0  0.1  0:01.61 init          │
│     2 root      20   0     0    0    0 S  0.0  0.0  0:00.00 kthreadd      │
│     3 root      RT   0     0    0    0 S  0.0  0.0  0:00.00 migration/0   │
│     4 root      20   0     0    0    0 S  0.0  0.0  0:00.00 ksoftirqd/0   │
│     5 root      RT   0     0    0    0 S  0.0  0.0  0:00.00 migration/0   │
│     6 root      RT   0     0    0    0 S  0.0  0.0  0:00.00 watchdog/0    │
│     7 root      20   0     0    0    0 S  0.0  0.0  0:00.09 events/0      │
│     8 root      20   0     0    0    0 S  0.0  0.0  0:00.00 cpuset        │
│     9 root      20   0     0    0    0 S  0.0  0.0  0:00.01 khelper       │
│    10 root      20   0     0    0    0 S  0.0  0.0  0:00.00 netns         │
│    11 root      20   0     0    0    0 S  0.0  0.0  0:00.00 async/mgr     │
│    12 root      20   0     0    0    0 S  0.0  0.0  0:00.00 pm            │
│    13 root      20   0     0    0    0 S  0.0  0.0  0:00.00 sync_supers   │
│    14 root      20   0     0    0    0 S  0.0  0.0  0:00.00 bdi-default   │
│    15 root      20   0     0    0    0 S  0.0  0.0  0:00.00 kintegrityd/0 │
└─────────────────────────────────────────────────────────────────────────┘
```

FIGURE 4.4    Example of top program output.

real name, UID, user name and group name, the terminal window responsible for launching the process, is available by issuing different options or commands.

The interaction with top, aside from refreshing itself based on an interval, is through single character input. The more useful controls for top are given in Table 4.1. In addition to these keystrokes, pressing the enter key will automatically update the image.

Many of the commands shown in Table 4.1 can also be specified from the command line as options when top is launched. These include –H to show threads, –i to show idle processes and –u username to show processes started by *username*. The user can specify –n # where # is an integer to indicate the number of refreshes that top should perform before terminating. For instance, top –n 10 will only update itself nine times after the

TABLE 4.1    Keystroke Commands for Top Instruction

| Keystroke | Meaning |
| --- | --- |
| A | Use alternate display mode |
| D | Alter interval time |
| L | Turn on/off load statistics |
| T | Turn on/off task statistics |
| M | Turn on/off memory statistics |
| f, o | Add/remove fields or alter display order |
| H | Show threads |
| U | Show specified user owned processes only |
| N | Show specific number of processes only |
| Q | Quit |
| I | Turn on/off including idle processes |

first display before it self terminates. If you do not specify –n, then you must stop top on your own. This can be done by either typing a q or pressing `control+c`.

The ps command is to some extent an inverse of top; it shows a detailed examination of the running processes as a snapshot (an instance in time). Unlike top, it is not interactive. It displays the information and exits.

The ps command by itself (with no options) displays only the running processes of the current user in the current terminal window. If you are typing from the command line, this means that any other processes are suspended or running in the background, aside from Bash itself.

Below is an example of what you might find when running ps. In this case, we see only two processes, the ps command itself and bash. By the time the output is displayed, the ps command will have terminated and the only active process is bash. The information displayed when using ps without options is minimal: the PID, the terminal window, the amount of CPU time being used by the process, and the command name itself.

```
PID     TTY      TIME CMD
16922   pts/0    00:00:00 bash
24042   pts/0    00:00:00 ps
```

There are a multitude of options available for ps although their proper usage can be confusing because of ps' history. The ps command originated from early Unix and has been modified for BSD Unix and the GNU's project. The result is that there are three categories of options available. The original version of ps required that options be preceded by hyphens, the BSD version required that options *not* be preceded by hyphens, and the GNUs version required that options be preceded by double hyphens (––). Therefore, it is important to remember which options require which notations.

In order to display all processes, you can use any of ps –A, ps –e or ps ax. The 'a' and 'x' options list all user processes and processes outside of the given terminal window respectively so their combination is all processes started by all users in all windows as well as processes run from the GUI. You can restrict those processes to a given list of users by using –u user,user,user where each *user* is either the user's ID or username (you can also use –U or –– User). Notice that each user is separated by a comma but no space. There are several ways to display threads along with processes using H, –L, m, or –m. In the latter two cases, the threads appear after the processes. The option r outputs only running processes.

You can also request the status of specific processes by listing the command names. This version of ps uses the option –C commandlist. As with the user list, the *commandlist* is a list of command names separated by commas (but no spaces). For instance, if you have several terminal windows open and are running processes like bash, emacs, man and find, you might issue the command ps –C bash,emacs,man,find to obtain information on just those processes. Notice that –C will find processes of all windows and all users.

Aside from specifying the processes that ps will examine, you can also specify the format of the output. The ps command with no options provides minimal information. You might use this to obtain a process' ID number or to see what is still running as you may have processes in the background that you have forgotten about. The l (lower case L) option is the long format. The s option provides signal format which provides information for the number of signals sent to each process. The option u provides user-oriented format and the option v provides virtual memory information. The option –f is the *full* format. The u option provides more detail than –f. The j (or –j) option provides the *jobs* format which includes the PID, the PPID (parent's PID), PGID (process group ID), SID (session ID), and TPGID (your terminal window's foreground process group).

You can specify your own formatting options using any of o, –o, or --format. In specifying your own formatting options, you can select the statistics that ps will output. For instance, you might issue ps -o pid,start,ppid to obtain the PID, start time, and PPID of the processes owned by you in the current terminal window. Note that with o (-o, --format) you only receive the specific fields requested.

You can also view the parent-child relationship between processes by displaying the processes in a process tree, or hierarchically. There are two options for this. The older option, –H, displays children commands (names) indented underneath the parent processes. The newer option, f, uses an approach called "ascii art," which uses \_ to indicate the parent–child relationship between related processes.

Another command, pstree, captures the parent–child relationship between processes. The hierarchical relationship is easier to view with the command pstree although this command is far less powerful in terms of available options and output statistics than ps.

Before we look at specific instances of ps commands, we need to understand the information that ps can provide. Each of the outputs has column headers describing the data of that column. Table 4.2 summarizes the more useful column headers, which are listed as abbreviations.

Let us focus on the STAT value. As shown in Table 4.2, the status will appear as one or more characters describing the status of the process. The most common values are S, R, or T. Running (R) and Stopped (T) mean that the process is running or has been stopped by the user via a control+z. Interruptible sleep (S) means that the process has entered a wait state, waiting for an event. The event might be input from the user or disk file or data to be provided to it from another process through what is known as interprocess communication. The process may also be in a wait mode for a prespecified amount of time. Uninterruptible sleep (D) is often a process invoked via a system call to the kernel. The process is waiting (asleep) as with (S) but in this case, it cannot be resumed until a specific event occurs.

Figure 4.5 illustrates several different views of the ps command. Each listing is truncated to just a few processes as the important point is to see the headers (and in the case of the hierarchical listing, how processes are arranged). The first portion demonstrates the options aux, showing several users (root, Student, postfix) from different terminal windows. The second portion demonstrates the options aef to display Ascii art illustrating parent–child relationships. The third demonstrates the "long" listing format, through ps al.

TABLE 4.2    Headers for Various ps Options

| Abbreviation | Meaning/Usage |
|---|---|
| STAT (or S) | Status: D (uninterruptible sleep), R (running), S (interruptible sleep), T (stopped), W (paging), X (dead), Z (defunct/zombie) |
| | May be followed by: < (high priority), N (low priority), L (locked pages), s (session leader), l (multi-threaded), + (foreground) |
| %CPU | CPU utilization of process |
| %MEM | Process' fraction of memory usage |
| COMMAND/CMD | Command including all parameters |
| START | Time process started |
| TIME | Accumulated CPU time |
| CLS | Class (used if process was scheduled) |
| ELAPSED | Elapsed time since process started |
| F | Flags associated with the process |
| GID, UID, PID | Group and user ID that owns the process, process ID |
| LABEL | Security label, primarily used in SE Linux |
| NI | Niceness value |
| PPID | Parent process' ID |
| RSP | Stack pointer |
| RSS (or RSZ) | Physical memory usage (does not include swap space) |
| SESS | Session ID |
| SZ (or VSZ) | Swap space size |
| P | Processor executing process (if multiprocessor) |
| TTY | Terminal window of process, ? if process is not from a terminal window |

```
USER       PID %CPU %MEM    VSZ   RSS TTY      STAT START   TIME COMMAND
root         1  0.0  0.1  19404  1460 ?        Ss   Oct11   0:01 /sbin/init
gdm       2009  0.0  0.0  20084   640 ?        S    Oct11   0:00 /usr/bin/dbus-launch --exit-with
root      2015  0.0  0.2  45152  2508 ?        S    Oct11   0:00 /usr/libexec/devkit-power-daemon
root      2056  0.0  0.3  49880  3960 ?        S    Oct11   0:00 /usr/libexec/polkit-1/polkitd
rtkit     2067  0.0  0.1 168504  1188 ?        SNl  Oct11   0:09 /usr/libexec/rtkit-daemon
root      2072  0.0  0.2 176940  2932 ?        S    Oct11   0:00 pam: gdm-password
Student  10572  0.0  0.3 150580  3448 ?        Sl   Oct22   0:00 /usr/bin/gnome-keyring-daemon --
Student  10582  0.0  0.6 247868  6296 ?        Ssl  Oct22   0:00 gnome-session
Student  10590  0.0  0.0  20084   636 ?        S    Oct22   0:00 dbus-launch --sh-syntax --exit-w
Student  10591  0.0  0.1  32340  1676 ?        Ssl  Oct22   0:00 /bin/dbus-daemon --fork --print-
```

```
  PID TTY      STAT   TIME COMMAND
25820 pts/3    Ss     0:00 /bin/bash ORBIT_SOCKETDIR=/tmp/orbit-Student HOSTNAME=CentOS6Templa
25835 pts/3    S+     0:00  \_ ftp www.nku.edu ORBIT_SOCKETDIR=/tmp/orbit-Student HOSTNAME=Cen
20810 pts/2    Ss     0:00 /bin/bash ORBIT_SOCKETDIR=/tmp/orbit-Student HOSTNAME=CentOS6Templa
27282 pts/2    R+     0:00  \_ ps aef ORBIT_SOCKETDIR=/tmp/orbit-Student HOSTNAME=CentOS6Templ
10995 pts/0    Ss+    0:00 /bin/bash ORBIT_SOCKETDIR=/tmp/orbit-Student HOSTNAME=CentOS6Templa
 1924 tty1     Ss+    0:37 /usr/bin/Xorg :0 -nr -verbose -auth /var/run/gdm/auth-for-gdm-NRcrX
 1908 tty6     Ss+    0:00 /sbin/mingetty /dev/tty6
 1902 tty5     Ss+    0:00 /sbin/mingetty /dev/tty5
 1898 tty4     Ss+    0:00 /sbin/mingetty /dev/tty4
 1896 tty3     Ss+    0:00 /sbin/mingetty /dev/tty3
 1894 tty2     Ss+    0:00 /sbin/mingetty /dev/tty2
```

```
F UID    PID  PPID PRI  NI    VSZ   RSS WCHAN  STAT TTY       TIME COMMAND
4   0   1894     1  20   0   4116   528 n_tty_ Ss+  tty2      0:00 /sbin/mingetty /dev/tty2
4   0   1896     1  20   0   4116   528 n_tty_ Ss+  tty3      0:00 /sbin/mingetty /dev/tty3
4   0   1898     1  20   0   4116   528 n_tty_ Ss+  tty4      0:00 /sbin/mingetty /dev/tty4
4   0   1902     1  20   0   4116   528 n_tty_ Ss+  tty5      0:00 /sbin/mingetty /dev/tty5
4   0   1908     1  20   0   4116   524 n_tty_ Ss+  tty6      0:00 /sbin/mingetty /dev/tty6
4   0   1924  1922  20   0 139392 36384 poll_s Ss+  tty1      0:37 /usr/bin/Xorg :0 -nr -verbose
0 501  10995 10993  20   0 108524  1888 n_tty_ Ss+  pts/0     0:00 /bin/bash
0 501  20810 10993  20   0 108524  1856 wait   Ss   pts/2     0:00 /bin/bash
0 501  25820 10993  20   0 108524  1844 wait   Ss   pts/3     0:00 /bin/bash
0 501  25835 25820  20   0 116312  1564 n_tty_ S+   pts/3     0:00 ftp www.nku.edu
0 501  27293 20810  20   0 108192   968 -      R+   pts/2     0:00 ps al
```

FIGURE 4.5  Output from various combinations of ps options.

## 4.5 MANAGING LINUX PROCESSES

By default, Linux runs processes and threads together by switching off between them. This is known as concurrent processing in that processes (and threads) overlap in execution. Because of the speed and power of modern processors, the user is typically unaware of the time elapsing as the processor moves from one process to another and back. Literally, the processor moves back to the first process in under a millisecond (thousandths of a second). Concurrent processing of processes is known as multitasking while concurrent processing of threads is known as multithreading. Linux does both.

We mentioned the difference between processes and threads in Section 4.2, but let us explore this again. A process is a stand-alone entity, it has its own code, data, and status information. A thread, sometimes called a lightweight process, shares its code and data (or at least some data) with other threads. Threads are in essence portions of a process. Threads will communicate with each other through shared data. For the processor to switch between threads, it does not have to load new program code into memory as it is already present. Switching between processes often requires additional overhead.

We should also define an application. As users, we run programs like Firefox or OpenOffice Writer. However, larger applications like these do not consist of single processes. Instead, there are many different processes that might run in support of such software. Some of these processes are services (parts of the operating system, covered later in the textbook). Some of these are multiple processes that support each other. In other cases, there may be a single process which spawns child processes or child threads.

Although by default, Linux executes processes and threads using multitasking and multithreading, it does not mean that all processes run in this fashion. As a user, you can dictate how processes run. The Linux command batch, for instance, forces a process to run in batch mode. A batch process is one that does not interact with the user. Therefore, any batch process must be provided its input at the time the process executes. Output might be sent to a file or to a terminal window. Batch processing in Linux defaults to executing only if system load permits it. This occurs when CPU utilization drops below a pre-set amount, usually 80%. Background processes that require I/O essentially are stopped so that they do not take up CPU time. In effect, if you are running entirely from the command line (no GUI at all), and you are not running background processes, then you are running in a single tasking mode.

Since the default is to run processes (and threads) concurrently, does the user have any control over how these are run? That is, can the user dictate that certain processes are more important than others? The answer is yes, by specifying priorities for processes. Priorities establish the amount of CPU time that processes are given as the CPU switches off between them. The higher the priority, the more attention the CPU provides the given process.

A process' priority is established by setting its *niceness* value. Niceness refers to how nice a process is with respect to other processes. A higher nice value means that the process is willing to offer some of its CPU time to other processes. Thus, the higher the nice value, the lower the priority. This might seem counter-intuitive in that we would expect a higher value to mean a higher priority, but in Linux, it is just the opposite.

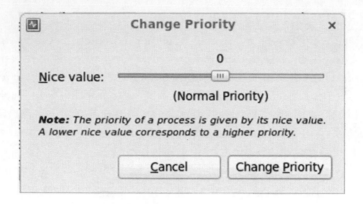

FIGURE 4.6  Changing process priority through GUI.

There are several ways to set or change a process' nice value. First, through the GUI System Monitor program, you can right click on a process (under the Process tab) and select Change Priority, which pops up the window in Figure 4.6. From this window, changing the niceness value is handled by adjusting the slider. As you can see in the figure, you are adjusting the process' nice value and so a lower nice value means a higher priority.

Alternatively, you can establish the nice value of a process when you launch it from the command line. This can be accomplished using the `nice` command. The nice command requires that you specify the actual command as one of its parameters, so you issue nice rather than the command itself, as shown here

```
nice -n # command
```

The value # is the niceness value, an integer which ranges between −20 and +19. The *command* is the command you want to issue. By default, the nice value is +10.

The following two examples illustrate how to launch instructions using nice. The find command operates with high priority while the script, myscript, runs with low priority. As both processes are launched in the background, we are able to run them both concurrently from the same terminal window.

```
nice -n -15 find ~ -name *.txt > found_files &
nice -n 15 ./myscript < inputfile > outputfile &
```

Once a process is running, you can change its nice value using the `renice` command. The renice command, at a minimum, requires the new priority and some specification of which process should be altered. You may either provide the process ID, multiple PIDs, or a list of users. If you select the latter, then all processes by those users are altered. What follows are several examples. The –n option is used to specify the new nice value. If multiple process IDs are provided, use –p. The –u option is used to denote users.

```
renice -n 5 18311
renice -n 19 -u foxr zappaf
renice -n -10 -p 23813 24113 26729
```

In the first and third of these examples, you are increasing the priority (lowering the niceness) of the processes. Unfortunately, Linux will not allow this unless you are root. While a user can give away CPU time from a process, the user is not allowed to take CPU time away from other processes even if that user owns those processes!

Notice in the first and third examples above that we are using the PID of the process(es) that we want to adjust. How do you obtain a PID of a process? As you have seen throughout this section, you can obtain it by searching through the list of entries in the Processes tab of the GUI System Monitor, or by looking at the entries listed by the top or ps command. But rather than searching through these lists, you can simplify the task for yourself.

You can use ps aux and pipe the result to grep. The grep program, which we will explore in Chapter 6, searches lines of text to match against a regular expression or a literal string. In our case, we want to match the response from ps aux to a string that describes the process(es) you are searching for. This might be your user name or it might be the name of the process, or even both. Below we see three examples. In the first, we are looking for all processes owned by user zappaf. In the second, we are looking for all bash processes. In the third, we are looking for all bash processes owned by zappaf. Note that the order of the two grep statements in the third example is immaterial. The result will be the same.

```
ps aux | grep zappaf
ps aux | grep bash
ps aux | grep bash | grep zappaf
```

Alternatively, we can use the command pidof. This command returns the PIDs of processes matching the command name given. For instance, pidof bash, will return the PID of every bash process running. Unlike ps which returns one process per line, this returns all of the PIDs as a single list. The parameter –s causes pidof to exit after it outputs the first found PID. You can supply pidof with a list of as many programs as desired. However, pidof does not differentiate between the programs found in its output.

Aside from renice, there are many other reasons for acquiring a process' ID. One such example is the lsof command. This command provides a list of open files attributed to the process. The command's structure is lsof –p *PID*. We visit another use of the PID in the next section.

## 4.6 KILLING PROCESSES

### 4.6.1 Process Termination

Typically, a Linux program runs until it reaches a normal termination point. In a program like ls, this occurs once the program has output all of the content that matches its parameters. In a program like cd, this occurs when the PWD and OLDPWD variables have been modified. In interactive programs, normal termination occurs when the user has specified that the program should exit. In top, this happens when the user enters a q, or for the vi editor, when the user enters :q. In a GUI, it happens when the user closes the window or selects Exit.

Some processes do not terminate normally. Some processes reach an error and stop functioning. In many Linux programs, an abnormal error causes a core dump to be saved. A *core dump* is an image or copy of the process' memory at the time the termination occurred. The core dump is produced to help a programmer debug the program. The core dump however can be a large file. As a user, if you write programs that terminate abnormally, you might find core dumps in your file space. Given a core dump, you can view its contents through a debugger program (we will not cover that in this text as this text does not concentrate on Linux programming). Note: as core dumps can be large, unless you plan on using them for debugging, you should delete them as they occur.

When a process does terminate, if it has spawned any children, this creates orphans. An *orphan* is a process whose parent has died. This creates a problem because, in many cases, a process reports back to its parent. Without a parent, the process cannot report. Therefore, upon a parent process terminating, the child process is re-parented, or adopted, by the init process.

The child process could be orphaned because of a number of different situations. First, the parent process may unexpectedly terminate (crash) because of some erroneous situation. Second, a child process may be launched to run in the background as a long-running job. In such a situation, it is not necessarily expected that the parent would run for the same duration. In this case, the parent is expected to terminate at some point while the child, which is now somewhat detached from the parent in that it is running independently, continues to exist. The third situation arises when the parent process creates the child process using the exec function call, in which case the parent dies and the child takes over for the parent, inheriting its PID. In this case, the child does not require re-parenting because the parent's parent automatically is inherited as the child's parent.

Another situation may arise in which a process has terminated, whether normally or abnormally, but remains in the system. In this case, the process is referred to as a *zombie*. In essence, such a process retains its PID but no longer receives attention from the processor. You can find zombie processes when you issue the ps command. Zombie process names will appear in ps with a STAT (state) of "Z."

The reason that zombies can arise is that some processes persist until their parent "cleans up after them." If the parent process is currently suspended or sleeping, a terminating process must wait for the parent to clean up. This clean up might simply be to report back its exit status or it may involve using data from the child. The parent itself might have placed itself into a waiting situation. If this is the case, the parent may resume only if a particular event arises or a specified amount of time has elapsed. For instance, the parent might issue a 24 hours wait and its child might terminate within an hour. The child would persist in the operating system for at least another 23 hours.

Even though zombies do not take up system resources, they can be annoying as they continue to appear in ps listings. Unlike normal processes which the user can kill (see below), the zombie process cannot be killed until the parent resumes. Therefore, if you find a lot of zombie processes and you want to get rid of them, restart the parent process(es) of the zombies.

This leads us to the discussion of killing processes. Why would you want to kill a process and how does this differ from exiting a process? In some cases, killing a process is the same as exiting a process. This is especially true if the process does not currently have any open resources.

Consider, for instance, using a text editor (e.g., vi or emacs). You can exit the program normally, which makes sure that any open files are first closed properly. If the program does not have an open file, then exiting the program and killing it leave the system in the same state. However, if you do have an open file and you kill the process, the file may become corrupted and so the system would be in a different state than had you exited the program properly.

Why then would you kill a process? Primarily, you kill processes that are not responding. This happens often in the Windows operating system but is less common in Linux. However, it still happens. In addition, a process launched into the background has no interface with the user and so could be stopped by killing it. Finally, if an emergency situation has arisen requiring that the system be shut down quickly, you or the system administrator may have no choice but to kill all running processes.

### 4.6.2 Methods of Killing Processes

To kill a process, the command is naturally `kill`. You must specify the PID of the process to kill. You can specify multiple PIDs in one kill command. You must however be the owner of the process(es) specified or root to kill it off. The kill command also accepts an interrupt signal. The signal indicates what action the application should take upon receiving the signal. The signal can specify that the software should try to recover from an error or that that the process should terminate immediately. There are nine signals available for the kill command, as shown in Table 4.3, although in fact there are 64 total signals available for programs in Linux.

Signal number 1 is also known as SIGHUP (signal hangup). This signal is used to terminate communication if a signal is lost. This can prevent a process from waiting indefinitely for a response via telecommunications. In the cases of signal numbers 2–8,

TABLE 4.3   Kill Signals

| Signal Number | Signal Name | Meaning |
| --- | --- | --- |
| 1 | SIGHUP | Hang up |
| 2 | SIGINT | Terminal interrupt |
| 3 | SIGQUIT | Terminal quit |
| 4 | SIGILL | Illegal instruction |
| 5 | SGTRAP | Trace trap |
| 6 | SIGIOT | IO Trap |
| 7 | SIGBUS | Bus error |
| 8 | SIGFPE | Floating point exception |
| 9 | SIGKILL | Kill |

an exception is raised so that, if handled by the software, the error can be handled. It is possible, based on the program, that after handling the exception, the program can continue executing. For instance, a floating point exception (overflow, division by 0) might be handled by the program by displaying a message that it could not perform the computation and request new input. In the case of signal number 9 (kill), or SIGKILL, the signal is never passed to the process. Instead, Linux immediately kills the process with no attempt to shut down properly or save the process. Thus, SIGKILL is the most powerful signal used in the kill command in that it assures immediate termination of the process.

When using the kill command, you include the signal as −s # where # is a number from 1 to 9. You might specify one of the following to kill the process with PID 12413.

```
kill -s 9 12413
kill -s SIGKILL 12413
kill -9 12413
```

Notice in the third example, we have omitted the –s in favor of using –9.

If you wish to kill all processes (other than the init process), you can use kill −1 (the number one). However, as most of the processes running are not owned by you the user, you would be unable to do this (the system administrator could). If you want to see all of the signals, use kill −1 (lower case L).

A variation of kill is the instruction killall. This instruction will kill all instances of the named process(es). For instance, you might enter

```
killall man vim
```

to kill all instances of man and vim that are running. As with kill, killall accepts a signal. However, if none is given, it defaults to signal 15, which is SIGTERM. The killall command has an interactive mode, specified with option –i, which pauses to ask the user before killing each process. Each process is presented to the user by name and PID. Aside from specifying programs by name, you can specify users in killall using the option −u userlist. This will kill all processes of the named user(s). Unless you are root, you would be unable to kill processes owned by other users.

You can also kill or exit processes using the Resource Manager tool. From the Processes tab, select a process, right click, and select one of End Process or Kill Process. The End Process selection will terminate the process normally while the Kill Process operates as if you pass -9 to the kill command.

### 4.6.3 Methods to Shut Down Linux

To wrap up this section, we look at commands that are used to shut down the Linux operating system. These are intended for the system administrators and not individual users. However, they are worth exploring here as we need to assure that the operating system is shut down properly. If the Linux operating system is shut down abnormally, such as by just turning the computer off, the result could be corruption of the file system.

The first command is shutdown. We specify a time unit in shutdown to send out a warning message to users. Additionally, 5 minutes before that time has elapsed, new logins become disabled. For instance, if time is +20, then the system will be shut down in 20 minutes and new logins disabled starting in 15 minutes. The time parameter can also be *now*, or a combination of hours and minutes using the notation hh:mm as in 15:30 to indicate 3:30 p.m. (military time). The shutdown command can include a string that is used as the shut down warning message. If no message is provided, a default message is used in its place.

The idea behind shutdown is to let users shut down their software themselves. For instance, a user might be editing a file in vi or emacs and this will allow the user to save the file and exit. When the time limit expires, shutdown automatically kills all running user processes and then system processes.

The shutdown command permits other options. The –r option indicates that after shutting down, the system should reboot automatically. The –c option forgoes a time unit so that it can shut down immediately and the –k option sends out the message and disables logins, but does not actually shut the system down.

The shutdown program actually leaves the computer running but has shut down the Linux GUI and most of the Linux services as well as most of the Linux kernel. You are provided a command line which permits limited input. Three of the commands available at this point are halt, reboot, and poweroff. The halt instruction terminates the operating system entirely. The reboot instruction causes Linux to reboot. Finally, power-off shuts down the hardware after halting the system. It should be noted that shutdown, halt, reboot, and poweroff are only available to root. The user can start the shutdown and reboot operations through the GUI. We discuss how users can execute system administrator operations later in the text when we introduce the sudo command.

## 4.7 CHAPTER REVIEW

Concepts and terms introduced in this chapter:

- Background—Processes which run when needed and when the CPU has time to run them. Because these processes are not running consistently as with multitasked processes, they cannot run interactively, so input and output must be done through files or offline.

- Batch processing—A form of process management whereby the operating system assigns a single process to the CPU to execute. Like background processes, batch processes do not run interactively so I/O must be done through files or offline.

- Child—A process spawned (started) by another process. In Linux, all processes form a parent–child relationship with only one process, init, having no parent.

- Context switch—The CPU switching from one process to another by swapping process status information.

- Core dump—The result of an abnormally terminating program is a file which stores a memory dump of the process at the time the error arose. Core dumps can help programmers debug programs but are often large files.

- EUID—The effective user ID of the running process. Typically, a process runs under the UID of the user who started the process, but it is sometimes useful for a process to run under a different UID instead. In such a case, the EUID differs from the UID.

- Exec—A system call to start a new process in place of the current process. The new process inherits the old process' PID and environment.

- FIFO—First in first out, or first come first serve, a simple scheduling strategy for ordering waiting processes (jobs) in a queue.

- Foreground—Processes which are being executed by the CPU (whether single tasking or multitasking) and currently have interactive access by the user.

- Fork—A system call to start a new process which is a duplicate of the parent process.

- Interrupt—A signal received by the CPU to suspend the current process and perform a context switch to an interrupt handler in order to handle the interrupting situation.

- Interrupt handler—A part of the operating system kernel set up to handle a specific interrupting situation.

- IRQ—An interrupt request made by a hardware device. Each IRQ is assigned to an interrupt handler.

- Multiprocessing—A form of process management in which processes are divided up by the operating system to run on different processors. Although this was reserved for parallel processing computers in the past, today's multicore processors can also accomplish this.

- Multiprogramming—In batch processing, if a process requires time-consuming I/O, the CPU idles during this time. In multiprogramming, the operating system causes a context switch so that the current process is moved to an I/O waiting queue while another process is selected to execute (or resume execution). This is also called cooperative multitasking.

- Multitasking—By adding a timer to count the elapsed clock cycles of an executing process, the operating system can force a context switch when the timer reaches 0 and raises an interrupt. In this way, we improve on multiprogramming so that no process can monopolize the CPU. This is also known as preemptive multitasking. Commonly, the processor maintains several running processes and switches off between them, visiting each in a round robin fashion.

- Multithreading—Multitasking over threads as well as processes.

- Niceness—The value that indicates the priority level of a Linux process. The higher the niceness, the more willing the process will voluntarily give some of its cycles to other processes, thus lowering it's a priority. Niceness values range from −20 (least nice, highest priority) to +19 (most nice, lowest priority).

- Orphan—An active process whose parent has terminated. As processes need to report before they terminate, any process requires a parent. An orphan is automatically adopted by the init process.

- PATH—An environment variable storing various directories. When issuing a command in a shell, the current directory and all directories in PATH are checked to find the executable program.

- PID—The process ID number, assigned when the process begins execution. PIDs are assigned sequentially so that with each new process, the PID increases by 1. PIDs are used in a variety of instructions including renice and kill.

- Priority—A value used by a scheduler to determine when a process should execute. In Linux, priorities determine the amount of time the CPU will focus on the process before moving on to the next process.

- Process—A running program which includes code, data, status (values for its registers), and resources allocated to it.

- Process management—How the operating system manages running processes. Older operating systems permitted only two processes to be in memory running at a time, the user process and the operating system. Later operating systems began using multiprogramming and multitasking. Today, most operating systems are multithreading and possibly multiprocessing.

- Process status—The collection of information that describes what a process is currently doing. This will be the important register values as well as allocated resources and locations in memory of such objects as the run-time stack and page table.

- Queue—A waiting line, used in operating systems to store processes waiting for access to the CPU or some device like a printer. Most queues store items in a FIFO manner but some use a priority scheme.

- Round robin—A scheduling algorithm where processes wait in a queue for service by the CPU which moves down the queue one process at a time until it reaches then end. It then resumes with the first process.

- Scheduling—The task of an operating system to organize waiting processes and decide the order that the processes will gain access to the CPU.

- Signal—A value or message passed to a process indicating the level of seriousness of the command. This is used in the kill instruction to indicate to what level the operating system should end a process.

- Single tasking—An old form of process management whereby the operating system only permits one process to be active and in memory at a time (aside from the operating system). In single tasking, the next process can only begin once the current process ends.

- Sleeping—A process' status which has moved itself into a waiting state for some circumstance to arise. This may be a specified amount of time (e.g., wait 1 hour) or for another process to complete its task.

- Sticky bit—An indicator, displayed in a directory's permissions, denoting that the directory is set up to allow other users to have write access to the directory but not to its contents.

- Suspended process—Using control+z from the command line, the current process enters a suspended state so that the user can gain control of the command line. The user can also use this to move a process from foreground to background.

- System monitor—A Linux GUI program that displays process status information, CPU, memory, swap space usage and network history for the last minute, and information about the file system.

- Threads—A variation of processes whereby a collection of threads share the same program code and possibly some of the same data but have their own unique data as well. A context switch between threads is faster than between processes so that multithreaded operating systems can be more efficient when running threads than a multitasking system.

- Time slice—The amount of time a process gets with the CPU until a context switch is forced.

- Zombie—A process which has terminated and freed up its resources but has not vacated the system because it has to report its end status to its parent, but its parent is currently unavailable (for instance, the parent may be suspended or sleeping).

Linux commands covered in this chapter:

- bg—To move a command-line issued process to the background.

- fg—To move a command-line issued process to the foreground.

- jobs—List the processes that are suspended or running in the background in this terminal window.

- kill—To terminate a (or multiple) running process.

- killall—To terminate a collection of processes all owned by the same user or group or of a particular name (e.g., killall bash).

- halt—When shutting down the Linux operating system, the shutdown command leaves the system in a text-based mode in which there are few commands available. Halt is used to stop the system whereas shutdown enters this particular mode.

- init—The first process run by the Linux operating system after kernel initialization so that init has PID of 1, and is used to start up the system and then become the parent of any orphaned process. The init process runs during the entire Linux session.

- nice—To alter a process' default priority, the higher the niceness the lower its priority.

- ps—Print the process snapshot of all running processes in the current terminal window (or through options, all processes of this user or all users).

- renice—Change the niceness of a running process.

- shutdown—Change the system to single user, text-based mode. From this state, the system administrator can issue a command of reboot, halt, or poweroff.

- top—A text-based program that displays process status and computer resource usage. It is interactive and updates itself every 3 seconds (although this duration can be altered).

## REVIEW QUESTIONS

1. From the desktop, you start a terminal window. Your default shell is Bash. From the terminal window, you type an mv command but add & after and then type ps. Show the tree of processes.

2. What registers make up a process' status?

3. Why is it important to save the PC value as part of the process' status?

4. How does batch processing differ from single tasking?

5. How does multiprogramming differ from batch processing?

6. How does multitasking differ from multiprogramming?

7. How does multithreading differ from multitasking?

8. Under what circumstances might a process give up access to the CPU?

9. What does the timer count? What happens when the timer reaches 0?

10. What is the ready queue and what does the ready queue store?

For questions 11–14, assume we have the following partial long listing of three executable programs:

| permissions | owner | group | filename |
| --- | --- | --- | --- |
| -rwsr-xr-x | user1 | group2 | foo1 |
| -rwsr-sr-x | user1 | group2 | foo2 |
| -rwxr-xr-x | user1 | group2 | foo3 |

11. If user foxr runs foo1, foo2, and foo3, which will run under EUID user1?

12. If user foxr runs foo1, foo2, and foo3, which will run under EGID group2?

13. Assume foo1, foo2, and foo3 all access files that are readable and writable only to user1. If foxr attempts to run all three programs, which can successfully read and write to this file?

14. If user user1 wishes to change the permissions of foo3 to match those of foo1, what four-digit value would user1 submit with the chmod command?

15. Assume that you have written an executable program, myprog, in your home directory/home/username. If you are currently in your home directory, what command would you issue to execute myprog? If you were instead in some other directory, what command would you issue to execute myprog using an absolute path to your home directory?

16. You have launched the program prog1 from the command line. This is an interactive program that will take several minutes to run. You wish to suspend it, start prog2 in the background, and resume prog1. How do you do this?

17. You have launched the program prog1 from the command line. This is an interactive program that will take several minutes to run. You wish to move it to the background and then start prog2. How do you do this?

18. You want to read the man pages for ps and top but you do not want to exit either man page as you wish to make some comparisons. How can you start both man commands and switch between them?

19. How does a job number, listed when you issue the jobs command, differ from a process' PID?

For questions 20–23, assume you have the following output from the jobs command:

```
[1]     Stopped     prog1
[2]-    Stopped     prog2
[3]+    Stopped     prog3
[4]     Running     prog4
```

20. Of the processes listed, which one(s) were invoked using & as in prog1 & ?

21. Which process would resume if you typed fg?

22. If you were to kill prog3, which process would resume if you then typed fg?

23. What happens if you type bg 1?

For questions 24–27, use the system monitor GUI (Applications --> System Tools --> System Monitor). Use the Processes tab for questions 24–26.

24. How many running processes are there? How many sleeping processes are there? What is a sleeping process?

25. What Nice values do the processes have?

26. What is the range of PIDs?

27. Click on the resources tab. How long is the history of CPU, memory, and network usage?

For questions 28–32, run top (hint: see the man pages).

28. In the display at the top of the output, what do CPU%us,%sy,%ni,%id,%wa,%hi,%si, %st each mean?

29. In the list of processes, what are VIRT, RES, and SHR?

30. For status, you might see S, R, D, Z. What do each of these represent?

31. With top running, type f followed by b. What column is added? What does this column tell you?

32. With top running, type f followed by g. What column is added? What does this column tell you?

33. What option(s) would you supply to ps to view the parent–child relationships between processes?

34. What option(s) would you supply to ps to view all running processes?

35. What option(s) would you supply to ps to view processes of only root?

36. How does ps av differ from ps au?

37. Using ps a, you will see the state of running processes. What do each of these stand for? S, Ss, S<, R, R+, D, X, Z?

38. Assume you want to run some process, foo, with a lower priority. How would you do this through the command line? Provide a specific instruction making up your own priority value.

39. Imagine a process, foo, is running. You want to lower its priority. Describe two ways to do this.

40. As a normal user, you want to increase one of your process' priorities. Are you allowed to?

41. What is the difference between SIGHUP and SIGKILL?

42. Provide an instruction that would stop the process whose PID is 18511.

43. You are running a script call myscript.sh but it seems to be hanging. Explain how you can stop this process step-by-step.

44. As a system administrator, you detect a user, foxr, is running processes that he should not be. You decide to end all of his processes. How will you do this?

45. What is the difference between shutdown and halt?

# CHAPTER 5

# Linux Applications

THIS CHAPTER'S LEARNING OBJECTIVES are

- Understanding the role of text editors in Linux and specifically how to use vi and/or emacs

- Learning about the available productivity software available in Linux

- Understanding the use of encryption in Linux and how to apply openssl

- Understanding the basic usage of the mail and sendmail programs

- Understanding the use of IP aliases and addresses and the domain name system (DNS) of the Internet

- Understanding the basic network commands available in Linux

## 5.1 INTRODUCTION

One of the strengths of the Linux platform is that much of it has been developed by the open-source community. Not only does this mean that the operating system itself is available for free, it also means that a lot of the application software for Linux is free. We briefly explore the open-source community and what it means to obtain the source code in Chapter 13. This chapter though is dedicated to examining some of the most popular and useful Linux applications software.

In this chapter, we will first focus on the two common text editors, vi and emacs. The vi editor is part of the Linux installation. The emacs editor may or may not be part of your installation. We also briefly look at the gedit text editor and the most popular of the Linux productivity software suites, OpenOffice. We limit our examination of OpenOffice because it should be software that you are able to figure out with little effort based on your experience using other productivity software such as Microsoft Office. We also briefly examine the text-based word processor LaTeX. LaTeX is very popular among Linux users as a powerful word processor. Most word processors are WYSIWYGs (what you see is what you get, or GUI based). LaTeX is a very different kind of word processor and a throwback to the earliest word processors where commands are embedded within the text itself.

We then focus on encryption software based on the openssl tool. In Section 5.5, we also examine what encryption is and why we might want to use it. The chapter ends with some of the network programs that a user may wish to use. There are other network programs that a system administrator will use and these are covered in Chapter 12.

## 5.2 TEXT EDITORS

As you are aware by now, operating Linux from the command line has a number of advantages over using the GUI. The primary advantage is the power that Linux commands offer through the use of options, which can be typically combined in a variety of ways. Although you can accomplish most tasks similarly through the GUI, it may take a greater amount of system resources, more time, and more operations (more clicks of the mouse) than you would face when entering commands. These same things can be said about using a GUI-based word processor over a text-based text editor.

In Linux, there are two common text-based text editors. These are vi (now called vim for vi improved) and emacs. The vi editor is the native editor, found in all Linux and Unix systems. The emacs editor is perhaps as commonly used although in most cases, it has to be first installed. Linux and Unix users typically consider the editor that they learn first to be their favorite of the two. Both have their own strengths and weaknesses, and both can be challenges to learn because they do not offer a traditional interface. A third editor that comes with the Gnome desktop is gedit. This editor only runs in the GUI environment though; so, relying on it may be a mistake as you might not always have access to the GUI.

You might wonder why you might use a text editor at all. In Linux, you will often find that you need a text editor for quick editing jobs. You will create text files for at least two reasons. First, you will write your own scripts (we examine scripts in Chapter 7). Scripts are written in text editors. Second, many Linux programs (including scripts) will use text-based data files for input. Some of the programs that operate on text files include grep, sed, awk (we look at these in Chapter 6), cat, less, more, diff, cmp, and wc (which we saw in Chapter 3) just to name a few. A third reason exists for system administrators. Many of the Linux-operating system services use configuration files that are text based.

Because of the extensive use of text files as a Linux user, and especially a Linux system administrator, you must be able to create and edit text files. GUI-based editors are simple enough to master but there are reasons why you might choose a text-based editor instead that echo the same strengths as using the command line over the GUI to control the operating system. These are efficiency, speed, and to a lesser extent, power. You might also find yourself in a situation where you only have access to the command line (e.g., when remotely logging in or if the GUI itself has failed). The cost of using these text-based text editors is the challenge of learning them. Both vi and emacs have very different styles of control (interface) over the GUI-based editors.

It should be noted that you can also use a word processor to create or edit a text file, but the word process brings a lot of overhead that is not needed, nor desired. Additionally, you have to remember to save a text-based file as a text file when done or else the word processor will save it under its own format such that it would not be usable by the Linux programs mentioned earlier.

Here, we look at vi and emacs in detail and then briefly look at gedit. Although emacs may require installation, the emacs editor is more powerful than vi and provides a great deal of support for programming. For instance, if you have the CTRL++ module in place, emacs performs brace and parenthesis matching. Or, if you have common lisp installed, you can actually run your common lisp code from inside the emacs browser. As mentioned above, Linux users who learn vi first tend to like vi far more than emacs while those who learn emacs first will favor it over vi.

## 5.2.1  vi (vim)

The first thing to know about vi is that the interface operates in one of the three modes. The main mode is known as the *command* mode. In this mode, keystrokes entered operate as commands rather than characters entered into the document. For instance, typing G moves the cursor to the end of the document and typing k moves the cursor up one line rather than entering the letter 'G' or 'k' into the document.

The other two modes are *insert* and *replace*. In insert, any character entered is placed at the current cursor position while in replace, any character entered replaces the character at the current cursor position. To move from insert or replace back to command, press the escape (<Esc>) key. What we will see below are the commands that control vi, that is, that are applied in the command mode. At any time, if you are in insert or replace, the keystrokes correspond to characters to be entered at the position of the cursor except for <Esc>.

The basic cursor movement commands are shown in Figure 5.1. The keys h, k, l, and j move the cursor left, up, down, and right on one position, respectively. The key 0 moves the cursor to the beginning of the current line and the key $ moves the cursor to the end of the current line. The ^ key moves the cursor to the first nonblank character of the line so that for a line starting with a nonblank, 0 and ^ do the same thing.

To reposition the cursor elsewhere, we have #G where # is a number. For instance, 10G repositions the cursor on line 10 and 500G repositions the cursor on line 500. If the number supplied is greater than the file's size in line numbers, then the cursor moves to the bottom of the file. In any case, the cursor is positioned at the start of the line. The key G by itself moves the cursor to the beginning of the last line in the file.

The key w moves the cursor to the beginning of the next word or punctuation mark (whichever comes first) while W moves the cursor to the beginning of the next word, ignoring any punctuation. The key e moves to the end of the word or a punctuation mark in the word while E moves to the end of the word. The key b moves to the beginning of the current word or punctuation mark in the current word while B moves to the beginning of the previous word.

H, M, and L, respectively, shift the cursor to the top, middle, and bottom lines in the window. Control+u (ctrl+u) and ctrl+b change the view of the document to be one screen back or one-half-screen back while ctrl+f and ctrl+d shift the view to be one screen forward or one-half-screen forward. You can also refresh the screen using ctrl+r or ctrl+l (lowercase L) although these may not be implemented.

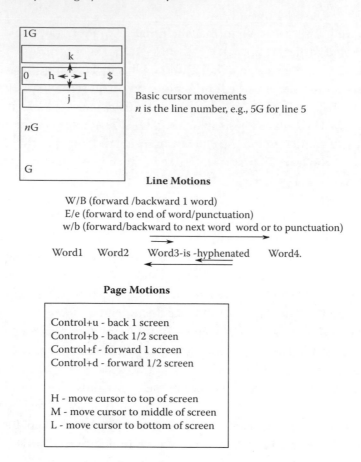

FIGURE 5.1   Cursor movements in vi.

Another means of movement to a new location in vi is to shift to a marked location within the document. To mark a location, use m*char*, where the *char* indicates a label for the mark, as in ma. To move to a marked location, use ʻ*char* as in ʻa. This lets you mark numerous locations in a document.

There are a variety of commands available to enter insert and replace modes. The most basic three are r (replace one character and then return to command mode), R (enter replace mode and stay there), and i (enter insert mode immediately before the location of the cursor). To move back to command mode from insert or replace, press <Esc>. The other commands enter insert mode but do so at different locations.

The key I moves the cursor to the beginning of the line to enter insert mode, a and A enter insert mode at the current cursor position and the end of the line, respectively. Finally, o and O insert blank lines and move the cursor to the new blank line, o creating the line immediately after the current line and O creating the line immediately before the current line. Figure 5.2 illustrates these insert locations.

There are many cut, copy, and paste commands available. To delete the character at the cursor, use x. To delete the word that the cursor lies in, use dw. To delete the characters in the word that lie before the cursor, use db. To delete the entire line that the cursor lies

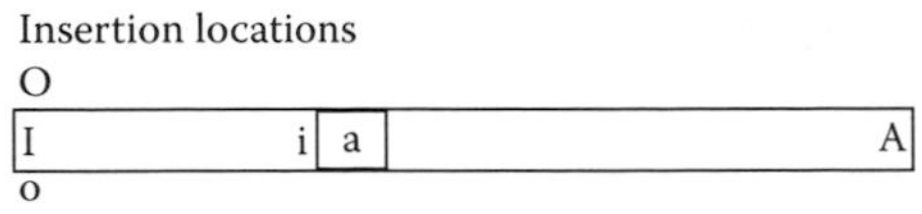

Insertion locations

O

| I | | i | a | | A |
|---|---|---|---|---|---|

o

FIGURE 5.2    Insertion commands in vi.

in, use dd. The commands dd, dw, and db cut the items into a buffer. The command yy copies the current line into a buffer. To retrieve the contents of the buffer, use p or P (put or paste) where p pastes in a line immediately above the line of the cursor's position and P pastes in a line immediately after the line of the cursor's position. The combination of dd/ dw/db and p/P perform cut and paste while yy and p/P perform copy and paste. Another interesting command is xp, which transposes the current and next character. In fact, it is a combination of the x command to delete the current character and p to insert the item in the buffer immediately after the current character. So, for instance, if you have abc and the cursor is under b, typing xp deletes b and pastes it after c. The command J joins the current line and the next line, thus removing the line break (\n) from the end of the current line.

There are several different search options. The search allows you to input either a string or a single character to search for. In the case of a string, the string can be literal characters or a regular expression (we cover regular expressions in the next chapter). To search forward from the cursor on for *string*, use /*string* <enter> and to search backward, use ?*string* <enter>. Once an instance of *string* has been found, you can continue to the search using / <enter> or ? <enter> for additional forward or backward searches, respectively.

You can search for a single character but only on the current line. The commands are f*char*, F*char*, t*char*, and T*char* where f and t search forward for the character *char*, positioning the cursor on the character or immediately before the character, respectively, and F and T search backward on the line for the character *char*, positioning the cursor on the character or immediately before the character, respectively. The command ; is used to repeat the last f, F, t, or T command. You can also search forward or backward in a document for a parenthesis. If the cursor is resting on an open or close paren, the % key will move the cursor to the corresponding close or open paren. For instance, if you have "the value of x (heretofore unknown) is not a concern" and the cursor is on the ')', pressing % will move it to the '('.

All file commands start with a : and are followed by one or more characters to denote the command. To open the file *filename*, use :r *filename*. To save a file, use :w if the file is already named (if unnamed, you receive an error message) and :w *filename* to save a file under the name *filename*. To exit vi, use :q. If the file has not been saved since last changed, you will be given an error. Use :q! to force vi to quit in spite of not saving the file. You can combine saving and quitting using :wq. You can also specify a command using :*command* as in :undo, :put, :sort. Finally, you can execute a shell command from within vi by using :!*command* as in :!wc (word count). This shifts your view from vi to the command line and then waits for you to hit <enter> to return to vi.

Finally, there are commands to repeat and undo. To repeat the last command, type period (.). To repeat a command multiple times, do *#command* where # is an integer number and *command* is the keystroke(s) to be repeated. As an example, 4dd will issue the dd command 4 times, or delete four consecutive lines. The command 12j will move the cursor down 12 lines. To undo the last command, use u. You can continue to use u to undo previous commands in reverse order. For instance, u undoes the last command and another u undoes the command before it. You can do #u where # is a number to undo a number of commands, as in 6u to undo the last six commands. The command U undoes all commands that have been issued on the current line of the text. Consider, for instance, deleting a character, moving the cursor one to the right on the line, using xp, moving the cursor further to the right on the line, and deleting a word using dw. Issuing U will undo all these commands.

### 5.2.2 emacs

The emacs text editor offers an option to vi. Some prefer emacs because it is more powerful and/or because the commands are more intuitive (in most cases, the letters to control cursor movement and related operations use the letters that start their names, such as control+f for "forward," control+b for "backward," etc.). Others dislike emacs because, upon learning vi, they find emacs to be either too overwhelming or dislike the commands.

Unlike vi, which has distinct modes, emacs has only one mode, insert. Commands are issued at any time by using either the control or escape key in conjunction with other keys. Therefore, any key entered that is not combined with control or escape is inserted at the current cursor position. In this section, we will describe these commands. We use the notation ctrl+*char* to indicate control+*char* as in ctrl+k, and esc+*char* to indicate escape+*char* as in esc+f. To perform ctrl+*char*, press control and while holding it down, press *char*. To perform esc+*char*, press the escape key and then after releasing, press the *char* key.

When you start emacs, you will be placed inside an empty (or nearly empty) buffer. The text you enter appears in this buffer as you type it. You are able to open other buffers and move between them even if a buffer is not visible at the moment. We will explore buffers later in this section.

As you enter the text, characters are placed at the position of the cursor. Upon reaching the end of the line, characters are wrapped onto the next line. If this occurs within a word, the word is segmented with a \ character. This is unlike a normal word processor that performs word wrap by moving the entire word that is segmented onto the next line. For instance, the text below might indicate how three lines appear.

Desperate nerds in high offices all over the world have been known to enact the m\
ost disgusting pieces of legislation in order to win votes (or, in places where they d\
on't get to vote, to control unwanted forms of mass behavior).

The above text represents one line. If the cursor appears anywhere inside this line, moving to the beginning of the line moves you to the 'D' in Desperate while moving to the end of the line moves you to the '.'

The most basic commands in emacs are the character movement commands. Figure 5.3 illustrates the most common and useful of these commands. You might notice, if you refer back to Chapter 2, that many of these commands are the same as the command-line editing commands found in Bash.

The basic commands shown in Figure 5.3 allow you to move forward, backward, up, and down using `ctrl+f`, `ctrl+b`, `ctrl+p`, and `ctrl+n` (forward, backward, previous, and next), and move to the beginning or end of the current line (`ctrl+a`, `ctrl+e`). The command `esc+<` and `esc+>` move the cursor to the beginning and end of the document, respectively.

The commands `esc+b` and `esc+f` move you to the beginning or end of the current word. Notice in Figure 5.3, the hyphen indicates a different word; so, `esc+b` takes you from 'l' to 'h', not 'e'. A separate `esc+b` is needed to move the cursor to the 'e'. You can move up and down one screen using `esc+v` and `ctrl+v`, respectively.

Editing commands are as follows:

- `ctrl+d`—delete the current character

- `ctrl+k`—delete from cursor to the end of the line (kill)

- `esc+d`—delete (kill) from cursor to the end of the current word

- `esc+<backspace>`—delete (kill) from this point backward to the beginning of the current word

- `ctrl+y`—retrieve all killed items and paste them at the cursor (anything killed but not deleted via ctrl+d)

- `ctrl+t`—transpose this and the next characters

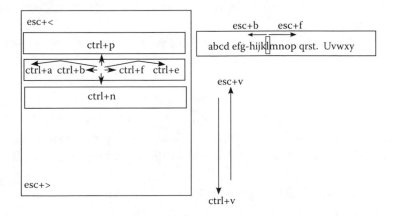

FIGURE 5.3   emacs cursor movement commands.

- esc+t—transpose this and the next word

- ctrl+x u—undo the last command (also ctrl + _ and ctrl + /)

- esc+u—convert the current word from this point forward into all uppercase letters

- esc+l—convert the current word from this point forward into all lowercase letters

- esc+c—convert this character into capital and the remainder into lowercase letters

Mastering the cut and paste via esc+d, esc+<backspace>, and ctrl+k followed by ctrl+y can be tricky. Any item killed is placed into a "kill buffer" and ctrl+y copies what is in the kill buffer to the current cursor position. However, as soon as you move the cursor using any of the cursor commands (see Figure 5.3), you end the kill buffer. You can still yank items back, but if you start deleting something new, you end the old kill buffer and replace it with a new one.

Consider, for instance, the following sentence in your buffer:

The lazy brown cow jumped over the man in the moon holding a silver spoon.

The cursor is currently over the 'l' in lazy. You perform the following operations: esc+d, esc+d to delete "lazy" and "brown," followed by esc+f, esc+f, and esc+f. Now, you do esc+d, esc+d, and esc+d. This deletes "the man in," leaving the sentence

The cow jumped over the moon holding a silver spoon.

You move the cursor back to before "cow" and do ctrl+y. What is yanked is just "the man in," not "lazy brown."

You can also cut a large region of text without using ctrl+k. For this, you need to mark the region. Marking a region is done by moving the cursor to the beginning of the region and doing either ctrl+<space> or ctrl+@. Now, move to the end of the region and use ctrl+w. Everything from the mark to the cursor's current position is deleted. Move to the new location and use ctrl+y to paste the cut region. If you use esc+w instead of ctrl+w, it performs a copy instead of a cut.

Commands can be repeated using esc+# command where # is a number, as in esc+5 ctrl+d. This will perform ctrl+d (delete character) 5 times in a row. You can also perform the previous operation any number of times by doing ctrl+x followed by z for each repetition. For instance, if you do ctrl+v to move the cursor forward one screen's worth, followed by ctrl+x zzz, the ctrl+v is executed 3 additional times to move the cursor forward by three more screens (four total). You can also program emacs to repeat a sequence of keystrokes. This is known as a *macro*. We briefly discuss recording and using a macro at the end of this section.

A buffer is merely a reserved area of memory for storing information. In emacs, you can have numerous open buffers. Using buffers, you can not only edit multiple documents at a time but you can also cut or copy and paste between them. You can even open multiple views into the same file by keeping each view in a separate buffer. Another interesting use of buffers is to have one buffer be a run-time environment, for instance, for the language

Common Lisp, and another buffer be a program that you are editing. You can then copy the program code from the second buffer into the first buffer to have it executed.

One special buffer is called the mini-buffer. This buffer appears in response to certain commands. The mini-buffer is a single line in size that is used to input information into emacs such as a command or a file name. One operation that calls for the mini-buffer is the search command. The command ctrl+s opens the mini-buffer (which is at the bottom of the emacs window). You enter your search term and press enter. The cursor is moved to the next instance of the search term. Note that search is not case sensitive. Continuing to press ctrl+s searches for the next instance of the search term last entered. Entering any other key (such as new characters to insert or new commands) causes the search term to be removed from the mini-buffer so that the next ctrl+s causes the mini-buffer to ask for a new search term. The command ctrl+r is a reverse search that again requests a search term in the mini-buffer.

The mini-buffer is also used with file operations. All file operations start with a ctrl+x followed by another control operation. These are shown below.

- ctrl+x ctrl+w—request new file name in mini-buffer, and save the file under this name.

- ctrl+x ctrl+s—save the document under the current name, and prompt for filename if the buffer has not yet been saved and is unnamed.

- ctrl+x ctrl+f—request file name in mini-buffer, and open that file in the main buffer.

- ctrl+x ctrl+c—exit emacs, if the file has not been saved since recent changes, you will be asked if you want to save it; if the file has never been saved, emacs exits without saving!

Note that entering a file name in the mini-buffer can use tab completion to complete the directory and file names.

If you have multiple buffers open, the buffer name will appear at the bottom of the window immediately above the mini-buffer. To move between buffers, use ctrl+x ctrl+<left arrow> and ctrl+x ctrl+<right arrow>. Or, to move to a named buffer, use ctrl+x b. The mini-buffer will prompt you for a name. If the buffer name you provide is not one of the open buffers, you are given a new buffer of that name. To list all buffers, use ctrl+x ctrl+b. To remove the buffer you are currently in, use ctrl+x k.

You can also open multiple buffers in the current window. The most common way to do this is to split the current buffer's screen into half. This is done through ctrl+x 2. With two buffers open, you can switch between them using ctrl+x o. You can continue to divide buffers by doing ctrl+x 2 although after a while, you will find that your buffers are very small. You can also collapse from two buffers to one using ctrl+x 1. See Figure 5.4 for an example of emacs with two main buffers and a mini-buffer.

Let us see how we might use two buffers. You start emacs and open the file file1.txt. You wish to copy contents of file1.txt into file2.txt. Using ctrl+s, you search forward for the beginning of the content that you wish to copy. Now, you do ctrl+x 2 to split the screen.

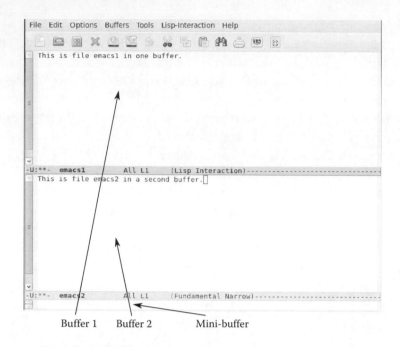

FIGURE 5.4    emacs with three buffers.

Use ctrl+x  o to move to the other buffer in the lower half of the screen. Here, you use
ctrl+x  ctrl+f to open file2.txt. Move the cursor to the proper position in this file to
insert. Do ctrl+x o to move back to the top buffer holding file1.txt. Using ctrl+<space>
and esc+w, you mark a region and copy it into a buffer. Again, you type ctrl+x  o to
move to the lower buffer. Now, you use ctrl+y to yank the copied content into place in
the second file. You save this file (ctrl+x  ctrl+s) and then kill this buffer using ctrl+x
k. You are back to file1.txt in one buffer.

The mini-buffer is available whenever you issue the command esc+x. This allows you to
enter a command in the mini-buffer. There are numerous commands available. A few are
listed in Table 5.1 below. Tab completion is available to help you complete the name of a
command. Or, if you want to see the commands available, press the tab key. This will list
all commands. If you want to see just those that start with the letter 'a', type an 'a' in the
mini-buffer followed by tab.

The doctor command runs a program called doctor. This is a variation of an older
artificial intelligence program called Eliza. Give it a try, you might find it intriguing! The
shell command opens a Linux shell inside the emacs buffer so that you can issue Linux

TABLE 5.1    Some emacs Commands

| append-to-file | check-parens | goto-line |
|---|---|---|
| auto-save-mode | copy-file | insert-file |
| calendar | count-lines-page | save-buffer |
| capitalize-region | count-lines-region | shell |
| capitalize-word | doctor | undo |

commands and capture all their outputs from within the text editor. In this way, you can literally save a session to a file for later examination (or reuse).

We wrap up our examination of emacs by considering macro definition and execution. A macro is essentially a recording of a series of keystrokes that you want to repeat many times. The keystrokes can comprise commands (e.g., ctrl+f, esc+b), characters that are being entered into the document, and characters or commands being entered into the mini-buffer. This allows you to perform some complex editing task and then have emacs repeat it for you. To define a macro, you type ctrl+x ( . At this point, everything you enter is recorded. Recording stops when you type ctrl+x ) . With a macro defined, you can execute it using ctrl+x e. Or, if you want to execute the macro repeatedly, you can use esc+# (where # is a number) followed by ctrl+x e.

Let us consider an example. You are editing a C program. The C program consists of a number of functions. Every function has the following format:

```
type name(type param, type param, type params, ...)
{
        instruction;
        instruction;
        ...
}
```

You want to convert from C to pseudocode. You will do this by rearranging the code to look like this instead:

```
name type param, type param, type param, ...
        instruction;
        instruction;
        ...
        return type;
```

This requires that you

1. Cut the type from the beginning of each function

2. Move the type to the end of the function and add it to the word return in the form return *type*

3. Remove the parens from the parameter list

4. Remove the {}

We will assume the cursor is currently at the beginning of the first function. To define this macro, do the following:

- ctrl+x (—start defining the macro

- esc+d—delete the first word (the type)

- `ctrl+a ctrl+d`—delete the blank space vacated by the type

- `esc+f ctrl+d [space]`—move passed the name of the function to the open paren and delete it and then add a blank space to separate the function name from the first parameter

- `ctrl+e ctrl+b ctrl+d`—move to the end of this line, back up, and delete the close paren

- `ctrl+a ctrl+n ctrl+d ctrl+d`—move to the beginning of the next line and delete the open curly brace; a second deletion deletes the end of the line character so that the first instruction now appears on the second line

- `ctrl+s }`—search for close curly bracket

- `ctrl+a [space] [space] [space] [space] [space] return ctrl+y; ctrl+k`—move to the beginning of the line with the close curly bracket, add several spaces (or use <tab>), insert the word `return` followed by yanking back the cut item from the second step (this is the type, note that since we did not use another esc+d or a ctrl+k, this will still be in the yank buffer), and finally delete the close curly brace

- `ctrl+s ( ctrl+a`—search forward for the next function that will be denoted by a parameter list with an open paren and move to the beginning of this line

- `ctrl+x )`—end the macro

Now, we can repeatedly execute it using `ctrl+x e`.

While this example may seem confusing, you can see the power of a macro. Once defined, this sequence of operations can be performed over and over to alter the format and the text within a document. Like vi, to understand and master emacs, you need to use it.

## 5.2.3 gedit

vi opens in the terminal window it was launched from, filling the window. emacs when launched from a terminal window will either open a new window in the GUI if this is available, or will fill the terminal window if you are restricted to a text-only environment. A third choice for a text editor is `gedit`. This program *only* runs in the GUI and so is not a choice if you have logged in through a text-only interface. However, you might find gedit far easier to use than either vi or emacs because you do not have to memorize any editing keystrokes. In spite of gedit's availability, you should learn vi and/or emacs for the occasions when you are operating in Linux without the GUI.

You can launch gedit from either the command line, typing `gedit`, or selecting it from the Accessories submenu of the Applications menu. An open file in gedit is shown in Figure 5.5. The typical user would interact with gedit through the mouse to position the cursor and to select the menu items. However, as with most GUI software these days, hot key keystrokes are available to control operations such as `ctrl+c` and `ctrl+v` for copy and paste or `ctrl+o` and `ctrl+s` to open or save a file.

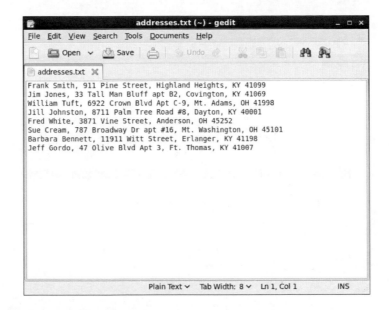

FIGURE 5.5 GUI-based gedit text editor.

## 5.3 PRODUCTIVITY SOFTWARE

Productivity software is a collection of software titles that enable the user to create documents whether they are word-processed documents, spreadsheets, presentation graphics documents, database relations, or graphic images. The most commonly used piece of productivity software is Microsoft Office. The open-source community has created both LibreOffice and OpenOffice as competitors. Both LibreOffice and OpenOffice have many of the same functions and features as Microsoft Office although they are not an exact match. When compared, LibreOffice is usually felt to have more features in common with Microsoft Office and is more powerful than OpenOffice. However, OpenOffice is able to import Microsoft Office files and export files to Microsoft Office format making it more flexible. Both LibreOffice and OpenOffice are open-source titles meaning that they are available in their source code format to allow for modifications. They are also both free. They are also available for most Linux platforms as well as Windows.

The titles of the individual software in LibreOffice and OpenOffice are the same. The word processor is called Writer, the presentation graphics program is called Impress, the spreadsheet program is called Calc, the database management program is called Base, the drawing software is called Draw, and there is a mathematical package called Math. Comparing LibreOffice or OpenOffice Writer to Microsoft Word, you find many of the same functions including wizards, tables, references and tables of content, spelling and grammar checkers, rulers, the ability to define macros, mail merge, footnotes/endnotes, headers, and footers.

Figure 5.6 displays the top portions of OpenOffice Writer, Impress, and Calc. Any user of Microsoft Office should have no difficulty learning the LibreOffice or OpenOffice equivalents although both LibreOffice and OpenOffice are more like Microsoft Office 2003 than more recent versions as they use dropdown menus rather than menus of button bars, as you will notice in Figure 5.6.

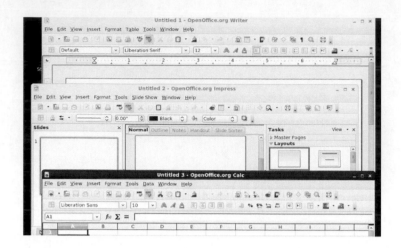

FIGURE 5.6    Several OpenOffice software titles.

## 5.4 LATEX

Another popular software title often used by Linux users is called LaTeX. LaTeX is a word processor, but unlike Microsoft Word and OpenOffice Writer, it is not a WYSIWYG. Instead, in LaTeX, formatting commands are embedded within the text using special character sequences, somewhat like html. For instance, to denote that a sequence of characters should be underlined, you would use \underline{text to be underlined} or to center justify a section of text, you would use \begin{center} text to be centered \ end{center}.

You place your text document, along with the LaTeX commands, in a text file and then process it using the latex command from the command-line prompt. LaTeX allows you to draw figures using postscript or include figures already stored in postscript. LaTeX also automatically generates bibliographies for you as long as you have set up a reference library by using the bibtex program.

Table 5.2 provides a description of some of the most useful formatting and document commands. Figure 5.7 provides an example of both a latex file and the document that latex generates from the file. The top portion of Figure 5.7 provides a short latex document to illustrate the use of the embedded commands. The bottom portion of Figure 5.7 is the finished document.

Aside from the commands shown in Figure 5.7, there are hundreds of other commands available in LaTeX. The font style commands use the format \style{text} such as \ em{hi there} or \bf{This is rather Bold!}. Styles for \bf, \em, \it, \rom, \sl, \sc, and \tt are shown below.

| |
|---|
| **Bold face** |
| *Emphasis* |
| *Italics* |
| Roman font face |
| *Slant* |
| CAPS |
| Teletype |

TABLE 5.2 Some LaTeX Commands of Note (not included in Figure 5.6)

| Command | Meaning | Example Usage |
|---|---|---|
| \documentstyle | Change type of document | \documentstyle[twocolumn]{book}<br>\documentstyle[11pt]{article} |
| \author{author names} | Author names are centered in smaller font under title | \author{Richard Fox and Frank Zappa} |
| \chapter{},<br>\section{},<br>\subsection{},<br>\part{} | Create a new portion of document with a title (size and numbering determined by type specified, e.g., chapter vs. section) | \chapter{How to Use LaTeX}<br>\section{Starting Your Document}<br>\subsection{The title page} |
| \begin{tabular}{...}...<br>\end{tabular} | Create a table where {...} indicates the justification for columns | \begin{tabular}{|l|lr|} {...}<br>\end{tabular}<br>Creates three columns: left justified, left justified and right justified, and the use of | indicates where table borders should be placed |
| \maketitle | Used to generate a title page | |
| \newline,<br>\linebreak,<br>\nolinebreak | Insert new line or permit a line break at this point, do not permit a line break | |
| \parindent{},<br>\textwidth{},<br>\textheight{} | Set paragraph indentation and margins | \parindent{.3in}<br>\textheight{1in} |
| \tiny, \scriptsize,<br>\footnotesize, \small,\large,<br>\Large,\LARGE, \huge | Various text size commands | |

To use LaTeX, you would first create your document. Normally, you would save this document as a .tex document. Tex is a more primitive program that LaTeX calls upon; so, you could actually use Tex commands but LaTeX is far easier to work with unless you had to work at a very primitive level of formatting.

Next, given your .tex file, issue your latex command: `latex filename.tex`. As long as there are no errors encountered, LaTeX then produces at least one file, *filename*.dpi. The dpi extension stands for "device independent" meaning that it could then be used by one of several other programs. To view your document or print it out, you must use an additional program. One such program is `dvi2ps` to create a postscript file. Today, it is more common to use `dvipdf` to create a pdf version.

Once the pdf is created, you can view and/or print out the file. LaTeX will also generate other files as needed, including *filename*.log and *filename*.aux. The log file contains comments generated by LaTeX while the aux file may contain any error messages generated.

Another useful feature of LaTeX is the easy ability to insert references. First, create one or more bibliography files, known as bibtex files. These files contain the reference information that you might refer to in any given document. A bibliographical reference in your bibtex file will contain information such as the authors, title, book or journal name, volume, year, pages, editor, and address. Each entry is preceded by the type of reference such as article, book, or collection. What follows is an example of an entry for a journal article.

```
\documentstyle{report}
\begin{document}
\begin{center}The title of the document\end{center}
\vspace{2pc}

This is a sentence. This is a second sentence. This is a
third sentence. This is a fourth sentence. This is the end
of a paragraph.

This is a new paragraph. The blank line between the two
paragraphs indicates a new paragraph. See how it
automatically indented? We can also prevent indentation.
\noindent We start a new paragraph without indentation.
This is another sentence for the non-indented paragraph.

And now we have another paragraph, indented. Lets look at a
list.
\begin{itemize}
\item The first item
\item The second item
\item The third item
\end{itemize}

This is much like html. We specify formatting around the
word(s) to be formatted such as {\it italicized text}, {\bf
bold faced font}, or \underline{underlined text}.
\end{document}
```

We start a new paragraph without indentation. This is another sentence for the
non-indented paragraph.

And now we have another paragraph, indented. Lets look at a list.

- The first item

- The second item

- The third item

This is much like html. We specify formatting around the word(s) to be
formatted such as *italicized text*, **bold faced font**, or underlined text.

FIGURE 5.7    LaTeX file (top) and resulting document (bottom).

The order of the entries (author, title, etc.) is unimportant and depending on the type of
entry, some fields are optional, such as volume for the article. Note that fox2010 is a label
that we will use in our LaTeX file.

```
@article{fox2010,
    author = {Fox, R. and Zappa, F.},
    title = {Using Linux to Dominate Mankind},
    journal = {The Linux Journal},
```

```
        volume = {March},
        year = 2010,
        pages = {85-93}
}
```

In your LaTeX document, you reference bibliography entries through the \cite command as in \cite{fox2010} or \cite{fox2010,keneally2008,mars2012}. You also have to specify the location of your bibliography through the \bibliography{*filename*} command. The filename is one or more bibtex files located in the same directory as your LaTeX file. These files are expected to end with the extension .bib as in myrefs.bib. You can also insert a \bibliographystyle{*type*} command to specify the *type* of bibliography style. The style type can include aaai-named, abbrev, abstract, acm, alpha, apa, bbs, chicago, kluwer, plain, and these (thesis) to name a few.

Once you have processed your LaTeX file, before you try to create your ps or pdf file, you need to process the bibliography. This is done with the command bibtex *filename* where *filename* is the name of your LaTeX file. Bibliography entries will appear in your document in square brackets ([]) and will either store a number or abbreviation of the author(s) depending on the format type selected. For instance, a reference might be [5] or [fox13].

## 5.5 ENCRYPTION SOFTWARE

In this section, we examine some of the encryption software available in Linux concentrating on openssl and other supporting tools. Before we look at the software, let us consider what encryption is and why and how we can use it.

### 5.5.1 What Is Encryption?

Encryption is a process of taking information in the form of a string of characters (whether ASCII, Unicode, binary, or other) and altering it by some code. The encoded message would make the information hard to understand if intercepted. Encryption and decryption are the translation processes of taking information and placing it into a coded form and taking a coded message and restoring the original form, respectively.

We want to apply encryption in our telecommunication to ensure that secure information such as passwords, credit card numbers, and even confidential text information cannot be viewed if intercepted during communication. This is necessary because the Internet is not set up to broadcast information in a secure form. We might also encrypt data files on our computer in case a computer was to fall into others hands.

Once encrypted, the information should hopefully be in a nearly unbreakable code. For a person to be able to view the original information without the code, the person would have to try to break the code. This might require trying all possible permutations of a code. This attempt at breaking a code is known as a *brute-force* approach. The number of combinations that such an attack might require depends on the encryption algorithm. The best encryption algorithms used today could take dozens of supercomputers more than $10^{18}$ (a billion billion) years to break.[*]

---

[*] As reported by *EE Times* at http://www.eetimes.com/document.asp?doc_id = 1279619.

There are two general forms of encryption algorithms used today: *symmetric key encryption* (also known as *private* key encryption) and *asymmetric key encryption* (also known as *public* key encryption). The *key* is a mathematical means of encrypting and decrypting information. The encryption algorithms used today apply a numeric key that might be 80, 128, or 256 bits in length. Various algorithms use different sized keys and apply them in different ways. Symmetric key encryption algorithms include advanced encryption standard (AES), the data encryption standard (DES, now out of date), and the triple data encryption algorithm (triple DEA which applies DES in three ways). Asymmetric key encryption algorithms include RSA (abbreviations of the three inventors, Ron Rivest, Adi Shamir, and Leonard Adleman), ElGamal, and the digital signature algorithm (DSA).

Figure 5.8 illustrates the use of symmetric key encryption (top half of the figure) and asymmetric key encryption (bottom half of the figure). In symmetric key encryption, there is a single key that handles both encryption and decryption. Since this key is the only protection to ensure the security of the encrypted data, it is known as a *private* key, one that should be held securely so that the data cannot be examined. You might use symmetric key encryption to encrypt your data files on your computer or to transmit data to another person over the Internet who knows the key.

For E-commerce, we need a mechanism, whereby a customer can encode confidential data such as a credit card number and submit that information over the Internet. The concern here is that if we were to use symmetric key encryption, then we would be giving the key to the customer who could potentially then use that key to decrypt other people's messages. So, a separate mechanism is needed, one where we have a public key that can be given to anyone and a private key. In asymmetric key encryption, the *public* key is used to encrypt messages but cannot be used to decrypt messages. The private key is used to decrypt messages as well as generate the public key. Therefore, the organization creates a private key and uses it to generate a public key. The private key is kept secure and the public key is provided to the customers.

You have no doubt used asymmetric, or public key, encryption yourself although you may be unaware of any of the actual mechanisms behind it because these are taken care of by your web browser. You can view the public key provided to you by an organization by

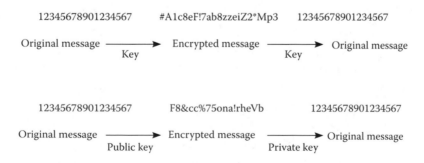

FIGURE 5.8 Symmetric (public) key encryption (top) and asymmetric (private) key encryption (bottom). (Adapted from Fox, R. *Information Technology: An Introduction for Today's Digital World,* FL: CRC Press, 2013.)

viewing the digital certificate stored by your web browser. Below, for instance, is a public key as generated by the RSA algorithm for a digital certificate.

```
b0 78 99 86 0e a2 73 23 d4 5a c3 49 eb b1 36 8c
7c ca 84 ae 3c af 38 88 28 99 8d 2d 58 13 b1 97
78 3e 52 20 67 ac 5b 73 98 6c 32 55 c9 70 d1 d9
aa 15 e8 2e 26 85 81 bc 56 e4 bc 80 63 db 4e d7
f5 02 be 51 63 1e 3c db df d7 00 5d 5a b9 e5 7b
6a ea 38 20 b2 3b b6 ee 75 54 84 f9 a6 ca 38 70
dd bf b0 ff a5 85 5d b4 41 fe dd 3d d9 2a e1 30
43 1a 98 79 93 a0 5f e0 67 6c 95 fa 3e 7a ae 71
7b e3 6d 88 42 3f 25 d4 ee be 68 68 ac ad ac 60
e0 20 a3 39 83 b9 5b 28 a3 93 6d a1 bd 76 0a e3
eb ae 87 27 0e 54 8f b4 48 0c 9a 54 f4 5d 8e 37
50 dc 5e a4 8b 6b 4b dc a6 f3 34 be 77 59 22 88
ff 19 2b 6d 76 64 73 da 0c 87 07 2b 9a 37 3a d0
e2 8c f6 36 32 6b 9a 79 cc d2 3b 93 6f 1a 4d 6c
e6 c1 9d 40 ac 2d 74 c3 be ea 5c 73 65 01 29 b1
2a bf 70 59 c1 ce c6 c3 a2 c8 45 5f ba 67 3d 0f
```

### 5.5.2 Openssl

Openssl is an open-source implementation of the SSL (secure socket layer) and TSL (transport layer security) protocols. With openssl, you can generate encryption keys (public and private), perform encryption operations (encrypt and decrypt messages or files), create certificates and digital signatures, and calculate message digests among other operations. Openssl works with a number of different encryption algorithms including ciphers AES, DES, IDEA, RC5, and triple DES, several cryptographic *hash* functions such as MD5, SHA-1, SHA-2, and MDC-2, and public key cryptography algorithms such as DSA and RDA.

A hash function is a mathematical function that translates a variable-length string into a fixed-length string, often applying some form of modulo (mod) operator. The mod operator performs a division, returning only the remainder. For instance, 38153 mod 101 gives us the remainder of 38153/101. Since 38153/101 = 377 and 76/101, 38153 mod 101 is 76. In this way, we took a longer value (38153) and reduced it in size. If we can convert a string into a number and mod it, we are able to reduce the string's size to a fixed-sized length. Using 101, our mod operator will return a number between 0 and 100. Hash functions are extensively utilized in many cryptographic algorithms to encrypt and decrypt messages.

You can see what openssl can do by typing `openssl help` or `openssl -h`. This will provide the standard commands available, message digest commands, and cipher commands. Let us examine how to encrypt and decrypt a file. The basic form of the command is

```
openssl enc cipher -in filename -out filename2
```

where *cipher* is the encryption algorithm you wish to apply. The two filenames listed, filename and filename2 are the name of the file to be encrypted and the name of the file to be created of the encrypted file, respectively. If you do not provide the output portion, by default, openssl dumps the resulting encrypted file to the terminal window, which will look like odd random symbols. To obtain a list of the encryption algorithms, type `openssl list-cipher-commands`.

The encryption command requires that you specify a passphrase. The passphrase is then needed to decrypt the file. Decryption uses the same exact instruction except for an added –d option after enc, as in `openssl enc -d` *cipher* `-in` *filename* `-out` *filename2*. When run, you will be asked to input the passphrase. You can avoid having to type the passphrase separately by adding `–pass pass:`*passphrase* to the command. One advantage of this is that you can remove the need for interactive access if, for instance, you are calling openssl from a shell script.

The encryption/decryption component of openssl uses a private key based on the algorithm selected and the passphrase. You can also use openssl to generate a public key. The public key is generated from the private key; so, you must first generate the private key. We will demonstrate this using RSA (you can also use DSA but it is more complicated). To generate a private key, we specify genrsa as the openssl command, the output file to store the key into, and the number of bits for the key. The default is 512 bits.

```
openssl genrsa -out mykey.key 2048
```

This produces a key stored in the file mykey.key.

If you look at the key created from the command, it will contain seemingly random letters, digits, and punctuation marks. For instance, the following was produced as a 128-bit key.

```
-------BEGIN RSA PRIVATE KEY-------
MGICAQACEQDNctpZY1qIZUDJWhrFdownAgMBAA ECEET6rv0cMqVgXwjwipD+L+EC
CQD6mQIPqBDcYwIJANHgrXHN21JtAggfZ49nkONU CwIJAIbd/Fb/VArZAggGaPTtzAuzkg==
-------END RSA PRIVATE KEY-------
```

We can also encode a passphrase into the key by adding an option such as –des or –des3. Alternatively, you can add the option `–passout` *value* where *value* is the passphrase. Now, given the private key, you can generate a public key using the following instruction.

```
openssl rsa -in mykey.key -pubout
```

This outputs the public key to the terminal window. You can redirect it to a file using

```
openssl rsa -in mykey.key -pubout > mykey.pub
```

This produced the following public key:

```
-------BEGIN PUBLIC KEY-------
MCwwDQYJKoZIhvcNAQEBBQADGwAwGAIRAMbq
Zr43A+UEI/Aio2r0dKkCAwEAAQ==
-------END PUBLIC KEY-------
```

With the public and private keys available, there are several things we can now do. We can generate a certificate, which you might use as part of a website to indicate that it is secure and can handle the HTTPS (secure HTTP) protocol, or you might use it for ssh or email. Here, we look at generating a *self-signed* certificate.

Normally, a certificate should be signed by a *certificate authority* to ensure its authenticity. This can cost thousands of dollars. A self-signed certificate may be used by an organization that requires a login to reach some specific web content and where the users know that the organization is trusted (e.g., employees of the company, students of a university).

We will use the x509 algorithm to generate a certificate. When generating a certificate, you can either have openssl generate a new private key or use an existing private key. We will look at both approaches. To generate a self-signed certificate with the key from above, you might specify the following:

```
openssl req -x509 -new -key mykey.key -days 365 -out mycert.pem
```

The 365 indicates the number of days that the certificate will be valid for. If the certificate is in use beyond this time period, it expires and any web browser is warned.

Issuing the above command places you into an interactive session where openssl queries you for information about the organization that this certificate will be used for:

```
Country Name (2 letter code) [XX]:
State or Province Name (full name) []:
Locality Name (eg, city) [Default City]:
Organization Name (eg, company) [Default Company Ltd]:
Organizational Unit Name (eg, section) []:
Common Name (eg, your name or your server's hostname) []:
Email Address []:
```

With your certificate generated, you would place it on your webserver in a directory requiring https access. Before doing so, you might want to test your certificate. You can view your certificate's key as it is in Ascii text. However, you might want to examine the entire contents, including the information you entered when generating the certificate. For that, use the operation

```
openssl x509 -text -in mycert.pem
```

Notice in this case that there is no hyphen preceding x509 unlike the previous instruction where you generated the certificate. Shown below is the informational portion of what is returned. You would also see the public key in hexadecimal notation and the certificate encoded using the public key (not shown here). Notice the expiration date that is 365 days after the creation date.

```
Certificate:
   Data:
        Version: 3 (0x2)
        Serial Number:
            99:d0:2f:8c:d2:ec:cd:a9
        Signature Algorithm: sha1WithRSAEncryption
        Issuer: C=US, ST=Kentucky, L=Highland Heights,
            O=ZappaFrank Industries, OU=Sales,
            CN=ns1.zappafrank.com/emailAddress=sales@zappafrank.com
        Validity
            Not Before: Feb 14 13:08:12 2013 GMT
            Not After : Feb 14 13:08:12 2014 GMT
        Subject: C=US, ST=Kentucky, L=Highland Heights, O=ZappaFrank
            Industries, OU=Sales,
```

CN=ns1.zappafrank.com/emailAddress=sales@zappafrank.com

As with the use of a passphrase, you can avoid the interaction in entering the country, state, city, and so on, by specifying all of the information from the command line. This is done by adding

```
-subj '/C = US/ST = Kentucky/L = Highland Heights/
     O = ZappaFrank Industries/OU = Sales/CN = ns1.zappafrank.com/
     emailAddress = sales@zappafrank.com'
```

If you wish to generate a new key when you generate your certificate, the syntax changes somewhat drastically. Instead of using –key *keyname*, you would specify

```
-new -newkey keyinformation -nodes -keyout keyname
```

where *keyinformation* includes the algorithm used to generate the key and the number of bits of the key, for instance, rsa:1024, and *keyname* is the name by which the key will be stored.

The openssl program calls upon numerous lesser Linux programs. These include genrsa and gendsa to generate private keys, rsa and dsa to generate public keys, enc to encrypt and decrypt messages given a key, and x509 to generate, test, and verify certificates. It also calls upon the program ca to sign certificates. Other Linux encryption tools are available, including gpg that is also open source and can generate keys, certificates and digital signatures, and can be used to encrypt and decrypt messages and files. The choice between openssl and gpg is commonly one of preference and efficiency. Both have their strengths and weaknesses.

## 5.6 EMAIL PROGRAMS

We briefly look at the standard email program of Linux in this section. Email is a complex and challenging topic. We make the assumption that email in your system is only between users on the local machine and not across the Internet. The email program we look at here is `mailx` although numerous other email programs exist for Linux including both text-based programs such as elm and Pine, and GUI-based programs such as Mozilla Thunderbird and Balsa. All of these programs are *email clients*, not servers. This means that you would use the given program to read your email messages and send out email messages or responses.

The *email server* has the responsibility of packaging the email up into a message that can be transported across the network and delivered to another email server. An email message waits on the server until the user accesses the new email via the client. While a computer can run both an email client and an email server, the two are usually separate and it is the client's responsibility to make contact with the server to access new emails or to send out new emails.

The first Unix mail program was called `mail` and was eventually in competition against a different mail program called `Mail`. Today, the most commonly used program is called `mailx`. To avoid confusion, both mail and Mail are symbolically linked to mailx so that the user could use any of these to start the mailx program. We will refer to the mailx program as mail in this section.

## 5.6.1 Sending Email Messages

To send an email message using mail, the command is `mail` *username(s)*. This invokes the interactive mailx program that first asks you to input a Subject. Then, you are dropped into a buffer to enter the body of your email. The buffer is not like an editor in that you are only able to use the backspace key and the enter key when editing your messages. You enter your text ending with control+d to exit the buffer and send your email. As we are assuming that your mail program will only permit emails to local users, the usernames are simply the account names of the users in the system (i.e., there is no extension such as @someplace.com).

A session may look like the following:

```
mail zappaf <enter>
Subject: missing document <enter>
We discovered a missing document in the file you gave us
this morning. It's the receipt from the sheet music you
said you purchased. Do you know where it is? <control+d>
```

The word Subject: appears after you press the enter key allowing you to enter a subject. You can press the <enter> key if you want to omit a subject for this email. You would receive a message indicating that the email had no subject but it would still be sent.

The mail program allows for a number of options to provide additional functionality to the email message you want to send. For instance, to attach a file to the email, use `-a` *filename*. Even with the attachment, you are still placed into the buffer to type a message, which you could leave blank by immediately pressing control+d. The option `-s` *subject* inserts *subject* into the subject line and thus avoids you having to enter the subject interactively. This can be useful if you are also including a file so that you can send an email without being placed into the buffer. The file included is not the same as attachment but instead is made available through –q *filename*.

To send carbon copies of the message to others, use –c *address(es)* and –b *address(es)* where addresses are separated by commas but no space and where –b sends a blind carbon (the names in –c listing will appear in the To: listing of the email but the names in the –b listing will not). You can use –r *address* to alter the From: address as received by the email recipient.

### 5.6.2 Reading Email Messages

If you omit the username(s) from the mail command, then you are requesting that mail check your mail spool for new emails. You will see either

```
No mail for username
```

or the list of new emails. Below is a listing in response to typing mail.

```
Heirloom Mail version 12.4 7/29/13. Type ? for help.
"var/spool/mail/username": 5 messages 2 new 2 unread
   1   marst    Fri   Jul  26   4:41    20/801    "weekend"
U  2   zappaf   Mon   Jul  29   10:53   20/726    "missing document"
U  3   marst    Mon   Jul  29   11:13   43/1831   "lunch?"
N  4   zappaf   Mon   Jul  29   12:41   22/1011   "missing document"
N  5   root     Mon   Jul  29   1:31    17/683
```

We see five emails listed above: one old message, two unread (U), and two new (N). The difference between unread and new is that new messages are those that have arrived since you last ran the mail. The unread messages are those that were available for reading the last time you ran the mail but you did not read them at that time. You would see an * prior to any email that has been read and saved.

With the above listing presented to you, you will see the mail prompt, a &. From here, you have many options. The most important are shown here.

- <enter>—read the first new email

- #—read the email numbered #, for instance, entering 2 would read the email from marst

- h—list all emails

- h #-#—list all emails numbered # to # as in h 100-120

- r/reply—reply to the most recently read email

- r/reply #—reply to email message with the given number

- s/save *filename*—save the most recently read email to *filename* in the current directory

- s/save # *filename*—save the email numbered # to *filename*

- list/help—list all commands available in mail

- fwd/forward *username*—forward the most recently read email to *username*

- fwd/forward # *username*—forward the email numbered # to *username*

- d/delete—delete the most recent email

- d/delete #—delete email numbered #

- q/quit—exit interactive mail

Upon exiting mail, you receive a message indicating the number of held messages. Messages are held in the file /var/spool/mail/*username*.

Any email message generated by mailx will start with a "header." The header consists of a number of lines of text that describe the sender, recipient(s), date, subject, MIME type, and a variety of other information. This information, aside from being useful to the reader of the email can also be used by the system administrator to troubleshoot the possible email problems. The header from the third email message from the previous list is shown below (without the message body). This email, from marst to foxr has the subject of lunch. The machine's name is localhost that has the default domain of localdomain.

```
Message 3:
From marst@localhost.localdomain Mon Jul 29 11:08:16 2013
Return-Path: <foxr@localhost.localdomain>
X-Original-To: foxr@localhost.localdomain
Delivered-To: foxr@localhost.localdomain
Date: Mon, 29 Jul 2013 11:13:16 -0400
To: foxr@localhost.localdomain
Subject: lunch?
User-Agent: Heirloom mailx 12.4 7/29/08
Content-Type: text/plain; charset=us-ascii
From marst@localhost.localdomain
Status: RO
```

To restate our initial assumption, you are only reading email sent locally on this machine. To establish email that comes from other machines, you need to connect your email client to an email server. We will not cover that topic. One last comment. To send and receive emails, the postfix service must be running. We discuss services in Chapter 11. Without the postfix service, any emails sent by mail will be queued until postfix is started. Once started, these queued messages can then be delivered.

## 5.7 NETWORK SOFTWARE

A computer network is a collection of computers and resources that can communicate together through some form of telecommunication media. Most commonly, these computers are physically connected together through some form of cable (although infrared and radio communication is also possible). Broadcast devices are used so that a message can be directed from one machine to another utilizing the message's destination address. Broadcast devices are hubs, switches, routers, and gateways. Typically, hubs and switches connect computers within a small network while routers steer messages between networks whether those networks are on one site, such as a university campus, or across the world. Gateways are routers that connect networks that use different communication protocols.

## 5.7.1 Internet Protocol Addressing

For a computer to communicate over a network, the computer must have an address. We typically use IP addressing because we utilize the TCP/IP protocol so that our computers can not only communicate over a local area network but also over the Internet. We will hold off on discussing TCP/IP in detail until Chapter 12, but here, we will examine IP addresses.

IP addresses come in two formats, IPv4 (version 4) that is a 32-bit binary value often represented as four integer numbers between 0 and 255, and IPv6 (version 6), which is a 128-bit binary value, often represented as 32 hexadecimal digits. Since IP addresses are hard to remember, we often instead refer to computers by names, or IP aliases. This requires that we have some form of IP alias to IP address translation process. This is where the DNS (domain name system) comes into play.

Across the Internet are a number of DNS name servers. A DNS server is a computer that contains DNS translation information. The server is set up to respond to translation requests from client computers across the Internet. The translation process usually takes place in two phases. First, a local DNS will attempt to map the network portion of the IP alias into the network portion of the IP address. For instance, google.com or nku.edu might be known by the local DNS server. If the local DNS server does not have the mapping information, the request is forwarded to another DNS server that is located "higher up" on the Internet. Thus, the DNS servers make up a hierarchy across the Internet. Eventually, a DNS server will have the translation information, or, you will receive an error that such an address cannot be found. The message can now be routed to that destination address.

At the destination, a local DNS server finishes the translation by mapping the full IP alias (including the machine's name) to the entry found in its table. In many cases, a destination may have subdomains that must also be mapped. For instance, nku.edu has a subdomain of hh and hh has numerous computers within its domain. The local DNS server at nku.edu finishes the translation of the alias to the IP address and now, the message can be routed on its way. The DNS process is mostly invisible (or transparent) to the user and takes milliseconds to at most a few seconds to complete.

The Internet (and other networks) support a number of protocols that handle the remainder of the message traffic. That is, the protocol(s) dictate how addresses are interpreted, how data are handled, the size of the data, which we call packets, error-handling information, and so forth. Among the protocols we deal with are TCP and UDP for handling packets, HTTP and FTP for handling file transfers, HTTPS, FTPS, SFTP, SSL, and others for transmitting data in a secure manner, and numerous others for handling email.

More information on TCP/IP can be found in Chapter 12 and we examine the Bind program for a DNS server in Chapter 15.

In this section, we look at Linux programs to support the user's ability to communicate over a network either by testing network operability (ping, traceroute), opening shells on other computers (telnet, ssh, and rsh), transferring files (ftp, sftp, and http), or communicating directly to a DNS server (nslookup, host, and dig). We will also briefly examine commands that tell you about your computer's network connectivity. We will concentrate only on command-line network instructions although some of these have similar GUI-based programs.

### 5.7.2 Remote Access and File Transfer Programs

One tool that Unix provided from its earlier incarnations was a means to connect to other computers. The basic tool is known as *telnet* and through telnet, you can use your current terminal window as if it was another computer's. The telnet program establishes a connection to that remote computer, sending your input to that computer and collecting that computer's output to display in your terminal window. In this way, telnet literally opens a window into another computer. You must first log in to the remote computer to gain access; so, you must have an account.

The `telnet` program requires that you include the remote computer's IP address. A number of options are available. These include –E to prevent escape characters from being interpreted, -a to attempt automatic login, -b to specify an IP alias instead of an IP address, -l to specify a username (this option is only useful if you are also using –a), and –x to turn on encryption. You can also specify a port number if you do not wish to use the default telnet port (which is typically port 23).

Communication between the two computers is handled by passing ASCII characters as streams between the two computers. This is unfortunate because it means that all messages that are sent to the remote computer are unencrypted. Among other things, this includes the password that you sent to log in to the remote computer. With techniques such as IP spoofing and packet sniffing, unencrypted messages can be fairly easily intercepted. Therefore, telnet is not secure and is largely unused today.

Another means of controlling a remote Unix or Linux computer is through *r-utilities*. This suite of network communication programs is intended to be used on a network of computers that share the same authentication server. The `rlogin` program lets you connect to the remote computer just as with telnet, except that you do not need to log in. As you have already logged into the local computer, since the remote computer shares the same authentication server, your account information is the same. So, the advantage of the r-utilities is that they save you from having to log in.

There are several other r-utilities. The `rsh` utility is used to run a command remotely on a networked computer. The command opens a shell, runs the given command, collects its output, and returns it back to the host computer for display. The rsh command cannot be used to run any interactive commands. As an example, you might issue a command such as `rsh 1.2.3.4 who` to obtain the list of users currently running on the computer 1.2.3.4. Alternatively, there is an `rwho` command that will do the same. Finally, `rcp` is a copy remote copy command that allows you to copy files from the remote computer to your host computer.

Although the r-utilities require that you have an account on the remote computer for authentication purposes, all interactions are handled in normal text, just as with telnet. Thus, any communication using either telnet or an r-utility is insecure. In place of telnet is `ssh`, the secure shell program. The ssh program uses public key encryption (refer back to Section 5.5). The local computer uses the public key, which is known to any or all computers. The public key encrypts messages into a code. Only the remote computer has the private key.

The ssh command is followed by the IP address or alias of the remote computer you want to log into. By default, ssh will attempt to log you in using your current account name. To override this, prepend the IP address/alias with username@, as in `ssh foxr@10.11.12.13` to log into the computer at 10.11.12.13 with the account name foxr.

Another old technology is FTP. FTP is file transfer protocol, and it was developed in 1971 for the ARPAnet (an early incarnation of the Internet). With FTP, a user makes a connection with a remote computer to transfer files. Notice how this differs from telnet and ssh that in essence opens a terminal window to the remote computer. With FTP, only files are moved between remote and host computers. Files can be sent to the remote computer, often referred to as uploading, and files can be sent from the remote computer to the local computer, often referred to as downloading. To access the remote computer in this way, you must either have an account or you must log in as an anonymous user. If the remote computer permits *anonymous* logins, this provides the user access only to the public area. In Linux, this is often a top-level directory called /pub.

The initial FTP command is like telnet, `ftp address` where *address* is the IP address or IP alias of the remote computer. You are then asked to log in. To log in anonymously, type anonymous for your user name and your email address for your password. Although the password is not used for authentication, it is captured in a log file so that others can see who has logged in anonymously (and so it is not truly anonymous!) Once connected, you have an `ftp>` prompt. From here, there are numerous commands available, as shown in Table 5.3.

Today, there are many GUI-based FTP programs. These typically have a drag-and-drop feel so that you can copy individual or multiple files very easily. You can change local and remote directories, view contents of local and remote directories, create and delete directories, and so forth. FileZilla is one of the most popular GUI-based FTP programs available in Linux.

FTP, like telnet, is insecure. This not only results in making FTP transfers (including any passwords sent) open to view by IP spoofers and packet sniffers, but also leaves open security holes on FTP servers. There are more secure forms of file transfer now available. These include secure FTP (not to be confused with SFTP) that uses various forms of security to protect the information being transmitted, FTPS that is FTP extended to include encryption, SFTP and SCP (secure copy), which use ssh to transfer files, and FTP over SSH, which requires that the user first logs in using ssh and then accesses the remote computer using FTP. Two drawbacks of FTP over SSH are that the remote computer must be able to handle both FTP and ssh and that the user must have an account on the remote machine because ssh does not permit anonymous logins.

TABLE 5.3   FTP Commands

| FTP Command | Meaning |
|---|---|
| ascii | Transfer files in ascii text (default mode) |
| binary | Transfer files in binary (necessary if files are binary files) |
| cd | Change directory on remote machine |
| close | Close connection but remain in ftp |
| delete | Delete remote file |
| get | Transfer file from remote computer to local computer (download) |
| help | Get the list of commands |
| lcd | Change directory on the local computer so that uploads originate from the specified directory, and downloads are saved to the specified directory |
| ls, mkdir, pwd | List contents of remote directory, create new directory on remote machine, and output current working directory on remote machine |
| mget | Mass get—used in conjunction with wildcards, for example, mget *.txt |
| mput | Mass put |
| open | Open a new connection to specified machine |
| put | Transfer file from local computer to remote computer (upload) |
| quit | Close connection and exit ftp |
| rmdir | Delete specified directory on remote computer |

FTP has largely become obsolete in the face of the newer HTTP and HTTPS protocols. HTTP and HTTPS are the protocols used in support of the World Wide Web. Commands to transfer files using these protocols can be entered at the command line or generated by software such as a web browser or a search engine spider (or crawler). HTTP has just a few commands, the most significant are GET (to request a file), HEAD (to request just the header of a file without the file's contents), OPTIONS (to see which commands the web server will respond to), PUT (to upload a file), and POST (to upload data to a server, such as a post to a discussion board). HTTPS is one of the secure forms of HTTP that involves public key encryption and the transfer of a certificate to ensure that the website is legitimate.

Both the wget and nc programs allow the user to transmit HTTP requests to servers from the command line. The wget program is a noninteractive and text-based form of access. The user enters the URL and the server responds with the file, which is stored locally to the hard disk rather than displayed in a web browser. In essence, wget serves the same purpose as FTP except that the remote computer is a web server and there is no log-in process.

The wget program supports a recursive download so that the command can be used to retrieve not only a target file by URL but also all the files that the target URL has links to, recursively. If not done cautiously, this could result in a download of an entire website! The recursive version of the command is useful for a web spider that attempts to accumulate all pages of a web server. There are GUI-based interactive versions of wget available. Below are three examples of wget, the first downloads a single file, the second uses a wildcard (the extension.jpg) to download a number of jpg files, and the third performs a recursive download.

- wget www.nku.edu/~foxr

- wget -A .jpg www.nku.edu/~foxr

- wget -r www.nku.edu/~foxr

The nc (netcat) program has been called the "Swiss Army Knife" of network programs. It contains a number of useful features that can be used to explore a network including its security. In our case, we are interested in using it to communicate with a web server. The basic command is nc `remotename port` as in nc `www.nku.edu 80`. The port 80 is the default port for HTTP communication. The nc command itself establishes an open connection to the remote computer. Once a connection has been made, the user is able to interact with the remote computer by sending it commands.

As we will focus on a web server, the commands we wish to send will all be HTTP commands. Therefore, we will want to send GET, OPTION, PUT, or HEAD messages. We must specify, along with the command, the URL and the version of HTTP in use. For instance, to obtain the file index.html, we might issue the command

```
GET /index.html HTTP/1.0
```

If the file is in a subdirectory, such as foxr's index.html, it would appear as

```
GET /foxr/index.html HTTP/1.0
```

With nc, we can specify preferences to the web server to perform *content negotiation*. A web server is set up to return the requested file, but in some cases, the web developers might provide multiple files to fulfill a variety of preferences. One type of preference is the file's language. Imagine that there are two versions of /foxr/index.html, one that ends with the extension .en and one with the extension .fr. These are most likely the same file but one is written in English (en) and one is written in French (fr). If a user wishes the French version, the request becomes

```
GET /foxr/index.html HTTP/1.0
Accept-Language: fr en
```

This request says "give me a version in French if available, otherwise English." You are able to establish preferences such as language, encoding (compression), character set, and MIME type in your web browser as well through the nc program.

### 5.7.3 Linux Network Inspection Programs

There are numerous other network programs available in Linux. Many of these are tools for the system administrator, such as `route`, `bind`, `sendmail`, `ss`, `ip`, and `netstat`. We will examine these in Chapter 12 when we focus on network configuration from the system administrator perspective. There are a few other programs worth noting though for both the administrator and the user. These are ip, ping, traceroute, host, dig, and nslookup. So, we wrap up the chapter by examining these.

As stated in the introduction, every computer on the Internet has a unique IP address. Many computers have two addresses, a version 4 address (IPv4), and a version 6 address (IPv6). The IPv4 address is stored as a single 32-bit binary number. If written as a decimal value, it is four individual numbers, each in the range 0–255 where each number is separated by a period such as 1.2.3.4, 10.11.12.13, or 172.31.251.3. Each number is called an *octet*.

The IPv6 address is a 128-bit binary number that is often written as 32 hexadecimal digits. The hexadecimal digits are grouped into octets of four digits long, and separated by colons. However, for convenience, leading zeroes are often omitted and if an octet is all zeroes, it can also be omitted. For instance, the IPv6 address

```
fe80::125:31ff:abc:3120
```

is actually

```
fe80:0000:0000:0000:0125:31ff:0abc:3120
```

To determine your computer's IP address, use the program ip. This program is stored in /sbin, which may not be in your path; so, you might have to issue this as /sbin/ip. The ip command provides many different types of information about your computer's network connections such as local router addresses and addresses of other local devices. The ip command expects the object you wish information on. For the address, use ip addr. This will supply you with both the IPv4 and IPv6 IP addresses if they are both available. To limit the response to just one of these addresses, use the option -f inet for IPv4 and -f inet6 for IPv6.

In Linux, you will have at least two interface devices to communicate over: an Ethernet device (probably named eth0) and a loopback device named lo. The eth0 device will be given an IPv4 and/or IPv6 address. The loopback device is used for your computer to communicate with itself rather than networked computers. The lo address is always 127.0.0.1 with an IPv6 address of all 0s.

Another way to obtain your IP address is with the older program ifconfig, also in /sbin. Both ifconfig and ip will display your IP addresses and both of them can be used to modify network address information. As these are tasks for a system administrator, we will not cover it here.

To test the availability of a remote computer, there are other available Linux programs. The easiest and most commonly used one is called ping. The ping program sends out continual messages to the destination address at 1 second intervals and reports on responses. For instance, ping 1.2.3.4 could result in the following output:

```
64 bytes from 1.2.3.4: icmp_seq=1 ttl=60 time=0.835 ms
64 bytes from 1.2.3.4: icmp_seq=2 ttl=60 time=0.961 ms
64 bytes from 1.2.3.4: icmp_seq=3 ttl=60 time=1.002 ms
```

Such messages repeat until you exit the program by typing control+c. Upon termination, you are given overall statistics such as

```
3 packets transmitted, 3 received, 0% packet loss,
time 2798 ms rrt min/avg/max/mdev = 0.835/0.933/1.002/0.071
```

As with most Linux commands, ping has a number of useful options. You can specify the number of packets transmitted using –c, as in `ping  -c  10` to force ping to exit after the 10th packet. The –f option outputs a period for each packet sent rather than the output shown above. You can establish a different interval for packet transmission using –i interval although intervals of < 0.2 seconds require system administrator privilege. The option –R records the route that the packet took to reach the destination. This is the output after the first packet's response. You can also specify your own route if you have knowledge of router addresses that would permit your message to be routed between your computer and the remote computer.

The `traceroute` command is like the ping command with the –R option. The traceroute command sends out packets to a remote computer and reports statistics on the route (or routes) that the packets took. While ping is useful to ensure that a computer is accessible, traceroute is useful in determining several things about a network. First, by examining several traceroutes, you can see what parts of your network are reachable and what parts might not be. Also, you can determine if any particular router or part of the network is overburdened because it is responding slowly. The newer traceroute6 (or traceroute -6) provides IPv6 addresses rather than IPv4 addresses.

The traceroute command sends out several packets, known as *probes*. To successfully reach the next network location, three probes must be received and a response must be received. Each probe is sent with a set time-to-live (ttl) value. This is initially a small value, but if a probe is unsuccessful in reaching the next network location, a probe is resent that has a larger ttl.

The traceroute output provides a list of every segment of the network that was successfully reached. In cases where a network segment was not reachable (a time-out is indicated if a response is not returned within five seconds), traceroute responds with * * *. The probes will reach a variety of types of devices en route to the remote computer. These will include your network's gateway to the Internet, perhaps an Internet service provider, routers at various Internet locations such as (or including) Internet backbone sites, the destination computer's gateway to the Internet, internal routers, and finally the destination computer. For each segment, the information provided includes the IP alias (if any), the IP address, and the time it took for the three probes to reach the device. If more probes reach the location, only the last 3 times are output.

The final programs of note are `host`, `dig`, and `nslookup`. These programs perform DNS lookups to translate IP aliases into IP addresses. A DNS server[*] contains one or more files that provide the address translation information, each file representing one domain (or subdomain) within that site.

---

[*] We provide only a cursory examination of DNS tables here. For more detail, see Chapter 15.

DNS servers tend to fall into one of the two categories, authorities that can be masters or slaves, and caches. An authority DNS is one that happens to store an SOA (start of authority) entry for a given domain. While this information can be propagated to other DNS servers including caches, only the SOAs can be modified to show changes to addressing information.

The difference between a master and a slave is that the master's DNS table is modified and then transmitted in regular intervals to all slaves. In this way, a system administrator who is maintaining a domain only has to modify a single computer's DNS table. As we will see below, information in this table specifies how often the slaves should be modified.

Other DNS servers may receive copies of address translations. In these cases, the servers will probably cache the data for some time. DNS servers can be set up to cache or not cache such information. Cache entries are fairly short lived so that out-of-date data are not kept. If a nonauthority DNS server is asked to perform address translation for a given domain, and it does not have the information in its cache, it will ask another DNS server. Ultimately, the request will be made of a DNS server that has the information in its cache, or is an authority (whether master or slave).

An authority's DNS table begins with a record about the authority itself. This appears in an authority record denoted as

```
domain-name IN     SOA           server-name(s) (
            values
)
```

If the domain name is already present in the file (prior to the SOA record), it can be replaced with an @. The entry IN indicates an Internet device. The entry SOA denotes that this record is the SOA information. The server names are any names that are aliased to this DNS server. If there is more than one server name, they are separated by spaces. The names should end with periods.

Inside of parentheses are five values that describe the authority. The first number is a serial number that is incremented by the administrator every time this file is edited. The serial number is used to compare master and slave records. If a slave receives an update from a master and the two records have different serial numbers, then the slave should update its own record. If the master and slave have the same serial number, then there is no reason for the update as it indicates that there has been no change.

The other four values are used to control how often updates are required. The second of the five values is the refresh rate, or how often a slave should attempt to contact the master. The third of the five numbers is the retry interval, used by the slave if it attempts to contact the master and the master does not respond. In such a case, the slave retries as specified by this interval.

For instance, the retry value might be anywhere from five minutes to two hours. The fourth value is the expiration period. This value indicates how long its own records should be good for. If this time period has elapsed, the slave should consider its records out of date and refuse to respond to requests for IP addresses of this domain. The final value is the minimum TTL, which is used by nonauthoritative DNS servers to retain cached entries.

The remainder of the DNS table lists the various device names found in the domain. These will include ordinary devices, mail servers, name servers, and other devices. These types are denoted by A, for devices that use IPv4 addresses, AAAA for devices that use IPv6 addresses, MX for mail servers, and NS for name servers. These entries consist of the machine name, IN, the type, and the IP address, for instance,

```
mymachine.hh.nku.edu     IN     A       1.2.3.4
```

Some entries have the type CNAME, a canonical name. A canonical name is the true name for an alias. In this case, there should be (at least) two entries in the table, one that maps the alias to the machine's true name, and one that maps the true name to its IP address. For instance, you might see

```
mymachine.hh.nku.edu     CNAME     machine1.hh.nku.edu
machine1.hh.nku.edu      A         1.2.3.4
```

so that mymachine is the alias for the true name machine1.

Now that we have some understanding of the DNS server, we can now explore the last three Linux programs. Of the three programs, `nslookup` is the oldest and most primitive, but this also makes it the easiest to use. The nslookup instruction expects as a parameter the IP alias of the computer whose IP address you wish to look up. Optionally, you can also specify a DNS server IP address to act as the server to perform the lookup. The format is

```
nslookup IP_alias [DNS_IP_address]
```

where the DNS_IP_address is optional.

The response from nslookup provides all IP addresses known for the IP alias and also lists the IP address of the DNS server used. Consider the three nslookup command responses in Figure 5.9. The first case is a request to look up www.nku.edu, sent to a DNS server that is the master for that domain and so, the response is authoritative. The second has a nonauthoritative response because the response did not come from a centos.com DNS authority. The third is also a nonauthoritative response, but is of interest because of the number of responses. We receive multiple IP addresses because google has several physical IP addresses to support their servers.

Both the dig (domain information groper) and host programs permit a number of options and provide more detailed feedback than nslookup. Both dig and host will return portions of the DNS server table.

With dig, the −t option allows you to specify the type of entry you are interested in. So, rather than returning information about a specific machine, you can query the DNS server for all devices in the domain that match the type. Consider `dig −t MX google.com`.

```
$ nslookup www.nku.edu
Server:        172.28.102.11
Address:       172.28.102.11#53

www.nku.edu canonical name = hhilwb6005.hh.nku.edu.
Name: hhilwb6005.hh.nku.edu
Address: 172.28.119.82

$ nslookup www.centos.com
;; Got recursion not available from 172.28.102.11,
trying next server
;; Got recursion not available from 172.28.102.13,
trying next server
Server:        10.11.0.51
Address:       10.11.0.51#53

Non-authoritative answer:
Name: www.centos.com
Address: 87.106.187.200

$ nslookup www.google.com
Server:        172.28.102.11
Address:       172.28.102.11#53

Non-authoritative answer:
www.google.com     canonical name = www.l.google.com.
Name: www.l.google.com
Address: 74.125.227.51
Name: www.l.google.com
Address: 74.125.227.49
Name: www.l.google.com
Address: 74.125.227.48
              (additional addresses omitted)
```

FIGURE 5.9   nslookup Command Responses

This will ask the google.com domain's DNS server to return all devices that are denoted as mail servers. The response provides several entries of the form

```
google.com  460   IN   MX   20     address
```

The value 460 indicates a TTL while 20 is used in load balancing since there are a number of entries (each with a different value).

Alternatively, try dig -t MX nku.edu. Here, you will find that the TTL value is static (does not change) from 3600. The command dig -t NS google.com gives much the same type of response except that the TTL is very large and there is no load-balancing value specified.

The dig command actually responds with several different sections. First, the dig command responds with a repeat of the command's arguments. It then summarizes the response as a header, a type of operation (Query), status (NOERROR), an ID number, flags, and the number of responses received from the DNS server. These are divided into a question (or query) section that in essence repeats the request, the number of items in an ANSWER section, number of items in an AUTHORITY section, and number of ADDITIONAL items. Figure 5.10 demonstrates the result from the query dig -t NS www.nku.edu. In this case, we are querying a DNS for a specific machine's information so there are fewer entries in the response.

Aside from –t, dig can also be queried using option –c to specify a class (IN is the only class we examine here), -p to specify a port, -6 to indicate that only IPv6 addresses should be used, and –b to send the dig command to a specified DNS server. Additionally, you can request responses of multiple machine and/or domain names. If preferred, you can place the request information in a file using dig -f *filename* address.

The host program by default responds with the IP address of the given alias. The option –d (or –v) provides more detail (or a verbose response). Like dig, host will respond with question, answer, authority, and additional sections. Unlike dig, host, when supplied with –d (or –v) responds with information from the SOA record. For instance, the

```
; <<>> DiG 9.7.3-P3-RedHat-9.7.3-8.P3.el6_2.1 <<>>
-t NS www.nku.edu
;; global options: +cmd
;; Got answer:
;; ->>HEADER<<- opcode: QUERY, status: NOERROR, id:
57418
;; flags: qr aa rd ra; QUERY: 1, ANSWER: 1,
AUTHORITY: 0, ADDITIONAL: 0

;; QUESTION SECTION:
;www.nku.edu.                    IN     NS

;; ANSWER SECTION:
www.nku.edu.       3600   IN     CNAME hhilwb6005.
hh.nku.edu.

;; Query time: 1 msec
;; SERVER: 172.28.102.11#53(172.28.102.11)
;; WHEN: Tue Aug 28 14:23:32 2012
;; MSG SIZE rcvd: 57
```

FIGURE 5.10  dig Command Response

instruction `host -d www.nku.edu` will provide the following in the AUTHORITY SECTION:

```
;; AUTHORITY SECTION:
hh.nku.edu.        3600  IN    SOA    nkuserv1.hh.nku.edu.
postmaster.exchange.nku.edu. 37368706 900 600 86400 3600
```

Notice here that we see the serial number, refresh rate, retry rate, expiration time, and minimum TTL. While this information is available via dig, it is not provided by default.

Figure 5.11 provides the results of `host -d www.nku.edu`. The information in this response is similar to that from dig, but here, we go beyond the single machine www.nku. edu to obtain information about the domain (nku.edu) as well.

Finally, with host, the –i option asks the DNS server to perform a *reverse IP lookup*. This is used to translate an IP address into an IP alias. You might wonder why anyone would want to perform a reverse IP lookup when the whole point of the DNS is to allow humans to use IP aliases so that they do not need to remember IP addresses even though the addresses are essential for Internet communication. But there are many approaches to attacking a computer by pretending to be an IP address that you are not. The reverse IP lookup can ensure that an incoming communication that claims to be from a particular IP address is correct.

## 5.8 CHAPTER REVIEW

Concepts and terms introduced in this chapter:

- ARPAnet—original form of the Internet, developed in the late 1960s through the early 1980s.

- Asymmetric encryption—a form of encryption that uses two separate keys, a public key to encrypt messages and a private key to decrypt messages. This form of encryption is primarily used to support E-commerce and secure communications between different users or sites on the Internet.

- FTP—file transfer protocol, for transferring files between computers over the Internet.

- HTTP/HTTPS—hypertext protocol transfer and the secure form, for transferring web pages from a web server to a web client.

- Insert mode—in vi, this mode inserts typed characters into the document rather than interpreting the characters as commands.

- Command mode—in vi, this mode interprets special-purpose keystrokes as commands to move the cursor, cut/copy/paste characters, perform file operations, or change from command to insert/replace mode.

- Computer network—a collection of computers and computer resources connected together by some medium to support resource sharing, document sharing, and electronic communication.

```
Trying "www.nku.edu"
;; ->>HEADER<<- opcode: QUERY, status: NOERROR, id: 2258
;; flags: qr aa rd ra; QUERY: 1, ANSWER: 2, AUTHORITY: 0,
ADDITIONAL: 0

;; QUESTION SECTION:
;www.nku.edu.                    IN     A

;; ANSWER SECTION:
www.nku.edu.        3600   IN     CNAME hhilwb6005.hh.nku.edu.
hhilwb6005.hh.nku.edu.   3600   IN     A      172.28.119.82

Received 73 bytes from 172.28.102.11#53 in 2 ms
Trying "hhilwb6005.hh.nku.edu"
;; ->>HEADER<<- opcode: QUERY, status: NOERROR, id: 84
;; flags: qr aa rd ra; QUERY: 1, ANSWER: 0, AUTHORITY: 1,
ADDITIONAL: 0

;; QUESTION SECTION:
;hhilwb6005.hh.nku.edu.          IN     AAAA

;; AUTHORITY SECTION:
hh.nku.edu. 3600  IN     SOA    nkuserv1.hh.nku.edu. postmaster.
exchange.nku.edu. 37369675 900 600 86400 3600

Received 104 bytes from 172.28.102.11#53 in 5 ms
Trying "hhilwb6005.hh.nku.edu"
;; ->>HEADER<<- opcode: QUERY, status: NOERROR, id: 15328
;; flags: qr aa rd ra; QUERY: 1, ANSWER: 0, AUTHORITY: 1,
ADDITIONAL: 0

;; QUESTION SECTION:
;hhilwb6005.hh.nku.edu.          IN     MX

;; AUTHORITY SECTION:
hh.nku.edu.         3600   IN     SOA    nkuserv1.hh.nku.edu.
postmaster.exchange.nku.edu. 37369675 900 600 86400 3600

Received 104 bytes from 172.28.102.11#53 in 3 ms
```

FIGURE 5.11   host Command Response

- Macro—a definition of keystrokes to be performed many times to save the user time, available in both vi and emacs.

- Private key—a key that can be used to encrypt and decrypt messages in symmetric (private key) encryption, or generate a public key and decrypt messages in asymmetric key encryption.

- Public key—a key made available to the public to encrypt messages but not decrypt messages, used in asymmetric (public key) encryption.

- R-utility—a suite of Linux network programs that permits a user of one computer in a Linux network to access the other Linux computers without having to log in.

- Symmetric encryption—a form of encryption in which one key, the private key, is used to both encrypt and decrypt messages. Mostly used when there is only one user such as when encrypting and decrypting a file system.

- Text editor—a program to create and edit text files (as opposed to a word processor which also permits formatting of text).

- WYSIWYG—an acronym for "what you see is what you get," a popular term expressing GUI-based software where you can see what the final product looks like before you print it. Early text-based word processors used embedded commands (see LaTeX in Section 5.4 for instance) so that the version of a file as seen during editing is not how the final product would look.

NOTE: See Chapters 12 and 15 for review terms on TCP/IP and DNS.

Linux commands introduced in this chapter:

- bibtex—generates the bibliography for a LaTeX document.

- dig—to query a DNS server about IP address information.

- emacs—a text-based text editor with many powerful features.

- ftp—a text-based program to transfer files between computers.

- host—like dig, to query a DNS server about IP address information.

- ifconfig—to obtain or reconfigure the IP address of a computer.

- ip—a network tool that supersedes many older network programs such as ifconfig (covered in detail in Chapter 12).

- latex—to process a .tex (LaTeX) file and create a neutral, device-independent (dvi) file.

- mail—also mailx, the built-in text-based Linux mail client program.

- nc—the netcat program offering a suite of network operations, including the ability to send HTTP requests to a web server without using a web browser.

- nslookup—like dig and host, but a more primitive DNS query tool.

- openssl—a collection of encryption algorithms that permit the user to generate public and private keys, encrypt and decrypt messages, and create digital certificates and signatures.

- ping—tests the availability and access of a networked resource.

- rlogin—one of the r-utilities, to remotely log into another Linux computer of the same network.

- rsh—one of the r-utilities, to open a shell on another Linux computer of the same network.

- rwho—one of the r-utilities, to execute who on another Linux computer of the same network.

- rcp—one of the r-utilities, to copy a file from another Linux computer of the same network to this computer.

- ssh—a secure form of telnet (see below) that uses public key encryption.

- telnet—a program to log into another computer across the Internet. Telnet is no longer used because messages, including passwords, are passed in an unencrypted form.

- traceroute—tests the route taken in communications between two networked computers.

- vi—a text-based text editor that comes with all Linux implementations.

- wget—a command-line noninteractive program to retrieve files from a web server without having to operate through a web browser.

## REVIEW PROBLEMS

1. In vi, what is the difference between the command 'r' and 'i'?

2. In vi, what is the difference between the command 'o' and 'O'?

3. In vi, what is the difference between the command 'w' and 'W'?

4. In vi, what is the difference between the command 'e' and 'E'?

5. In vi, you are editing a document and currently on line 21. How would you reach line 1? Line 12? The last line of the document?

6. In vi, to copy the current line and paste it below the current line, you would do yy followed by p. How would you do this in emacs?

7. As a Linux user, what feature of Bash helps you learn how to control emacs?

8. What is a macro? Provide an example of why you might want to define one.

9. Research LibreOffice and OpenOffice. Create a list of five features available in LibreOffice that are not available in OpenOffice.

10. Why might you want to use LaTeX rather than a WYSIWYG word processor?

11. Given the following LaTeX definition for a table, draw how it would appear in a document.

```
\begin{tabular}{|l||l|}
  Item 1 & Item 2 \\
  Item 3 & Item 4 \\
  Item 5 \\
  & Item 6 \\
\end{tabular}
```

12. What is Tex? How does it differ from LaTeX?

13. If I have a public key to your private key and you send me an encrypted message, can I decrypt it? If I have a private key to your public key and you send me an encrypted message, can I decrypt it?

14. Using openssl, what are the steps that you would need to generate a private key?

15. Using openssl, what are the steps that you would need to generate a public key?

16. Using openssl, what are the steps that you would need to generate a digital certificate?

17. Assume a digital certificate has the following information:

```
Issuer: C = UK, L = London, O = Bobs Books, OU = Sales,
CN = tardis.bobsbooks.com.uk
```

What do each of C, L, O, OU, and CN stand for?

18. How do you control the valid dates when generating a digital certificate in openssl?

19. What is the difference between an email client and an email server?

20. How does the command `mail` differ from the command `mail` *address*?

21. What is the difference between `mail -b` *address* and `mail -c` *address*?

22. By default, where are emails stored until they are read?

23. What must be true of networked Linux computers for you to be able to use r-utilities between them?

24. How does ssh differ from telnet?

25. When using ftp, how does cd differ from lcd?

26. When using ftp, what command allows you to upload a file to the remote computer?

27. How does wget differ from ftp?

28. Explain why you might want to use ping. Explain why you might want to use traceroute instead of ping.

29. Explain why you might want to use nslookup.

30. How does nslookup differ from dig and host?

31. What is the equivalent in host of the command dig (i.e., the version of host that provides detailed information like dig)?

32. How would you specify using dig that you only want to obtain information for the listings of NSs?

33. What is an SOA?

# CHAPTER 6

# Regular Expressions

THIS CHAPTER'S LEARNING OBJECTIVES are

- To understand the use of regular expressions

- To understand the usage of the metacharacters of regular expressions

- To understand how to apply each of the metacharacters

- To be able to use grep/egrep to search for files for strings

- To be able to use the basic forms of sed and awk to solve search problems

## 6.1 INTRODUCTION

A regular expression (regex) is a string that expresses a pattern used to match against other strings. The pattern will either match some portion of another string or not. We can use regular expressions to define pattern matching rules which can then be used by programs. For instance, regular expressions are applied in software that performs spam filtering and natural language understanding.

Let us look at a concrete example to illustrate the problem that a regular expression can solve. You want to build a spam filter. You need to come up with a way to filter out any email message that contains the word Viagra. Searching for the literal string "Viagra" or "viagra" is easy enough. But, clever spammers will try to disguise the word by using substitution letters, adding characters or removing letters in the word. We might expect any of the following variations:

- v.i.a.g.r.a

- vlagra

- vi_ag_ra

- vi@gr@

- ViAgRa

- Viagr

To build our spam filter, we would not want to list every possible potential appearance of Viagra. For the six examples listed here, there are thousands of others that someone might come up with. Instead, we could define a single regular expression that could cover many or most of the possibilities.

Let us consider another example. We want to define a regular expression to match any person's full name (we will assume just first and last name). A name starts with a capital letter and is followed by lower-case letters. This defines any capitalized word. To differentiate, a name will consist of two of these with a space in between. We can define this regex as consisting of an upper-case letter followed by some number of lower-case letters followed by a space followed by a capital letter followed by some lower-case letters. Note that our regex does not ensure that the names found are actually people's names. With the regular expression defined, we can use a Linux program, for instance grep, to scan through a file to locate all of the strings that are names.

To define a regular expression, you write a pattern that consists of *literal* characters and *metacharacters*. The literal characters are characters that must match precisely, in the order provided and in the specified form (e.g., an upper-case letter 'A' must match an upper-case letter 'A'). The metacharacters are characters that are not compared but instead are applied to alter the interpretation of literal characters. For instance, the '.' (period) metacharacter represents "any character." The regex b.t will match any string that has the letter 'b' followed by any single character followed by the letter 't.' The regular expression b.t combines two literal characters, 'b,' and 't,' and one metacharacter, '.'. This regex then will match such strings as

- bat

- but

- bit

- b3t

- b+t

- b.t

- b t

The regex will not match the string bt because the . must match a character. In the string bt, there is no character between the 'b' and 't' to match. The regex b.t will also not match b12t, baat, or b%$#!t because there is not *exactly* one character between the 'b' and 't.'

There are several very powerful tools available in Linux that utilize regular expressions. Because of this, it is very useful for the Linux user to understand regular expressions and be able to utilize them through these tools. For the system administrator, these tools are often utilized in a number of ways to simplify their tasks.

Aside from examining regular expressions (and numerous examples), this chapter presents three useful tools: grep, sed, and awk. The grep program searches a text file for instances of the given regular expression, returning any lines that contain a matching string. This can be very useful for searching multiple files for specific content such as people's names, IP addresses, or program instructions. The egrep program is a variation of grep which allows for more metacharacters and so we will primarily examine it when we look at grep in Section 6.4. Regular expressions are also used in the programs sed and awk although we will see that these programs are far more complex than grep/egrep. We examine sed in Section 6.5 and awk in Section 6.6. Regular expressions can also be used in both vi and emacs when searching for strings. If you are a programmer, you will also find that many modern programming languages have regular expression facilities. One of the earliest languages to make use of regex was perl, but we find it in such languages as Java, Python, Ruby, PHP, and .Net platform languages (C++, C#, J#, ASP, VB, etc).

## 6.2 METACHARACTERS

In this section, we present the regex metacharacters available in Linux. Each of these will be discussed in detail and illustrated through a number of examples. We build upon the metacharacters in that we will find some metacharacters can be applied in conjunction. That is, some metacharacters not only modify literal characters but can also modify other metacharacters. You might find this section to be challenging as this is a complicated topic. Working through the examples presented should help.

Table 6.1 lists the various metacharacters that are part of the standard and extended regular expression set. In this section, we examine their usage with numerous examples.

The first thing you might notice from the characters in Table 6.1 is that some of them are the same as Linux wildcard characters. Unfortunately, the * and ? differ between their use as wildcards and their use in regular expressions. So we need to understand the context in which the symbol is being used. For instance, the *, when used as part of a regular expression in grep differs from the use of * when it is used in the ls instruction. We will consider this difference in more detail in Section 6.3.

### 6.2.1 Controlling Repeated Characters through *, +, and ?

The first three metacharacters from Table 6.1 are all used to express a variable number of times that the character that precedes the metacharacter can appear. With *, the preceding character can appear 0 or more times. With +, the preceding character can appear 1 or more times. With ?, the preceding character can appear 0 or 1 times exactly.

Consider the following strings:

1. aaaabbbbcccc

2. abcccc

3. accccc

4. aaaaaabbbbbb

TABLE 6.1    Regular Expression Metacharacters

| Metacharacter | Explanation |
| --- | --- |
| * | Match the preceding character if it appears 0 or more times |
| + | Match the preceding character if it appears 1 or more times |
| ? | Match the preceding character if it appears 0 or 1 times |
| . | Match any one character |
| ^ | Match if this expression begins a string |
| $ | Match if this expression ends a string |
| [chars] | Match if the next character in the string contains any of the chars in [] |
| [char$_i$-char$_j$] | Match if the next character in the string contains any characters in the range from char$_i$ to char$_j$ (e.g., a-z, 0–9) |
| [[:class:]] | Match if the next character in the string is a character that is part of the :class: specified. The :class: is a category like upper-case letters (:upper:) or digits (:digit:) or punctuation marks (:punct:). Table 6.2 lists the classes available |
| [^chars] | Match if the next character in the string is not one of the characters listed in [] |
| \ | The next character should be interpreted literally, used to escape the meaning of a metacharacter, for instance \$ means "match a $" |
| {n} | Match if the string contains n consecutive occurrences of the preceding character |
| {n,m} | Match if the string contains between n and m consecutive occurrences of the preceding character |
| {n,} | Match if the string contains at least n consecutive occurrences of the preceding character |
| \| | Match any of these strings (an "OR") |
| (…) | The items in … are treated as a group, match the entire sequence |

The regular expression a*b*c* would match all four strings because this regular expression will match any string that has 0 or more a's followed by 0 or more b's followed by 0 or more c's. String #3 has no b and string #4 has no c but * permits this as "0 or more."

The regular expression a+b+c+ would only match the first two because each letter, a, b, and c, must appear at least one time. The regular expression a?b?c* will match the second and third strings because the letters a and b must appear 0 or 1 times apiece and the letter a appears once in each and the letter b appears once in the second string and 0 times in the third string. The first and fourth strings have too many occurrences of a and b to match a?b?c*.

Notice in the previous examples, order must be maintained. The string aaaaacccccbbbbbb would not match a+b+c+ because the b's must appear before the c's. Would a*b*c* match this new string? It seems counter-intuitive to say yes it matches because the c's appear before the b's in the string. But in fact, a*b*c* does match for a reason that is not obvious. When we use the * metacharacter, we are saying that the string must contain "0 or more" instances of the given literal character. In this case, we are requiring that the string contain at least 0 a's, at least 0 b's, and at least 0 c's, in that order. It does because the string contains for instance aaaaa which means that the a* portion matched. The b* and c* portions also match because immediately following the a's are 0 b's and 0 c's. In fact, when we use the * metacharacter, we have to be careful because, being able to interpret it as "0 or more" means it can match just about anything and everything. The same can be said of the ? metacharacter as it can also match 0 of the preceding literal characters. In Section 6.2.3, we will see how to better control our regular expressions.

## 6.2.2  Using and Modifying the '.' Metacharacter

The '.' metacharacter is perhaps the easiest to understand. It is a true wildcard metacharacter meaning that it will match any single character. So, b.t will match any string that has 'b,' something, 't,' no matter what that something is. The . will match any character no matter if that character is a letter, digit, punctuation mark, or white space. Consider the regular expression . . . which will match any three-character sequence, whether it is abc, a b, 123, 3*5, or ^%#.

We can combine the . metacharacter with the previously described three metacharacters *, +, and ?. If *, +, or ? appear after the . then the *, +, or ? modifies the . itself. For instance, .* means "0 or more of any character" whereas .? means "0 or 1 of any character."

The expression b.?t means 'b' followed by 0 or 1 of anything followed by 't.' This would match bit, bat, but, b3t, b+t, as well as bt in which case the ? is used to match against 0 characters. If we use .+ instead, we are requiring that there be one or more of any character. If there is more than one character in the sequence, those characters do not have to be the *same* character. For instance, b.+t will match a 'b' followed by any sequence of characters ending with a 't' including any of the following:

bit    boat   baaaaat      b1234t        b+13*&%3wert

but not bt. The regular expression b.*t allows there to be 0 or more instances of characters and so it is the same as b. +t except that bt also matches.

How do the regular expressions

b*t    b+t    b?t

differ from the expressions

b.*t   b. +t   b.?t

In the former three cases, the metacharacters *, + , and ? modify the b in the expression while in the latter three cases, the metacharacters *, + , and ? modify the . in the expressions. In the first three expressions, only the 'b' is variable in nature. These three expressions match against any number of b's followed by a t, at least one b followed by a t, and 0 or 1 b followed by a t respectively. In the latter three cases, the *, +, and ? are applied to the . so that there are any number of characters between the b and t, at least one of any characters between the b and t, and either 0 or 1 character between the b and t respectively. So, for instance, b*t will match any of bbbbbt, bt, bbbbbbbbt, and t (0 b's) while b.*t will match babcdt, b12345678t, bbbbbt, and bt.

### 6.2.3 Controlling Where a Pattern Matches

Let us reconsider a problem that we mentioned earlier. Recall that a regular expression will match a substring of a string. Now consider the string bbbbbbbattttttt. The regular expression b?.t+ will match this string even though it only matches a substring. See Figure 6.1 where we can see that b?.t+ matches the last b in the string (b? requires 0 or 1 b's), followed by any single character (the a), followed by 1 or more t's. Even though the string has seven b's, an a, and seven t's, the b? does not have to match all seven b's. The period matches the single a and the t+ matches the 7 t's. We do not require that the regular expression match the *entire* string. Note also in the figure is ^b?.t+. We describe this later.

Unfortunately, the lack of control of where an expression matches can lead to a significant problem. What if we *want* a regular expression to match the entire string? How do we specify that there must not be multiple b's? Before we describe how we can do this, let us define what we mean by *substring* and *string*. We will consider a string to be a line of text. A substring is any (0 or more) consecutive characters of a string. The line of text that makes up our string will typically be one line of a textfile. The line ends with an end-of-line character, often represented as \n. We write a regular expression to match any portion of that string. If there is a match, then we match the string on the line. If we want our expression to match the entire line (or string), we have to specify this in our expression.

By this new definition, a regular expression that matches the empty string (no characters at all) will match that line. For instance, the expression a* will match 0 or more a's, so if there are no a's on the line, it still matches.

In fact, the regular expression a* would match any line. This is why it is important to control *where* a regular expression is allowed to match within a line. This leads us to the next two metacharacters: ^ and $. The ^ metacharacter indicates that the regular expression can only match from the start of the string (the beginning of the line). The $ metacharacter indicates that the regular expression can only match at the end of the string.

If, for instance, we have the expression ^b?.t+ (the earlier expression with a ^ added to the front), then this will match a string that *starts with* 0 or 1 b followed by anything followed by 1 or more t's. This will not match the earlier string of bbbbbbbattttttt. Why not? The ^b forces the regular expression to begin matching this string from the beginning. Therefore, b?. will match the first b (0 or 1 b) and the second b (the period matches any single character) but now the regular expression expects the remainder of the string to be one or more t's and since we have another b, there is no match. Refer back to Figure 6.1 where you can see that the t+ portion of the regex fails to match.

To obtain a complete match of a regular expression to a string, use both ^ and $. For instance, ^b?.t+$ will match any string that starts with 0 or 1 b, has any character, and

FIGURE 6.1  Controlling where an expression matches.

ends with 1 or more t's. So this regex will match the following strings: batttttttt, bbtttttt, and atttttttt (in this case there are 0 b's).

Let us examine some more examples using ^ and/or $. We start with the regex ^0a*1+. This regex says that to match a string, it must start with a 0. This is followed by 0 or more a's followed by 1 or more 1's. This string will therefore match any of 0a1, 0aaaa1111, 011111, and 01. It can also match strings where there are characters after the 1 because we are not restricting this regex to "end the string" using $. So it will also match 0a1a and 011110. What it cannot match are strings like 0a (there must be at least one 1), 0000aaaa1111 (after the single 0 at the start). We must see a's if any and at least one 1 before any other characters or aaaa1111 because there must be a 0 to start.

Now consider the variation ^0a*1+$. With the closing $, we are stating that a string must not contain any characters after the 1 (or 1's). Thus, 0aaaa1111 and 01 will match but 0a1a and 011110 will not.

One last example to finish this subsection is the regex ^$. Since the ^ indicates that the regex must match at the start of the string and the $ indicates that the regex must match at the end of the string, this regex must match the full string. But as there are no literal characters or metacharacters between the two, this says "match any string that has nothing in it" which is known as the *empty string*. Thus, this regex only matches the empty string.

## 6.2.4 Matching from a List of Options

So far, our expressions have allowed us to match against strings that have a variable number of characters, but only if the characters appear in the order specified. For instance, a+b+c+ can match any number of a's, b's, and c's, but only if they appear in that order. What if we want to match any string that contains any number of a's, b's, and c's, but in no particular order? If we want to match any three character sequence that consists of only a's, b's, and c's, as in abc, acb, bca, and so forth, we will need an additional metacharacter.

The [ ] metacharacters, often referred to as brackets, straight brackets, or braces, allow us to specify a list of options. The list indicates that the *next* character in the string can match *any single character* in the list.

Inside of the brackets we can specify characters using one of three notations: an enumerated list as in [abcd], a range as in [a-d], or a *class* such as [[:alpha:]]. The class :alpha: represents all alphabetic characters. Obviously we would not use :alpha: if we only wanted to match against a subset of letters like a-d. Alternatively, we could use a-zA-Z to indicate all letters. Notice when describing a class, we use double brackets instead of single brackets.

Consider the regular expression [abc][abc][abc]. This expression will match any string that contains three consecutive characters that are a's, b's, or c's in any combination. This expression will match abc, acb, and bca. And, because we are not restricting the number of times any character appears, it will also match aaa, bbb, aab, aca, and so forth. We could also use a range to define this expression, as in [a-c][a-c][a-c].

We can combine [ ] with *, +, and ? to control the number of times we expect the characters to appear. For instance, [abc]+ will match any string that contains a sequence of 1 or more characters in the set a, b, c while [abc]* will also match the empty string. In this latter case, we actually have a regular expression that will match anything because any string can contain 0 a's, b's, and c's. For instance, 12345 contains no a's, b's, or c's, and so it can match [abc]* when * is interpreted as 0.

Now we have a means of expressing a regular expression where order is not important. The expression [abc]+ will match any of these four strings that we saw earlier that matched a*b*c*:

- aaaabbbbcccc

- abccccc

- acccccc

- aaaaaabbbbbb

This expression will also match strings like the following.

- abcabcabcabc

- abacab

- aaaaaccccc

- a

- cccccbbbbbbaaaa

We can combine any characters in the brackets as in [abcxyz], [abcd1234], or [abcdABCD]. If we have a number of characters to enumerate, a range is more practical. We would certainly prefer to use a range like [a-z] than to list all of the letters. We can also combine ranges and enumerations. For instance, the three sequences above could also be written as [a-cx-z], [a-d1-4], and [a-dA-D] respectively. Now consider the list of all lower case consonants. We could enumerate them all as [bcdfghjklmnpqrstvwxyz] or we could use several ranges as in [b-df-hj-np-tv-z].

While we can use ranges for letters and digits, there is no range available for the punctuation marks. You could enumerate all of the punctuation marks in brackets to capture "any punctuation mark" but this would be tedious. Instead, we also have a class named :punct: which is applied in double brackets, as in [[:punct:]]. Table 6.2 provides a listing of the classes available in Linux.

Let us now combine all of the metacharacters we have learned with some examples. We want to find a string that consists only of letters. We can use ^[a-zA-Z]+$ or ^[[:alpha:]]+$. The ^ and $ force the regex to match an entire string. Thus, any string that contains nonletters will not match. If we had used only [a-zA-Z]+, then it could match any string that contains letters but could also have other characters that precede or succeed the letters such as abc123, 123abc, abc!def, as well as ^#!$a*%&. Why do we use the + in this regex? If we had used *, this could also match the empty string, that is,

TABLE 6.2    Classes Defined for Regular Expressions

| Class | Meaning |
|---|---|
| :alnum: | Any letter or digit |
| :alpha: | Any letter |
| :blank: | Space and tab |
| :cntrl: | Any control character |
| :digit: | Any digit |
| :graph: | Any nonwhitespace |
| :lower: | Lower-case letter |
| :print: | Any printable character |
| :punct: | Any punctuation mark including [ and ] |
| :space: | Any whitespace (space, tab, new line, form feed, carriage return) |
| :upper: | Any upper-case letter |
| :xdigit: | Any digit or hexadecimal digit |

a string with no characters. The + insists that there be at least one letter and the ^ and $ insist that the only characters found are letters.

We could similarly match a string of only binary digits. Binary digits are 0 and 1. So instead of [a-zA-Z] or [[:alpha:]], we use [01]. The regex is ^[01]+$. Again, we use the ^ and $ to force the expression to match entire strings and we use + instead of * to disallow the empty string. If we wanted to match strings that comprised solely digits, but any digits, we would use either ^[0-9]+$ or ^[[:digit:]]+$.

If we want to match a string of only punctuation marks, we would use ^[[:punct:]] + $. Unlike the previous examples, we would not use [...] and enumerate the list of punctuation marks. Why not? There are too many and we might (carelessly) miss some. There is no range to indicate all punctuation marks, such as [!-?], so we must either list them all, or use :punct:.

If we want to match a string that consists only of digits and letters where the digits precede the letters, we would use ^[0-9] + [[:alpha:]] + $. If we wanted to match a string that consists only of letters and digits where the first character must be a letter and then can be followed by any (0 or more) letters and digits, we would use ^[[:alpha:]][0-9a-zA-Z]*$.

### 6.2.5 Matching Characters That Must Not Appear

In some cases, you will have to express a pattern that seeks to match a string that does not contain specific character(s). We might want to match a string that has no blank spaces in it. You might think to use [. . .]+ where the . . . is "all characters except the blank space." That would require enumerating quite a list as it would have to include every letter, every digit, and every punctuation mark. In such a case, we would prefer to indicate "no space" by using the notation [^ ]. The ^, when used inside of [] means "do not match" against the characters listed in the brackets. The blank space after ^ indicates that the only character we do not want to match against is the blank.

Unfortunately, our regex [^ ] will have the same flaw as earlier expressions in that if it locates any single nonblank character within the string, it is a match to the string. If our string is "hi there," the [^ ] regex will match the 'h' at the beginning of the string because it is not a blank, and so [^ ] matches the string. To express that the string should contain no blanks, we need to first indicate that we are matching the entire string using ^ and $, and then we have to indicate that there can be many nonblanks using either * or +. We could solve this problem through the regex ^[^ ]+$. This states that the entire string, from start to finish, must contain characters that are not blanks. If you want to also match the empty string, use ^[^ ]*$.

### 6.2.6 Matching Metacharacters Literally

All of our metacharacters are punctuation marks. What happens when the regular expression requires one of these punctuation marks as a literal character? For instance, you want to match a number that contains a decimal point. Your first inclination might be to express this as [0-9]+.[0-9]+, meaning that the number has some digits followed by a period followed by some digits. But recall that the decimal point or period (.) means "any character." This would cause the regular expression to match 123.567 but also 123a567 as well as 1234567 where the period matches the 'a' and '4' respectively. To match a punctuation mark that is one of the metacharacters, we could use [[:punct:]], but this would match any punctuation mark, not the one(s) we expect.

There are two solutions to this problem. First, we could specify the punctuation mark inside of [] as in [.]. This will work for most punctuation marks although it would not work for [ or ]. Alternatively, we can specify that the punctuation mark should be treated literally and not as a metacharacter. To accomplish this, we precede the punctuation mark with a \ character. The \ means "treat the next character literally," or "escape the meaning of the next character." This is why the \ is known as an *escape* character. Using \, we see that [0-9]+\.[0-9]+ forces the period to be treated literally rather than as a metacharacter. So this revised regex will match any sequence of one or more digits followed by a period followed by one or more digits.

For another example, let us define an expression to match a simple multiplication problem of the form 3*4=12. Recall that the * means "0 or more occurrences," but here we would want the * to be treated as a literal character. We could define our regular expression in a couple of different ways:

```
[0-9]+[*][0-9]+=[0-9]+
[0-9]+\*[0-9]+=[0-9]+
```

The above expressions define a pattern of "at least one digit followed by an asterisk followed by at least one digit followed by an equal sign followed by at least one digit." Notice that there is no way to ensure that the multiplication problem is correct, so this expression can match 3*4 = 7 as well as 3*40 = 12. The following regular expression would not work because the * is being used to modify the +, which is not legal syntax.

```
[0-9]+*[0-9]+=[0-9]+
```

NOTE: since = is not a metacharacter, we do not need to precede it with \ although \= is equivalent to = (i.e., the \ does not impact the =).

We can omit the \ when expressing some of the metacharacters under limited circumstances. Since the $ is used to indicate "ends the expression," if a $ does not appear at the end of the regular expression, it is treated literally and therefore does not require the \. Similarly, the ^ is expected to appear in only two positions, at the beginning of an expression or as the first character inside the []. If a ^ appears anywhere else, it is treated literally. On the other hand, the characters [ and ] must be preceded by the \ if you wish to treat them literally. This is unlike other metacharacters which could be expressed in [] because we would not be able to specify [[]] to indicate literal [ or ] marks.

## 6.2.7 Controlling Repetition

Through the use of * and +, we can specify repetition of one or more characters. But * and + may not be precise enough for circumstances where we want to *limit* the number of repetitions. To control the number of repetitions expected, we use an additional set of metacharacters, {}, sometimes called the curly brackets or curly braces. There are three ways that we can apply the curly brackets as shown below. In each, n and m are integer values.

- $\{n\}$ – exactly n occurrences

- $\{n, m\}$ – between n and m occurrences where n < m

- $\{n,\}$ – at least n occurrences

For instance, [0-9]{5} will match exactly five digits while [0-9]{5,} will match five or more digits. The notation [0-9]{1,5} will match anywhere from one to five digits.

Keep in mind that the regular expression can match any substring of a string. Consider the following string:

```
1234567890abc
```

The regular expression [0-9]{5} will match this string because there is a substring that contains exactly five digits. In fact, [0-9]{5,} will also match this string. To define a regular expression that contains exactly five consecutive digits, we would have to include ^ and $. However, ^[0-9]{5}$ will *only* match a string that is exactly five digits long. It would not match the string above. If we wish to match strings that contain five consecutive digits but can have other, nondigit, characters, we have to enhance our regular expression.

The expression [^0-9]*[0-9]{5}[^0-9]*says "match a string that contains of 0 or more nondigits followed by five digits followed by 0 or more nondigits." This regular expression will match any of the following strings:

- abc12345abc

- 12345abc

- abc12345

- 12345

The expression would not match 1234567 or abc123def because neither of these strings contains exactly five digits in sequence. It would match 1ab23456de7 though because the five digits are surrounded by nondigits. How does ^[^0-9]*[0-9]{5}[^0-9]*$ differ?

## 6.2.8 Selecting between Sequences

Now that we can control the exact number of repetitions that we expect to see, let us define a regular expression to match a zip code. If we consider a five-digit zip code, we can use [0-9]{5}. If we want to match a five-digit zip code that is in a string that contains other characters, we might use the previously defined expression ^[^0-9]*[0-9]{5}[^0-9]*$. In fact, if we know that the zip code will always follow a two-letter state abbreviation followed by a blank space, we could be more precise, as in [A-Z]{2} [0-9]{5}[^0-9]*$.

We also have nine-digit zip codes. These are zip codes that consist of five digits, a hyphen, and four digits, as in 12345-6789. We would define this sequence as [0-9]{5}-[0-4]{4}. Now we have a new problem. Which expression should we specify? If we specify both, as in

[0-9]{5} [0-5]{5}-[0-9]{4}

we are stating that the string must have a five-digit sequence followed by a space followed by a five-digit sequence, a hyphen, and a four-digit sequence. We need to be able to say "or." Recall from earlier that we were expressing "or" using [list]. However, the items in [] indicated that any single character should match, not that we want to match an entire sequence.

We use another metacharacter to denote "or" in the sense that we have two or more expressions and we want to match either (any) expression against a string. This metacharacter is | (the vertical bar). Now we can express a zip code using the following.

[0-9]{5}|[0-5]{5}-[0-9]{4}

In the above expression, the | appears between the two definitions: the five-digit zip code and the nine-digit zip code. That is, the regex will match a five-digit number OR a five-digit number followed by a hyphen followed by a four-digit number.

Let us consider another example. The Cincinnati metropolitan region extends into three states, Ohio, Kentucky, and Indiana. If we want to define a regular expression that will match any of these three states' abbreviations, our first idea might be to express this as [IKO][NYH]. This will match any of IN, KY, and OH, so it seems to solve the problem. However, there is no way to control the ideas that "if the first character in the first list matches, then only use the first character in the second list." So this expression could also match any of IY, IH, KN, KH, ON, or OY. By using the | we can avoid this problem through IN|KY|OH.

The final metacharacters are the parentheses, (). These are used when you want to encapsulate an entire pattern of metacharacters, literal characters, and enumerated lists inside another set of metacharacters. This allows you to state that the trailing metacharacter applies to the entire pattern rather than the single preceding character.

For instance, we want to match against a list of words. Words will consist of either capitalized words or lower case words and will be separated by spaces. A single word is indicated using `[A-Za-z][a-z]*`, that is, any upper or lower-case letter followed by 0 or more lower-case letters. To express the blank space, we will follow the above pattern with a blank space, giving us '`[A-Za-z][a-z]* `'. The quote marks are shown to clearly indicate the blank space. Now, to express that there are several of these words in the string, we will want to add `{2,}`. However, if we place the `{2,}` after the blank space, it will only modify the blank space. We instead want the `{2,}` to modify the entire expression. Therefore, we place the expression in () giving us `([A-Za-z][a-z]* ){2,}`. Now we have an expression that will match two or more sequences of "upper- or lower-case letter followed by 0 or more lower-case letters followed by a blank."

If we expect to see between two and five words in the string, we would express this as

```
([A-Za-z][a-z]* ){2,5}
```

To ensure that the two to five words makes up the entire string, we might enclose the expression within `^` and `$` marks as in

```
^([A-Za-z][a-z]* ){2,5}$
```

However, there is a flaw in our expression. We might assume that the final word in the string does not end in a blank space. How can we say "two to five words *separated* by blank spaces?" The word "separated" means that there is a blank between words, but not before the first or after the last. What we need here is to present the last word as a "special case," one that does not include the space. How?

```
^([A-Za-z][a-z]* ){1,4}[A-Za-z][a-z]*$
```

Now our expression says "one to four words followed by spaces followed by one additional word." Thus, our words are separated by spaces but there is no space before the first or after the last words.

## 6.3 EXAMPLES

In this section, we reinforce the discussion of Section 6.2 by providing numerous examples and descriptions.

- `[0-9]+` Match if the string contains at least one digit.

- `^[0-9]+` Match if the string starts with at least one digit.

- `^[0-9]+$` Match if the string only consists of digits. The empty string will not match. Use `*` in place of `+` to also match the empty string.

- `[b-df-hj-np-tv-z]+[aeiou][b-df-hj-np-tv-z]+` Match a string that includes at least one vowel that is surrounded by at least one consonant on each side. Add `^` and `$` around the regex to specify a string that consists exactly of one vowel with consonants on either side.

- `[A-Z]{4,}` Match if the string contains at least four consecutive upper-case letters.

- `[A-Z]{4}` Although this appears to say "match if the string contains exactly four consecutive upper-case letters," it will in fact match the same as the preceding example because we are not forcing any characters surrounding the four upper-case letters to be nonupper case characters.

- `[^A-Z][A-Z]{4}[^A-Z]` Match if the string contains exactly four upper-case letters surrounded by other characters. For instance, this would match "abcDEFGhi" and "Hi There FRED, how are you?" but not "abcDEFGHijk." It will also not match "FRED" because we are insisting that there be nonupper-case letters around the four upper-case letters. We can fix this as shown below.

- `[^A-Z][A-Z]{4}[^A-Z]|^[A-Z]{4}$`

- `^$` Match only the empty string (blank lines).

- `...` Match a string that contains at least three characters of any type. Add `^` and `$` to specify a regex that matches a string of exactly three characters.

- `[Vv].?[Ii1!].?[Aa@].?[Gg9].?[Rr].?[Aa@]` This regex might be used in a spam filter to match the word "Viagra" along with variations. Notice the `.?` used in the regex. This states that between each letter of the word viagra, we can accept another character. This could account for such variations as `Via!gra` or `V.I.A.G.R.A.` The use of 1, !, @, and 9 are there to account for variations where these letters are replaced with look-alike characters, for instance @ for a.

- `([A-Z][[:alpha:]]+ )?[A-Z][[:alpha:]]+, [A-Z]{2} [0-9]{5}$` This regex can be used to match the city/state/zip code of a US postal address. First, we expect a city name. A city name should appear as an upper-case letter followed by additional letters. The additional letters may include upper-case letters as in McAllen. Some cities are two names like Los Angeles. To cover a two-word city, we expect a blank space and another name. Notice the `?` that follows the close parenthesis to indicate that we would expect to see this one or zero times. So we either expect a word, a space and a word, or just a word. This is followed by a comma and space followed by two upper-case letters to denote the state abbreviation and a space and five digits to end the line. We might expect two spaces between state and zip code. We need to include an optional space. We could use either of `[]{1,2}` or `[][]?`. We can also add `(-[0-9]{4})?` to indicate that the four-digit zip code extension is optional.

- [A-Za-z_] [A-Za-z_0-9]* In most programming languages, a variable's name is a collection of letters, digits, and underscores. The variable name must start with a letter or underscore, not a digit. In some languages, variable names can also contain a dollar sign, so we can enhance our regex by adding the $ character in the second set of brackets. In some languages, variable names are restricted in length. For instance, we might restrict variables to 32 or fewer characters. To denote this, we can replace the * with {0,31}. We use 31 instead of 32 because we already have one character specified. Unfortunately, this would not prevent our regex from matching a 33-character variable name because we are not specifying that the regex not match 33 characters. We could resolve this by placing delimiters around the regex. There are a number of delimiters such as spaces, commas, semicolons, arithmetic symbols, and parentheses. Instead, we could also state that before and after the variable name we would not expect additional letters, digits, or underscores. So we could improve our regex above to be

- [^A-Za-z_0-9][A-Za-z_][A-Za-z_0-9]{0,31}[^A-Za-z_0-9]

- ([([] [0-9]{3}[)] )?[0-9]{3}-[0-9]{4} In this expression, we describe a US phone number. A phone number consists of three digits, a hyphen, and four digits. If the number is long distance, we include the area code before the number. The area code is three digits enclosed in parentheses. For instance, a phone number can be 555-5555 or (123) 555-5555. If the area code is included, after the close paren is a space. In the above regex, the two parens are placed in [] to differentiate the literal paren from the metacharacter paren as used to indicate a sequence. We could have also used \( and \). Some people will write the 10-digit phone number (with area code) without the parens. We can include this by adding a ? after each paren as in [(]? and [)]? however, this would cause the regex to match if only one of the two parens are supplied as in (123 555-5555. Alternatively, we can provide three different versions of the regex with an OR between them as in

- [(][0-9]{3}[)] [0-9]{3}-[0-9]{4}|[0-9]{3} [0-9]{3}-[0-9]{4}|[0-9]{3}-[0-9]{4}

- [0-9]+(.[0-9]+)? This regex will match a numeric value with or without a decimal point. We assume that there must be at least one digit and if there is a decimal point, there must be at least one digit to the right of the decimal point. By placing the ? after the sequence of period and a digit, we are saying that if one appears the other must appear. This allows us to have 99.99 or 0.0 but not 0. with nothing after the decimal point.

- $[0-9]+\.[0-9]{2} Here, the $ indicates that we seek a dollar sign and not "end of string." This is followed by some number of digits, a period, and two digits. This makes up a dollar amount as in $123.45. We have three problems with this regex if we want to match any dollar amount. First, we are discounting a dollar amount that has no cents such as $123. Second, the regex would not prevent a match against something like $123.45678. We would not expect to see more than two digits after the decimal point but our regex does not prevent this. Finally, if the dollar amount contains commas, our regex will not match. To resolve the first problem, we provide two versions:

- `$([0-9]+|[0-9]+\.[0-9]{2})` Now we can match either a dollar sign and digits or a dollar sign, digits, a period, and two digits. To resolve the second problem, we have to embed our regex such that it is not followed by digits.

- `$([0-9]+|[0-9]+\.[0-9]{2})[^0-9]` This in itself would require that a dollar amount not end a line. So we could enhance it as follows:

- `$([0-9]+|[0-9]+\.[0-9]{2})[^0-9]|$([0-9]+|[0-9]+\.[0-9]{2})$` Although this looks a bit bizarre with two dollar signs around the latter portion of the regex, the first is treated literally and the last means "end of string." The commas can be more of a challenge and this is left to an end of chapter exercise.

Now that we have introduced regular expressions (and no doubt confused the reader), we will examine in the next three sections how to use regular expressions in three common pieces of Linux software, grep, sed, and awk. Numerous examples will be offered which will hopefully help the reader understand regular expressions.

## 6.4 GREP

The name grep comes from *global regular expression print*. The idea behind grep is to search one or more files line-by-line to match against a specified regular expression. In this case, the string being matched against is the entire line. If the regular expression matches any part of the line, the entire line is returned. The grep program can search multiple files and will return all matching lines. This is a convenient way to search files for specific types of content.

### 6.4.1 Using grep/egrep

The format for grep is

```
grep [options] regex file(s)
```

The *regex* is any regular expression that does not include the extended set of metacharacters (such as {}). To use the extended regular expression set, either use grep –E or egrep. The –E denotes "use egrep," so they are identical. The egrep program can apply all metacharacters, so it is preferred to always use egrep (or grep –E). We will examine the other options later in this section.

In addition, it is good practice to place your regex inside of ' ' as in egrep 'regex' file(s). This is to prevent confusion between the metacharacters that can appear as wildcards, such as * and +. Recall that the Bash interpreter performs filename expansion before executing an instruction. This could result in an * or + being applied by the Bash interpreter, replacing the wildcard with a list of matching files, and thus not being part of the regular expression. We will examine this problem in more detail at the end of this section. For now, all regular expressions will be placed inside of single quote marks.

We have a series of files, some of which include financial information. Assume that such files contain dollar amounts in the form $number.number as in $1234.56 and $12.00. If we want to quickly identify these files, we might issue the grep instruction

```
egrep '$[0-9]+\.[0-9]{2}' *
```

As another example, we want to locate files that contain an address. In this case, let us assume the address will contain the zip code 41099. The obvious instruction is

```
egrep '41099' *
```

However, this instruction will match any file that contains that five-digit sequence of numbers. Since this is part of an address, it should only appear within an address, which will include city and state. The 41099 zip code is part of Highland Heights, KY. So we might further refine our regular expression with the following instruction:

```
egrep 'Highland Heights, KY 41099' *
```

In this case though, we might have too precise a regular expression. Perhaps some addresses placed a period after KY and others have only one space after KY. In yet others, Highland Heights may not appear on the same line. We can resolve the first two issues by using a list of "or" possibilities, as in KY |KY |KY\. and we can resolve the second problem by removing Highland Heights entirely. Now our instruction is

```
egrep '(KY |KY|KY\.) 41099' *
```

The use of the () makes it clear that there are three choices, KY with a space, KY without a space, and KY with a period, followed by a space and 41099.

Let us now turn to a more elaborate example. Here, we wish to use egrep to help us locate a particular file. In this case, we want to find the file in /etc that stores the DNS name server IP address(es). We do not recall the file name and there are hundreds of files in this directory to examine. The simple approach is to let egrep find any file that contains an IP address and report it by name. How do we express an IP address as a regular expression?

An IP address is of the form 1.2.3.4 where each number can range between 0 and 255. We will want to issue the egrep command:

```
egrep 'regex-for-ip-address' /etc/*
```

where *regex-for-ip-address* is the regular expression that we come up with that will match an IP address. The instruction will return all matching lines of all files. Included in this list should be the line(s) that matched the file that we are looking for (which is resolv.conf).

An IP address (version 4) consists of four numbers separated by periods. If we could permit any number, the regular expression could be

```
[0-9]+.[0-9]+.[0-9]+.[0-9]+
```

However, this is not accurate because the . represents any character. We really mean "a period" so we want the period interpreted literally. We need to modify this by using \. or [.]. Our regular expression is now

```
[0-9]+\.[0-9]+\.[0-9]+\.[0-9]+
```

The above regular expression certainly matches any IP address, but in fact it can match against any four numbers that are separated by periods. The four numbers that make up the IP address must be within the range of 0–255. How can we express that the number must be no greater than 255? Your first thought may be to use [0-255]. Unfortunately, that does not work nor does it make sense. Recall that in [] we enumerate a list of *choices* to match *one* character in the string. The expression [0-255] can match one of three different sets of single characters: a character in the range 0–2, the character 5, and the character 5. This expression is equivalent to [0-25] or [0125] as that second 5 is not needed. Obviously, this expression is not what we are looking for.

What we need to do is express that the item to match can range from 0, a single digit, all the way up to 255, three digits long. How do we accomplish this? Let us consider this solution:

```
[0-9]{1,3}
```

This regular expression will match any sequence of one to three digits. This includes 0, 1, 2, 10, 11, 12, 20, 21, 22, 99, 100, 101, 102, 201, 202, and 255 so it appears to work. Unfortunately, it is too liberal of an expression because it also matches 256, 257, 258, 301, 302, 400, 401, 500, 501, 502, 998, and 999 (as well as 000, 001, up through 099), none of which are permissible as parts of an IP address.

If we do not mind our regular expression matching strings that it should not, then we can solve our problem with the command

```
egrep '[0-9]{1,3}\.[0-9]{1,3}\.[0-9]{1,3}\.[0-9]{1,3}' /etc/*
```

We are asking grep to search all of the files in /etc for a string that consists of 1–3 digits, a period, 1–3 digits, a period, 1–3 digits, a period and 1–3 digits. This would match 1.2.3.4 or 10.11.12.13 or 172.31.185.3 for instance, all of which are IP addresses. It would also match 999.998.997.996 which is not an IP address.

If we want to build a more precise regular expression, then we have some work to do. Let us consider again the range of possible numbers in 0–255. First, we have 0–9. That is easily captured as [0-9]. Next, we have 10–99. This is also easily captured as [1-9][0-9]. We could use the expression [0-9]|[1-9][0-9].

What about 100–255? Here, we have a little bit more of a challenge. We cannot just use [1-2][0-9][0-9] because this would range from 100 to 299. So we need to enumerate even more possible sequences. First, we would have 1[0-9][0-9] to allow for any value from 100 to 199. Next, we would have 2[0-4][0-9] to allow for any combination from 200 to 249. Finally, we would have 25[0-5] to allow for the last few combinations, 250–255.

Let us put all this together to find a regex for any legal IP address octet. Combining the parts described above, we have

[0-9]|[1-9][0-9]|1[0-9][0-9]|2[0-4][0-9]|25[0-5]

This gives us each of these options

- 0–9        [0–9]
- 10–99      [1–9][0–9]
- 100–199    1[0–9][0–9]
- 200–249    2[0–4][0–9]
- 250–255    25[0–5]

Now we can get back to our regular expression. The expression must express the above list of options, followed by a period, followed again by the list of options and a period, followed again by the list of options and a period, followed finally by the list of options one more time. Thus, we want to have the list of options followed by a period occur three times exactly, followed by the list of options a fourth time. This can be represented as

(([0-9]|[1-9][0-9]|1[0-9][0-9]|2[0-4][0-9]|25[0-5])\.){3}
([0-9]|[1-9][0-9]|1[0-9][0-9]|2[0-4][0-9]|25[0-5])

Now our grep instruction is

```
egrep '(([0-9]|[1-9][0-9]|1[0-9][0-9]|2[0-4][0-9]|
25[0-5])\.){3}([0-9]|[1-9][0-9]|1[0-9][0-9]|2[0-4]
[0-9]|25[0-5])'/etc/*
```

## 6.4.2 Useful egrep Options

If you were to issue the previously defined egrep instruction to search for IP addresses among the files of /etc, you would obtain many responses. Of course the file we were looking for was resolv.conf, but another file, ntp.conf would also have matching lines. Let us examine the output from the egrep statement. We receive output like that of Figure 6.2.

```
networks:loopback 127.0.0.0
networks:link-local 169.254.0.0
ntp.conf:restrict 127.0.0.1
ntp.conf:#restrict 192.168.1.0 mask 255.255.255.0 nomodify notrap
ntp.conf:#broadcast 192.168.1.255 autokey # broadcast server
ntp.conf:#broadcast 224.0.1.1 autokey     # multicast server
ntp.conf:#multicastclient 224.0.1.1       # multicast client
ntp.conf:#manycastserver 239.255.254.254  # manycast server
ntp.conf:#server  127.127.1.0 # local clock
openct.conf:       # >= linux-2.6.27.14
openct.conf:       # >= linux-2.6.28.3
pam_ldap.conf:host 127.0.0.1
pam_ldap.conf:#uri ldap://127.0.0.1/
pam_ldap.conf:#uri ldaps://127.0.0.1/
Binary file prelink.cache matches
resolv.conf:nameserver 172.28.102.11
resolv.conf:nameserver 172.28.102.13
```

FIGURE 6.2   Partial Result of egrep Searching for IP Addresses.

Notice among the responses shown in Figure 6.1 that many files contain IP addresses (many of which are for localhost, 127.0.0.1) or define netmasks (we discuss netmasks in Chapter 12). One interesting match is for the file prelink.cache. This file is a binary file and really should not have matched. To ignore binary files, we add the option –I.

There are many other useful options available in grep. The option –c responds with the total number of matches rather than the matches themselves. If grep executes on a number of files, each filename is listed along with the number of matches in that file, including 0 if no lines in the file matched the regular expression.

The –n option will insert line numbers before the file name. If we wish to discard the filenames, we can use –h. The option –H forces grep to display the filenames. The option –h is used by default if grep is only searching a single file while –H is used by default if grep is searching multiple files. Figure 6.3 shows the output from the same command as shown in Figure 6.2, but with the –H, –I, and –n options applied.

The option –i forces egrep to ignore case. This allows you to specify letters in your regular expression without having to differentiate upper versus lower case. For instance,

```
      egrep '[ABCabc]' somefile
```
and   `egrep -i '[abc]' somefile`

will accomplish the same task.

Recall the challenge of properly using [^...] to denote "NOT." A far easier solution is to use the –v option, which means "invert the match." For instance, if we want to search for all strings that do not contain a period, we could try to write a regular expression that conveys this, but we could also use the command

`egrep -v '\.' somefile`

```
2:loopback 127.0.0.0
3:link-local 169.254.0.0
14:restrict 127.0.0.1
18:#restrict 192.168.1.0 mask 255.255.255.0 nomodify notrap
26:#broadcast 192.168.1.255 autokey # broadcast server
28:#broadcast 224.0.1.1 autokey          # multicast server
29:#multicastclient 224.0.1.1            # multicast client
30:#manycastserver 239.255.254.254       # manycast server
31:#manycastclient 239.255.254.254 autokey # manycast client
35:#server   127.127.1.0 # local clock
17:host 127.0.0.1
25:#uri ldap://127.0.0.1/
26:#uri ldaps://127.0.0.1/
4:nameserver 172.28.102.11
5:nameserver 172.28.102.13
```

FIGURE 6.3   Results from Sample Grep Instruction.

A regular expression that will match a string without a period is not as simple as [^\.] because this only says "match if any portion of the string does not contain a period." Thus, the expression [^\.] will match any string that is not solely periods.

The option –m allows you to specify a number so that grep stops searching after it reaches that number of matches. This could be useful if you are only looking for a few or even one match. For instance, if you want to know if a particular expression matches anywhere in a file, you could use –m 1.

If you only want grep to output the actual matches (i.e., the portion of the line that matched), use –o. Finally, –l (lower case "L") and –L will output only file names that contain matches or file names that do not contain matches respectively rather than the matched lines.

### 6.4.3  Additional egrep Examples

Let us now concentrate on a few examples that search the standard Linux dictionary for words that fit certain patterns. If you examine the contents of this dictionary (stored in /usr/share/dict/words), you will see words that are numbers, words that contain numbers, words that start with upper-case letters, words that are all lower case, and some words that are all upper case, all upper-case letters and numbers, and words with punctuation marks.

We start by looking for all 30-letter words.

```
egrep '[[:alpha:]]{30}' /usr/share/dict/words
```

The output of this command is two words:

```
dichlorodiphenyltrichloroethane
pneumonoultramicroscopicsilicovolcanoconiosis
```

Just in looking at this output you can see that they are different length words. They are not both 30-letters long. In fact the first word is 31 letters and the second is 45. To obtain exactly 30-letter words, we have to match a string from the start of the line to the end. We enhance our regular expression to be `^[[:alpha:]]{30}$` and find that there are no 30 letter words.

Let us look for 20 letter words that start with a capital letter.

```
egrep '[A-Z][[:alpha:]]{19}' /usr/share/dict/words
```

We use 19 in curly brackets because A-Z represents the first letter, so we are now looking for words that, after the first capital letter, have 19 additional letters. This regular expression has the same drawback as our previous grep command's expression in that we did not enforce that the regex should precisely match the entire line. By adding `^` and `$` to our regular expression we wind up with exactly five words:

```
Archaeopterygiformes
Biblicopsychological
Chlamydobacteriaceae
Llanfairpwllgwyngyll
Mediterraneanization
```

Let us try another one. If you look through the file, you will see many words with various punctuation marks. Some words are hyphenated, some have periods (e.g., abbreviations), and some use / as in AC/DC. There are a handful of words that contain other punctuation marks. We want to find all words that contain punctuation marks aside from hyphens, periods, and slashes. How can we do this? The regular expression for any punctuation mark is easy, `[[:punct:]]`. Here, we want to limit the expression to not include three forms. We cannot do so with :punct:, so instead, we need to enumerate the remaining punctuation marks. We will not include ' and " because these have special meanings in grep. We will also not use \ (the escape character) or [], as we will need to enumerate our list of punctuation marks inside of []. This leaves us with ``[`~!@#$%^&*()_=+{}|;:<>,?]``. Our grep command is

```
egrep '[`~!@#$%^&*()_=+{}|;:<>,?]' /usr/share/dict/words
```

which results in the following list.

```
2,4,5-t
2,4-d
A&M
A&P
AT&T
&c
hee-hee!
he-he!
IT&T
```

Can we simplify the above regex? Yes, we examine how to do so below.

To this point, we have used grep to search files for instances of strings. By piping the results of one command to grep, we can also reduce the output of other commands. Two common uses are to pipe the result of `ls -l` to `grep` and `ps` to `grep`. You might recall from earlier chapters that `ps ax` will list all processes irrelevant of user or terminal window. This creates a substantial amount of output. This can be viewed by piping the result to less, but if we were interested in only particular processes, we could also pipe the result to grep.

Consider wanting to view active processes that have utilized CPU time beyond 0:00. First, we want to issue `ps ax`. From this list, we want to find all processes whose Time is not 0:00. We will search using `egrep '0:00'` and then invert the match using –v. Our command is

```
ps ax | egrep -v '0:00'
```

Notice that the grep command does not include a filename. Why not? Because the input to grep is being redirected from a previous command.

Similarly, imagine that you want to view any files in /etc that are readable and writable. We want to perform `ls -l /etc`. Now, we pipe the result to grep. What regular expression do we seek? First, we want files, not directories. So we expect the permissions to start with a hyphen. Next, we want files that are both readable and writable. Thus, the first three characters in the long listing are expected to be '-rw.' We want to ensure that these are the first three characters of the line, so we add '^' to the beginning of the regular expression. This leaves us with the command:

```
ls -l /etc | egrep '^-rw'
```

Another example is to find the files in /etc that are not both owned by root and in root's private group. That is, which files do not have 'root root' in their long listing? For this, we want to specify "anything but 'root root'," so we will again use –v. Our instruction is

```
ls -l /etc | egrep -v 'root root'
```

Let us reconsider our earlier solution to list all words in the Linux dictionary that contained punctuation marks other than the hyphen, period, and slash. We can base this on the above strategy of piping the result of a Linux instruction to egrep –v. In this case, we will use `egrep '[[:punct:]]' /usr/share/dict/words` to obtain all words that contain a punctuation mark. Now we pipe the result to `egrep -v '[./-]'` to rule out all of those that contain one of the hyphen, period, and slash. The entire statement is

```
egrep '[[:punct:]]' /usr/share/dict/words | egrep -v '[./-]'
```

You might wonder about the order that the three punctuation marks are placed in the brackets. The hyphen, when used in brackets, indicates a range. We have to make sure that the hyphen is at the end of anything listed in brackets when we want to indicate that the hyphen is a character to match and not part of a range. The period and slash can be placed in either order.

### 6.4.4 A Word of Caution: Use Single Quote Marks

To wrap up this section, let us revisit the decision to place our regular expressions in quote marks as we have done throughout this section when using grep. As it turns out, in most cases, those quotes are not needed. For instance, ls -1 /etc | egrep ^-rw will work as well as the instruction listed earlier. But the instruction ls -1 /etc | egrep -v root root does not work correctly. The quotes are needed here because of the blank space between root and root. So our command should be

```
ls -1 /etc | egrep -v 'root root'
```

But there is another reason why we need to enclose the regular expression in quote marks. This reason has to do with the Bash interpreter and filename expansion, or globbing. Recall that ls * will list all items in the current directory. Similarly, an instruction like wc * will apply the wc operation to all files in the current directory. The Bash interpreter performs filename expansion before the instruction is executed. So wc * is first converted into wc *file1 file2 file3 file4* ... where the files are all files found in the current directory. Then, the wc command is applied to the list of files.

Now consider the instruction egrep abc* foo.txt. The * seems to indicate a regular expression wildcard. You might interpret this instruction as saying "search the file foo.txt for lines that contain an a followed by a b followed by 0 or more c's." Unfortunately, the Bash interpreter does not treat it this way. Instead, globbing takes place prior to the execution of the egrep command. Let us assume that the current directory has a file named abcd.txt (and no other entries whose name starts with ab). When you enter the above egrep instruction, globbing takes place before egrep executes. Globbing will produce one match, the file abcd.txt. The result of globbing is that anything that matched replaces the string with the wildcard. So the instruction becomes egrep abcd.txt foo.txt. That is, the regex abc* is replaced by the regex abcd.txt. Now, egrep executes. In this case, egrep will look for strings in foo.txt that match the regex abcd.txt where the period is the metacharacter to match any single character. If foo.txt happened to have the string abc-detxt, it would match. However, if foo.txt had the string abcccc, it would not match!

To force the Bash interpreter to not perform globbing mistakenly on the regular expression, you place the regular expression inside of quote marks. Aside from the *, recall that globbing operates on ? and []. In addition, the Linux interpreter applies a different interpretation to $ (variable name), {}, and several other characters. So, between the possibility that globbing will interfere with the command and the possibility that the regular expression contains a blank space, you should always use the single quote marks. What about using the double quote marks, ""? Recall that the Bash interpreter will perform operations on an item placed in double quote marks, such as "$NAME," that it would not perform when the item is placed in single quote marks. For this reason, use the single quote marks and not the double quote marks.

## 6.5 SED

Sed is a stream editor. A stream editor is a program that takes a stream of text and modifies it. This can be useful if you want to perform mass substitutions on a text file without having to open the file to perform a search and replace operation. In this section, we take an introductory look at how to use sed for fairly simple editing tasks. A full description of sed is well beyond the scope of this text.

### 6.5.1 Basic sed Syntax

With sed, you specify a regular expression which represents a pattern of what you want to replace. You then specify a replacement string to take the place of any strings that match the pattern. The generic form of a sed command is

```
sed 's/pattern/replacement/' filename
```

where *pattern* is the regular expression that describes the strings you wish to replace and *replacement* is the replacement string.

By default, sed outputs the resulting file with the replacements to the terminal window. You might wish to then redirect the output to a new file. The format of the sed command then becomes

```
sed 's/pattern/replacement/' filename > filename2
```

You would use a second file with redirection because the first file, filename, is currently open and so cannot be written to. You could later use mv *filename2 filename*.

There is only one replacement string in our sed command while the pattern could potentially match many strings. Thus, any string that matches the pattern has the matching portion of the string (the substring) replaced by the one replacement string. However, the replacement string can reference the matched string in such ways that you can, with effort, have different replacement strings appear. We will explore these ideas later in this section. First, we start with a simple example.

Let us look at some examples using a file of addresses called addresses.txt. As is common, people who live in apartments might list their apartment with the abbreviation apt or Apt. We want to convert these to read "Apartment." We can do so with a sed command as follows:

```
sed 's/[Aa]pt/Apartment/' addresses.txt > revised_addresses.txt
```

In this instruction, we are searching each line of the addresses.txt file for either Apt or apt and replacing the string with Apartment.

People may also indicate an apartment using #, Number or number. To incorporate these into our regular expression, we must use | to separate each possibility. However, | is a character that cannot be used in a sed command as is. We need to precede it with a \ in our regular expression. Our previous instruction becomes

```
sed 's/[Aa]pt\|[Nn]umber\|#/Apartment /' addresses.txt
   > revised_addresses.txt
```

This expression reads "replace Apt, apt, Number, number and # with Apartment."

Unfortunately, our solution is not perfect. While we might see an address in the form Apt 3a or Number 7, when people use #, it is typically with the format #12. The difference is that with Apt and Number there is a space before the apartment number, yielding the replacement in the form of "Apartment 3a" or "Apartment 7." But the replacement for "#12" is "Apartment12." How can we enforce a blank space after the word "Apartment?" One way to resolve this is to change our instruction to be

```
sed 's/[Aa]pt\|[Nn]umber\|#/Apartment/' addresses.txt
   > revised_addresses.txt
```

We are inserting a blank space after the word "Apartment" in the replacement string. This however leads to these strings as replacements: "Apartment  3b," "Apartment  7," and "Apartment 12." That is, when Apt, apt, Number, or number are used, a blank space appears after Apartment as part of our replacement string but there was already a blank space between Apt/apt/Number/number and the apartment number leading us to having two blank spaces rather than one. Only when # is used do we receive a single blank space.

We resolve this problem by specifying several /pattern/replacement/ pairs. We do this by providing sed with the option –e. The format of such a sed command is:

```
sed -e 's/pattern1/replacement1/'
   -e 's/pattern2/replacement2/' ... filename
```

The . . . indicates additional -e 's/pattern/replacement/' pairs. Notice that –e precedes each pair.

We will want to use a different replacement for the pattern #. Our new sed instruction is

```
sed -e 's/[Aa]pt\|[Nn]umber/Apartment/'
   -e 's/#/Apartment /' addresses.txt > revised_addresses.txt
```

Here, we see that the replacement string differs slightly between the patterns that match Apt/apt/number/Number and #. For the latter case, the replacement string includes a space while in the form case, the replacement string does not contain a space.

Continuing on with our example of a file of addresses, assume each address includes a zip code in the form #####-#### where each # would be a digit. We want to eliminate the -#### portion. Let us assume that four-digit numbers following hyphens only appear in zip codes (for instance, we would not see a street address in the form 12–3456 Pine Street).

In order to accomplish this task, we want to search for a hyphen followed by four consecutive digits and replace that string with nothing. To denote the pattern, we could use the regular expression -[0–9]{4}. Our command should then be

```
sed 's/-[0-9]{4}//' addresses.txt > revised_addresses.txt
```

This instruction specifies that the pattern of a hyphen followed by four digits is replaced by nothing.

Unfortunately, we are using {} in this regular expression. Remember from Section 6.2 that {} are part of the extended regular expression set. If we wish to use the extended set, we have to add the option –r. We have two alternatives then. First, add –r. Second, use -[0–9][0–9][0–9][0–9] as our regular expression. Obviously, the –r is easier. Our new command is

```
sed -r 's/-[0-9]{4}//' addresses.txt
   > revised_addresses.txt
```

This instruction will now convert a string like

```
Ruth Underwood, 915 Inca Road, Los Angeles, CA 90125-1234
```

to

```
Ruth Underwood, 915 Inca Road, Los Angeles, CA 90125
```

Recall that grep matched the regular expression to each line in a file. If any line matched the regular expression, the line was output. It did not really matter if the expression matched multiple substrings of the line. As long as any single match was found, the line was output. With sed, we are looking to replace the matching string with a replacement string. What if there are multiple matching strings in one line of the file? Will sed replace each matched string?

The answer is not normally. By default, sed will stop parsing a line as soon as a match is found. In order to force sed to continue to work on the same line, you have to specify that the search and replace should be "global." This is done by adding the letter 'g' after the final / and before the close quote. Let us imagine a file that contains a number of mathematical problems where the words "equals," "times," "plus," and "minus" are used in place " = ," "*," " + ," and "-." We want sed to replace all of the words with the operators. Further, since any particular line might have multiple instances of these words, we need to do a global replace. Our sed command might look like this (assuming that the file is called math.txt):

```
sed -e 's/equals/=/g' -e 's/times/*/g' -e 's/plus/+/g'
   -e 's/minus/-/g' math.txt
```

### 6.5.2 Placeholders

While using –e is a convenient way to express multiple patterns and their replacement strings, this will not work in a more general case. Let us consider a situation where we want to place any fully capitalized word inside of parentheses (I have no idea why you would want to do this, but it illustrates the problem and the solution nicely). Our regular expression is simply [A-Z] + , that is, any collection of one or more capital letters. If we had a specific list of words, for instance ALPHA, BETA, GAMMA, we could solve the problem easily enough using the –e option as in

```
sed -e 's/ALPHA/(ALPHA)/g' -e 's/BETA/(BETA)/g'
  -e 's/GAMMA/(GAMMA)/g' ... filename
```

NOTE: the ... indicates additional uppercase words that we want to parenthesize.

Just how many additional –e /pattern/replacement/pairs would we need? The problem is that we want to replace any fully capitalized word. So in this case, the ... would indicate every possible upper case word. This would be tens of thousands of entries! (or if we allowed any combination of upper-case letters, it could be an infinite list). It is clearly ridiculous to try to enumerate every possibility. And in fact, we can easily represent "upper case words" with the single regular expression [A-Z] +.

What we need is a mechanism in sed that can be used to represent the matched string, or the portion of the string that matched the regular expression. If we could denote the matched portion with a placeholder, we could then use that placeholder in the replacement string to indicate "use the matched string here." sed does have such a mechanism, the & character.

We can now solve our problem much more simply using a single /pattern/replacement/ pair. The pattern is [A-Z]+. The replacement is to put the string in parentheses, so we will use (&). This indicates a replacement of "open paren, the matched string, close paren." Our sed command is now much simpler.

```
sed 's/[A-Z]+/(&)/g' filename
```

There are modifiers that can be applied to &. These modifiers include \U, \u, \L, and \l to entirely upper case the string, capitalize the string (the first letter), entirely lower case the string, or lower case the first letter of the string respectively. For instance

```
sed 's/[A-Z][a-z]*/\L&/g' filename
```

will find any capitalized words and lower case them. Notice that we could also use \l since we are only seeking words that start with a capital letter and therefore there is only one letter to lower case. The modifiers \U, \u, \L, \l only impact letters found in the matched strings, not digits or punctuation marks.

Aside from the &, sed offers us another facility for representing portions of the matched string. This can be extremely useful if we want to alter one portion of a matching string, or rearrange the string. To denote a portion of a string, we must indicate which portion(s) we are interested in. We denote this in our regular expression by placing that portion of the pattern inside \( and \) marks. We can then reference that matching portion using \1. In sed, we can express up to nine parts of a pattern and refer to them using \1, \2, up through \9.

Let us assume the file names.txt contains information about people including their full names in the form *first_name middle_initial last_name*, such as Homer J. Simpson or Frank V. Zappa. Not all entries will have middle initials, such as Michael Keneally. We want to delete the period from any middle initial when found. To locate middle initials, we would seek [A-Z]\. to represent the initial and the period. We embed the [A-Z] in \(\) to mark it. This gives us the pattern

```
\([A-Z]\)\.
```

Although this is somewhat ugly looking, it is necessary. Now to remove the period, the replacement string is \1. The \1 refers to whatever matched \([A-Z]\). The rest of the match, the \., is not part of the replacement and so is omitted. Our sed command is

```
sed 's/\([A-Z]\)\./\1/' names.txt
```

This sed instruction says "find any occurrences of a capital letter followed by a period in a line and replace it with just the capital letter." Note that if a person's first name was also offered as an initial, as in H. J. Simpson, this replacement would result in H J. Simpson. If we want to ensure that we delete all periods from initials, we would have to add 'g' after the last / in the sed command for a global replacement. Alternatively, if we want to ensure that only a middle initial's period is deleted, we would have to be more clever. A revised expression could be

```
sed 's/\([A-Z][a-z.]+ [A-Z]\)\./\1/' names.txt
```

Let us now consider an example of referencing multiple portions of the pattern. We want to rearrange the names.txt file so that the names are changed from the format

*first_name middle_initial. last_name*

to the format

*last_name, first_name*

while omitting the middle initial entirely. This would change Homer J. Simpson to be Simpson, Homer. We need to remember all three parts of the name.

For a first name, we would expect [A-Z][a-z]+, for the middle initial, we would expect [A-Z]\., and the last name would be the same as the first. We would embed each of these in \(\) marks giving us

```
\([A-Z][a-z]+\) \([A-Z]\.\) \([A-Z][a-z]+\)
```

Our replacement string would be \3,    \1. That is, we would put the third string first followed by a comma followed by a space followed by the first string, and no second string. The sed command is

```
sed 's/\([A-Z][a-z]+\) \([A-Z]\.\) \([A-Z][a-z]+\)/
\3, \1/' names.txt
```

Unfortunately, the above sed command will not work if the first name is just an initial or if the person's name has no middle initial. To solve these problems, we can use multiple pattern/substitution pairs using the –e option. In the first case, we replace [A-Z][a-z]+ with [A-Z]\. giving us –e  's/\([A-Z]\.\)  \([A-Z]\.\)  \([A-Z][a-z]+\)/\3, \1/'. For the second case, we remove the middle portion of the expression and renumber \3 to \2 since there are only two patterns we are remembering now, giving us –e  's/\ ([A-Z][a-z]+\)  \([A-Z][a-z]+\)/\2, \1/'

As you can see, sed commands can become both complicated and cryptic very quickly. We end this section with a brief listing of examples. Assume names.txt is a file of names as described above. The first example below capitalizes every vowel found. The second example fully capitalizes every first name found. Notice the use of * for the lower-case letters. This allows us to specify that the first name could be an initial (in which there are no lower-case letters). The third example removes all initials from the file. The fourth example replaces every blank space (which we assume is being used to separate first, middle, and last names) with a tab. The fifth example matches every line and repeats the line onto two lines. In this case, the replacement is & (the original line) followed by a new line (\n) followed by & (the original line again).

- sed 's/[aeiou]/\u&/g' names.txt

- sed 's/[A-Z][a-z]*/\U&/' names.txt

- sed 's/[A-Z]\.//g' names.txt

- sed 's//\t/g' names.txt

- sed 's/[A-Za-z.]+/&\n&/' names.txt

## 6.6 awk

The awk program, like sed and grep, will match a regular expression against a file of strings. However, grep returns the matching lines and sed replaces matching strings, the awk program allows you to specify actions to perform on matching lines. Actions can operate on the substring(s) that matched, other portions of the line, or perform other operations entirely. In essence, awk provides the user with a programming language so that the user can specify condition and action pairs to do whatever is desired. We also see in this section that awk does not need to use regular expressions for matches but instead can use different forms of conditions.

Overall, awk (named after its authors Aho, Kernighan, and Weinberger) is a more powerful tool than either grep or sed, giving you many of the same features as a programming language. You are able to search files for literal strings, regexes, and conditions such as if a particular value is less than or greater than another. Based on a matching line, awk then allows you to specify particular fields of a matching line to manipulate and/or output, use variables to store values temporarily, and perform calculations on those variables, outputting results of the calculations.

Unlike grep and sed, awk expects that the text file is not just a sequence of lines but that each line is separated into fields (or columns). In this way, awk is intended for use on tabular information rather than ordinary text files. While you could potentially use awk on a textfile, say a text document or an email message, its real power comes into existence when used on a file that might look like a spreadsheet or database, with distinct rows and columns. Figure 6.4 illustrates this idea where we see two matching lines with awk operating on the values of other fields from those lines.

### 6.6.1 Simple awk Pattern-Action Pairs

The structure of a simple awk command will look something like this:

```
awk '/pattern/ {print ...}' filename
```

The pattern is a string, regular expression, or a condition. The action, in this case, is a print statement which specifies what to print from the matching line. The ... would need to be replaced with the content to output. Given such a command, awk examines filename line-by-line looking for *pattern*.

awk considers each line to be a series of data separated into fields. Each field is denoted by $n where n is the column of that field where the leftmost (first) field of a line is $1. The notation $0 is reserved to indicate "all fields in the line" which we could use in the print statement.

Let us reconsider our file from Section 6.5 names.txt which contained first names, middle initials, and last names. We want to print out the full names of those with middle initials. We could use /[A-Z]\./for the pattern and {print $0} for the action (assuming that each line only contains names, otherwise $0 would give us the full line which might be more information than we want to output). If we want to ensure that only the names are output, and assuming that the names appear in the first three fields of each line, we could use {print $1 $2 $3}. The full awk command is as follows.

```
awk '/[A-Z]\./ {print $1 $2 $3}' names.txt
```

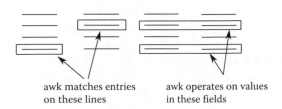

awk matches entries          awk operates on values
on these lines               in these fields

FIGURE 6.4   awk operates on fields within rows of a file.

Here, we see that $1 represents the first field (first name) of the row, $2 represents the second field (middle initial) of the row, and $3 represents the third field (last name) of the row. If we do not want to output the middle initial, we could use the following command.

```
awk '/[A-Z]\./ {print $1 $3}' names.txt
```

What is a field? Fields are indicated by a *delimiter* (a separator). For awk, delimiters are either spaces or tabs (indicated with \t). Therefore, whether a file is clearly in a tabular form or just text with spaces, awk is able to be used on it. However, if it is a textfile, we may not know the exact number of items (fields) per line as the number of items (words) can vary line-by-line.

The simple awk structure and example above only specify a single pattern. As with sed, you are able to specify any number of patterns for awk, each with its own action. The structure of a more elaborate awk command is

```
awk '/pattern₁/ {action₁}
     /pattern₂/ {action₂}
     /pattern₃/ {action₃}
     ...
     /patternₙ/ {actionₙ}' filename
```

This instruction is interpreted much like a nested if-then-else statement in a programming language. Working line-by-line in *filename*, if pattern$_1$ matches, execute action$_1$, else if pattern$_2$ matches, execute action$_2$, else if ..., else if pattern$_n$ matches, execute action$_n$. As soon as a pattern matches and the corresponding action is executed, awk moves on to the next line.

Let us consider some more interesting uses of awk. Assume we have a textfile, sales.dat, which contains sales information. The file consists of rows of sales information using the following fields:

```
Month    Salesman   Sales   Commission amount       Region
Jan      Zappa      3851    .15                     CA, OR, AZ
```

Aside from the first row, which is the file's header, each entry is of sales information for a given employee. There may be multiple rows of the same month and/or the same salesman. For instance, another row might contain

```
Feb      Zappa      6781    .20                     CA, OR, WA
```

First, let us compute the salary earned for each salesman whose region includes AZ. Here, we have a single pattern, a regular expression to match AZ. In fact, since there is no variability in the regular expression, our pattern is literally the two characters "AZ". For each matching line, we have a salesman who worked the Arizona region. To compute the salary earned, we need to multiply the sales by the commission amount. These two amounts are fields 3 and 4 respectively. We want to print these values. Our awk command could be

```
awk '/AZ/ {print $3*$4}' sales.txt
```

Unfortunately, this will only provide us with a list of sales values, but not who earned them or in which month. We should instead have a more expressive output using {print $1 $2 $3*$4}. This would give us JanZappa577.65! We need to explain to awk how to format the output. There are two general choices for formatting. First, separate the fields with commas, which forces awk to output each field separated by a space. Second, between each field, we can use "\t" or " " to indicate that a tab or a blank space should be output.

Next, let us compute the total amount earned for Zappa. The awk command

```
awk '/Zappa/ {print $1 "\t" $3*$4}' sales.txt
```

will provide one output (line) for each Zappa entry in the file. This will not give us a grand total, merely all of Zappa's monthly sales results. What we need to do is accumulate each value in some running total. Fortunately, awk allows us to define and use variables. Let us use a variable named *total*. Our action will now be {total = total + $3*$4}. We can also print out each month's result if we wish, so we could use {print $1 "\t" $3*$4; total = total + $3*$4;} or if we want to be more efficient{temp = $3*$4; print $1 "\t" temp; total = total + temp}. Our new awk command is

```
awk '/Zappa/ {temp = $3*$4; print $1 "\t" temp;
    total = total + temp}' sales.txt
```

Notice that we are not outputting total in the above awk statement. We will come back to this in the next subsection.

While awk is very useful in pulling out information from a file and performing computations, we can also use awk to provide specific results from a Linux command. We would do this by piping the result of an instruction to an awk statement. Let us consider a couple of simple examples.

Let us output the permissions and filenames of all files in a directory. The ls –l long listing will provide 10 characters that display the item's file type and permissions. This first character should be a hyphen to indicate a file. If we have a match, we then want to output the first and last entries on the line ($1 and $9). This can be accomplished as follows.

```
ls -l | awk '/^-/ {print $1, $9}'
```

Notice that the awk instruction does not have a filename after it because its input is coming from the long listing. The regular expression used as our pattern, ^-, means that the line starts with a hyphen.

In another example, we want to obtain process information using ps of all running bash shells. This solution is even easier because our regex is simply bash. We print $0 to output the full line including for instance the PID and statistics about each bash shell's processor usage.

```
ps aux | awk '/bash/ {print $0}'
```

### 6.6.2 BEGIN and END Sections

Our earlier example of computing Zappa's total earnings computed his total pay but did not print it out. We could change our action to be {temp = $3*$4; print $1 "\t" temp; total = total + temp; print total}. This would then explicitly output the value of temp for each match. But this will have the unfortunate effect of outputting the total for every row in which Zappa appears; in addition, the total will increase with each of these outputs. What we want to do is hold off on printing total until the very end of awk's run.

Fortunately, awk does have this capability. We can enhance the awk command to include a BEGIN section and/or an END section. The BEGIN section is executed automatically before awk begins to search the file. The END section is executed automatically after the search ends. The BEGIN section might be useful to output some header information and to initialize variables if necessary. The END section might be useful to wrap up the computations (for instance, by computing an average) and output any results. We enhance our previous awk instruction to first output a report header and then at the end, output the result.

```
awk 'BEGIN {print "Sales results for Zappa"; total = 0}
        /Zappa/ {temp = $3*$4; print $1 "\t" temp;
             total = total + temp}
    END {print "Zappa's total sales is $" total}' sales.txt
```

The above instruction works as follows. First, the BEGIN statement executes, outputting the header ("Sales results for Zappa") and initializes the variable total to 0. This initialization is not necessary as, in awk, any variable used is automatically initialized to 0. However, initializing all variables is a good habit to get into. Next, awk scans the file line-by-line for the pattern Zappa. For each line that matches, temp is set to the values of the third and fourth columns multiplied together. Then, awk outputs $1 (the name), a tab, and the value of temp. Finally, temp is added to the variable total. After completing its scan of the file, awk ends by output a closing message of Zappa's total. Note that if no lines contained Zappa, the output would be simply:

```
Sales results for Zappa
Zappa's total sales is $0
```

Now, let us combine the use of the BEGIN and END sections with a multipatterned instruction. In this case, let us compute the total salaries for three employees. We want to have, as output, each employee's total earnings from sale commissions. This will require maintaining three different totals, unlike the previous example with just the total for Zappa. We will call these variables total1, total2, and total3.

```
awk 'BEGIN {total1 = 0;total2 = 0;total3 = 0}
        /Zappa/ {total1 = total1 + $3*$4}
        /Duke/ {total2 = total2 + $3*$4}
        /Keneally/ {total3 = total3 + $3*$4}
     END {print "Zappa $" total1 "\n" "Duke $" total2 "\n"
        "Keneally $" total3}' sales.txt
```

As with our previous examples, the regular expression exactly matches the string we are looking for, so it is not a very challenging set of code. However, the logic is slightly more involved because we are utilizing three different running totals.

### 6.6.3 More Complex Conditions

Let us look at an example that requires a greater degree of sophistication with our patterns. In this case, let us obtain the number of salesmen who operated in either OH or KY. To specify "or," we use /pattern1/|||/pattern2/ where the notation || means "or." If we have a matching pattern, we want to increment a counter variable. In the END statement, we will want to output this counter's value. We omit the BEGIN statement because we do not need a header in this case (the END statement outputs an explanation of the information that the command computed for us and the variable, counter, is automatically initialized to 0).

```
awk '/OH/|||/KY/{counter = counter + 1;}
        END {print "Total number of employees who serve OH or KY: "
            counter}' sales.txt
```
If we wanted to count the number in OH and KY, we would use && instead of ||.

Let us consider a different file, courses.dat, to motivate additional examples. Imagine that this file contains a student's schedule for several semesters. The file contains fields for semester (fall, spring, summer, and the year as in fall12 or summer14), the course which is a designator and a course number as in CSC 362.001 (this is divided into two separate fields, one for designator, one for course number), number of credit hours, location (building, room), and time. For instance, one entry might be

```
fall12 CSC  362.001 3 GH 314 MWF 9:00-10:00 am
```

Let us create an awk command to output the courses taken in a particular year, for instance 2012. We would not want to use the pattern /12/ because the "12" could match the year, the course number or section number, the classroom number, or the time. Instead, we need to ensure that any 12 occurs near the beginning of the line. We could use the expression /fall12/|||/spring12/|||/summer12/. A shorter regular expression is one that finds 12 in the first field. Since the first field will be a string of letters representing the season (fall, spring, summer), we can denote this as [a-z]+12. To indicate that this must occur at the beginning of the line, we add ^ to the beginning of the expression. This gives us the command

```
awk '/^[a-z]+12/{print $0}' courses.dat
```

An awk command to output all of the 400-level courses should not just contain the pattern /4/ nor /4[0-9][0-9]/ because these could potentially match other things like a section number, a classroom number, or in the case of /4/, credit hours. Instead, we will assume that all course designators are three-letter combinations while all classroom buildings are two-letter combinations. Therefore, the course, as indicated as 4[0-9][0-9] should follow after [A-Z][A-Z][A-Z]. Since we require three letters in a row, this would not match a building. Our awk command then would look like this:

```
awk '/[A-Z][A-Z][A-Z] 4[0-9][0-9]/{print $0}' courses.dat
```

We can compute the number of hours earned in any particular semester or year. Let us compute the total hours for all of 2012. Again, we will use ^[a-z]+12 to indicate the pattern as we match the "12" only after the season at the beginning of the line. But rather than printing out the entry, we want to sum the value of hours, which is the fourth field ($4). Our awk command will be as follows.

```
awk '/^[a-z]+12/ {sum = sum + $4}
     END {print "Total hours earned in 2012 is " sum}'
     courses.dat
```

The metacharacter ^ is used to denote that the regular expression must match at the beginning of the line. To indicate "not," we use ! before the pattern. We would use !/*pattern*/ to match if pattern does not occur on the line. For instance

```
awk '!/^fall/{print $0}' courses.dat
```

would output all of the lines that do not start with "fall."

Aside from matching a /*pattern*/, awk allows for comparisons of field values against other values. These comparisons use the relational operators (<, >, = =, ! =, < =, > =). As with the print statement, you reference a field's value using $*n* where *n* is the field number. For instance, to print all of the courses that are 3 credit hours from your courses.txt file, you would use

```
awk '$4 = = 3 {print $0}' courses.txt
```

In this instruction, awk examines the file line-by-line comparing each line's fourth field to the value 3. On a match, it outputs that entire row. We could have instead written a pattern to find entries with a 3 in the fourth field. For instance, the credit hour will always appear with a space before and after it, unlike any other occurrence of a number, so the pattern /3/ would have also worked.

While regex-based patterns are very expressive, there are limitations to what you can do, at least easily. So the conditions that utilize relational operators are often easier. Returning to obtaining a list of all 400-level courses, we could more easily accomplish this with the following awk statement.

```
awk '$3 >=400 {print $0}' courses.dat
```

Notice that we are comparing a field that includes a decimal point against an integer number.

Let us consider a harder, more elaborate example. We have a payroll.dat file that lists the person's name, number of hours worked, and hourly wages as three fields on each line. We could use the following to output the pay of anyone who earned overtime.

```
awk '$2 > 40 {print $1 "\t $" 40 * $3 +
    ($2 - 40) * $3 * 1.5}' payroll.dat
```

This statement compares the hours field ($2) with 40 and if greater, then computes the pay, including overtime (overtime pay is normal wages for the first 40 hours plus the hours over 40 times wages times 1.5 for time and a half). Notice that this will *only* output the pay for people who have earned overtime. We will see below how to output everyone's pay, no matter if they worked normal hours or overtime.

You can combine comparisons by using && or ||. For instance, to find any employee who worked fewer than 35 hours and earns more than $20 per hour, you might use

```
awk '($2 < 35 && $3 > 20) {print $1}' payroll.dat
```

As another example, let us compute and output all of the entries of Zappa when he worked overtime. This combines two different conditions: the name is Zappa (field 1) and the hours is greater than 40 (field 2). We could achieve this in two ways, through the combination of a pattern and a condition /Zappa/ && $2 > 40 or through two conditions $1 =="Zappa" && $2 > 40. In either case, we output the full line, {print $1}. Here are the two versions of this awk command.

```
awk '/Zappa/&& $2 > 40 {print $0}' payroll.dat
awk '$1 =="Zappa" && $2 > 40 {print $0}' payroll.dat
```

What happens if you want to perform different operations for different patterns? There are two possible approaches. First, we might have different actions for different patterns. Second, we might use an if-else statement (covered in the next subsection).

We will want to compute either overtime pay or normal pay. Below is the awk command for doing this.

```
awk '$2 > 40 {print $1 "\t $" ($2-40)*$3*1.5 + 40*$3}
    $2 <= 40 {print $1 "\t $" $2*$3}' payroll.dat
```

The first pattern searches for any line where the employee earned overtime pay. The second pattern searches for any line where the employee earned normal pay.

Let us enhance our awk command to compute and output the *average* pay that was earned by the group of employees. We will also output each person's individual pay. To compute an average, we need to amass the total pay of all employees and divide it by the number of employees. To amass the total pay, we add each individual's pay to the total after we compute each individual's pay. We will add two variables, total_pay, and count. We will also use a third variable, current_pay, to shorten our statements.

We use BEGIN to initialize total_pay and count to 0 (recall that we do not need to initialize variables to 0, but this is good programming form, so we will). We have two patterns, one for normal pay and one for overtime pay. For either match, we compute and store in current_pay that employee's pay and add that value to total_pay while also outputting the result. We also increment count. When we are done, we use an END statement to compute the average pay and output the result.

```
awk 'BEGIN {total_pay = 0.0; count = 0}
    $2 > 40 {current_pay = ($2-40) * $3 * 1.5 + 40 * $3;
        total_pay += current_pay; count++;
        print $1 "\t $" current_pay}
    $2 <=40 {current_pay = $2*$3; total_pay += current_pay;
        count++; print $1 "\t $" current_pay}
    END {print "Average pay is $" total_pay/count}'
    payroll.dat
```

Note that count++ is an increment instruction equivalent to count=count+1. The ++ notation originated in C and is available in many languages like C, C++, perl, and Java.

Now that we have seen how to compute averages in awk, let us consider how to use this idea where the input is coming from a pipe. Specifically, how can we find the average file size for a given directory? As we saw earlier when piping ls  −1 to awk, we want to search for files so our regex is ^-. In this case, we want to add up the file sizes, which is the 5th field in the ls  −1 as well as count the number of files. Thus, we need two variables; we will use total and count. Finally, in an END clause, we will compute and output the average.

```
ls -1 | awk 'BEGIN {total = 0; count = 0}
            /^-/{total +=$5; count ++}
            END {print total/count}'
```

There is one problem with our instruction. What if none of the items in the current directory are files? We would wind up with no matches resulting in the computation of 0/0, which would yield an error. We can resolve this problem with additional logic as described next.

## 6.6.4 Other Forms of Control

The awk instruction provides a number of other operations making it like a full-fledged programming language. These operations include control statements like selection statements and loops. The selection statements are an if statement and an if-else statements. These statements are similar in syntax to those found in Java and C. In Java and C, the statements look like:

The if statement is: if(condition) statement;

The if-else statement is: if(condition) statement; else statement;

For awk, the entire statement is placed inside of {} marks and each clause (the if statement and the else statement) are placed in {} as well. There is also a nested if-else statement in Java and C and similarly in awk. In Java and C, the syntax is

if(condition1) statement1; else if(condition2) statement2; else statement 3;

Again, place the entire statement and each clause (statement1, statement2, statement3) in {}.

The role of the if statement is the same as our previous /pattern/{action} pair. We present a condition. The condition is tested line-by-line and if it is true for a given line, the associated statement(s) executes on that line. If the statement is an if-else and the condition is false for a given line, then the else action(s) executes. With the if-else structure, we do not need any /pattern/{action} pairs. Here is an example illustrating our previous payroll solution.

```
awk 'BEGIN {total_pay = 0.0; count = 0}
     {if ($2 > 40) {current_pay = ($2-40) *$3*1.5 + 40*$3;
           total_pay +=current_pay; count++;
           print $1 "\t $" current_pay}
     else {current_pay = $2*$3; total_pay += current_pay;
           count++; print $1 "\t $" current_pay}
     }
     END {print "Average pay is $" total_pay/count}'payroll.dat
```

Here, in between the BEGIN and END sections is a statement {if ... else ...}. The statement inside of the {} is executed on every line. There is no pattern to match but instead the if-else statement always executes. The if-else statement has a condition to determine whether to execute the if clause or the else clause.

We can also solve the earlier problem of computing the average file size of a directory if the directory contains no files. In this case, we do not want to perform total / count if count is 0. We add an if-else statement to the END clause.

```
ls -l | awk 'BEGIN {total = 0; count = 0}
             /^-/{total +=$5; count++}
             END {if(count > 0) {print total/count}
                  else print {"no files found!"}'
```

There are many other features available in awk including loops, arrays, built-in variables, and input statements to name a few. The use of loops is somewhat limited or irrelevant because awk already loops by performing the code between the BEGIN and END clauses for each line of the file. However, you might find a use for a loop in performing some computation within one of the actions.

The other features flesh out awk so that it can be used as a command line programming language. That is, features like input, arrays, and loops provide the power to use awk in place of writing a program. And since awk is a command line program, it allows the knowledgeable Linux user to write code on the command line that can operate in a similar way as a program (or shell script).

These two sections will hopefully whet your appetite to learn more about these tools. The primary reason for covering them in this text was to provide further examples and uses of regular expressions. As with anything Linux, check out the man pages for sed and awk. You can also learn a great deal about these programs on the Internet where there are numerous websites dedicated to sed and awk tutorial and user guides.

## 6.7  CHAPTER REVIEW

Concepts and terms introduced in this chapter:

- Character class—an abbreviated way to specify a list of all characters within a grouping such as alphabetic (:alpha:), upper-case letters (:upper:), digits (:digit:), and punctuation marks (:punct:).

- Character range—an abbreviated way to specify a list of characters that are expressed as the smallest and largest separated by a hyphen. Ranges are permissible for letters and digits but not other types of characters.

- Escape character—used to indicate that the given metacharacter should be treated literally and not as a metacharacter, for instance by forcing the period to act as a period and not "any character."

- Enumerated list—specifying all of the options of characters by listing them in [].

- Metacharacters—special characters that are not interpreted literally but are used to express how preceding character(s) should be matched against.

- Regular expression—a string comprising literal characters and metacharacters for use in pattern matching.

- Stream editor—a program that can search for and replace strings without opening the file in an editor.

- String—any set of characters (including the empty string); we refer to string as the master string for a regular expression to compare against.

- Substring—any subset of consecutive characters of a string (including the empty string).

Linux commands covered in this chapter:

- awk—powerful program that searches a file line-by-line for matching patterns and applies operations to the items in that line. Operations include output and assignment statements containing arithmetic operations.

- grep/egrep—tool to search for matches of a regular expression to every line in one or multiple files.

- sed—a stream editor to search each line of a file for one or multiple matching strings of a regular expression and replace all matched items with a replacement string.

## REVIEW QUESTIONS

1. What is the difference between [0-9]+ and [0-9]*?

2. What is the difference between [0-9]? and [0-9]+?

3. What is the difference between [0-9] and [^0-9]?

4. How would .* be interpreted?

5. How would . + be interpreted?

6. Is there any difference between [0-9] and [[:digit:]]?

7. We want to match against any sequence of exactly five digits. Why does [0-9]{5} not work correctly?

8. Is there any difference between [0-9]{1,} and [0-9]+?

9. How does .?.?.?.?.? differ from .{1,5}?

10. Interpret [0-999]. If we truly wanted to match any number from 0 to 999, how would we express it correctly?

11. Imagine that we want to match the fractional value of .10 (i.e., 10%). What is wrong with the expression .10?

12. Write a regular expression to represent two vowels in a row.

13. Write a regular expression to represent two of the same vowel in a row.

14. We want to match against any arithmetic expression of the form X op Y = Z where X, Y, and Z are any numbers and op is any of +, -, *, or /. Write the proper regular expression.

15. To find four words in a sentence, we might use ([[:alpha:]]+ ){4}. Why is this incorrect? How would you fix it?

For questions 16–19, imagine that we have a file that lists student information, one row per student. Among the information for each student is every state that the student has lived in. For instance, we might have one entry with OH and another with OH, MO, NY. Further, assume that the only use of commas in the entire line will be to separate the states.

16. We want to find every student who has lived in either OH or KY. Write such a regular expression.

17. Why does [^OK][^HY] match students who have not lived in either OH or KY?

18. We want to find a student who has lived in OH or NY. Why is [ON][HY] not proper?

19. We want to find all students who have lived in at least three states. Write such a regular expression.

For questions 20–23, assume a file contains a list of information about people, row by row where each row starts with a person's first name, a comma, and the person's last name followed by a colon.

20. Write a regular expression to find anyone whose last name starts with either F, G, H, I, or J.

21. Write a regular expression to find anyone whose first and last names are both exactly six letters long.

22. Write a regular expression to find anyone whose first and last names do not contain an 'a.' Both first and last names must not contain an 'a.'

23. Write a regular expression to find anyone whose last name contains two upper-case letters as in McCartney. The second upper-case letter can occur anywhere in the name.

For questions 24–30, use the Linux dictionary found in /usr/share/dict/words.

24. Write a regular expression to find words that have two consecutive punctuation marks.

25. Write a regular expression to find any words whose entries contain a digit with letters on either side of it.

26. Write a regular expression to find all words that begin with the letter a and end with the letter z.

27. Write a regular expression to find all five letter words that end with a c.

28. Write a regular expression to find any entries that contain a q followed by a non-u (as in Iraqi). The q can be upper or lower case.

29. Write a regular expression to find all entries that contain two x's somewhere in the word.

30. Write a regular expression to find all words that contain the letters a, b, c, and d in that order in the word. The letters do not have to appear consecutively but they must appear in alphabetical order.

31. In this chapter, we saw how to write a dollar amount that could include dollars or dollars and cents but we did not try to include commas. Provide a regular expression that can contain any amount from $999,999.99 down to $0.00 with the comma properly included.

32. Following up on number 31, can you come up with a regular expression that permits any number of digits to the left of the decimal point such that the comma is correctly inserted?

33. What is wrong with the following grep instruction?

```
grep abc* foo.txt
```

34. What is the difference between using –H and –h in grep?

35. Assume we have a grep statement of the form grep 'someregex' *.txt. Should we use the option –H? Explain.

36. When using the –c option, will grep output a 0 for a file that contains no matches?

37. Explain the difference between grep '[^abc]' somefile and grep –v '[abc]' somefile.

38. What regular expression metacharacters are not available in grep but are available in egrep?

39. Write a sed command to remove every line break (new line, \n) in the file somefile. These should be replaced by blank spaces.

40. Explain what the following sed command will do:

```
sed 's/ //' somefile
```

41. Explain what the following sed command will do:

```
sed 's/ //g' somefile
```

42. Explain what the following sed command will do:

```
sed 's/[[:digit:]]+\([[:digit:]]\)/0/g' somefile
```

43. Explain what the following sed command will do:

```
sed 's/^\([[:alpha:]]+\)\([[:alpha:]]\)\([[:space:]]\)/
\2\1\3/g' somefile
```

44. What does the & represent in a sed command?

45. What is the difference between \U and \u when used in a sed command?

46. Write a sed command to reverse any capitalized word to start with a lower-case letter and whose remaining characters are all upper case. For instance, Dog becomes dOG while cat remains cat.

47. Write a sed command to replace every occurrence of '1' with 'one,' '2' with 'two,' and '3' with 'three' in the file somefile.

For questions 48–53, assume the file payroll.dat contains employee wage information where each row contains

first_name        last_name        hours             wages             week

where week is a number from 1 to 52 indicating the week of the year. Multiple employees can occur in the file but they will have different weeks.

48. Write an awk command to output the first and last names of all of all employees who worked during week 5.

49. Write an awk command to compute and output the total pay for employee Frank Zappa assuming no overtime.

50. Revise #49 so that the computation includes overtime at 1.5 times the wages specified.

51. Write an awk command to compute the average number of hours worked for the weeks that Frank Zappa appears.

52. Compute the number of times any employee worked overtime.

53. Compute the average wage of all records in the file.

# CHAPTER 7

# Shell Scripting

T HIS CHAPTER'S LEARNING OBJECTIVES are

- To be able to write Bash shell scripts

- To know how to use variables and parameters in shell scripts

- To be able to write input and output statements in shell scripts

- To understand and write conditional statements in Bash

- To be able to properly apply selection statements in shell scripts

- To be able to properly apply loop statements in shell scripts

- To understand how to use arrays and string operations in Bash

- To understand the role of the function and be able to write and call functions in Bash

- To understand the differences between Bash and C-shell scripts

## 7.1 INTRODUCTION

Most computer programs are written in a high-level language (e.g., C++, Java) and then compiled into an executable program. Users run those executable programs. A script is a program that is *interpreted* instead of compiled. An interpreter takes each instruction in the script one at a time, translates it to an executable statement, and executes it. This is far less efficient for the end-user because the end-user must then wait during instruction translation. Compiled programs have already been translated so the end-user runs the program without waiting for translation.

Many Linux users write and run both types of programs. The developers will write code in C or C++ (or Java). These tend to be larger programs where compilation is a necessity so that the executable code can be distributed to end-users to run. However, when it comes to small programs, many Linux users and system administrators will write interpreted programs instead, called *scripts*. There are many scripting languages available for Linux including Python, Ruby, PHP, and Perl. Each shell type has its own scripting language as well. In this chapter, we concentrate on the Bash scripting language although at the end of the chapter we briefly compare it to the C-shell scripting language.

The reason that we need to learn how to write scripts is that we need to produce code to automate Linux system tasks. Many of these scripts are the glue of the Linux operating system, holding together the already written portions of the system. We can get away with using an interpreted language here because our scripts will commonly be small programs whose execution time is not impacted much by the need to be interpreted during execution.

The advantage of scripting is that you can experiment with individual instructions to see what they do first, while putting together your program. Also, the interpreted environment provides the user with the ability to have a "session" by which he or she can refer back to previous definitions and results. Thus, as the session goes on, the user/programmer can add more and more to it and recall the items already defined.

As each shell has its own interpreter, you can enter shell instructions from the command line. The shell session then creates the environment whereby definitions such as aliases, variables, functions, and the history list persist. You can also capture the shell instructions in a file to be recalled at later times for convenience. If the tasks that make up this file are commonly used, the user can then schedule this file to be executed at regular intervals. Thus, the shell script is a mechanism for automation.

The Bash shell scripting language is a programming language that includes instructions much like those found in most programming languages such as loops, selection statements, input and output statements, and assignment statements. A Bash script can therefore combine such instructions with Linux commands.

Bash also permits the definition of functions and then function calls among the script code. Additionally, variables are available and can include environment variables (e.g., HOME, PWD, and PATH) along with any variables defined in the script or those defined from the command line and then exported. Parameters can also be supplied by the user when the script is executed. These parameters can then be utilized in the script.

In this chapter, we introduce shell scripting as a programming tool. We will examine most of the constructs available in the Bash shell. At the end of the chapter, we will also briefly examine the constructs found in the Csh shell.

Note: It is assumed that the reader already has some knowledge of programming and so concepts such as variables, logic, loops, and so forth are not described in any detail.

## 7.2 SIMPLE SCRIPTING

### 7.2.1 Scripts of Linux Instructions

Every shell script must start with a comment that specifies the interpreter which should run the script. For instance, you would start your script with one of the following lines, depending on which shell you are operating on (or which shell interpreter you wish to call upon). Here are examples for each of Bash, Csh, and Tcsh.

```
#!/bin/bash
#!/bin/csh
#!/bin/tcsh
```

The script file must also be executable. We will use permission 745, or alternatively 755.

Among the simplest scripts we can write are those that perform a sequence of Linux operations. For instance, as a user, you might want to start off each login with a status report. This report might include the current time/day, your disk utilization, and a list of any empty files found. We can write a script that executes commands such as: `date`, `du -s ~`, `find ~ -empty`. Our script might look like this:

```
#!/bin/bash

date
du -s ~
find ~ -empty
```

The blank line after `#!/bin/bash` is not needed and if we wanted, we could separate each of the other commands with blank lines. We can also add comments to the script to describe the instructions. Comments always follow the # symbol as in

```
date # output the current date and time for the report
```

### 7.2.2 Running Scripts

Assume that we save the previous script under the name `start`. After changing the permissions of the file `start` to 745 (using chmod), we can execute this script by entering `./start`. The `start` script outputs information to your terminal window.

Consider what might happen if you have a lot of empty files under your home directory. The output may stream through the terminal window too rapidly for you to see the full report. To make the output more readable, we might enhance this script in two ways. First, we can add echo statements that output literal messages that explain what the output is such as "your disk utilization is" and "your empty files are." We can also add line breaks to add blank lines in the output to separate the three types of output being generated. Second, we can send some or all of the output to a disk file. This would allow the user to view the output using more or less or in a program like vi.

The output statement is echo, as we saw from Chapter 2. To redirect the output of a statement, we use > *filename* (or >> *filename* to append to an existing file).

```
#!/bin/bash

date
echo          # output a blank line
du -s ~
echo          # output a blank line
find ~ -empty >> empty_report.txt # redirect output
```

Here, the choice of using the redirection operator >> to send the result of find to a file allows us to accumulate this information every time we run the script, rather than replace it. If we also redirect date to empty_report.txt, we can track the files that are empty over time.

An alternative to performing redirection within the script is to redirect the script's output. Consider the following script: called start:

```
#!/bin/bash
date
echo
du -s ~
echo
find ~ empty
```

Rather than executing start simply as ./start, we could redirect the script's output by using

```
./start >> start_up_output.txt
```

What should we do with this script once it is written? While we could just try to remember to run it from time to time, we could schedule its execution. Later in the text, we look at the scheduling commands at and crontab. A user might want to execute this script every time they start a bash shell or log in. Recall from Chapter 2 that your .bashrc file is executed every time you open a bash shell and your .profile file is executed every time you log in. The user could add ./start into the user's .bashrc or .profile file. Alternatively, the user could place the contents of the script (date, echo, du –s ~, etc.) directly in the .bashrc or .profile file.

### 7.2.3 Scripting Errors

There are many reasons why a script might generate an error. Unfortunately for us as programmers, the errors are only caught at run-time. That is, you will only know if an error exists when you try to run the script. This differs from compiled programming languages where the compiler will discover many or most of the errors that might exist in a program when you compile the program. Here, there are no syntax errors but instead run-time errors that might be caused by syntactic issues (such as misspelling a reserved word) or execution issues (such as dividing by zero).

Unfortunately the error messages provided by the Bash interpreter may not be particularly helpful. What you will see in response to an error is the line number(s) in which an error arose. Multiple error messages can be generated because in most cases, if an error occurs, the Bash interpreter continues to try to execute the script.

Let us consider a couple of errors that we might come across. The following script uses an assignment statement and an output statement (echo). Since we have already seen these in Chapter 2, you should have no difficulty understanding what this script is supposed to do. Assume this script is stored in the file error.

```
#!/bin/bash
NAME = Frank Zappa
echo $NAME
```

When run, we receive the message:

```
./error: line 3: Zappa: command not found
```

This message indicates that when Bash attempted to run echo $NAME, something went wrong. Specifically what went wrong was that $NAME is storing two values with no quotes around it, so retrieving the value from $NAME yields an error. We did not get an error with the second instruction though. The solution to this error is to place "" around Frank Zappa in the second line.

In the following case, we have a script that performs a series of operations for us. You can see we are using wc to obtain information from the /etc/passwd, /etc/group, and /etc/shadow files. These three files store user account information. In this case, we are just outputting the number of lines of each file. Let us call this script account_report.

```
#!/bin/bash
wc -l /etc/passwd
wc -l /etc/group
wc -l /etc/shadow
```

Running this script generates output as we might expect for the first two instructions: the number of lines of the two files. However, for the third line, we get this error:

```
wc:/etc/shadow: Permission denied
```

Our error message indicates that it was not our script that caused an error but instead the wc command from within the script. The error arises because the shadow file is not accessible to any user other than root.

We will explore other errors that may arise as we introduce Bash programming instructions in later sections of this chapter.

## 7.3 VARIABLES, ASSIGNMENTS, AND PARAMETERS

### 7.3.1 Bash Variables

A shell script can use variables. There are two types of variables: environment variables that, because they have been exported, are accessible from within your script or from the command line and variables defined within your script that are only available in the script. All variables have names. Names are commonly just letters although a variable's name can include digits and underscores (_) as long as the name starts with a letter or underscore. For instance, you might name variables x, y, z, first_name, last_name, file1, file_2, and so forth. You would not be able to use names like 1_file, 2file, or file 3 (the space is not allowed). Variable names are case sensitive. If you assign the variable x to have a value and then use X, you are referencing a different variable. Note that all reserved words (e.g., if, then, while) are only recognizable if specified in lower case. Unlike most programming languages, variable names can be the same as reserved words such as if, then, while, although there is no reason to do this as it would make for confusing code.

Variables in Bash store only strings. Numbers are treated as strings unless we specify that we want to interpret the value as a number. If we do so, we are limited to integer numbers only as Bash cannot perform arithmetic operations on values with decimal points (floating point numbers). Bash also permits arrays, which we examine later in this chapter.

## 7.3.2 Assignment Statements

In order to assign a variable a value, we use an assignment statement. The basic form of the assignment statement is

$$VARIABLE = VALUE$$

No spaces are allowed around the equal sign. Unlike programming languages like C and Java, the variable does not need to be declared.

The value on the right hand side can be one of four things:

- A literal value such as Hello, "Hi there," or 5

- The value stored in another variable, in which case the variable's name on the right hand side of the assignment statement is preceded by a dollar sign as in $FIRST_NAME or $X

- The result of an arithmetic or string operation

- The result returned from a Linux command

Literal values should be self-explanatory. If the value is a string with no spaces, then quote marks may be omitted. If the value is a string with spaces, quote marks must be included. There are two types of quote marks, single quote marks (') and double quote marks (""). These are differentiated below. Here are several examples of assignment statements and their results.

- X = 5

  - X stores the string 5 (this can be interpreted as an integer value, as will be demonstrated later)

- NAME = Frank

  - NAME stores *Frank*

- NAME = "Frank Zappa"

- NAME stores *Frank Zappa*, the quote marks are required because of the space between Frank and Zappa

- NAME = Frank  Zappa

  - An error arises because of the space, the Bash interpreter thinks that Zappa is a command, which does not exist

- NAME = FrankZappa

  - No quote marks are required here because there is no space, so in fact the literal value is really one string of 10 characters

- NAME = ls  *

  - An error should arise as Bash tries to execute *, which is replaced by all of the entries in the current directory. If all of the entries in the directory are executable programs, this will result in all of these programs executing!

- X = 1.2345

  - X stores 1.2345, but the value is a string, not a number

The instruction X = 1 would yield the error Fatal server error: Unrecognized option:  =. Unlike the error messages shown in Section 7.2, we do not even get a line number informing us of where the error arose!

The assignment statement can copy the value of one variable into another. To retrieve the value stored in a variable, you must precede the variable name with a $ as in $NAME or $X. However, the $ is not applied if the variable is enclosed in single quote marks (' '). It is applied if the variable is enclosed in double quote marks (" "). Let us look at some examples. Assume FIRST_NAME stores the string Frank and LAST_NAME stores the string Zappa. Here are several different assignment statements and the result stored in NAME.

- NAME = "$FIRST_NAME $LAST_NAME"

  - NAME stores *Frank Zappa*

- NAME = '$FIRST_NAME $LAST_NAME'

  - NAME stores *$FIRST_NAME $LAST_NAME*

- NAME = $FIRST_NAME$LAST_NAME

  - Name stores FrankZappa, notice that since there is no space on the right hand side of the assignment statement, we did not need to place $FIRST_Name $LAST_NAME in quote marks

- NAME = $FIRST_NAME $LAST_NAME

  - Results in an error because there is a space and no quote marks

- NAME = "FIRST_NAME LAST_NAME"

  - NAME stores *FIRST_NAME LAST_NAME* because there is no $ preceding the variable names

- NAME = 'FIRST_NAME LAST_NAME'

  - NAME stores *FIRST_NAME LAST_NAME*

Once a variable has been given a value, there are several ways to alter it. To assign it a different value, just use a new assignment statement. For instance, if X was set to 0, we can do X = 1 to change it. To remove a value from a variable, you can either assign it the NULL value which is accomplished by having nothing on the right hand side of the equal sign, for instance as

$$X=$$

Alternatively, you can use the unset command as in unset X. Either way, X will no longer have a value.

### 7.3.3 Executing Linux Commands from within Assignment Statements

The VALUE portion of the assignment statement can include executable statements. Consider the assignment statement DATE = date. This literally sets the variable DATE to store the string *date*. This would not differ if we enclose date in either "" or '' marks. However, if date is placed within either ` ` marks or $() marks, then Bash executes the Linux command. So, we revise our previous assignment statement to be DATE = `date` or DATE = $(date). This instruction causes the Linux date command to execute and rather than sending the output to the terminal window, the output is stored in the variable DATE. Notice that we did not use "" or '' marks here. If the right hand side includes multiple parts such as literal text, values in variables, and the results of Linux operations, we would embed the entire right hand side in "" or '' marks. For instance

DATE = "Hello $FIRST_NAME, today's date and time is $(date)"

The single quote mark after s is permissible within the double quote marks. We would not want to use the single quote marks around the entire right hand side because it would treat both $NAME and $(date) literally. For instance

DATE = 'Hello $NAME, todays date is `date`'

will cause DATE to store *Hello $NAME, todays date is `date`*. Additionally, by using the single quote marks, we would not be able to place the single quote after the s in today's.

Let us consider another example of executing a command on the right hand side of an assignment statement. What will happen when executing an ls command? The command LIST = $(ls) would result in LIST storing the list of all files and subdirectories in the current directory. However, unlike the actual list produced by ls, each of these items, listed as a string, would be separated by spaces on a single line and not on separate lines. Similarly, LIST = $(ls −l) would cause LIST to store the long listing of the items in the directory where each portion of the long listing is separated by a single space. This might look something like:

```
total 44 −rwxr−− r−−. 1 foxr foxr 448 Sep 6 10:20
addresses.txt −rwxr−−r−−. 1 foxr foxr 253 Aug  6 08:29
computers.txt −rw-rw-r−−. 1 foxr foxr 235 Sep 11 14:23
courses.dat . . .
```

which is a total jumble of information. If the item enclosed inside ` ` or $() is not a Linux command, an error will arise.

In addition to applying commands, we can also take advantage of Bash's filename expansion using wildcards. The instruction LIST = *.txt would store in the variable LIST the list of all filenames with the .txt extension. The instruction LIST = file?.txt would store in the variable LIST the list of all filenames that start with file followed by one character and end with the .txt extension, for instance file1.txt and file2.txt but not file.txt or file21.txt.

We can use various operations on the right hand side of our assignment statement. In general, we think of two types of operations, string operations and arithmetic operations. The most commonly used string operation is string concatenation, that is, the conjoining of multiple strings to create one larger string. Concatenation is accomplished merely by listing each string on the right hand side of the assignment statement. If the strings are to be separated by blank spaces, then the entire sequence must be enclosed within quote marks. The items concatenated can be any combination of variables, literal characters, and Linux operations embedded in ` ` or $() notation. We saw several examples earlier such as

- NAME="$FIRST_NAME $LAST_NAME"

- NAME=$FIRST_NAME$LAST_NAME

- LIST=$(ls)

We examine some other string operations later in this section.

You need to make sure that any command placed inside ` ` or $() notation is a legal command which the user who is running the script has access to. For instance,

```
Greeting = "Hello $NAME, today is $(dat)"
```

yields an error of dat: command not found.

If you do choose to declare a variable, the statement is `declare [options] name[= value]`. This notation indicates that options and the initial value are optional. Options are denoted using + to turn an attribute of the variable off and − to turn it on. The attributes are a (array), f (function), i (integer), r (read-only), and x (exported). For instance, `declare −rx ARCH = 386` will create a read-only variable, ARCH, with the value 386 and export it beyond the current shell. You cannot use +r as by default, a variable is not read-only.

### 7.3.4 Arithmetic Operations in Assignment Statements

For arithmetic operations, we have to denote that the values stored in variables are to be interpreted as numbers and not strings. We annotate such operations using one of two notations. First, we can precede the assignment statement with the word let, as in `let a = n + 1`. Second, we embed the arithmetic operation inside the notation $(()), as in a = $((n + 1)). Notice in both cases, n, a variable on the right hand side of the assignment statement, does not have a $ immediately precede it. You are allowed to include the $ before n in either case but it is not necessary. You can also use the prefix or postfix increment/decrement operators ++ and -- as found inC, Java, and similar languages. The syntax in this case is ((var++)), ((++var)), ((var --)), and (( --var)). The variable must currently be storing a numeric value. If it is storing a string, the variable's string value is lost and replaced by 1 (for ++) or −1 (for --).

The Bash interpreter can perform many arithmetic operations on integer values. It cannot however perform arithmetic operations on fractional values (any value with a decimal point). The operators available are shown in Table 7.1.

Let us consider some examples.

- N=1

- N=$((N+1))

  - N is now 2

TABLE 7.1 Bash Arithmetic Operators

| Arithmetic Operators | Meaning |
| --- | --- |
| +, −, *, / | Addition, subtraction, multiplication, division |
| ** | Exponent (e.g., x**2 is $x^2$) |
| % | Modulo (division retaining the remainder only) |
| ~ | Negation of 1's complement value |
| <, >, <=, >=, ==, <>, != | Relational operators, the last two are available for not equal[a] |
| <<, >>, &, ∧, \| | Bit operations (left/right shift, AND, XOR, OR) |
| &&, \|\|, ! | Logical AND, OR, NOT |
| = | Assignment |
| +=, −=, *=, /=, %= | Reassignment (e.g., x = x + y can be written as x+ = y) |
| &=, ∧=, \|=, <<=, >>= | Reassignment using &, ∧, \|, <<, >> |

[a] Note that we do not use <, <= , and so on in conditional statements. We cover this in Section 7.4.

- `let N+=1`

  - Reassign N to be 1 greater, N is now 3, note that without the "let" statement, N+=1 is interpreted instead as string concatenation

- `let Y=$N+1`

  - Y is 4, the $ in front of N is not needed so this could be let Y = N + 1

- `let Z=Y%N`

  - Z is 1 (4/3 has a remainder of 1)

- `Y=$(($Y+1))`

  - Y is now 5, we can omit the $ before Y

- `let Y=$((Y << 2))`

  - Y is now 20. << is a left shift which results in a number being doubled. The number 2 in this instruction indicates that the left shift should be done twice. As Y was 5, the left shift doubles Y to 10 and then 20.

- `((Y++))`

  - Y is now 21

- `((Q--))`

  - As Q had no value, it is now −1

- `N=$N+1`

  - N has the value of 3 + 1 (literally a 3, a plus, and a 1) because $N+1 is treated as a string concatenation operation instead of an arithmetic operation

You must specify the arithmetic operations correctly. For instance, a statement like X = $((Y + Z*)) will yield an error because the operation is lacking an operand. The specific error will look like: `Y + Z*: syntax error: operand expected (error token is "*")`.

You can specify multiple assignment statements on the same line. This is accomplished by separating each assignment with a space. For instance, rather than assigning three variables on separate lines as in

```
X = 1
Y = 2
Z = 3
```

we could do this in one instruction as X = 1  Y = 2  Z = 3.

The Bash interpreter also has a command called `expr` which performs and returns the result of a string or arithmetic operation. This is not part of an assignment statement; instead the result is printed to the terminal window. You might use this if you need to see the result of a quick computation. Here are some examples. Assume N is 3.

- `expr $((N+1))`
  - This returns 4 without changing N
- `expr $N+1`
  - This is treated as string concatenation and returns the string 3 + 1
- `X=`expr $((N+1))``
  - The expr command computes and returns 4, which is then stored in X, so X stores 4 when this instruction concludes

### 7.3.5 String Operations Using `expr`

The `expr` command is commonly used with string operations. Three string operations are `substr`, `index`, and `length`. The format of the expr statement using a string operation is

```
expr command string params
```

where command is one of the commands (substr, index, length), *string* is either a literal string or a variable storing a string (or even a Linux operation that returns a string), and params depend on the type of command as we will see below.

The substr command performs a substring operation. It returns the portion of the string that starts at the index indicated by a start value whose length is indicated by a length value. We specify both start and length in the substr command using the syntax

```
expr substr string index length
```

where *string* is the string, *index* is an integer indicating the position in the string to start the substring, and *length* is an integer indicating the length of the substring. For instance, `expr substr abcdefg 3 2` would return the string `cd`.

Any of string, index, and length can be stored in variables in which case we would reference the variable using $VAR. If NAME stores "Frank Zappa," start is 7 and length is 3, then `expr substr "$NAME" $start $length` would return `Zap`. Notice that $NAME is placed inside of "". This is necessary only because NAME contains a blank. If $NAME instead stored FrankZappa, the quotes would not be necessary.

The index command is given two strings and it returns the location within the first string of the first character in the second string found. We can consider the first string to be the master string and the second string to be a search string. The index returned is the location in the master string of the first matching character of the search string. For instance, `expr index abcdefg qcd` returns 3 because the first character to match from qcd is c at index 3.

The length command returns the length of a string. Unlike substr and index, there are no additional parameters to length other than the string itself. So for instance, you might use expr length abcdefg which would return 7 or expr length "$NAME" which would return 11 using the value of NAME from a couple of paragraphs back.

Each of these string operations will only work if you execute it in an expr command. As with using expr with arithmetic expressions, these can be used as the right hand side of an assignment statement in which case the entire expr must be placed inside ` ` marks or inside the notation $(). Here are some examples. Assume PHRASE stores "hello world"

- X=`expr substr "$PHRASE" 4 6`
  - X would store "lo wor"
- X=`expr index "$PHRASE" nopq`
  - X would store 5
- X=`expr length "$PHRASE"`
  - X would store 11

### 7.3.6 Parameters

Parameters are specified by the user at the time that the script is invoked. Parameters are listed on the command line after the name of the script. For instance, a script named a_script, which expects numeric parameters, might be invoked as

```
./a_script 32 1 25
```

In this case, the values 32, 1, and 25 are made available to the script as parameters.

The parameters themselves are not variables. They cannot change value while the script is executing. But they can be used in many types of instructions as if they were variables. Parameters are accessed using the notation $n where n is the index of the parameter (for instance, $1 for the first parameter, $2 for the second).

As an example, the following script returns substrings of strings. The user specifies the index and length of the substrings as parameters. The script then outputs for each string defined in the script the corresponding substring.

```
#!/bin/bash
STR1 = "Hello World"
STR2 = "Frank Zappa Rocks!"
STR3 = "abcdefg"
expr substr "$STR1" $1 $2
expr substr "$STR2" $1 $2
expr substr $STR3 $1 $2
```

If this script is called sub and is executed as ./sub  2  3, the output is

```
ell
ran
bcd
```

Notice in the first two expr statements that $STR1 and $STR2 are in "" because the strings that they store have spaces. This is not necessary for $STR3.

Once a parameter has been used in your script, if you do not need it, you can use the instruction shift which rotates all parameters down one position. This would move $2 into $1. This can simplify a script that will use each parameter in turn, one at a time so that you only have to reference $1 and not other parameters. The proper usage of shift is in the context of a loop. We will examine this later in this chapter.

Four special notations provide us not the parameters but information about the script. The notation $0 returns the name of the script itself while $$ returns the PID of the script's running process. The notation $# returns the number of parameters that the user supplied. Finally, $@ returns the entire group of parameters as a list. We can use $# to make sure that the user supplied the proper number of parameters. We can use $@ in conjunction with the for statement (covered in Section 7.5) to iterate through the entire list of parameters. Both $# and $@ are discussed in more detail in Sections 7.4 and 7.5.

## 7.4 INPUT AND OUTPUT

### 7.4.1 Output with echo

We have already seen the output statement, echo. The echo statement is supplied a list of items to output. These items are a combination of literal values, treated as strings, variables (with the $ preceding their names), and Linux commands placed inside ` ` or $() marks. You cannot use the expr statement though in this way.

Here are some examples of output statements. Assume that any variable listed actually has a value stored in it.

- echo Hello $NAME, how are you?

- echo Your current directory is $PWD, your home is $HOME

- echo Your current directory is `pwd`, your home is ~

- echo The date and time are $(date)

- echo The number of seconds in a year is $((365*24*60*60))

The echo statement also permits a number of "escape" characters. These are characters preceded by a back slash (\). You must force echo to interpret these characters, if present. This is done through the –e option. Table 7.2 displays some of the more common escape characters available. To use the escape characters (with the exception of \\), the sequence including any literal characters that go with the escape characters must be placed within "" marks.

TABLE 7.2    Some Useful Escape Characters in Bash

| Escape Character | Meaning |
| --- | --- |
| \\ | Backslash |
| \a | Bell (alert) |
| \b | Backspace |
| \n | New line |
| \t | Horizontal tab |
| \v | Vertical tab |
| \0### | ASCII character matching the given octal value ### |
| \xHH | ASCII character matching the given hexadecimal value HH |

Here are some examples of using the escape characters and what they output

- `echo -e "Hello World!\nHow are you today?"`
  - `Hello World`
  - `How are you today?`
- `echo -e "January\tFebruary\tMarch\tApril"`
  - `January    February    March    April`
- `echo -e "\x40\x65\x6C\x6C\x6F"`
  - `Hello`
- `echo -e Hello\\World`
  - `Hello\World` (recall that the quote marks are not needed for \\)

The use of "" and '' in echo statements is slightly different from their use in assignment statement. As we see below, the quote marks appear to do the same thing as we saw with assignment statements. The "" return values of variables and Linux commands whereas in '', everything is treated literally. Assume FIRST stores Frank and LAST stores Zappa.

- `echo $FIRST $LAST`
  - `Frank Zappa`
- `echo "$FIRST $LAST"`
  - `Frank Zappa`
- `echo '$FIRST $LAST'`
  - `$FIRST $LAST`
- `echo "$FIRST \n$LAST"`
  - `Frank \nZappa`

- `echo -e $FIRST \n$LAST`
  - `Frank nZappa`
- `echo -e "$FIRST \n$LAST"`
  - `Frank`
  - `Zappa`
- `echo -e "abc\bdef"`
  - `abdef`
    - the \b backs up over the 'c'
- `echo -e "\0106\0162\0141\0156\0153"`
  - `Frank`
- `echo -e "$FIRST\t$LAST"`
  - `Frank          Zappa`

But there is a difference as seen with a variable such as `LIST = *.txt`. Imagine that there are three text files in this directory, file1.txt, foo.txt, and someotherfile.txt. Examine the output of these three commands.

- `echo $LIST`
  - `file1.txt foo.txt someotherfile.txt`
- `echo "$LIST"`
  - `*.txt`
- `echo '$LIST'`
  - `$LIST`

The second and third cases, are what we might expect, the value stored in $LIST and the literal string $LIST. However, the first example does not return $LIST or the value it stores, but instead the value $LIST stores is interpreted and we see all of the matching files. The "" are preventing filename expansion from occurring whereas without the "", the value of $LIST (*.txt) is itself being evaluated.

By default, the echo statement will cause a line break at the end of the line's output. You can prevent echo from outputting the line break by using the option –n. This is useful when wanting to prompt the user for input.

### 7.4.2 Input with read

The input statement is `read`. The read statement is then followed by one or more variables. If multiple variables are specified they must be separated by spaces. The variable names are listed without $ preceding them. The read can be used to input any number of variable values. The following are some examples of read statements.

- `read FIRST`
- `read FIRST LAST`
- `read x y z`

The second of these examples expects two input values, separated by a space and third expects three input values, each separated by a space. If the user does not provide the number of values as expected, any remaining variables are given NULL values. For instance, if the user were to input 5 10 <enter> in response to the third read statement, x would be 5, y would be 10, and z would have the value NULL. If the user were to input more values than needed, the remaining values are grouped together in a list so that the last variable receives all of these values together. For instance, if the user inputs 5 10 15 20 <enter> for the third read statement, then x is assigned 5, y is assigned 10, and z is assigned the list of 15 20.

Executing the read statement results in the script pausing while the cursor blinks in the terminal window. This would not inform the user of the expectations behind the input statement. Therefore, it is wise to always output a prompting statement that explains what the user is to do. There are two ways to accomplish this. The first is through an echo statement preceding the input statement. For instance, we might use:

```
echo Enter your name
read NAME
```

With this notation, the echo statement's output would appear and the cursor would then move to a new line so that the terminal window might look like the following.

```
Enter your name
Frank
```

The −n option for echo causes the output to not conclude with a new line. If we modify the echo statement to use −n, then when we run the program, the prompting message and the text appear on the same line.

```
echo −n Enter your name
read NAME
```

When run, we might see the following after we enter our name.

```
Enter your nameFrank
```

Obviously, this does not look very good, so we should probably alter the echo statement again as follows.

```
echo -n "Enter your name "
```

Alternatively, the read statement has an option, -p, that outputs a prompting message prior to the input. The format is

```
read -p "message" var1 var2 ...
```

The message must be enclosed in quote marks unless the message itself has no blank spaces. The quote marks are preferred though because you can add blank spaces after it. For instance:

```
read -p "Enter your name " NAME
```

A few other options that you might be interested in are as follows. With –N, you specify an integer number. The read statement then accepts input as usual but if the input exceeds the integer number in characters, only that number are used. For instance, `read -N 10` will only read up to the first 10 characters. The –N option will only store the input into one variable, so you would not use it if your input was for multiple variables.

The –t option allows you to specify a number of seconds. If the input is not received from keyboard within that number of seconds, the instruction times out and no value is stored in the variable (or variables if the instruction has multiple variables). A timeout of 0 is used if you wish to obtain input from a file. In such a case, read will indicate a successful input by returning the number 0. We examine input from files below.

The -s option has the read statement execute in silent mode. In this mode, any input characters are not echoed to the terminal window. This can be useful when inputting a password or other sensitive information. Finally, if you perform read with no variables, the Bash interpreter uses the environment variable REPLY to store the input in to.

Notice that by default, the read statement accepts input from the keyboard. Recall from Chapter 2 that one of the forms of redirection is <. Through this form of redirection, the Linux command accepts input from a file rather than keyboard. Since most Linux commands, by default, expect their input to come from disk file, the < form of redirection is not often used. However, if we execute a script and redirect input from a file, then the read statement accepts input from that file rather than keyboard. Consider the following script, we will assume is called `input.sh`.

```
#!/bin/bash
read X
read Y
read Z
echo $((X+Y-Z))
```

We can run this script in two ways, `./input.sh`, in which case X, Y, and Z are input from the keyboard, or `./input.sh < datafile.txt`, in which case X, Y, and Z obtain the values of the first three rows of data from the file datafile.txt. Any data that follows the first three rows is unused.

If the input from the user contains multiple data on one line, the script will yield an error because the value stored in X, Y, or Z is not a single datum. For instance, if the user were to input

```
5 10 <enter>
15 <enter>
20 <enter>
```

then X is storing 5 10, not just a single number. The statement `$((X + Y-Z))` then generates an error. Similarly, if we run the script by redirecting the input from datafile.txt and this file does not contain exactly one datum per line, we receive the same error. Alternatively, if datafile.txt has fewer than three rows of data, the remaining variables in the program are given the value NULL, which will have no impact on the arithmetic operation. For instance, if datafile.txt has the numbers 10 and 20, the echo statement would output 30 because Z was not given a value.

## 7.5 SELECTION STATEMENTS

Selection statements are instructions used to make decisions. Based on the evaluation of a condition, the selection statement selects which instruction(s) to execute. There are four typical forms of selection statement:

- The if-then statement—if the condition is true, execute the then clause.

- The if-then-else statement—if the condition is true, execute the then clause, otherwise execute the else clause.

- The if-then-elif-else statement—here, there are many conditions; execute the corresponding clause. If this sounds confusing, it is. We will examine it below.

- The case statement—in some ways, this is a special case of the if-then-elif-else statement. The case statement enumerates value-action pairs and compares a variable against the list of values. When a match is found, the corresponding action is executed.

In essence, these variations break down into one-way selection, two-way selection, and n-way selection (for the last two in the list).

### 7.5.1 Conditions for Strings and Integers

Before we explore the various forms of if-then statements, we need to understand the condition. A condition is a test that compares values and returns either true or false. The comparison is typically a variable to a value, a variable to another variable, or a variable to the value of an expression. For instance, does x equal y? is x greater than 5? does y + z equal a − b? In Linux, there are three forms of conditions

- compare two strings together using == or != (not equal to),

- compare two integers together using –eq, -ne, -lt, -gt, -le, -ge (equal to, not equal to, less than, greater than, less than or equal to, greater than or equal to respectively),

- compare a file against an attribute (for instance, is the file readable?).

Notice that string comparisons do not include less than, greater than, and so on.

Conditions can be written using one of two types of syntax. The more common approach is to use [] marks although the syntax has greater restrictions as we will see. The other approach uses (()). We first examine the use of [].

For [], the components that make up the condition must all be separated by spaces. To test the value in the variable NAME to the value Frank, use [ $NAME == Frank ]. Without spaces around $NAME, =, and Frank, you will receive an error message. In the case of [$NAME==Frank] will result in the error [Frank: command not found].

Comparing strings can be tricky. Imagine $NAME above stored "Frank Zappa" (recall, if a string has a space in it, we must insert it in quote marks). The natural comparison would be

```
[ $NAME == "Frank Zappa" ]
```

This will also cause an error. In this case, the evaluation of $NAME contains a blank space causing Bash to misconstrue that $NAME is not a single item. The error message is somewhat cryptic stating bash: too many arguments. The message conveys that there are too many arguments in the comparison. To enforce Bash to treat the values in $NAME as a single string, we enclose it in quote marks, as we did the right hand side of the comparison. Thus, we use "$NAME" instead of $NAME. Our comparison becomes

```
[ "$NAME" == "Frank Zappa" ]
```

Numeric comparisons are made using the two-letter abbreviations, such as –eq or –lt (equal to, less than). On either side of the comparison you can have variables, literal values, or arithmetic expressions. Arithmetic expressions should appear within the proper notation as described in Section 7.3. Let us assume X, Y, Z, and AGE all store numeric values. The following are possible comparisons.

- [ $X -ne $Y ]

- [ $Z -eq 0 ]

- [ $AGE -ge 21 ]

- [ $((X*Y)) -le $((Z+1)) ]

The alternate syntax for a condition is to use ((condition)) as in (($x > 0)). For this type of condition, you may omit any dollar sign prior to a variable name and utilize the relational operators (>, >=, <, <=, ==, !=) in place of the two-letter abbreviated comparison operators. You may also omit the blank spaces around each portion of the condition. For instance, you could replace [ $x -gt 0 ] with any of (($x>0)), (( $x > 0 )), ((x>0)), (( x>0 )), or (( x > 0 )). The earlier examples can be rewritten as follows.

- ((X!=Y))

- ((Z==0))

- ((AGE > 21))

- ((X*Y < Z + 1))

It should be noted that using the relational operators (e.g., <, >=) instead of the two-lettered versions in conditions with brackets can lead to erroneous behavior. We explore an example in the next subsection. Therefore, use the two-letter operators when comparing numeric values in the [] notation and use the relational operators in the (()) notation.

Compound conditions are those that perform multiple comparisons. In order to obtain a single true/false value from a compound condition, the comparisons are joined together using one of two Boolean operators, AND or OR (see the appendix if you are unfamiliar with the use of AND and OR). In Bash programming, AND is denoted as && and OR as ||. For instance, $X -ne $Y && $Y -eq $Z is true if the value in X does not equal the value in Y and the value in Y is equal to the value in Z. To further complicate the syntax, a compound conditional must be placed not in [] but in [[]], that is, double brackets. Alternatively, a compound conditional can also be placed in (()) symbols. Examples follow.

- [[ $X -eq 0 && $Y -ne 0 ]] —true if X is 0 and Y is not 0

- [[ $X -eq $Y || $X -eq $Z ]] —true if X equals either Y or Z

- [[ $X -gt $Y && $X -lt $Z ]] —true if X falls between Y and Z

- ((X > Y || X < Y–1)) —true if X is greater than Y or less than Y-1

Many introductory programmers make a mistake with compound conditions in that they think they can express a condition involving one variable in a shorthand way. For instance, if we want to test to see if X is equal to both Y and Z, the programmer may define this as

[[ $X -eq $Y -eq $Z ]]

or

[[ $X -eq $Y && $Z ]]

These are both incorrect. We must specify both X$ -eq $Y and $X -eq $Z, so the proper condition is

```
[[ $X -eq $Y && $X -eq $Z ]]
```

You can indicate NOT through the exclamation mark. We do not need to use NOT when comparing strings or numbers because we have "not equal to" available to us in the form of != and –ne. The use of ! is more useful when performing file comparisons.

Another variation for a condition is to use the test command. The test command is followed by the condition to be tested, but in this case, without either [] or (()) symbols. If your condition is a compound conditional, then use –a and –o in place of && and || respectively. Here are some examples.

- `test $AGE -ge 21`

- `test $AGE -gt 12 -a $AGE -lt 30`

- `test -z $AGE`

### 7.5.2 File Conditions
One feature missing from most programming languages that is somewhat unique of Bash's scripting language is the ability to test files for properties. This can be extremely useful in preventing run-time errors. Consider a script that is set up to run other scripts such as

```
#!/bin/bash
./script1 > output.txt
./script2 >> output.txt
./script3 >> output.txt
```

If one of the three script files does not exist or is not an actual file (it might be a directory) or is not executable, then running the script results in a run-time error that might leave the user wondering what went wrong. By testing file properties, the above script can be improved. We could improve the above script by testing each file and only executing it if it exists, is a regular file, and is executable.

The file comparisons use two formats

```
[ comparison filename ]
```

and

```
[ ! comparison filename ]
```

The comparisons are shown in Table 7.3. The use of ! allows you to negate the result. For instance, if you want to test if a file *does not* exist, you would use `[ ! -e filename ]`.

TABLE 7.3   File Operators

| Comparison Operator | Meaning |
|---|---|
| −e | File exists |
| −f | File is regular (not a directory or device type file) |
| −d | File is a directory |
| −b | File is a block device |
| −c | File is a character device |
| −p | File is a named pipe |
| −h, −L | File is a symbolic link |
| −S | File is a domain socket |
| −r, −w, −x | File is readable, writable, executable |
| −u, −g | User ID/Group ID set (the 's' under executable permissions) |
| −O, −G | You or your group owns the file |
| −N | File has been modified since last read |

In addition to the comparisons in the table, you can also compare two files to see if one is older than the other using −ot, one is newer than another using −nt, or two files are the same using −ef (useful in the case of links). For instance, the condition [ f1 −ot f2 ] is true if f1 is older than f2.

Here are some examples and their meanings.

- [ −f file1.txt ] is true if file1.txt exists and is a regular file.

- [ −h file1.txt ] is true if file1.txt is a symbolic link to another file.

- [[ −r file1.txt && −w file1.txt ]] is true if file1.txt is both readable and writable.

- [ ! −e $FILENAME ] is true if the file whose name is stored in the variable FILENAME does not exist.

- [ −O file1.txt ] is true if file1.txt is owned by the current user.

- [ ! file1.txt −nt file2.txt ] is true if file1.txt is not newer than file2.txt.

- [ $FILE1 −ef $FILE2 ] is true if the file whose name is stored in the variable FILE1 is the same as the file whose name is stored in the variable FILE2. Notice that since FILE1 and FILE2 are variables, we could test [ $FILE1 == $FILE2 ] for the same physical string, but this would not be true if the two filenames are of files who are linked together by a hard link.

There are also comparisons to test a variable. You can test to see if a variable is storing a value or stores the special value null. To test for null, use [ −z $var ] and to test to see if it has a value (not null), use [ −n $var ]. You can also test for nonnull using [ $?variable ].

These tests are useful to avoid run-time errors when you attempt to perform an operation on a variable that may not yet have a value.

### 7.5.3 The If-Then and If-Then-Else Statements

Now that we have viewed the types of comparisons available, we can apply them in selection statements. The first type of selection statement is the if-then statement, or a one-way selection. If the condition is true, the *then clause* is executed, otherwise the *then clause* is skipped. The syntax for the if-then statement is as follows.

```
if [ condition ]; then action(s); fi
```

The syntax is somewhat bizarre. First, there must be a semicolon after the ] after the condition. Second, if there are multiple actions, they must be separated by semicolons. Finally, the statement must end with the word fi. However, the use of the semicolons is only needed if we place the entire instruction on a single line. If we separate each part onto a separate line, the semicolons are not needed. Here are several examples.

- `if [ -e file1.sh ]; then ./file1.sh; fi`
  - if the file exists, execute it
- `if [ $X -gt 0 ]; then COUNT=$((COUNT+1)); fi`
  - if the value in X is greater than 0 add 1 to COUNT
- `if [ $NAME == $USER ]; then echo Hello master, I am ready for you; fi`
  - specialized greeting if the value in NAME is equal to the user's login
- `if [[ $X -ne 100 && $USER != Frank ]]; then X=0; Y=100; NAME=Joe; fi`

Let us rewrite the last two of these examples as they might appear in a script file.

```
if [ $NAME == $USER ]
    then
        COUNT=$((COUNT+1))
    fi
if [[ $X -ne 100 && $USER != Frank ]]
    then
        X=0
        Y=100
        NAME=Joe
    fi
```

Notice that the indentation above is strictly for our own readability and not required by the Bash interpreter. The following three sets of code are equivalent as far as the Bash interpreter is concerned.

- `if [ $X -gt $Y ]; then echo greater; X=0; fi`

- ```
  if [ $X -gt $Y ]
       then
             echo greater
             X=0
  fi
  ```

- ```
  if [ $X -gt $Y ]
  then
  echo greater
  X=0
  fi
  ```

Recall from the last subsection, we noted that using the relational operators (e.g., <, >=) in place of the two-letter abbreviations can yield an improper output. Consider the following script.

```
#!/bin/bash
X=0
Y=1
if [ $X > $Y ]; then echo greater; fi
if [ $X == 0 ]; then echo zero; fi
```

This script, when run, outputs

```
greater
zero
```

There is no logical reason why greater should be output as it is clearly incorrect. For $X, we are testing it to 0 using the string comparison == (rather than the numeric comparison –eq). However, X does store 0, which can be interpreted as a string or a number, so the output of zero does make sense.

In each of the examples covered so far, if the condition is true, the action(s) is performed and if not, the statement ends and the interpreter moves on to the next instruction. In many situations, we have an alternative action that we would like executed if the condition is false. For this, we need a two-way selection statement, or an if-then-else* statement. The difference between the if-then and if-then-else statements is the inclusion of the word "else" and an *else clause*. Here are a few examples.

- `if [ $SCORE -ge 60 ]; then GRADE=pass; else GRADE=fail; fi`

  - The variable GRADE is assigned either pass or fail based on the value in SCORE.

---

\* In some languages, the word "then" is omitted and so the statements are referred to as an if-statement and an if-else statement. This is true in such languages as C, C++, Java, and Perl to name a few.

- `if [ $AGE -ge 21 ]; then echo You are allowed to gamble, good luck; else echo You are barred from the casino, come back in $((21-AGE)) years; fi`

  - this statement selects between two outputs based on comparing your age to 21, notice the use of $((21-AGE)) in the echo statement, this computes the number of years under 21 that the user is

- `if [ $X -le 0 ]; then echo Warning, X must be greater than 0; else SUM=$((SUM+X)); fi`

  - we are adding X to SUM if the value in X is greater than 0.

### 7.5.4 Nested Statements

The need for an n-way selection statement arises when there are more than two possible outcomes. For the if-then and if-then-else, there is a single condition that is either true or false. But there are many situations where we will need to use multiple comparisons. For instance, when assigning a letter grade based on a student's numeric score on an assignment, one typically uses a 90/80/70/60 grading scale. This can be expressed using an n-way selection statement by saying "if the grade is 90 or higher, then assign an A, else if the grade is 80 or higher, then assign a B, else if the grade is 70 or higher, then assign a C, else if the grade is 60 or higher, then assign a D, else assign an F." Notice how this reads in English with words if, then, else if, and else. These words inform us as to how to express this using what is called a nested if-then-else statement.

In Bash, we use a slight variation on the expression "else if" by using the term "elif" in its place. Thus, our full instruction is referred to as an if-then-elif-else statement. The if statement and each elif statement will have a condition associated with it. If the condition is true, the corresponding Then clause is executed. The else clause at the end of the statement is optional depending upon whether it is needed or not. Here, we see a full example comparing the numeric value in the variable SCORE to each of 90, 80, 70, 60 to determine what letter grade to store in the variable GRADE.

```
if [ $SCORE -ge 90 ]
     then GRADE=A;
elif [ $SCORE -ge 80 ]
     then GRADE=B;
elif [ $SCORE -ge 70 ]
     then GRADE=C;
elif [ $SCORE -ge 60 ]
     then GRADE=D;
else GRADE=F;
fi
```

The indentation differs a little from the previous examples. Again, indentation is immaterial to the Bash interpreter and only used here for readability. We could also write it as shown below.

```
if [ $SCORE -ge 90 ]
   then GRADE=A;
   elif [ $SCORE -ge 80 ]
        then GRADE=B;
        elif [ $SCORE -ge 70 ]
             then GRADE=C;
             elif [ $SCORE -ge 60 ]
                  then GRADE=D;
                  else GRADE=F;
fi
```

Also notice that any if-then-elif-else statement will have (no more than) one else clause and exactly one fi statement (always required to end any if/if-then/if-then-elif-else statement). Without the else clause, an if-then-elif statement may not execute any action. If, for instance, we omit the final else clause in the example above and SCORE was 59, then none of the conditions would be true and therefore none of the then clauses would execute. The result is that GRADE would have no value.

Let us consider a more complicated example. We input three values stored in X, Y, and Z. We will assume they are all numeric values. We want to find the largest of the three.

```
#!/bin/bash
read -p "Enter three numbers: " X Y Z
if [ $X -gt $Y ]
    then
        if [ $X -gt $Z ]
            then Largest=$X
            else Largest=$Z
        fi
    elif [ $Y -gt $Z ]
        then Largest=$Y
    else Largest=$Z
fi
```

This code contains a single if-then-elif-else statement. However, the then clause contains its own if-then-else statement. Thus, there are two if statements involved in this instruction, one to end the inner if-then-else statement and one to end the outer if-then-elif-else statement. The idea of an "inner" and "outer" statement is what yields the term *nested* when we talk about nested statements. Here we have a nested if-then-else statement.

Let us put all of these ideas together and write a script. We want to write a generic calculator script. The script expects three inputs: two numbers and an arithmetic operator, input in infix notation (as in 5 + 3 or 29 * 21). We will reference the three input parameters as $1, $2, and $3. We will also test to make sure that there are three parameters otherwise the user did not provide a legal arithmetic statement. We test for three parameters by comparing the parameter value $# to 3. We will assume the arithmetic operators permissible are +, −, *, /, %. Our script might look like this.

```
#!/bin/bash
if [ $# -ne 3 ]
    then echo Illegal input, did not receive 3 parameters
    elif [ $2 = "+" ]; then echo $(($1+$3))
    elif [ $2 = "-" ]; then echo $(($1-$3))
    elif [ $2 = "*" ]; then echo $(($1*$3))
    elif [ $2 = "/" ]; then echo $(($1/$3))
    elif [ $2 = "%" ]; then echo $(($1%$3))
    else echo Illegal arithmetic operator
fi
```

If this script is stored in the file calculator, we might invoke it as

```
./calculator 315 / 12
```

or

```
./calculator 41 + 815
```

The logic behind this script should be fairly self-explanatory. There is one logical flaw with the script that has nothing to do with the code itself. You might recall from Chapter 2 that the symbol * when used in Linux commands causes the Linux interpreter to perform filename expansion prior to the execution of the script. Imagine that you are running this script in a directory that contains five files. The command ./calculator 5 * 3 <enter> is expanded to be ./calculator 5 *file1 file2 file3 file4 file5* 3 where file1, file2, and so on are files found in the current directory. Now we are providing the script with seven parameters instead of three and so obviously we get the "Illegal input" message rather than the result that we were looking for! To avoid this problem, we will use 'm' instead of '*.' This would change the third elif statement into

```
elif [ $2 = "m" ]; then echo $(($1*$3))
```

### 7.5.5 Case Statement

The case statement is another form of n-way selection statement which is similar in many ways to the if-then-elif-else statement. The instruction compares a value stored in a variable (or the result of an expression) against a series of enumerated lists. Upon finding a match, the corresponding action(s) is(are) executed. The case statement (known as a switch in C/C++/Java) is more restricted in other languages than it is in Bash. For instance, in C/C++/Java, the expression can only be a variable storing an ordinal value.[*] Also, most languages require that an enumerated list be just that, an enumerated list. Thus, if we wanted

---

[*] An ordinal value is one that has a specific predecessor and a specific successor found with integers and characters. For instance, 5's predecessor is 4 and its successor is 6 and 'c' has a predecessor of 'b' and successor of 'd.' On the other hand, a number like 2.13513 or a string like "Hello" has no specific predecessor or successor.

to implement the previous grading script using a case statement, in most languages we would have to enumerate all of the values between 90 and 100, between 80 and 89, between 70 and 79, and so forth. The Bash version though gives us a bit of a break in that we can employ wildcard characters to reduce the size of an enumerated list.

The Bash form of the case instruction is

```
case expression in
     list1) action(s);;
     list2) action(s);;
     ...
     listn) action(s);;
esac
```

Notice the syntax. First, each action ends with two consecutive semicolons. Second, the enumerated list ends with the close parenthesis sign. Finally, the entire statement ends with an esac statement.

The expression can be a value from a variable, as in $X, the result of an arithmetic operation such as $((X*5)), or the result from a Linux command. The enumerated lists can be one or more literal strings and can combine Bash wild cards. For instance, * by itself would match everything and can be used as a default clause in place of the final enumerated list. Any of the actions can contain multiple instructions, in which case the instructions are separated by semicolons.

Revising our previous calculator program using the case statement yields the following script.

```
#!/bin/bash
if [ $# -ne 3 ]
    then echo Illegal input, did not receive 3 parameters
    else case $2 in
         +) echo $(($1+$3));;
         -) echo $(($1-$3));;
         m) echo $(($1*$3));;
         /) echo $(($1/$3));;
         %) echo $(($1%$3));;
         *) echo Illegal arithmetic operator;;
    esac
fi
```

We see that the last enumerated list item is *) which is the default clause so that if none of the other lists match the value of $2, then this action is performed. Therefore, * is not used to indicate multiplication but instead "everything else."

Aside from the * wildcard, you can also use [...] and [!...] to specify a pattern in a list that contains any of the enclosed characters or does not contain any of the enclosed characters respectively. For instance, [Pp][Cc] would match PC, Pc, pc, pC, and [A-Z][!A-Z] would match two characters where the first is an upper-case letter and the second is anything other than an upper-case letter. We can also use | to separate full strings such as spring|summer|fall|winter.

Let us consider another example. We ask the user for their favorite color and based on the response, output a message.

```
echo What is your favorite color?
read color
case $color in
     red) echo Like red wine ;;
     blue) echo Do you know there are no blue foods? ;;
     green) echo The color of life ;;
     purple) echo Mine too! ;;
     *) echo That's nice, my favorite color is purple ;;
esac
```

We tend to see case statements in use in shell scripts used to control system services, in situations like the following. Here, we have a script that will perform an operation as submitted by the user to start or stop a script.

```
if [ $# -eq 0 ]; then echo Error, no command supplied
else
     case $1 in
          start)/sbin/someservice -s ;;
          stop)/sbin/someservice -k ;;
          restart)/sbin/someservice -k
                  /sbin/someservice -s ;;
          status)/sbin/someservice -d ;;
          *) echo Usage:/etc/init.d/someservice [start |
             stop | restart | status] ;;
esac
fi
```

First, we make the assumption that the user has issued the command /etc/init.d/someservice *command* where someservice is a service name. Second, we make the assumption that the service program is itself stored in /sbin and will receive the option –s to start, -k to stop, and –d to display its status. We also assume that the user entered a *command* which in this case should be one of start, stop, restart, or status. Without a command, the script responds with the error message that no command was supplied and with any command other than start, stop, restart, or status, the script responds with the default usage statement explaining how to use this script.

Note that the action corresponding to restart has two instructions, to call /sbin/someservice –k and then /sbin/someservice –s. These could be placed on a single line, in which case they would need to be separated by semicolons. The * enumerated item is our default case, used to output to the user how to use the instruction. We explore this style of code in Chapter 11 when we explore one of the service scripts in detail.

The following examples demonstrate further use of the wildcards. In the first case, we are testing various possible Yes/No answers for the user. In the second case, we are comparing a file name against a list of possible extensions to determine how to handle the file (in this case, we are attempting to compile the program).

```
echo -n Do you agree to the licensing terms of the program?
read answer
case $answer in
    [yY] | [yY][eE][sS]) ./program ;;
    [nN] | [nN][oO] ) echo I cannot run the program ;;
    *) echo Invalid response ;;
esac

if [ $# -eq 0 ]; then echo Error, no directory provided
else
    for file in $1; do
        case $file in
            *.c) gcc $file ;;
            *.java) javac $file ;;
            *.sh)./$file ;;
            *.*) echo $file is of an unknown type
            *) echo $file has no extension to match
        esac
    done
fi
```

Finally, you can alter the case statement's behavior by using ;& instead of ;; to end an action. With ;; the action, after executing, causes the case statement to terminate. Instead, if an action executes and ends with the notation ;&, then the case statement continues by comparing the remaining lists. Thus, such a case statement could conceivably execute multiple actions. In the following code skeleton, any combination of the first three could match. The first case represents that the variable has a lower-case letter stored in it. If so, it executes its action(s) but continues on to the second case because of the ;& ending the instruction. The second case represents that the variable has an upper-case letter stored in it. If so, the second set of action execute and again, the instruction continues on. The third case represents that the variable has a digit in it. In this situation, if it matches it executes the action(s) and then the case statement terminates. If this third case does not match, the default executes. Thus, by mixing the ;; and ;&, we can control how far into the case statement we continue if we have matches. With only ;; we end the case statement as soon as we find a match.

```
*[a-z]*) ... ;&
*[A-Z]*) ... ;&
*[0-9]*) ... ;;
*) ...
```

## 7.5.6 Conditions outside of Selection Statements

There is one other form of selection statement to explore which is a short-hand notation for an if-then statement by using *short circuiting*, which can be applied by using && or ||. With respect to &&, if the first half of the condition is false, there is no need to compute the second half because it is immaterial. Similarly for ||, if the first half of the condition is true then there is no need to compute the second half. For instance, consider

```
A > B  &&  B > C
```

If A is not greater than B, then there is no need to compare B to C because the entire expression must be false. Alternatively, in

```
D > E  ||  F > G
```

If D is greater than E, there is no need to compare F and G because the entire expression is true.

In Linux, we apply this concept to pair up a condition and an executable statement as follows.

- `[ some condition ] && some action`
- `[ some condition ] || some action`

In the first case, if the condition is false, the action is not performed. In the second case, if the condition is true, the action is not performed. How might we put this to use?

The most common situation is to combine a condition that tests a file to see if it is accessible with an instruction that accesses the file. Consider, for instance, that we expect to find the script somescript.sh, which should be a regular file. If the file is found and is regular, we want to execute it. We could use an if-then statement of the form

```
if [ -f somescript.sh ]; then ./somescript.sh; fi
```

Or we could use the short circuiting approach

```
[ -f somescript.sh ] && ./somescript.sh
```

Another example might test a variable for a value. If the variable does not have a value, we should not use it.

```
if [ -r $var ]; then echo Variable var has no value; fi
```

Can be rewritten as

```
[ -r $var ] || echo Variable var has no value
```

## 7.6 LOOPS

### 7.6.1 Conditional Loops

The Bash scripting language has three types of loops: *conditional* loops, *counter-controlled* loops, and *iterator* loops. A conditional loop is a loop controlled by a condition, as described in Section 7.4. There are two versions of the conditional loop: a while loop and an until loop. They are both *pre-test* loops meaning that the condition is tested before the loop body is executed. The difference between the two is in the semantics of the condition. The while loop iterates *while* the condition is true. The until loop iterates *until* the condition becomes true.

The syntax of the two loops is shown below.

```
while [ condition ]; do
        action(s);
done

until [ condition ]; do
        action(s);
done
```

As with other Bash syntax, if you separate the word do from the condition, then the semicolon can be omitted. Additionally, if there are multiple actions, the semicolons separating them can be omitted if the actions are on separate lines. The conditions must conform to the conditions we discussed in Section 7.4.

Let us examine an example illustrating both loops. The following two loops will input numbers from the user and sum those values together. The condition to exit the loop occurs when a negative number is entered. The while loop's condition is [ $VALUE -ge 0 ] while the until loop's condition is the opposite, [ $VALUE -lt 0 ].

```
SUM=0
read -p "Enter the first number, negative to exit" VALUE
while [ $VALUE -ge 0 ]; do
     SUM=$((SUM+VALUE))
     read -p "Enter next number, negative to exit" VALUE
done

SUM=0
read -p "Enter the first number, negative to exit" VALUE
until [ $VALUE -lt 0 ]; do
     SUM=$((SUM+VALUE))
     read -p "Enter next number, negative to exit" VALUE
done
```

Conditional loops can be controlled by a number of different types of logic. For instance, in the above, the user controls when to exit the loop by entering a negative number. This is sometimes referred to as a *sentinel* loop in that a sentinel value is used as a terminating condition. Here, any negative number is the sentinel value. We can also control a conditional loop by asking the user to respond to a prompt of whether to continue or to exit. Below is a similar set of code to the two previous examples but requires the extra step of asking the user if there are more numbers to input.

```
SUM=0
read -p "Do you have numbers to sum?" ANSWER
while [ $ANSWER == Yes ]; do
    read -p "Enter the next number" VALUE
    SUM=$((SUM+VALUE))
    read -p "Do you have additional numbers?" ANSWER
done
```

Another way to control a conditional loop is to base the condition on the result of a computation. Below is a loop that will compute and output all of the powers of 2 less than 1000. When the loop exits, VALUE will be greater than or equal to 1000 (in fact, it will be 1024), which is not printed out because it is greater than the largest value that the code is to print out.

```
VALUE=1
while [ $VALUE -lt 1000 ]; do
    echo $VALUE
    VALUE=$((VALUE*2))
done
```

### 7.6.2 Counter-Controlled Loops

Another type of loop in Bash is the counter-controlled loop. This loop is a variation of the C/Java for loop. In this loop, you specify three pieces of information: a variable *initialization*, a loop *continuation condition*, and a variable *increment*. The format of this instruction is

```
for ((initialization; condition; increment));
    do
        loop body
    done
```

When this form of for loop executes, it first performs the initialization. This will initialize the variable that is to be used in both the condition and increment. The condition is then tested. If the condition is true, the loop body executes and then the increment instruction is performed. The condition is then retested and if true, the loop body executes followed by the increment instruction. This repeats until the condition becomes false. For those of you unfamiliar with C or Java, this for loop can be a challenge to use both syntactically and logically.

What follows are several examples of the for loop statement (without the loop body or do/done statements). The notation i++ is a shortcut for saying i=i+1, or incrementing i by 1.

- `for ((i=0; i < 100; i++))` – iterate from 0 to 99
- `for((i=0; i < 100; i=i+2))` – iterate from 0 to 99 by 2s
- `for ((i=100; i > j; i--))` – iterate from 100 down to j, whatever value j is storing

The syntax for the initialization, condition, and increment are similar to the syntax that we saw earlier in conditions that used (()) instead of []. In this case, you can forego the $ preceding any variable name, you do not need to insert blanks around each portion of the statement, you can use the relational operators (<, >=, etc) rather than the two-letter conditions, and you are free to use the shortcut operators such as ++ or +=.

## 7.6.3 Iterator Loops

The final loop available in Bash is another for loop. In this case, it is an iterator loop. An iterator loop is one that executes the loop body one time per item in a list. The iterator loop uses the keyword for, but is more like a for each loop as found in C# and perl. The Bash for loop has the form

```
for VAR in LIST do; action(s) done
```

The variable, VAR, takes on each element of the LIST one at a time. You can use the VAR in the action(s). The variable VAR is often referred to as the loop variable.

The LIST can be an enumerated list such as (1 2 3 4 5), the result of a Linux command which generates a list (e.g., ls, cat), a list generated through filename expansion (for instance *.txt would expand into all file names of.txt files in the current directory) or the list of parameters supplied to the script using $@.

The following example will sum up the numbers passed to the script as parameters. The variable VALUE obtains each parameters supplied by the user, one at a time. That value is then added to the SUM. If this script were called adder, we might execute it using ./adder 581 32 115 683 771 where the five numbers are used for the list in place of $@.

```
#!/bin/bash
SUM=0
for VALUE in $@; do
     SUM=$((SUM+VALUE))
done
echo $SUM
```

In this example, notice how the loop variable, VALUE, is used in the loop body. If the list that the loop variable iterates over consists of file names, we could apply file commands.

We might, for instance, wish to see the size of a series of files. The wc command will provide us this information.

```
#!/bin/bash
for filename in $@; do
     wc $filename
done
```

To ensure that all of the parameters were actually files, we should enhance this script to test each filename. We would use [ -e $filename ] to make sure the file exists. Our new script is

```
#!/bin/bash
for filename in $@; do
     if [ -e $filename ]; then
          wc $filename
     fi
done
```

Earlier, we introduced the instruction shift. Now that we have seen the use of $@, we will reconsider shift. The easiest way to iterate through the parameters in a list is through the for loop iterating over $@. For instance,

```
for value in $@; do . . .
```

However, we can also use shift and the counter-controlled for loop.

Below, we convert the earlier example of summing up the values in $@ to use the counter-controlled for loop and the shift instruction. Recall that shift will shift each parameter down one position. So, if there are four parameters, stored in $1, $2, $3, and $4, shift will cause $2 to shift into $1, $3 into $2, and $4 into $3. The original value of $1 is gone and now we have three values. Executing shift again will move $2 into $1 and $3 into $2. In the script below, we first obtain the number of parameters, as stored in $#. We now use a for loop to iterate that many times, using the value in $1 as the current parameter. The shift instruction then moves each parameter down so that there is a new value in $1.

```
#!/bin/bash
NUMBER=$#
SUM=0
for ((i = 0; i < NUMBER; count + +)); do
     SUM=$((SUM+$1))
     shift
done
echo $SUM
```

### 7.6.4 Using the Seq Command to Generate a List

We can also use the iterator version of the for loop as a counter-controlled loop. This is done by replacing the list with a sequence generated by the seq command. The seq command returns a list of numbers based on between one and three parameters. The variations available for seq are shown below.

- *seq last*—generate the list of numbers from 1 to *last* by 1
- *seq first last*—generate the list of numbers from *first* to *last* by 1
- *seq first step last*—generate the list of numbers from *first* to *last* by *step*

The values of first, last, and step can be literal values such as seq 1 10 or they can be values stored in variables, as in seq $FIRST $LAST, or they can be values returned from Linux commands. The result of seq is a list so for instance seq 5 would return (1 2 3 4 5) while seq 3 10 3 would return (3 6 9).

We can use seq to control a for loop in one of two ways. First, we can store the result of seq in a variable, such as LIST='seq 1 10' and then reference the variable in the iterator for loop. Second, we can insert the 'seq' instruction in the iterator for loop itself. The following loop iterates from 1 to 10 by 2, printing each value. This will produce the list of odd numbers from 1 to 10, printing each on a separate line.

```
#!/bin/bash
for number in 'seq 1 2 10'; do
    echo $number
done
```

If we replace the echo statement with echo -n "$number", we would produce the list on one line.

### 7.6.5 Iterating over Files

One of the more common uses of the Bash iterator for loop is to iterate through the files in a directory. The following example takes each .txt file and examines whether it is writable or not. If it is writable, its filename is output.

```
#!/bin/bash
for filename in *.txt; do
    if [ -w $filename ];
        then echo $filename
    fi
done
```

A more meaningful example is to execute several scripts in a directory. Commonly, startup scripts are placed into a single directory. We might use a for loop to iterate over the items in the directory, and if they are regular files, execute them one at a time. We would not want to execute items that were not regular files (such as directories). The script below also reports on the number of scripts started.

```
#!/bin/bash
TOTAL = 0
for filename in *; do
     if [ -f $filename ]; then
          ./$filename
          TOTAL=$((TOTAL+1))
     fi
done
echo $TOTAL scripts started
```

Let us consider a script to sum up the size of the normal files in a directory. We can obtain each file's size in a number of ways, but we will use the du command. We will iterate through the files in the directory using a for loop and then use an assignment statement like

```
TOTAL=$((TOTAL+`du $filename`))
```

Unfortunately, this will not work because du returns two values, the file size and the file name. To solve this problem, we can also employ awk.

To simplify the logic, we will divide the computation step into three parts. First, use du to obtain the file size and name. Second, pipe the result from du to awk where we will print out the file size. We will take the result of these two combined instructions and place it into a variable, TEMP. Third, we will add TEMP to TOTAL. The script is as follows.

```
#!/bin/bash
for file in *; do
     if [ -f $file ]; then
          TEMP=`du $file | awk '{print $1}'`
          TOTAL=$((TOTAL+TEMP))
     fi
done
echo Total file size for regular files in ~ is $TOTAL
```

We need to be careful with the types of operations that we might place inside ` ` marks in the for loop. Consider the following example.

```
for filename in `ls -l`; do
     if [ -w $filename ]; then ...
     fi
done
```

We might think that, as shown earlier, this would iterate through all of the files in the current directory and perform some operation on the writable files. However, with ls –l, we do not merely receive a list of filenames. Instead, the long listing of the entire directory is supplied. Consider this short excerpt of ls –l from the /etc directory:

```
-rw-r--r--. 1 root root   58 Dec  8  2011 networks
-rw-r--r--. 1 root root 3390 Dec  7  2011 nfsmount.conf
-rw-r--r--. 1 root root 2144 Aug 13 09:51 nscd.conf
```

As a list, returned from ls –l, the for loop would iterate over each of these items. Thus, the variable, filename, in the above code excerpt takes on each of the following items one at a time. As most of these are not filenames, the condition, [ -w $filename ], will yield an error.

```
-rw-r--r--.
1
root
root
58
Dec
8
2011
networks
-rw-r--r--.
1
root
root
3390
Dec
7
2011
nfsmount.conf
-rw-r--r--.
1
root
root
2144
Aug
13
09:51
nscd.conf
```

Let us put together everything we have seen about the iterator for loop and write a script, called word_match.sh, to count the number of occurrences of a particular string in a given file. The script will expect to receive two parameters, a filename and a string. If we did not get two parameters, or the first parameter is not an existing, regular file, the script will output an error message. Otherwise, it will use cat to provide a listing of the contents of the file. A for loop will then iterate over that listing and compare each item (string) to the second parameter. We will count the number of matches and output a summary at the end.

```
#!/bin/bash
count=0
if [ $# -ne 2 ]; then
    echo ERROR, illegal number of parameters
elif [ ! -f $1 ]; then
    echo ERROR, $1 is not a regular file
else
    for word in `cat $1`; do
        if [ $word = $2 ];
            then count=$((count+1))
        fi
    done
    echo The number of occurrences of $2 in $1 is $count
fi
```

You invoke the above script as ./word_match.sh *filename* word where *filename* is the file you are examining and *word* is the string you are searching for.

### 7.6.6 The While Read Statement

Earlier, we visited the while loop. A variation of the while loop is known as the while read loop. This loop combines the while statement and the read statement. With this version of the while statement, there is no condition. Instead, the loop continues to execute while there are still data to read. The form of this instruction is

```
while read varlist; do
    loop body
done
```

The *varlist* is a list of variable names. The loop expects to input a value for each variable in this list for each iteration. When there are no more data received, the loop terminates. No more data is either indicated by the user entering control+d if the input is coming from keyboard, or EOF (end of file) if input is from a file.

The following script, called sum, inputs pairs of numbers multiplying the two of each pair together, summing the resulting products into a SUM variable. For instance, if the input data is the list: 10 20 30 40 50 60, the script would compute 10 * 20 + 30 * 40 + 50 * 60 = 4400.

```
#!/bin/bash
SUM=0
while read X Y; do
    SUM=$((SUM+X*Y))
```

```
done
echo $SUM
```

Now to run this script, you could either invoke it normally, as in ./sum, or by redirecting input from an input file as in ./sum < data.txt.

There are some challenges in ensuring that this works correctly. You will need a data file that contains the pairs of numbers. The format of the data in the file must match the read statement. In this case, we are reading X and Y in a single statement, so X and Y must appear on the same line in the file. If each pair was on a separate line, the script would obtain the datum on the line to store in X and assign 0 for Y since there is no second datum on the line. The result would be a sum of 0.

Alternatively, if the user did not redirect input from a file, then input would come from the keyboard. However, there is no prompting message so the user may not know what is expected of them. Further, the user is not supposed to just enter a single datum but pairs of data. In order to end input, the user would have to type control + d.

An alternative to requiring that the user redirect input to the script is to handle the redirection within the script itself. This is accomplished with the done statement that ends the while read loop. You would specify done < inputfile. Our previous script becomes the following.

```
#!/bin/bash
SUM=0
while read X Y; do
     SUM=$((SUM+X*Y))
done < data.txt
echo $SUM
```

The disadvantage of this approach is that we are hardcoding the input file's name in the script. This removes control from the user who might want to run the script on several different input files. One solution to this is to replace done < data.txt in the script with done < $1 and expect that the user will enter the data file name as the first (or only) parameter to the script. For instance, we could invoke the script as ./sum data.txt.

## 7.7 ARRAYS

Arrays are data structures that store multiple values. They are a form of container structure. In most languages, an array must be a *homogenous* type meaning that all of the elements stored in the array must be of the same type. But in Bash, with only two types, strings and integers, an array could store either or both types of values.

As arrays store multiple values, how do you indicate the value you want to access? This is done through an array *index*. The index is the numeric position in the array of the element. Like C/C++ and Java, numbering starts at 0, so the first element is at position 0, the second element is at position 1, and the nth element is at position n − 1.

### 7.7.1 Declaring and Initializing Arrays

Unlike most languages, array variables do not need to be declared in a Bash script. Instead, you just start using the array. To place values into an array, you can either place individual values into array locations or you can initialize the entire array. We look at the second approach first. In these examples, we will use a variable called ARRAY to be our array.

```
ARRAY=(apple banana cherry date)
```

This assigns ARRAY to have four values. The values are stored in individual array locations with apple at position 0, banana at position 1, cherry at position 2, and date at position 3. Although the above approach is simple, there are several other ways to place values into the array.

Another way to initialize the array is to use the following notation.

```
declare -a ARRAY=(apple banana cherry date)
```

In some versions of Linux and Unix, the declaration is required. We will assume it is not. You can also declare an array using ARRAY=(). This creates an initially empty array.

The second way to insert values into an array is to place values into individual array locations. We reference an array location using the notation ARRAY[$i$] where $i$ is the position. Recall that if our array stores n elements, then $i$ will range between 0 and n − 1. We can therefore insert the four pieces of fruit into our array using

```
ARRAY[0]=apple
ARRAY[1]=banana
ARRAY[2]=cherry
ARRAY[3]=date
```

We can also initialize the array using this notation.

```
ARRAY=([0]=apple [1]=banana [2]=cherry [3]=date)
```

You might wonder why this last approach is worthwhile since it is obviously longer and more typing than using ARRAY=(apple banana cherry date). Imagine that you want to insert values into some locations but not consecutive locations. You want room to insert other items into positions in between each of these. You could accomplish this through one of the following approaches. First, we could initialize each array location individually.

```
ARRAY[0]=apple
ARRAY[2]=banana
ARRAY[4]=cherry
ARRAY[6]=date
```

Alternatively, we could combine these assignments into one statement using the following notation.

```
ARRAY=([0]=apple [2]=banana [4]=cherry [6]=date)
```

In either case, ARRAY[1], ARRAY[3], and ARRAY[5] are currently empty.

## 7.7.2 Accessing Array Elements and Entire Arrays

Now that we can insert items into an array, how do we get values out of the array? For any other variable, the approach is to reference $variable. Can we use $ARRAY or $ARRAY[0] for instance? The answer, unfortunately, is no. To retrieve an item from an array is slightly more cumbersome. We must wrap the ARRAY[] inside of ${}. Therefore, to obtain a value from ARRAY, we use ${ARRAY[0]} and ${ARRAY[1]}, and so on. To obtain the ith value in the array, where i is a variable storing an integer number, we use ${ARRAY[i]}. Notice that we do not precede i with a $.

In most languages, you are not able to inspect an entire array with one instruction. In Bash, the notation ${array[@]} will return the entire array as a list. This allows you to use the array in an iterator for loop much like we saw with the $@ parameter. The notation ${array[*]} will also return the array as a list. There is a distinction between these two notations though if you place them in quote marks. Imagine that ARRAY stores (apple banana cherry date). Then, we see the following behavior:

- "${ARRAY[@]}" returns the list ("apple" "banana" "cherry" "date")

- "${ARRAY[*]}" returns the list ("apple banana cherry date")

So, with @, you receive each array item individually quoted while with *, you receive the entire array quoted as a single list. You can also obtain the number of elements in the array through either ${#ARRAY[@]} or ${#ARRAY[#]}.

We will typically control access to our array through either a counting for loop or an iterator for loop. The iterator for loop is easier as there is less code to deal with. For the counter for loop, we will use ${#ARRAY[@]} to obtain the number of elements in the loop and iterate from 0 to ${#ARRAY[@]}-1. If our loop variable is i, then we access each element using ${ARRAY[i]}.

In summary, we have the following ways to access an array.

- ${array[i]}—access the ith element of array
- ${array[@]}—return the entire array as a list
- ${array[*]}—return the entire array as a list
- "${array[@]}"—return the entire array as a list where each item is quoted
- "${array[*]}"—return the entire array as a list where the entire list is quoted

- ${#array[@]}—return the size of the array

- ${#array[*]}—return the size of the array

Here are two example loops with an array called a:

- for value in ${a[@]}; do ...$value... ; done

  - $value stores each array value

- for ((i = 0; i < ${#a[@]}; i++)); do ...${a[i]}... ; done

  - ${a[i]} stores each array value

## 7.7.3 Example Scripts Using Arrays

Let us take a look at how you might use an array. In the following script, an array of URLs is presented to the wget program to download each file. We test each file afterward to see if it worked. The script outputs warning messages for each URL not found.

```
#!/bin/bash
list=(www.nku.edu/ ~ foxr/CIT371/file1.txt
      www.uky.edu/cse/welcome.html
      www.uc.edu/cs/csce310/hw1.txt
      www.xu.edu/welcome.txt www.cs.ul.edu/ ~ wul/music/lyrics.txt)
for i in ${list[@]}; do
      wget $i
      if [ ! -e $i ]; then
          echo Warning, $i not found!
      fi
done
```

What follows is an alternate version of the code. Here, we use the other style of for loop. Notice here we reference ${#list[@]} which tells us how many elements are in the array (four in this case) leading the loop to be: for ((i = 0; i < 4; i++)); do.

```
#!/bin/bash
list=(www.nku.edu/ ~ foxr/CIT371/file1.txt
      www.uky.edu/cse/welcome.html
      www.uc.edu/cs/csce310/hw1.txt
      www.xu.edu/welcome.txt
      www.cs.ul.edu/ ~ wul/music/lyrics.txt)
for ((i=0; i < ${#list[@]}; i++)); do
      url=${list[i]}
      wget $url
      if [ ! -e $url ]; then
          echo Warning, $url not found!
      fi
done
```

As another example, let us consider a script that we wish a set of users to be able to access. Although we could resolve this through permissions and a special group, we could also use this approach. The script first defines the legal users in an array. Next, the script iterates over the array, testing each one against the current user's USERNAME. Upon a match, we set isLegal to 1 to indicate that the user is a legal user of the script. After testing all users in the array, if the user is not legal, they are not allowed to run the script. Notice that the ... is the main portion of the script, omitted here as the code that the script will execute is immaterial.

```
#!/bin/bash
legalUsers=(foxr zappaf underwoodi dukeg marst)
isLegal=0
for user in ${legalUsers[@]}; do
    if [ $user == $USERNAME ]; then isLegal=1; fi
done
if [ $isLegal -eq 1 ]; then
    . . .
    else echo you are not authorized to run this script!
fi
```

Finally, we present an example that will monitor specific file modifications. Assume we have a list of files that we are interested in tracking. We place those file names in an array. The stat command, among other information, replies with last modification date and time. Using the option -c "%Z", stat returns the number of seconds since the Epoch (midnight, January 1, 1970) that the file was last modified. We pass to this shell script the number of seconds past the Epoch that we are interested in. Any file modified since then will be stored to a log file.

```
#!/bin/bash
time=$1
files=(. . .)          //list the files here
for file in ${files[@]}; do
    if [ `stat -c%Z $file` -lt $time ]; then
        echo $file >> modified.log; fi
done
```

As stated earlier, we could similarly accomplish these tasks using a list rather than an array.

## 7.8 STRING MANIPULATION

### 7.8.1 Substrings Revisited

Earlier in the chapter, we examined three string operations: length, index, and substring. We invoked each of these using the notation `expr command string params`. There is another command we could use called match. But before we look at match, there is a simpler way to obtain a substring. We do this through *${string:start:length}* where *start* and *length* are integer numbers or variables storing integer numbers. This operation returns the substring of *string* starting at *start* with *length* characters. As with the array, but unlike using the expr substr command, the first character of a string is at location 0 so that ${string:2:4} would actually provide the third, fourth, fifth, and sixth characters of the string, or the equivalent of `expr substr $string 3 4`.

Imagine that we have three variables, First, Middle, and Last, which are a person's first name, middle name, and last name. We want to form a variable, Initials, to store the initials of the three names. We could do this in one of two ways.

```
Initials="`expr substr $First 1 1`
    `expr substr $Middle 1 1``expr substr $Last 1 1`"
```

or

```
Initials=${First:0:1}${Middle:0:1}${Last:0:1}
```

When using the *${string:start:length}* notation, if the value for *start* is greater than the length of the string, nothing is returned but it does not yield an error. If the value for *length* is larger than the remaining length of the string, it returns the rest of the string. For instance, if name="Frank Zappa" then ${name:6:15} returns "Zappa." Note that if you omit length, it defaults to the remainder of the string, for instance ${name:0} returns "Frank Zappa" and ${name:3} returns "nk Zappa."

You can also use a negative value for *start*. This indicates that counting will start from the right end of the string rather than the left end. For this to work, you must either place the negative value in parentheses or add a space after the colon. For instance, ${name: -3} returns "ppa" and ${name:(-7)} returns "k Zappa." If you fail to include the space or parenthesize the negative number, the entire string is returned. You can combine the negative number with a length, which operates as you would expect; it returns *length* characters starting from the right side as indicated by the negative start value. For instance, ${name: -7:3} returns "k Z" and ${name:(-4:3)} returns "app."

Just as the notation ${string:start:length} is a shortcut for `expr substr`, we can employ a shortcut for `expr length $string`. In this case, the notation is ${#string}, which is similar to the notation used to obtain the number of items in an array. Using name from the previous examples, echo ${#name} will return 11 (11 characters in the string "Frank Zappa").

### 7.8.2 String Regular Expression Matching

Let us return to the expr command and look at another string operation known as match. This allows us to compare a string to a regular expression. The expr operation then returns the portion of the string that matched the regular expression. Unlike grep, match will only begin matching from the left hand side of the string. The notation is

```
`expr match "string" '\(regex\)''
```

for some regular expression *regex*. Here are some examples including what is returned. The return value is followed by a comment to explain it.

- `echo `expr match "abcdefg" '\([a-dg]*\)''`

    - `abcd`

    - our regex only matches, in order, a, b, c, d, and g, so the match stops after d

- `echo `expr match "abcdefg" '\([a-dg]*$\)''`

    - matches nothing because the $ requires that it match at the end of the string but the match begins at the front of the string and the g is separated from abcd by two characters that are not part of the regex

- `echo `expr match "abcdef123" '\([a-z]*\)''`

    - `abcdef`

    - cannot match the digits

- `echo `expr match "abcdef123" '\([a-z]*[0-9]\)''`

    - `abcdef1`

    - only matches one digit

- `echo `expr match "abcdef123" '\([a-z][0-9]\)''`

    - will not match because there is no digit in the second position

To match from the right end of the string, add a .* immediately after the opening single quote around the regex as in

```
echo `expr match "abcdefg123" '.*\([0-9]\)''
```

which will return 3, the last matching character. Note that this notation may not work as you might expect. For instance

```
echo `expr match "abcdefg123" '.*\([0-9]*\)''
```

returns nothing while

```
echo `expr match "abcdefg123" '.*\([0-9][0-9][0-9]\)''
```

returns 123.

If we omit the \( \) around the regex, then the operation returns the number of matching characters. Here are a few additional examples.

- `echo `expr match "abcdef" '[a-z]*''`

  - 6

  - 6 characters matched the regex

- `echo `expr match "abcdef123" '[a-z]*''`

  - 6

  - the last 3 characters do not match

- `echo `expr match "abcdef123" '[a-z]*[0-9]''`

  - 7

  - the letters and the first digit match

- `echo `expr match "abcdef123" '[A-Z]*''`

  - 0

  - nothing matched

- `echo `expr match "abcdef123" '[A-Z]*$''`

  - 0

  - nothing matches since the regex begins at the beginning of the string and the string is not solely composed of letters

We can replace the word match in these operations by placing a colon immediately after the string as in

```
echo `expr "abcdef": '[a-z]*''
```

There are other tricks we can play with regular expressions. In this case, we will use the ${string} notation but here we use one of the following:

- `${string#regex}`—deletes shortest matching substring from beginning of string

- `${string##regex}`—deletes longest matching substring from beginning of string

- `${string%regex}`—deletes shortest matching substring from end of string

- `${string%%regex}`—deletes longest matching substring from end of string

By "deletes," this means that the string is returned with the matching substring omitted. The string itself, if stored in a variable, is unaffected.

Here are a few examples followed by their output. In these examples, assume string=abcdef123.

- `echo ${string#[a-z]*[a-z]}`
  - `cdef123`
- `echo ${string##[a-z]*[a-z]}`
  - `123`
- `echo ${string%[a-z0-9]*[a-z0-9]}`
  - `abcdef1`
- `echo ${string%%[a-z0-9]*[a-z0-9]}`
  - outputs nothing (empty string)

## 7.9 FUNCTIONS

A function is a stand-alone piece of code that can be called from different program units. The function is a type of *subroutine*. All programming languages have some subroutine mechanism whether it is known as a function, procedure, or method. The subroutine is a means of breaking programs into smaller chunks of code. The subroutine supports a concept called *modularity* (writing your code in modules). Through modularity, it is often easier to design, implement, and debug your program code.

Although functions are not mandatory, or even required, the use of functions in Bash provides a mechanism to support *reusable* code. You might for instance define a number of functions in a file and then load that file into any script that needs to call upon any of those functions. If you were to define a function that searched a directory for bad permissions of files, you could call upon this function from many other scripts where you felt this operation would be useful. Thus, defining the function once allows you to reuse it time and time again.

### 7.9.1 Defining Bash Functions

In Bash, the function is somewhat different from the ordinary subroutine for several reasons. First, since Bash is interpreted, the placement of the function definition can be anywhere whether in a file by itself, amid other instructions in a shell script, defined at the command line, or in a file of other functions. Once the function has been defined to the interpreter, it can be called.

Second, functions receive parameters. The typical notation in most programming languages for defining parameters is to enumerate them inside of parentheses immediately after the function's name. For instance, a C function which expects three parameters might be called by `someFunction(a, b, 10);` and the function's header might be described as

```
void someFunction(int a, int b, int c)
```

But in Bash, you do not pass parameters by the parentheses. Instead, you pass them similar to passing parameters to a script by placing the parameters after the function name on the same line. Then, rather than receiving the parameters in a header as shown with the C function someFunction above, the parameters are accessed using $1, $2, and so on.

Note that parameters passed in Bash all qualify as *optional* parameters. This means that the function does not require parameters or that the number of parameters passed to the function can vary. Most programming languages have a mechanism for optional parameters, but it is not the default. For instance, the previous example of a C function expects exactly three parameters. If you were to write a function call that contained only two parameters or had four parameters, you would receive a syntax error when you compiled that C code because it did not meet the expectations of the function. In C, you would have to specifically indicate if any parameters were optional. In Bash, all parameters are optional. If, in your function, you reference $1 and no parameters were passed to the function, then $1 has the NULL value.

When you define your function, you may but do not have to precede the function's name with the word function. If you do use the word function, then you omit the parentheses. For instance, we could define someFunction either as

```
someFunction() {
    . . .
}
```

or as

```
function someFunction {
    . . .
}
```

You can also specify the function's body on the same line as the function header itself. If you do so though, you must follow every instruction with a semicolon, including the last instruction. By separating your function body instructions on separate lines, you do not need the semicolons. Here are three versions of the same, simple function:

```
foo() {
    echo Function foo
    echo Illustrates syntax
}

function foo {
    echo Function foo
    echo Illustrates syntax
}

foo() {echo Function foo; echo Illustrates syntax;}
```

## 7.9.2 Using Functions

The placement of a function is not necessarily important as long as the function is defined *before* it is called. If you place a function in a file of functions, say functions.sh, then in a script which will use those functions, you will want to execute the functions.sh file first. You can do this in one of two ways.

- `./functions.sh`

- `source functions.sh`

If your function is in the same file as a script which calls upon the function, the function *must* be defined in the file earlier than the function call. However, the script can place that function after other script code. The following skeleton of a script file illustrates the function's placement.

```
#!/bin/bash
//script instructions
//function f1 defined
//more script instructions
//call to f1
//more script instructions
```

Functions can but do not have to return a value. Most functions in Bash scripts tend to receive parameters and operate on those parameters, outputting results (if any) but not returning values. If you return a value, it is often used to specify some error code. Functions for the most part then are similar to ordinary scripts. They can contain input, output, assignment, selection, and iteration instructions. Mostly, functions will be short.

Here is an example of a simple function. It expects to receive a file name as a parameter. If the file exists and is a regular file, and it is executable, the file is executed. Otherwise, an error message is output.

```
runIt() {
    if [[ -f $1 && -x $1 ]]; then ./$1
            else echo Error, $1 does not exist or not executable
    fi
}
```

Now we have the code that utilizes `runIt`.

```
for file in $@; do
    runIt $file
done
```

Obviously, there is no need to have the function runIt. We could have just as easily (or more easily) placed the if statement inside of the for loop and replaced $1 with $file.

The function `runIt` might be useful though if we feel that other shell scripts might utilize it.

### 7.9.3 Functions and Variables

*Local* variables in a function are declared, unlike the variables of a normal script. This is done through

```
local var1 var2 var3 . . .
```

The reason to define local variables is to ensure that the variables no longer exist outside of the function once the function ends. Otherwise, if a variable used in a function happens to be the same name as a variable outside of the function, the value of the outer variable may be altered. Consider the following.

```
X=0

foo() {
        X=$1
        X=$((X+1))
        echo $X
}

echo $X
foo 5
echo $X
```

In this code, X starts at 0. The function foo is defined but not yet invoked, so any code in foo does not yet impact the script. The value of X is output, which is 0. Now, foo is called, passing the parameter 5. In foo, X stores the parameter, so X now stores 5. Next, X is incremented to 6. The value of X is output, 6. After foo terminates, we resume in the script immediately after the instruction `foo 5`. This leads to the last instruction, `echo $X`. This outputs 6 because X had changed in foo. So we see that X has changed from 0 to 5 to 6.

Now consider changing foo by adding the following line as the first line.

```
local X=$1
```

The script behaves somewhat differently now because the variable X exists in two places. First, X in the script (but not the function) is set to 0 and then output. Next, foo is called, being passed the parameter 5. The second X is now created *local* to foo. This variable is initialized to 1 and then changed to 5 in the instruction X=$1. This version of X is incremented to 6 and output. The `echo $X` in the function outputs 6. When the function terminates, the local version of X is removed from memory. The `echo $X` at the bottom of the script outputs 0 because this version of X was never affected by the function.

The above example illustrates that a local variable exists only within the confines of the function in which it was declared. The above example then uses two variables. They both happen to be named X, but they are accessible in different locations.

Without the local statement in foo, the X in foo is the same as the X outside of foo. With local, there are two X's. What about a variable that is defined in a function which was not first initialized prior to the function's call? This leads to yet another situation. In this case, what we will see is that a variable used in a function that did not exist prior to the function's call will continue to persist after the function terminates.

Let us consider a simple swap function. You might use this to sort an array of numbers. In our case, we will simply define swap to exchange two variable values, X and Y. In this example, prior to calling the function, we define X and Y. The value TEMP is first defined in the function and so did not exist prior to calling the function.

```
#!/bin/bash

X=5
Y=10
swap() {
        TEMP=$X
        X=$Y
        Y=$TEMP
}

swap
echo $X $Y
```

The script first initializes X to 5 and Y to 10. It then defines the swap function. Now, swap is called. The swap function knows of X and Y since they were defined in the code prior to the function call. In swap, we use TEMP to store X, we move Y into X and then TEMP into Y. Now X has 10 and Y has 5. TEMP also has 5. After the function terminates, we output $X and $Y giving us 10 5. Interestingly at this point, TEMP is still 5. Had we changed the echo statement to be

```
echo $X $Y $TEMP
```

we would see 10 5 5. Had we made TEMP a local variable in swap by declaring `local TEMP`, then TEMP would not have a value outside of swap and this latter echo statement would output a blank (null) for TEMP.

It should be noted that the initialization of X and Y could be moved to after the function's definition. By moving them after the } in the function and before swap, the function still receives the values of X and Y since they were defined before the function call. Thus, the script would still provide the same output.

### 7.9.4 Exit and Return Statements

All functions return an exit status. If the function executes without error, the exit status by default is 0. We can modify the exit status through the command exit *n* where *n* is an integer. You define the exit status to indicate the error. If you examine the man page of various instructions, exit codes are sometimes provided. For instance, useradd has numerous error codes, 0 being success, 1 being an error trying to update the /etc/passwd file, 2 being invalid command syntax, 3 being an invalid option, and so forth. You can obtain the status of a function's exit with the notation $?.

In addition, functions can also return an integer value. You would use this option of a function for one of two reasons. First, you may wish to write a function to compute and return a value. For instance, you might write a function, maximum, to return the largest value in a list. Second, the function could return a value in response to some situation within the function. In this way, the function does not *have* to return a value but *may* if the situation merits it. Recalling the runIt function, we might want to return the number of files passed to the function that did not match the condition, returning nothing if all files matched the condition.

To specify a return value, use the instruction return *n* where *n* is an integer. This is similar to exit *n*. In the case of exit, n should be a 0 or a positive number. This is not necessarily the case with return. If you use return, then the return value is also stored in $? Thus, we can utilize this value after the function terminates. Consider the following example script.

```
#!/bin/bash

maximum() {
    if [ $# -eq 0 ]; then exit 9999
    else
        max=$1
        shift
        for num in $@; do
            if [ $num -gt $max ]; then max=$num; fi
        done
    fi
    return $max
}

maximum 5 1 2 6 4 0 3
echo $?
```

This script defines the function maximum and then the first executable instruction is the call to maximum. We pass the function the list 5, 1, 2, 6, 4, 0, and 3. In maximum, if no parameters are received, it immediately exits with an error code of 9999. Otherwise, it sets max to $1 which is 5. The shift instruction shifts all parameters down one position. Now it iterates through the rest of this list (1, 2, 6, 4, 0, 3) and for each number, if it is larger than max, it replaces max. Once the for loop terminates, the if statement terminates, and the function returns the value in max, which in this case is 6. The last instruction in the script outputs $? which is the value provided either by the exit instruction or the return instruction. In this case, it will be 6, the value from the return.

## 7.10 C-SHELL SCRIPTING

The Bash scripting language is somewhat based on the syntax of an old language known as Algol. Newer features have been added such as the (()) notation to avoid the challenging syntax of conditions using [] and to permit easy arithmetic computations like ((i++)) and the use of the relational operators like ((i > j)). The C-shell (csh) language, as its name might imply, bases its scripting language's syntax on the C programming language. Here, we compare the csh syntax to that of bash. Note that TC-shell (tcsh) will use the same syntax.

We will not cover all of the programming language constructs like we did in the previous sections of this chapter but instead show comparisons. If you have learned Bash scripting, converting to csh scripting should be easy. Alternatively, if you had already known csh scripting, learning Bash is also easy. Both scripting languages have many similarities and it's primarily their syntax that differs. There are a few notable exceptions though where a related construct is not available in the other scripting language, which will be stated here as we provide the comparison.

### 7.10.1 Variables, Parameters, and Commands

As with bash, csh is case sensitive. Variable names when assigned or used, must match the name of the variable as previously assigned or declared, both in spelling and case. Similarly, all reserved words are only recognized in lower case.

Variables, as with bash, can only store strings although can also be interpreted as storing integers. Arrays are also available as with bash. To assign variables, the notation is set *VARIABLE = VALUE*. As with bash, the right hand side can be a literal value, a variable preceded by a $, or the result of an arithmetic operation. The csh language also utilizes the various shortcut operators (e.g., ++ and +=).

There are four significant differences here between csh and bash. First, we precede the assignment statement with the command set. Second, if the VALUE contains an arithmetic operation where we want to treat the values as numbers, we use the symbol @ preceding the assignment statement in place of the word set. The third difference is that we can insert spaces around the equal sign as well as around the arithmetic operators if the instruction uses @. Fourth, spaces can exist before and/or after the equal sign.

Here are some examples.

- set  X = 5

  - X is treated as a string

- set  X = 5

  - X is treated as a string

- `set  Y = Hello`
- `set  Z = "$X  $Y"`
  - Z takes on the value "5 Hello"
- `set  Z = $X  $Y`
  - although this does not cause an error, the lack of quote marks leaves Z storing only 5 rather than "5 Hello"
- `@  Z = $X + 1`
  - Z is now 6
- `@  Z++`
  - Z is now 7
- `@  Z+=5`
  - Z is now 12
- `set  Z = $X + 1`
  - this yields an error, set: Variable name must begin with a letter. because the csh interpreter is confused by attempting string concatenation without quote marks
- `set  Z = "$X + 1"`
  - Z stores "5 + 1" literally

There are three ways to unset a variable. As with bash, you can use the `unset` command. You can also accomplish this by either specifying set *VARIABLE* or set *VARIABLE* = ''. A variation of the set command is `setenv`. This is used to assign a variable a value and export it. Thus, only use setenv when defining environment variables.

Parameters are referenced the same as with bash, using $1, $2, and so on. $# is the number of parameters passed, $0 is the name of the script, and $$ is the PID of the script process. You can also obtain the values of the arguments through the notation $argv[1], $argv[2], and so forth with $argv[*] being the entire list of parameters. There is no $argv equivalent for $0 or $#. The variable $< stores the most recently entered input. If the user has entered multiple data on a single line, the entire list is stored in $<.

### 7.10.2 Input and Output

The echo statement is the same as we saw with bash, with the same options and with quote marks serving the same purposes. Input however differs as there is no explicit input statement. Instead, you must use $<. Reference to $< causes the interpreter to pause execution until the user presses enter. The following illustrates an example of this form of input.

```
#!/bin/csh
echo Enter your name
set name = $<
echo Nice to meet you $name
```

You do not need the variable name in the above example. Omitting the set statement, we could instead replace $name with $< in the second echo statement.

### 7.10.3 Control Statements

Conditions appear in single parens. Comparisons between values are made using only the relational operators (<, <=, >, >=, ==, !=). String comparisons, as with bash, are limited to equal to and not equal to (==, ! =). File tests are available using the same condition codes (e.g., -e for file exists, -f for file exists and is regular, -r for readable). To test a variable to see if it is initialized, use $?VAR as in $?x to see if x has a value. Compound conditions do not require any special syntax, unlike with bash where [[]] are used. Boolean operators of AND, OR, and NOT are the same as in bash (&&, ||, !).

There are four forms of if statements. The first is much like the C if statement. It is a one-way statement in which the then clause must be self-contained all on one line. The syntax is

```
if (condition) action(s)
```

Notice the omission of the word then. You are unable to use this form of statement if you have an else clause or if the actions are placed on separate lines. An optional endif statement may end the line if desired.

There is a second one-way statement which is the if-then statement similar to what we saw in bash earlier. The syntax here is

```
if (condition) then
    then-clause action(s)
endif
```

In this case, the then-clause must appear on a separate line from the word then. If there are multiple instructions in the then-clause, they can be on one or multiple lines but multiple instructions on one line must be separated by semicolons. The word endif must occur on a line by itself and is required.

The two-way statement is an if-then-else statement. This is the same syntax as the if-then statement with the exception that after the then-clause and before the endif, an else-clause appears. The else clause is the word else followed by one or more statements. Oddly enough, the word else and its statements do not have to occur on separate lines and can in fact occur on the same line as part of or all of the then-clause. The entire construct ends with an endif statement.

There are two forms of n-way selection statements, the nested if-then-else and the switch. The nested if-then-else is very much like bash' if-then-elif-fi but in this case, the elif is replaced by the two words else  if. Below is an example that compares the bash example (left) to the csh example (right).

```
if [ $x -gt $ y ]; then        if ($x > $y) then
    z=$((x+y))                      @ z = $x + $y
elif [ $x -gt $z ]; then       else if ($x > $z) then
    y=$((x+z))                      @ y = $x + $z
else x=$((y+z))                else @ x = $y + $z
fi                             endif
```

The switch statement is similar in syntax to the switch statement from C. The format is

```
switch (test)
    case pattern1:
        action(s)
        breaksw
    case pattern2:
        action(s)
        breaksw
    case pattern3:
        action(s)
        breaksw
    ...
    default:
        action(s)
endsw
```

The test is usually a variable but could also be an instruction which returns a value. The case portions can have multiple patterns which would appear on separate lines as

```
case pattern1a:
case pattern1b:
case pattern1c:
    action(s)
```

Each pattern is a string where wildcards of *, ?, |, and [] are permitted as in [Yy] or y*, which might be used to express different forms of the answer "yes." The default is optional but allows you to specify a default case much like using *) in bash' case statement. The breaksw statements are also optional but the way the switch statement is written, without them, the switch statement continues to compare the test value to the patterns even if a previous pattern matched. Omitting breaksw is like using ;& in bash' case statement.

There are two types of loops available for csh scripting, foreach and while. The while loop is the same as bash' except for the syntax. The foreach loop is the same as bash' iterator for loop. There is no equivalent of a counting loop in csh but can be simulated through either the while or the foreach. The syntax for each type of loop is given below.

```
while (condition)
     action(s)
end
```

The *condition* must evaluate to a Boolean (true or false). The actions must appear on a separate line from the word while and the expression. The actions can be combined onto one line if separated by semicolons, and the end statement can also appear on the same line as the actions.

```
foreach VARIABLE (arguments)
     action(s)
end
```

The *arguments* must be a list or an operation that returns a list. This could include, for instance, $@ (or $argv) for the list of parameters, the response from a Linux command such as `ls *.txt`, or an enumerated list such as (1 3 5 7 9). To simulate the counter for loop, use `seq first step last` to generate the values to iterate over. Neither *first* nor *step* is required and default to 1. For instance, `seq 5` generates the list (1 2 3 4 5) while `seq 3 7` generates the list (3 4 5 6 7).

### 7.10.4 Reasons to Avoid csh Scripting

With this brief overview of csh scripting, there are several reasons to not use csh for scripting. In spite of the more C-like syntax, the csh scripting is very incomplete and therefore there are many things that you can easily do in bash scripting that you either cannot do or cannot do easily in csh. Here we just look at a few.

- There is no facility to write or call functions in csh.

- There is no specific input statement relying instead on the use of $<. The $< operation obtains input from the standard input used by the script. By default, standard input is the keyboard. However, when you run the script you can change this to a file, for instance by running your script as ./script < file. In this case, $< obtains input from the file, one line at a time. If you want to obtain just one datum from a file it cannot be done. You would have to parse the list in $< to break it into individual items. Additionally, you cannot receive input from both keyboard and file in the same script!

- The substring notation ${string:first:last} does not exist in csh although the expr substr operation is available.

- The one-line if statement in csh does not perform short-circuiting. In bash, the following condition will work correctly whether X has a value or not:

  ```
  [[ -n $X && $X -gt 0 ]]
  ```

  but in csh, the equivalent condition

  ```
  ($?X -a $X > 0)
  ```

  yields an error if X is not defined.

- In many cases, you need to specify statements in multiple lines. This prevents you from easily issuing such commands either from the command line or in an alias statement.

- Although the bash scripting language's syntax has its own peculiarities and oddities, the need to separate the operators in an arithmetic expression by inserting spaces can be both annoying and troubling if you forget to do so.

- There are peculiarities with the interpreter leading to errors in some unexpected situations. One of which deals with quoting and the use of escape characters like \" and \'.

- There are also challenges with respect to properly piping some instructions together in csh.[*]

Neither bash nor csh present the most easily usable scripting languages. If you hope to develop your own software and run it in Linux, you are better served by learning a more complete and usable language. Linux has its own interpreters and compilers for perl, python, php, ruby, and other languages and its own compiler for C/C++. If they are not available in your Linux installation, they are available as open source software and easily downloaded using yum (see Chapter 13).

## 7.11 CHAPTER REVIEW

Concepts and terms introduced in this chapter:

- Array—a data structure that can store multiple values. To reference one specific value you need to specify the index (location) of the value of interest.

- Assignment statement—a programming instruction that assigns a variable a value. The value can be a literal value, the value stored in another variable, or the result of some expression such as an arithmetic operation or a Linux command.

- Compound condition—a condition which has multiple parts that are combined through Boolean operators AND and OR.

- Condition—a comparison that evaluates to true or false. Typically, the comparison is between a value stored in a variable and either a literal value, a value stored in another variable, or the result of a computation.

- Counter loop—a loop which iterates a set number of times based on a sequence of values such as 1 to 10 or 10 to 50 by 2.

---

[*] Tom Christiansen, a Unix developer, has produced a famous document for why you should not use csh for scripting. You can find a copy posted at http://www.faqs.org/faqs/unix-faq/shell/csh-whynot/.If you are an introductory programmer or unfamiliar with scripting, you might find many of the code examples to be very cryptic.

- Else clause—the set of code corresponding with the else statement in a two-way selection statement. This clause executes if the condition is false.

- Error code—functions can return an integer number that describes the exit status of the function indicating success or error.

- File condition—a set of conditions that test a file for type or access permission or other.

- Function—a subroutine; or a set of code that can be invoked from different areas of a script or program. Functions support modularity.

- Iterator loop—a loop which iterates once per item in a list.

- Local variable—a variable that is accessible only from within a function.

- Loop statement—a programming language instruction which repeats a body of code based on either a condition, the length of a list, or the sequence of counting values supplied.

- N-way selection—a selection statement which has n conditions to select between one of n groups of actions.

- One-way selection—the simplest form of selection statement which tests a condition and if true, executes the associated action(s).

- Parameter—a value passed to a function that the function might utilize during execution.

- Script—a small program executed by an interpreter.

- Selection statement—a programming language instruction which evaluates one or more conditions to select which corresponding action(s) to execute.

- Short circuiting—a technique when evaluating a compound condition such that if the entire condition can be determined without evaluating the entire statement, it ends. This occurs when two conditions are ANDed or ORed together. If the first condition is false, there is no need to evaluate the remaining condition(s) if ANDed together. If the first condition is true, there is no need to evaluate the remaining condition(s) if ORed together.

- Subroutine—a piece of stand-alone code that can be invoked from other code. Used to support modularity. In bash scripting, the unit to implement a subroutine is the function.

- Substring—a portion of a larger string.

- Then-clause—the set of code corresponding with the then statement in a one-way or two-way selection statement. This clause executes if the condition is true.

- Two-way selection—a selection statement with one condition that executes one of two sets of code based on whether the condition is true or false.

- Variable condition—tests available in Linux to determine if a variable is storing a value or is currently null.

Linux (bash, csh) commands covered in this chapter:

- case—n-way selection statement in bash.

- declare—used to declare a variable and change its attributes in bash.

- echo—output statement in Linux.

- end—used to denote the end of a while loop, foreach loop, or switch statement in csh.

- endif—used to denote the end of an if, if-then, if-then-else, or nested if-then-else statement in csh.

- elif—used to specify an else if condition in a nested if-then-else statement so that the statement has more than one possible condition.

- else—used to specify a second option of actions in Linux for the two-way selection statement.

- exit—used in bash functions to indicate that the function should terminate with a specified error code. 0 is the default for no error, a positive integer is used to denote an error.

- expr—a Linux instruction which performs either an arithmetic or string operation.

- for—in bash, the instruction used for either a counter loop or iterator loop.

- foreach—the iterator for loop available in csh.

- fi—in bash, the end of an if-then, if-then-else, or if-then-elif-else statement.

- if—used to begin any selection statement in both bash and csh (excluding the case/switch).

- length—a string operation to return the number of characters stored in a string.

- let—used in bash scripts to assign a variable the result of an arithmetic computation.

- local—used to declare a variable to be local to a bash function.

- read—the bash input statement. There is no corresponding statement in csh.

- return—used in bash functions to exit a function and return an integer value. In the case of return, the value can be positive or negative (unlike exit).

- seq—generate a list of the numeric sequence given the first and last values in the sequence and a step size. First and step size are optional and default to 1.

- set—used in csh to express an assignment statement.

- switch—the csh version of case, an n-way selection statement.

- then—reserved word in if-then statement to indicate the start of the then-clause.

- unset—used in both bash and csh to remove a value (uninitialized) from a variable.

- until—conditional loop in bash.

- while—conditional loop in bash and csh.

## REVIEW QUESTIONS

Assume for all questions that you are using a Bash shell.

For questions 1–8, assume X = 5, Y = 10, Z = 15, and N = Frank for each of the problems. That is, do not carry a value computed in one problem onto another problem.

1. What will echo $X $Y $Z output?

2. What will X store after executing X=$Y+$Z?

3. What is wrong with the following statement? Q=$X

4. Does the following result in an error? If not, what value does X obtain? X=$((Y+N))

5. Does the following result in an error? If not, what value does X obtain? let X=$N

6. What is the value stored in X after the instruction ((--X)) ?

7. What is the value stored in N after the instruction ((N++)) ?

8. What is the result of the instruction echo $((Y++)) ? Would echo $((++Y)) do something different? NOTE: if you are unfamiliar with C or Java, you will have to research the difference between the prefix and postfix increment.

9. Which of the following are legal variable names in bash? VARIABLE A_VARIABLE VARIABLE_1 1_VARIABLE variable A-VARIABLE while WHILE.

For questions 10–14, assume A=one and B=two.

10. What is output with the statement echo $A $B?

11. What is output with the statement echo "$A $B"?

12. What is output with the statement echo '$A $B'?

13. What is output with the statement echo $A$B?

14. What is output with the statement echo "A B"?

15. What is the difference in the output of the following two statements?

    ```
    ls
    echo `ls`
    ```

16. You have executed the script foo.sh as ./foo.sh 5 10 15 20. What are the values of each of the following? $0, $1, $2, $3, $4, $5, $#, $*.

    For questions 17–22, assume X="Hello World", Y=3, Z=5, and S="acegiklnor".

17. What is output from the statement `echo `expr substr "$X" $Y $Z`?

18. What is output from the statement `echo `expr substr "$X" $Z $Y`?

19. What is output from the statement `echo `expr substr $S $Y $Z`?

20. What is output from the statement `echo `expr index "$X" $S`?

21. What is output from the statement `echo `expr length "$X"`?

22. What is output from the statement `echo `expr index $S "$X"`?

23. Write an echo statement to output on one line the current user's username, home directory, and current working directory.

24. Write an echo statement to say "the current date and time are" followed by the result of the date command.

25. Assume we have variables location and work which store the city name and company name of the current user. Write an echo statement to output a greeting message including the user's username, his/her city name, and the company name.

26. Locate a listing of all ASCII characters. Using the notation \xHH, write an echo statement to output the string "Fun!"

27. What does the following instruction do assuming NAME stores your name?

    ```
    echo Hello $NAME, how are you today? >> greeting.txt
    ```

28. The script below will input two values from the user and sum them. What is wrong with the script? (there is nothing syntactically wrong). How would you correct it?

    ```
    #!/bin/bash
    read X Y
    SUM=$((X+Y))
    echo The sum of $X and $Y is $SUM
    ```

29. What is the –p option used for in read?

30. Write a script to input five values from the user (keyboard) and output the average of the five values.

31. What would you need to do so that the script from #30 would input the data from the disk file numbers.dat?

32. Write a conditional to test to see if HOURS is greater than or equal to 40.

33. Write a conditional to test to see if the user's name is not zappaf.

34. What is wrong with the following conditional? There several errors.

    ```
    [  $X > $Y  &&  $Y > $Z  ]
    ```

35. Write a conditional statement to test if the variable NUM is between 90 and 100.

36. Write a conditional statement to test to see if the file foo.txt does not exist. The condition should be true if foo.txt does not exist.

37. Write a condition to determine if the file foo.txt is neither readable nor executable.

38. Write a condition to determine if the file foo.txt is owned by both you the user and you the group.

39. What is a then-clause?

40. If a person is 21 years or older, they are considered an adult, otherwise they are considered a child. Write a statement given the variable AGE storing the word adult or child in the variable STATUS.

41. Write a script which inputs a user's name and state (location). If the user lives in either AZ or CA, they are charged a fee of $25. If the user lives in either OH, IN, or KY, they are charged no fee. Otherwise, they are charged a fee of $10. If their name is Zappa, Duke, or Keneally, their fee is doubled. Output their name and fee.

42. Assume the program diskcheck  -r returns the number of bytes remaining free in the current file system and diskcheck  -a returns the total capacity of the file system. You will have to use both of these to obtain the percentage of disk space remaining to determine what the user should do. NOTE: you cannot just divide the amount remaining by the total because this will be a fraction less than zero. Instead, multiply the amount remaining by 100 first, and then divide by the total to get a percentage remaining. Write a program to determine the amount remaining and output this percentage and the proper response based on the following table:

| Amount Available | Response |
| --- | --- |
| 100–61% | No need for any action |
| 60–41% | Keep an eye on the system |
| 40–26% | Delete unnecessary software |
| 25–16% | Implement disk quotas on all users |
| 15–6% | Add a new disk drive |
| 5–0% | Panic |

For instance, if `diskcheck -r` returns 127713651 and `diskcheck -a` returns 1000000000, then the program should output 12%, Add a new disk drive.

43. Revise the calculator program (that uses the if-then-elif-else clauses) to add the operations ** (exponent), << (left shift), and >> (right shift). Since **, << and >> can all be interpreted by the interpreter as Bash wildcards or redirection, you need to use some other symbol(s) for the parameters.

44. Explain why you would not want to rewrite the SCORE/GRADE if-then-elif-else statements from Section 7.4.4 using a case statement.

45. Rewrite the "favorite color" case statement from Section 7.4.5 using an if-then-elif-else statement.

46. If the variable FOO has a value, we want to output it. Write such a statement using the [ condition ] &&/|| action syntax.

47. Redo #46 to set FOO to 1 if it currently has no value.

48. Write a while loop which will input numbers from the user until they enter 0 and count and output the number of times they enter the number 100.

49. Redo #48 where the user supplies a parameter to the script and the number that they pass will be the number you count. For instance, if this script is called as `./count.sh 50` then the script will count the number of times the input numbers equal 50.

For questions 50–53, assume X=5 and Y=10. How many times will each loop iterate?

50. while [ $X –lt $Y ]; do ((X++)); done

51. while [ $X –gt $Y ]; do ((X++)); done

52. while [ $X –lt $Y ]; do ((X++)); ((Y++)); done

53. while [ $X –lt $Y ]; do ((X++)); ((Y––)); done

54. Rewrite the following iterator for loop as a counter for loop.

```
for num in (1 2 3 4 5); do . . . done
```

55. Rewrite the following iterator for loop as a counter for loop.

```
for num in ('seq 5 3 15'); do . . . done
```

56. Rewrite the following counter for loop to count downward instead of upward.

```
for ((i = 0;i < n;i++)); do . . . done
```

57. Write a script which will receive a list of parameters and will iterate through them and count the number of times the parameter is greater than 0. The script will output the number greater than 0, or an error message if no parameters are supplied.

58. An infinite loop is a loop which will never reach its exit condition. We can write infinite loops using while, until, or counter for loops (see the example below), but not iterator for loops. Why could an iterator for loop never be an infinite loop? For both cases below, assume x is initial equal to 0.

    Infinite while loop: `while [ $x -lt 10 ]; do echo $x; done`

    Infinite for loop: `for ((i = 0;x < 10;i++); do echo $i; done`

59. A palindrome is a string which reads the same forward as backward, as in radar or madamimadam (Madam, I'm Adam) but not abacab. Write a script to determine if a parameter is a palindrome and output the parameter and whether it is or not.

60. Repeat #59 so that the script receives a list of strings and outputs all of those that are palindromes. Output an error message if the script receives no parameters.

61. Write a script which will receive a list of parameters, compute, and output the average of the list, or an error if no parameters are supplied.

62. Write a script which will receive a list of parameters and output the number that is the largest. If no parameters are supplied, output an error message.

63. Assume array users stores a list of usernames. Write a loop to iterate through the usernames and count the numbers who are currently logged in. You can determine if a user, x, is logged in by using `who | grep $x` and seeing if the response is null or a value.

64. Assuming array is a variable storing an array of values, what is the difference (if any) between `${array[@]}` and `${array[*]}`? Between `"${array[@]}"` and `"${array[*]}"`?

65. Write a loop to output each letter in the string str using the notation `$str:i:j`.

66. Write a function which receives a list of values as parameters and computes and outputs their sum.

67. Write a function which receives a list of values as parameters and computes and returns their sum.

68. Write a function which receives a list of values as parameters and computes and returns their average.

69. Revise your function from #68 to return an error code of 9999 if the function receives no parameters and so cannot compute an average, otherwise return the average.

70. Rewrite your palindrome script from #60 so that the palindrome checking code is in a function which returns a 1 if the parameter passed to it is a palindrome and a 0 if it is not a palindrome. Then, call the function using a for loop to iterate over a list of strings and based on the return value, output the total number of strings that were palindromes.

71. Redo #70 so that the strings are all stored in an array.

# CHAPTER 8

# Installing Linux

THIS CHAPTER'S LEARNING OBJECTIVES are

- To understand the issues involved in operating system installation
- To understand the components that make up the Linux-operating system
- To understand the role of the kernel and the types of kernel design
- To know how to install Linux (both CentOS and Ubuntu)
- To understand what partitions are and how to establish them during installation
- To understand the concept of virtual memory and the need for a swap partition
- To understand the role of SELinux

## 8.1 INTRODUCTION

Most computers come with an operating system already installed. This is true of all Macintosh computers and most computers under the general umbrella of IBM PC compatibles (or "windows machines"). You might install a Linux-operating system if

1. You purchase hardware that does not have an operating system already installed
2. You want to dual boot your computer between Linux and another OS (most likely Windows)
3. You want to replace your existing operating system (e.g., Windows) or
4. You want to add Linux to a Windows system by installing Linux inside a virtual machine

If your choice is #2 or #3 above, you must be careful. Dual booting is a straightforward operation once both operating systems are installed. However, installing the second operating system must not conflict with the initial operating system. One of the steps during a Linux installation is partitioning the file system. You must make sure, if another operating system is present, that you do not wipe out any portion of that operating system when installing the new system. Deleting the native operating system means destroying all data currently present. If you have saved that data to an external storage device, then you should be able to restore the data to another computer.

### 8.1.1 Installation Using Virtual Machines

The safest way to install an operating system is into a virtual machine (refer back to Chapter 1 to read about virtual machines). A virtual machine is not a computer but instead software that mimics another computer. This would require a virtual machine software package (e.g., Virtual Box by Sun or VMware Client by VMware). Once the VM program is installed and running, you can create a new VM.

Creating a VM requires obtaining the target-operating system's installation program (usually on optical disk) and installing the OS inside the VM. This VM is stored on the disk until you are ready to run it. Now, when your computer is running in the native operating system, you start the VM program and then run the VM itself. Your computer, through the running VM, mimics a different computer as if you had a portal into another computer running its own hardware, operating system, and application software. See Figure 8.1. Although a VM will require heavy resources (processor, memory, and disk space), the installation will not compromise the native operating system such that you are free to experiment without risk. Additionally, you can install multiple VMs of different operating systems.

### 8.1.2 Preinstallation Questions

When installing an operating system, you should answer these questions:

1. Why are you installing the operating system?

2. Where are you installing the operating system?

3. How are you installing the operating system?

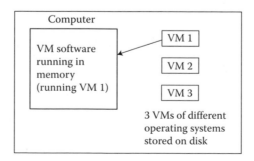

FIGURE 8.1    Computer running VM.

The "why" is generally answered because you want to explore the new operating system. This may be a business decision to see if productivity increases with the new OS. This may be a personal choice because you are tired of problems of the current OS (e.g., security holes) or because specific software only runs in another OS and you want to use that software. It may simply be a hobbyist decision.

The "where" goes back to the earlier list of choices—are you installing the OS on new hardware, dual booting, wiping out the old OS, or installing in a VM? This dictates whether you have to worry about overwriting the current OS on the disk. Only if you are installing for a dual-booting computer do you have to worry about the current contents of the hard disk. You might establish the second operating system on a second disk. Alternatively, you will have to partition the current hard disk into two or more pieces, one with the native OS and one with Linux.

The "how" also goes back to the earlier list of choices although there are more options now. First, are you installing the OS anew or are you updating/upgrading the OS? Second, are you installing the OS on new hardware (i.e., hardware with no native OS) or are you wiping out the current OS, or are you establishing a dual boot? Alternatively, are you installing the OS in a VM? Third, are you installing from an optical disk, flash drive, or over the network? Finally, are you installing the OS or are you just using a *live disk*? The live disk option allows you to boot directly from an optical disk (or flash drive) and thereby bypass the native OS. In doing so, your computer accesses all software, including the OS itself, off of the optical disk. Unfortunately, since the optical disk will be read-only, you would not be able to make permanent changes to the OS and you would have to save files either on a special location on your hard disk or on flash drive. Linux (and Unix) is one of the few operating systems that can be booted and run directly from an optical disk. MacOS X, some versions of Windows (but not Windows 7), and DOS are also live bootable with restrictions.

In this chapter, we examine Linux installation. We look at two installations, Red Hat Linux using the CentOS 6 distribution, and Ubuntu version 12. We make the assumption that you are installing in a virtual machine for simplicity. The same steps would be used to install on new hardware, or if you were deleting the native OS. Similar steps would be used in creating a dual-boot system. However, these steps are not identical and you would be best served to research how to set up a dual-boot system with respect to specifying disk partitions so that you do not damage or destroy the native operating system.

Before you begin any installation, you must understand the hardware that you are installing the OS onto. You should know your processor type, amount of memory, hard-disk capacity, network card or MODEM type, whether you have USB ports, and so forth. You should also know the type of hard disk (IDE or SATA/SCSI) and if you have multiple hard disks, their types and which hard disk(s) you plan to install Linux onto.

If you are installing Linux to dual boot with an existing operating system, you should perform a complete backup of the current operating system and data files first. You should also create a DOS-bootable disk in case the master boot record becomes corrupt. If you make a mistake during the installation of Linux such that your native OS is damaged, you can then attempt to recover any damaged or lost files. While you could not reinstall the native OS from a backup, you should have or be able to obtain that operating system's installation CD so that you could reinstall it if necessary and restore the remainder of your files from backup.

One aspect of installing Linux (at least Red Hat Linux) is that you identify a root password. To this point of the book, we have been examining Linux as a user. As we move past this chapter, we examine Linux from a system administration point of view.

Before we look at Linux installation, we will first examine the Linux-operating system: its components and what to expect from it. In Section 8.2, we also introduce the concept of the operating system kernel and talk about how the Linux kernel differs from those of other popular operating systems.

## 8.2 THE LINUX-OPERATING SYSTEM

We defined the operating system in Chapter 1 as software that is used to manage our computer system for us. Although we do not require an operating system to use our computer, the operating system makes computer usage much easier. For, without the operating system, a program that required the use of resources, such as disk access or memory movement, would have to contain machine instructions that could handle such operations. In the earliest computers, when resources were limited to perhaps tape access and when programmers regularly moved program code in the memory using a scheme called *memory overlays*,* adding such code to your program was common. Today, the operating system not only simplifies the task of the computer programmer, but it also greatly enhances the abilities of the user to control the computer and all of its resources.

The operating system is not a single program but a collection of programs that perform varying types of operations. Central to the operating system is a single program called the *kernel*. We examine the kernel in detail shortly. The operating system also contains *device drivers*, which are programs that allow the operating system to communicate with different types of hardware devices. For each device added to the computer, a device driver is required. Many common device drivers come with the operating system but others must be installed when the device is being installed. The operating system also contains *utility programs* that add to the capability of the operating system in terms of performance, monitoring, and so forth. For instance, disk defragmentation and backup programs are utilities.

### 8.2.1 Operating Systems and Modes of Operation

It is the operating system that manages system resources for us. This means that the user is not allowed to directly access system resources. This might seem odd because it is the user who has all the power and control, right? Not so. When the user selects Open from the File menu and selects a file through the pop-up window, the user is not actually causing the computer to open the file. Instead, the user is specifying an action for the application software to perform. The application software then requests the action to be performed by the operating system.

---

* As the main memory sizes were very small in earlier computers, programmers would often have to split their programs into two or more segments. They would write the code so that, when the first segment had been used, the program would then load the next segment over the first one in the memory, thus creating an overlay.

It is the operating system that decides if and how the action should be carried out. First, the operating system must decide if the request is permissible. Does the user have adequate access rights to the resource? Second, the operating system will decide how the item is to be treated. A file, for instance, can be opened with full access or as a read-only file. Third, the operating system issues the command(s) to the resource(s) by communicating through the appropriate device driver(s).

This leads us to a discussion of how access is controlled in the computer. Most computers use two modes of execution: *user mode* and *privileged mode*. User mode should be self-explanatory in that this is the mode for all users and in this mode, the user can only make requests. The privileged mode, also known as administrator mode, monitor mode, system mode, or supervisor mode, is able to control all aspects of the computer from memory movement to disk access to network access to handling of interrupting situations. In user mode, any such operation is a request requiring that the operating system kernel determine if the requesting user has sufficient access to the requested resource.

Given a request, the operating system first takes control of the computer. It does so by switching modes from user to privileged mode and then examines the request and the user or application that generated the request. The mode is indicated by a flag, usually stored as 1 bit in a control register of the CPU. Upon completion, the operating system changes mode back to user mode and relinquishes control back to the user. All this is hidden from the user.

In some operating systems, there are several different levels of user modes. The Multics operating system, for instance, had a hierarchical set of privileges. The Intel processors starting with the 80286 introduced a protected mode to handle address space issues. Another mode developed for Intel starting with the 286 was the real mode, used to differentiate whether a program's code was using newer instructions made available starting with the 286 processor, or older code written for the original 8086 processor.

Along with the division of operating modes is a division in memory. There is the user's address space (or users' address spaces) and the operating system's address space. Requests to access memory outside one's address space typically lead to a *memory violation* that can either cause a terminating error in the program or at least an interrupt to the operating system to decide how to handle it.

The operating system, on the other hand, is free to access any or all address spaces as needed. Within the operating system, there may be further divisions so that, for instance, the kernel has one address space and other utility programs have their own address spaces. In this way, errors caused by some subcomponent of the operating system may not interfere with the kernel.

## 8.2.2 System Calls

How do applications communicate with the operating system to place requests? This is handled through a mechanism known as a *system call*. A system call is implemented either as a function call or a message passed to the kernel. In Linux, system calls are implemented as C functions.

Depending on the version of Linux you are working with, there are between 300 and 400 system calls available. The idea is that application software is programmed to place system calls so that, when the program needs to invoke the kernel, it does so by calling a kernel function. The system call invokes the kernel that then switches mode and executes the system call. During the execution of the system call, if the operation requested is not appropriate for the software or user, the system call is not completed and either an error arises or the application is informed via an error code that describes the problem.

One drawback of the system call is that it requires a good deal of overhead because the system is switching from one process (the application) to another (the operating system), from one mode (user) to the other (privileged), and from one address space to another. Further, it is possible that this switch will also require the use of virtual memory.

Some examples of Linux system calls are presented in Table 8.1. Although you as a Linux user may never have to directly deal with system calls, it is useful in understanding this concept. As a Linux developer (whether operating system or application software), you would have to understand system calls.

## 8.2.3 The Kernel

The kernel is a single program in charge of basic system operations such as process execution, memory management, resource (device) management, and interprocess communication. There are three schools of thought in designing operating system kernels: monolithic kernels, microkernels, and hybrid kernels.

A *monolithic* kernel is one in which the kernel is a single program that operates solely within the privileged mode and in its own address space. Communication between the user

TABLE 8.1    Examples of Linux System Calls

| Name | Use |
| --- | --- |
| accept, bind | Accept a network socket connection, bind a name to a socket |
| access | Check user's permission of a file |
| chdir | Change working directory |
| creat | Create a file (or device) |
| create_module | Create a loadable module entry |
| execve, execp, execl, and so on | Execute a program |
| exit | Terminate the calling process |
| fork, vfork | Create a child process, vfork further blocks the parent |
| getpriority/setpriority | Return or set process-scheduling priority |
| getuid/setuid | Return or set a user ID |
| kill | Send a signal to a process (depending on the signal, it might terminate the process or might cause exception handler to be invoked) |
| mknod | Create a file (or device or named pipe) |
| mmap | Map (or unmap) files/devices into memory (in essence, place a new process into an area of memory, or remove it from memory) |
| mprotect | Set protection on an area of memory |
| open, close | Open, close a file |
| pause | Suspend a process until signaled |
| pipe | Create an interprocess channel for communication |
| read, write | Read from or write to an open file |
| wait | Wait for the process to change state before continuing |

side (running applications) and the kernel side is handled through system calls. Through the system calls, running processes are able to communicate requests for process management, memory management, I/O, and interprocess communication to the kernel.

Early operating systems used a monolithic kernel but one that was rather small because computer memory was limited and operating systems of that era were not asked to do a lot. As computer capabilities grew, operating systems became larger and monolithic kernels became more and more complex.

A reaction to the complexity of the monolithic kernel was the development of the *micro-kernel*, starting in the 1980s. In this approach, the operating system contains a smaller kernel, which operates within privileged mode and in its own address space. To support the expected range of functions, the operating system is extended by a number of subsystems referred to as *servers*. Servers operate with user applications using user mode and the user section of memory's address space.

Communication between application software and kernel is much like with the monolithic kernel, but the kernel then calls upon servers to handle many of the operations. Thus, the microkernel involves a far greater amount of system calls as the servers are separate from the kernel.

Among the server components are file system servers, network servers, display servers, user interface servers, and servers that can directly communicate with device drivers. Remaining in the kernel are the process scheduler, memory manager (including virtual memory), and the interprocess communication between operating system components. One of the most prominent operating systems to use the microkernel is Mach, Carnegie Mellon University's implementation of Unix.

The *hybrid* kernel compromises between the two extremes where the kernel is still kept small, but server-like components are added on. Here, the server components run in kernel mode but often in the user's address space. In this way, processes can call upon the servers more easily than with the microkernel approach and thus bypass some of the time-consuming system calls. The small kernel might handle tasks such as interrupt handling and process and thread scheduling. The servers, also called *modules* (as discussed shortly) might handle virtual memory, interprocess communication, the user interface, I/O (device drivers), and process management.

Windows NT, and later versions of Windows, including Windows 7 and 8 use forms of hybrid kernels. The MacOS X and the iOS used on Apple mobile devices combine the microkernel of Mach with components from FreeBSD and NetBSD (both descendants from Berkeley Standard Distribution) that use a modular kernel and a monolithic kernel, respectively. Thus, the resulting kernel, known as XNU (X is not Unix), is a hybrid kernel.

You might wonder what the fuss is about. Those who work with either microkernels or hybrid kernels cite that the monolithic kernel is too complicated and large. As the monolithic kernel is essentially one large piece of code, making a minor modification to one portion might have completely unexpected impacts on other portions. In fact, it is possible that errors may arise that are very hard to identify and locate in the code. This could lead to an unstable operating system that yields errors and system crashes for reasons that have little to do with the actual modified code. There are also concerns that the monolithic kernel will be inefficient because of its size.

Imagine, for instance, that a programmer modifies a few lines of code that deal with an interrupt handler. Unknowingly, the code change impacts some aspect of memory management. Now, a user is running a program and when virtual memory kicks in, the error arises. The programmers cannot identify the cause because there is nothing wrong with the memory manager. Tracing it to the changed interrupt-handling code may take a huge effort.

Linux was developed around a monolithic kernel. Andrew Tanenbaum, a computer scientist and author of many textbooks, initially criticized Linus Torvald's Linux because he felt that a monolithic kernel would not succeed with modern computers. This has led to a continued debate over which approach is better.

Given the problems with a monolithic kernel, why would anyone want to produce one? To improve on the efficiency and reduce the impact of errors of the monolithic kernel, the modern monolithic kernel includes numerous *modules*, each of which implements one or more of the core kernel's responsibilities. Modules are loaded either at kernel initialization time, or on demand. Thus, the kernel can itself be modularized. The efficiency of a monolithic kernel's execution is in part based on the number of modules loaded. Loading fewer modules leads to a more efficiently executing kernel. Through modules, the Linux kernel can be kept relatively small.

Figure 8.2 illustrates a rudimentary difference between the monolithic kernel and microkernel. On the top, the monolithic kernel is large and handles all kernel operations. System calls are limited to just the applications software invoking the kernel. On the bottom, the microkernel is smaller and simpler but system calls exist now between the kernel and servers as well as between the application software and servers.

### 8.2.4 Loading and Removing Modules

As a system administrator, you are able to tune the kernel by adding and removing modules. This is of course something you would only want to do if you find your Linux-operating

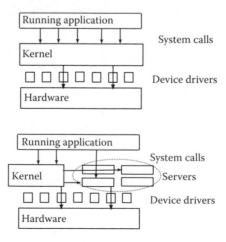

FIGURE 8.2 Monolithic kernel (top) versus microkernel (bottom).

system is not running efficiently leading you to remove modules or is alternatively lacking functionality leading you to add modules. To see the modules that are loaded by default, use the command `lsmod`. This will tell you the modules that are loaded, their size, and what software or service might use that module. You can also inspect each individual module by using `modinfo modulename` as in `modinfo ext4` to view information about the specified module. Information available includes the filename and location of the module (stored under `/lib/modules/version/kernel` where *version* is the version number for your version of Linux such as 2.6.32-279.el6.x86_64), the license, description, authors, version number, and other modules that depend on this module.

If you decide that a loaded module is not required, you can remove it from the kernel through `rmmod`. To add a module to the kernel, use `insmod`. Notice that these two commands will alter the modules that are loaded when you next boot the system; they do not impact your system as it is currently running. You can however add a module on demand using `modprobe`. The format of the command is `modprobe modulename options` where *options* are module specific. With this command, you can also remove a current module with the –r option, list the loaded modules with the –l option, and view dependencies with the –D option.

We will explore the loading and initialization of the Linux kernel in Chapter 11. Next, we look at the installation of the Linux-operating system in two forms, CentOS 6.0 (Red Hat) and Ubuntu 12 (a variation of Debian Linux).

## 8.3 INSTALLING CENTOS 6

CentOS is a variation of the Red Hat Linux distribution. The CentOS 6 installation is very straightforward and fairly quick. The only difficult choice you will face comes when you must decide on the hard-disk partitioning.

### 8.3.1 The Basic Steps

To install CentOS in a VM, start your VM software and select the option to create a new VM. Insert the CentOS installation disk in the optical drive. If you are instead installing CentOS directly onto your hard disk, turn on your computer and as your system boots, enter the ROM BIOS selection (usually this requires hitting the F2 key). Insert the installation disk in the optical drive. Select the option to boot from optical disk rather than hard disk from the ROM BIOS menu. Whichever approach you use, the selection should result in the CentOS installation menu being displayed (see Figure 8.3).

You will want to select the first option "Install or upgrade an existing system." The first step in the installation process asks if you want to test the media before installation. Testing the media will confirm that the optical disk is accessible and has an installation program on it. Skipping this test will save time and should be selected if you are certain that the media works. To skip the media test, press the <tab> key and then <enter>. See Figure 8.4.

You will then see the first installation screen, introducing CentOS 6 and your only choice is to click the Next button. See Figure 8.5.

You will then be taken to numerous screens to select installation options. These are as follows. Select the language. There are over 60 languages available. The default is English.

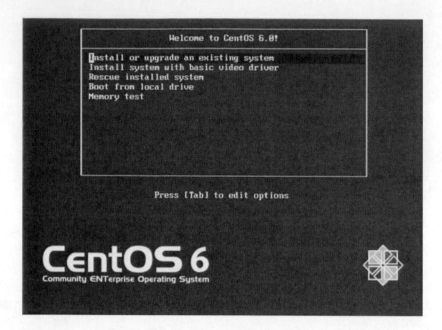

FIGURE 8.3    CentOS installation menu.

Once selected, click the Next button. The next screen contains the keyboard format. U.S. English is again the default. There are several versions that utilize English letters aside from U.S. English such as U.S. International and U.K. English and several different versions of French, German, Italian, and others. Click on Next for the next screen.

The next choice is the type of device(s) for your installation. You have two choices, basic storage devices and specialized storage devices. The typical installation is the former as the latter is primarily used for storage area networks. See Figure 8.6.

Having selected basic storage, you are now given a storage device warning pop-up window. This will explain that expected partitions were not found on your default storage

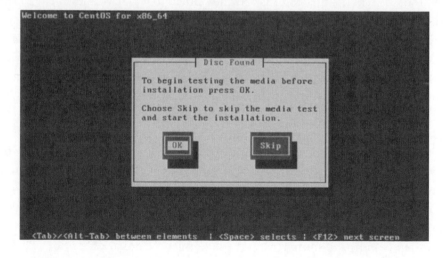

FIGURE 8.4    Media test screen.

FIGURE 8.5 CentOS installation welcome screen.

device and how to proceed. At this point, you will want to partition your hard disk. If you are establishing a dual-boot computer, you will want to retain the current partition(s) such as a Windows partition; otherwise, you can discard any found data files. See Figure 8.7.

Moving forward, you are next asked to provide a hostname for your computer. This hostname will be used as your machine's alias, appended by your network name. For instance, if you name your machine linux1 and your network is known as somecomputer.com, then the full IP alias for this machine would be linux1.somecomputer.com. By default, the hostname is called localhost.localdomain. Obviously, this name is not sufficient if your computer will play a significant role on the Internet. Also available at this step is a button that will allow you to configure your network connection. You can do this now or later.

Selecting this button brings up the Network Connections pop-up window as shown on the left side of Figure 8.8. You will select your network interface(s) as wired, wireless, mobile broadband, VPN, or DSL. In this case, the computer has a wired interface (an Ethernet card). You might notice that this device is listed as never used. Selecting this device and clicking on Edit… brings up the window on the right side of Figure 8.8. Here, we can adjust its settings. For instance, you can change the name of this device although eth0 is the default name and we will assume that you retain this name.

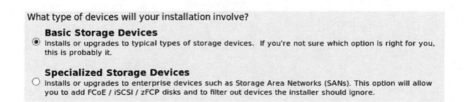

FIGURE 8.6 Storage device selection.

FIGURE 8.7    Storage device warning.

Under the Wired tab, we see the device's MAC address and its MTU size (defaulting to automatically set). We can also set up its security and any IPv4 and IPv6 settings such as whether this machine will receive an IP address from a DHCP server or have a static IP address. You can also specify your default router(s). You might notice in Figure 8.8 that the checkbox "Connect automatically" is not selected. This means that upon booting, your network connection will be unavailable. We want to make it available; so, select "Connect automatically." You should also make sure "Available to all users" is selected. When done, click Apply… and then Close the Network Connections window.

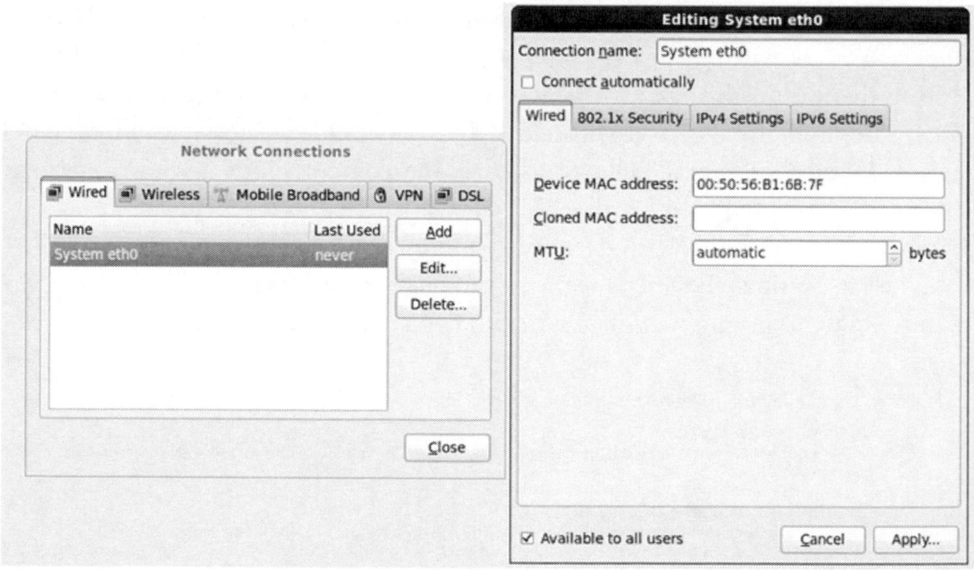

FIGURE 8.8    Initializing your network connection.

Note that the steps described in the previous two paragraphs are not necessary at this point. After successfully installing Linux, booting to Linux, and logging in, you can bring up the same Configure Network window and perform these steps at that time.

At the next screen, you are asked to select your time zone. This will be used for establishing the date and time. The default is the eastern time zone in the United States (New York). You can change the selection either through the drop down box or by clicking on the desired city in the map provided. Additionally, you can select whether your computer's system clock will use UTC (the coordinated universal time) or not. It defaults to using UTC.

At this point, you are asked to enter a root password. This password will be used any time anyone wishes to log in as root. The password should be a strong password, one that you will remember and one that you probably do not want to tell other people about (except for other people who will serve as system administrators on the same computer). You will enter the password twice to confirm that you typed it in correctly. If your password is too weak, you will be warned and asked if you want to use it anyway. Once done, click Next.

## 8.3.2 Disk-Partitioning Steps

We now reach the portion of installation where you are able to customize the installation onto the disk. Your choices are to use all space on the storage device, replace an existing Linux system, shrink the current system, use free space, or create a custom layout. If you are setting up a dual-boot computer, you must select "Use Free Space" to retain the files that already exist. If you are repairing or replacing your Linux system, you would select "Replace Existing Linux System(s)." Selecting this option would remove the Linux partitions that previously existed but does not remove non-Linux partitions. The selection "Shrink Current System" performs repartitioning of your already-existing Linux system. You would use this if your original partitioning has not worked out. We discuss partitions in more detail below and again in Chapter 10.

For our selection, we will select "Use All Space" (if we are installing Linux in a VM or on new hardware) or "Use Free Space" (if we are installing Linux as a dual-boot system on a computer which already has an operating system installed). We make the assumption that we can use all space here. You can either use the default-partitioning layout or select your own. In most cases, as a system administrator, you will want to create your own partitioning layout.

At the bottom of the current screen, you will find two checkboxes, "Encrypt system" and "Review and modify partitioning layout." We will do the latter (you can also encrypt the system if you desire, but we will skip that step). Once the "Review and modify partitioning layout" checkbox is selected, and having selected "Use All Space," click on Next. This brings up a warning pop-up window alerting you that you might be deleting or destroying data files in changing the partitioning layout. Select "Write changes to disk" to continue. This brings up a review of the default partitions, as shown in Figure 8.9.

Let us take a closer look at Figure 8.9. First, we see two different drives being represented, the LVM Volume Groups and the Hard Drives. The Hard Drives list is of the physical hard disk(s) available to us. We have one hard disk in this case, denoted as sda. This stands for

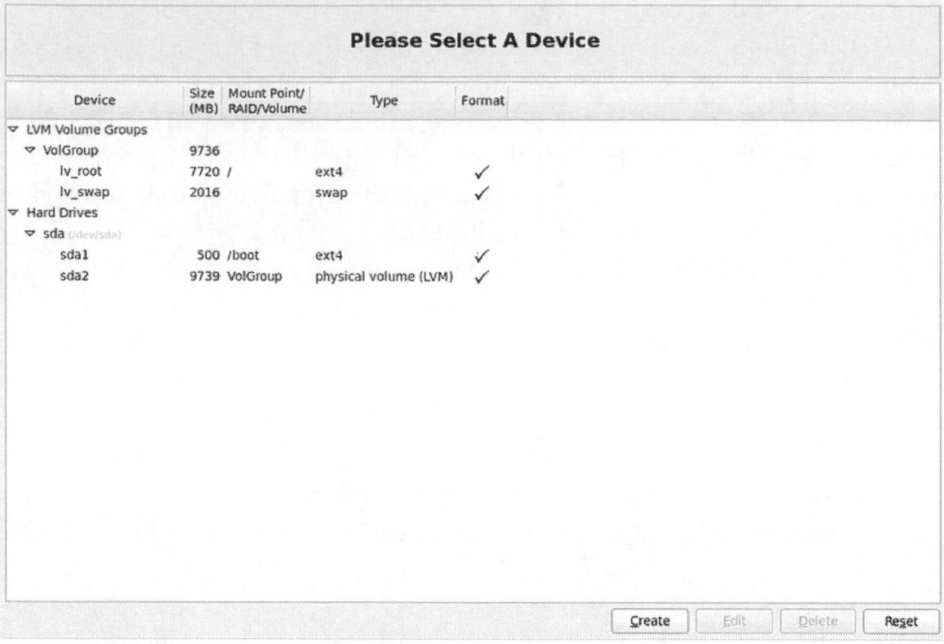

**Please Select A Device**

| Device | Size (MB) | Mount Point/ RAID/Volume | Type | Format |
|---|---|---|---|---|
| ▽ LVM Volume Groups | | | | |
| ▽ VolGroup | 9736 | | | |
|    lv_root | 7720 | / | ext4 | ✓ |
|    lv_swap | 2016 | | swap | ✓ |
| ▽ Hard Drives | | | | |
| ▽ sda *(/dev/sda)* | | | | |
|    sda1 | 500 | /boot | ext4 | ✓ |
|    sda2 | 9739 | VolGroup | physical volume (LVM) | ✓ |

[ Create ]  [ Edit ]  [ Delete ]  [ Reset ]

FIGURE 8.9    Default partitions.

a SATA hard drive. An IDE hard drive would be denoted as hda. Under sda, which will be denoted in Linux as /dev/sda, there are two file systems, sda1 and sda2. Thus, we would see them in the /dev directory as /dev/sda1 and /dev/sda2. The first of these file systems stores the boot partition (containing the /boot directory), which is 500 MB in size. The /boot directory contains the boot loader program (GRUB by default) and the Linux kernel stored in a partially compressed form. The second file system stores VolGroup, a logical volume, and is 9739 MB in size.

You might notice that VolGroup is defined itself as having two partitions, lv_root and lv_swap of 7720 and 2016 MB, respectively (note that the disk sizes reflected here are of a VM established with 10 GB of disk space, and you might find a much larger size during your installation because you are using an entire or a large fraction of an internal hard drive). The lv_root partition is denoted as / and stores the root of the Linux file system. In essence, all subdirectories under / are stored in this partition with the exception of /boot. The swap partition is treated separately from the rest of the file system. We will explore swap later in this chapter. Notice that both / and /boot are of type ext4. This is one of the most popular types for Linux file systems. We explore ext4 versus other types of file systems in Chapter 10. The swap file system is of type swap and VolGroup is of type LVM.

LVM is the *logical volume manager*. This is a means of implementing one or more disk partitions in a logical rather than physical way so that partitions can be easily resized. This means that part of the file system is managed by software. The software specifically keeps track of the partitions' file usage so that it can modify the "boundaries" of the partitions.

This allows you to have an adaptable set of partitions whereby they grow as needed. Using physical disk partitions, you are locked into the established size of the partitions. If you make a partition too small, you might find that you do not have enough disk space for Linux to run at all, or at least to run effectively. Or, you might find that you have run out of space and you have to delete files. At the other extreme, if one partition is too large, then other partitions might eventually suffer from too small a size. Now, these other partitions may fill up more quickly than desired. Resizing or otherwise changing partitions requires making physical changes to the disk that has the potential of destroying data. We examine LVM in more detail in Chapter 10.

Let us imagine that we do not want to use this default setup of partitions and want to create instead partitions for / (root), /var, and /home. These are *mount points* that provide an entry to the partition from the root file system. That is, the mount point is the logical location as we reference it from the operating system. All these partitions will be managed by LVM so that we can ensure that as they grow, as long as there is disk space available, the partitions will not run out of space. First, we need to delete the current setup. Select the two rows of lv_root and lv_swap (one at a time) and select the Delete button. You would wind up with a single entry underneath VolGroup that says Free followed by the amount of free space (9736 in this case).

Now, we can create our new partitions. Click on Create and you will see the Create Storage window as shown in Figure 8.10. You would select LVM Logical Volume underneath the

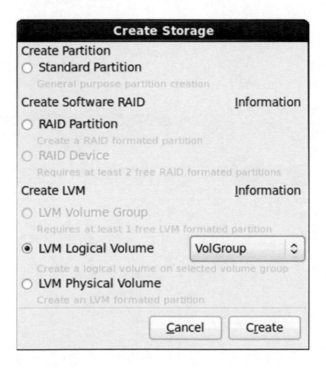

FIGURE 8.10   Creating a partition.

Create LVM section and then select the appropriate Volume Group (VolGroup is the only one available in our example).

Clicking on the Create button from this pop-up window presents you with two new pop-up windows, as shown in Figure 8.11. In the Make Logical Volume window, select the Mount Point, File System Type, and Size. By default, you are provided ext4 for the type, which is the best choice. Mount Point for our first partition will be / (root) and we will use a size sufficient for / to leave us space for other partitions. A size of 4000 MB is satisfactory. With this filled in, click on OK and both windows disappear leaving us with two entries under VolGroup, LogVol00 whose size is 4000 MB, whose mount point is / and whose type is ext4, and Free which is now 5736. We repeat this process to create /var (in this case, we will give /var 4000 MB).

For the swap space, instead of selecting a mount point, we select under File System Type "swap". Swap does not have a mount point, so the Mount Point selection becomes <Not Applicable>. Here, we select a size of 2000 MB because we are limited based on the size of the virtual machine (10 G). The common size for swap space is often cited as 1–2 times that of main memory. However, Linux is a very efficient operating system and you may never need more than about 2 GB worth of swap space. Our last partition has a mount point of /home and will use the remaining space, 736. Note that we could have chosen to encrypt any of these file systems during this partitioning.

At this point, we would have the following list of partitions specified:

```
LVM Volume Groups
    VolGroup          9736
        LogVol03       736     /home      ext4
        LogVol01      4000     /var       ext4
        LogVol00      4000     /          ext4
        LogVol02      1000                swap
```

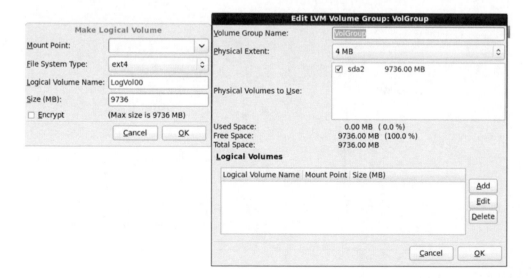

FIGURE 8.11 Creating logical partitions.

```
Hard Drives
   sda
      sda1      500         /boot          ext4
      sda2      9739        VolGroup
                    physical volume (LVM)
```

This summary tells us that our physical hard disk is divided into two file systems, sda1 and sda2 that is storing the boot partition (/boot) and VolGroup, an LVM. VolGroup itself consists of four logical groups, /, /home, /var, and swap that has no mount point. We can now proceed with formatting our disk with the above layout. Click on Next and a Format Warnings pop-up window appears. Click Format and then a Writing storage configuration to disk-warning window appears and click on Write changes to disk. A number of windows appear indicating the formatting and creation of the file systems as specified. Now, we can continue with the installation.

The main advantage of LVM is that a partition is not restricted to the size specified when we formatted the disk. Instead, a partition grows by grabbing space from the available disk, sda2 in this case, as needed and as available. Only if sda2 runs out of space would a partition not be able to grow further. By using the LVM, we, for instance, could potentially use more space for /home than the 736 MB specified while using less space for say /var.

However, the flexibility from LVM comes with some costs. One reason why you might want to forego the use of an LVM is that it adds a layer of complexity to the file system, resulting in both a reduction in efficient access and a complication with processes such as system booting and disaster recovery. Again, refer to Chapter 10 for a greater discussion on LVM.

In the meantime though, let us look at creating partitions with physical sizes by using the Standard Partition selection from the Create Storage window. With this selection, we get an Add Partition window rather than a Make Logical Volume window. See Figure 8.12.

In creating our own physical partitions, we specify much of the same information as we did with the LVM approach. We need to specify the mount point, file system type, and size. For the size specification, we can provide the size in the Size box that indicates a fixed size, we can select Fill all space up to and then specify a maximum size, or we can select Fill to maximize allowable size that uses the remainder of this drive's free space. You can also select whether this partition will be a primary partition and whether it should be encrypted. A primary partition is one with a boot program stored in it. We use a similar set of partitions as we did earlier except that we will forego a separate /boot partition, instead placing /boot with the root partition (/). Here are the values we might select:

- Partition 1: / (mount point), ext4 (type), 4000 MB, Fixed Size

- Partition 2: /var, ext4, 4000 MB, Fixed Size

- Partition 3: <blank> (mount point), swap, 2000 MB, Fixed Size

- Partition 4: /home, ext4, Fill to maximum size

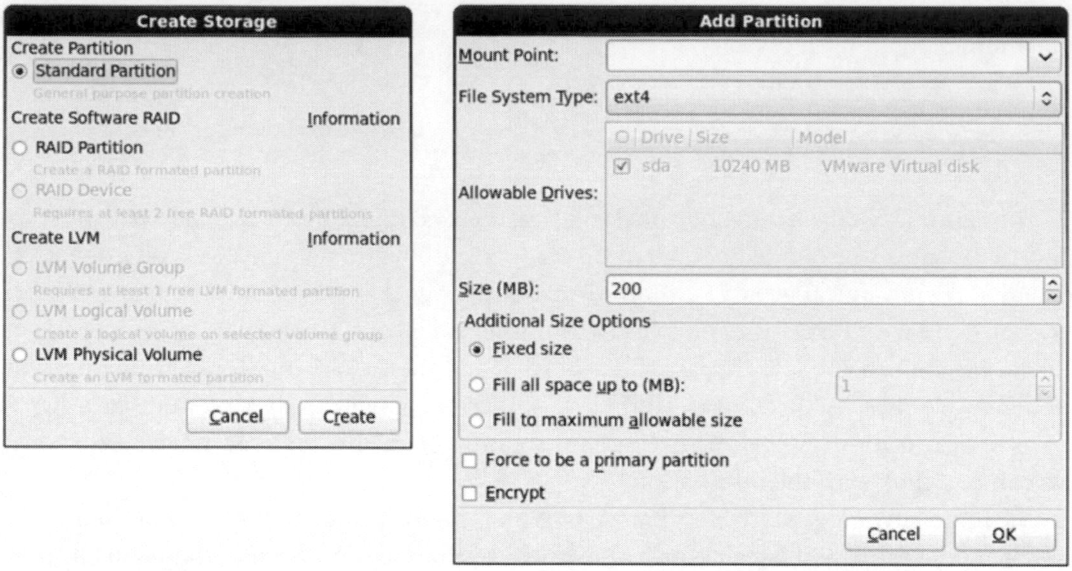

FIGURE 8.12  Creating a partition.

As before, swap space does not have a mount point. All the other partitions have mount points.

Before partitioning your hard disk, you need to consider what you are going to use your computer for and make a wise choice. Once partitioned, while it is possible to repartition or resize your partitions, it is not necessarily a good thing to do given that you can easily destroy data in the process. We examine the file system and partitioning in more detail in Chapter 10. For now, we will continue with our installation.

With your partition information filled out, select Next. Two pop-up windows will appear consecutively. The first warns you that the changes you are making will delete data, click on Format. The second states that you are writing configuration information to the disk. Select Write changes to the disk. You will then see formatting taking place on /dev/sda (or /dev/hda) as the partitions are created.

The next window allows you to specify the location of the boot loader program and the boot image(s) of your operating system. For our first partition using LVM, /boot was explicitly placed on /dev/sda1. Without using LVM, the /boot directory is located on the root (/) partition that is again placed on /dev/sda1. You should not have to do anything with the bootloader specification as the default settings should be correct. You can also establish a boot loader password if desired. Click Next to continue.

### 8.3.3 Finishing Your Installation

We have reached the last screen before installation finally begins. We must specify the type of installation. The type of installation dictates the software to be installed. Your choices are Desktop that provides access to the GUI, Minimal Desktop that provides only some GUI facilities, Minimal that provides only text-based access (and is the default), Basic Server, Database Server, Web Server, Virtual Host, and Software Development Workstation. The last five categories cause different application software to be loaded. If you were intending your computer to be used for one of these latter categories, you do not necessarily have to select it now as you could install the software later (see Chapter 13). You can also add additional software packages on this page. The default is to "Customize later." Select Desktop and then Next. At this point, CentOS 6 is unpacked and installed. Depending on the speed of your processor, hard disk, and optical drive, this will take several minutes, perhaps as many as 20. The last step after installation is to reboot your computer (the VM).

Upon rebooting, you will see a welcome screen. From here, you are nearly done with the installation. First, you will have to agree to the license. In this case, it is the CentOS license under the GPL. The license states that CentOS 6 comes with no guarantees or warranties and also states that individual software packages will come with their own licenses.

At the next window, you are required to provide your first user's account. It is important to create this first user account so that, when you are to log in for the first time, you log in as a user and not as root. Logging in as root through the GUI is never advisable because you have control over all files in the file space and so it is too easy to delete or destroy data accidentally. For the first user's account, you will provide the username, the user's full name, and the user's initial password. As with the root password, this password will be tested to see if it is a strong password, and if not, you will be warned. You can also select a network login mechanism. The Advanced... button will bring up the User Manager window, which we will explore in Chapter 9.

The next window allows you to set the date and time or alternatively synchronize this over the network. Finally, the last window provides you the ability to establish the kdump kernel-dumping mechanism including whether kdump should be enabled, the size of the dump, and the kdump configuration file. Alternatively, you can leave it disabled or not alter the default. After completing this step, you are provided with the CentOS 6 GUI login screen and you are done! Log in and start using Linux.

Although you now have a desktop available, you will probably find that there are a number of additional configuration steps required. These include, at a minimum, the creation of user accounts (we cover this in Chapter 9), the installation of additional software (we cover this in Chapter 13), and ensuring that your network is configured correctly (we cover this in Chapter 12). Other aspects of configuring your Linux environment include establishing and configuring services (covered in Chapter 11), setting up scripts to automate tasks such as backups, and setting up proper security such as the firewall (also covered in Chapter 12).

## 8.4 INSTALLING UBUNTU

Ubuntu is a descendant of Debian Linux. Here, we discuss the installation of Ubuntu 12. As with the CentOS install, much of it is simplified. The start-up window asks if you want to try Ubuntu (via the Live or bootable CD) or install. You would select Install and the language of choice (English is the default). See Figure 8.13.

You are cautioned that you should have at least 4.5 GB of available hard-drive space and Internet access. You can install updates during the installation as well as install third-party software. You are then asked whether to erase the hard disk and install Ubuntu, or create a separate partition for Ubuntu to the current disk setup (you can also resize existing

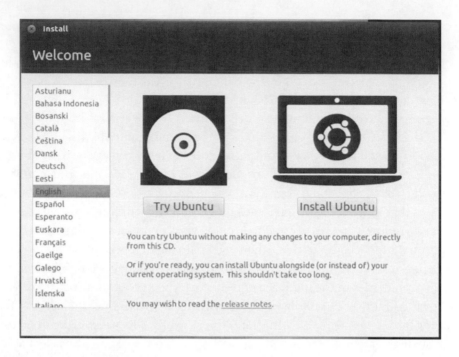

FIGURE 8.13   Ubuntu install window.

Ubuntu partitions). Unless you specify the latter, you will be given the default Ubuntu partitioning. You are asked to select the disk drive to use.

If you wish to partition the disk yourself, you would select partitions similar to what we explored with our CentOS installation. Figure 8.14 demonstrates this process through the install window and the new partition window. First, you select your hard disk (/dev/sda in this case) and select New Partition Table… A pop-up window warns you that you might be destroying previous partitions (as we have selected an entire hard disk to partition). Now, we see (as shown in Figure 8.14) that /dev/sda consists solely of free space (17,179 MB in this case). Click on Add… to create a new partition and the pop-up window shown on the right

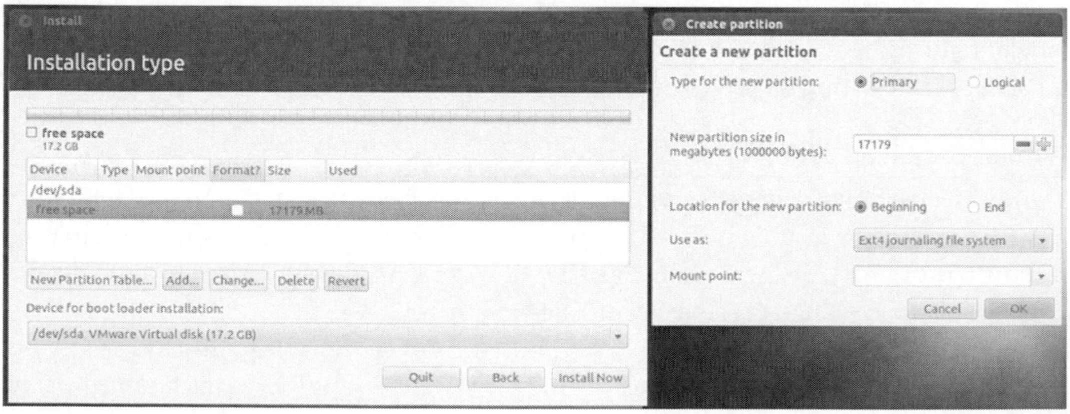

FIGURE 8.14   Creating Ubuntu partitions.

side of Figure 8.14 appears. You now fill out the new partition information by specifying the size, type, and mount point. Types are the same as with CentOS but you have a few additional types available: ResierFS, btrfs, JFS, Fat16, and Fat32. You can also specify if this will be a primary partition, a logical partition, and whether the new partition should start at the next available free location on the disk or end at the last remaining free location on the disk.

At this point, as with CentOS, you are asked to specify your time zone, keyboard layout, and establish an initial user account. Unlike in CentOS, this happens concurrently while the installation package is being unpacked and copied. After establishing your initial account, installation occurs without interruption. However, you have the option of scrolling through information about what Ubuntu has to offer. The installation process only takes a few minutes. You are then required to restart your computer (or VM).

At this point, you are ready to use Ubuntu. Unlike the CentOS installation where there are two separate accounts whose passwords you specified during installation (root and an initial user), here, you have only specified one password, that of the initial user. In fact, root is given an initially random password. To then accomplish any system administrator tasks, this would normally leave the original user in a bind. You do not know the root password and yet you need to be able to accomplish system administrative tasks. To get around this problem, Ubuntu establishes this initial user account with some administrative capabilities. This is similar to the Windows 7 operating system. From the GUI, this administrator user has access to most system administrator programs. From the command line, the user would have to accomplish the tasks through the program sudo. The sudo command is used by one user to carry out commands as another user. The typical usage is to carry out tasks as root without having to log in as root. We will explore sudo in detail in Chapter 9. For now though, when you issue a system administrator command from the command line, you would precede it with the word sudo as in:

- `sudo chown foxr file1.txt`—change the owner of the given file to foxr

- `sudo more /etc/shadow`—view a file that is otherwise unreadable by nonroot

- `sudo useradd -m newuser`—create account for *newuser* (the useradd instruction is covered in Chapter 9)

- `sudo /etc/init.d/network restart`—restart the networking service

- `sudo visudo`—edit the sudo configuration file to establish further sudo rights for other users

Although it might seem inconvenient to not be able to directly access the root account, there are advantages to this approach. It simplifies the initial user's life in two ways. The user does not have to remember multiple passwords and the user does not have to specify a root password. It also saves the possible confusion of a Linux user who switches to root to carry out a task and then forgets to switch back to their normal user account. Here, you are never root, but to carry out a root task, you must use sudo. Further, every sudo command is logged so that a history of commands can be explored at a later time.

You could change the root password using sudo although this is strongly discouraged. Doing so would require the command

```
sudo passwd root
```

You would then be prompted to enter the new password and confirm it.

## 8.5 SOFTWARE INSTALLATION CHOICES

While we focus on software installation in Chapter 13, there are a number of options available when installing Linux. In CentOS, for instance, the OS installation process allows you to specify additional packages to install. Only portions of these packages are installed depending on the type of installation selected (e.g., Desktop, Minimal Desktop, etc.). The packages are broken into 14 categories including Applications, Base System, Databases, Desktops, Development, Servers, System Management, and Web Services. Below, we focus in more detail on some of the options to these packages (other packages such as Languages, Load Balancer, etc., are not discussed here).

Application packages include Emacs, Graphics Creation Tools, Internet Applications, Internet Browsers, and the Office Suite. Within each of these packages, there are multiple programs and subpackages available. For instance, under Graphics, you can find the GNU Image Manipulation Program (GIMP) along with GIMP plugins, help files, and support files. There is a vector-drawing program and a program for displaying and manipulating images in X Windows. There is the SANE program and software to support SANE. The Internet browser package includes versions of Firefox and Firefox plugins. The Office Suite has individual programs for word processing, presentation graphics, spreadsheets and databases, as well as support programs.

Base systems include packages for backups, client management, console tools for remote login over the Internet, debugging tools, support for various programming languages (Java, Perl, and Ruby), network tools, and scientific computation applications. All the database packages revolve around MySQL (client and server) and PostgreSQL (client and server).

Another set of packages provide different desktop programs and environments. The default desktop for CentOS is Gnome. If you install a minimal desktop during installation, you are only provided minimal Gnome programs. The full desktop installation provides all of Gnome. The desktop packages include the KDE desktop, the X windows system, extra fonts, and remote desktop client programs.

The development packages provide programming tools for software developers. These include Ant for Java, a version of Python, the Yacc tool to construct your own language compilers, C++ support added to the GNU's C compiler, Fortran 77, Ada 95, Object-C, subversion for version control, and other useful programs.

The server packages include a backup server, network storage and file servers, an email server, an FTP server, a print server, and server-oriented system administration tools. The web server is located under the Web Services packages. The default web server for CentOS is Apache. Other packages available under Web Services are the LDAP authorization module for Apache, the Kerberos authentication module for Apache, SSL software, the Squid proxy server, and language support for Apache including PHP and Perl.

## 8.6 VIRTUAL MEMORY

During the installation of both CentOS and Ubuntu, we created a swap partition. We might find our system performing very poorly without this partition. Let us explore what the swap partition is and what it is used for.

With our modern multitasking operating systems, it is desirable to run many programs at one time. Programs require memory space to store both the code and the data. Even with large amounts of RAM like 8 GB, it is possible and even likely that you will run out of available memory space while running programs. Virtual memory extends the main memory to the hard disk. The idea is that whatever cannot fit into the memory will be placed in virtual memory for the time being. Virtual memory is implemented by creating swap space on the hard disk. By having its own swap partition, swap space is treated differently from the rest of the file space that can lead to improved performance over ordinary hard-disk access.

The implementation details for virtual memory will differ from operating system to operating system. Here, we look at the typical strategy. First, we divide the program's code and data space into fixed-sized units called *pages*. The number of pages that a program will require differs from program to program. One program may consist of four pages and another may consist of 4000 pages.

The operating system is in charge of loading pages into the memory as are needed. Memory is itself divided into fixed-sized units called *frames*. One page precisely fits into one frame. That is, the page size and the frame size are the same.

Let us consider an example of a program that comprises 12 pages. Upon starting the program, the operating system copies the entire program into swap space. Now, it decides to load the first set of pages of this program into the memory. Let us assume it decides to load four pages of this program. The operating system finds four available frames and loads the pages from hard disk to memory. Then, if or when the program moves beyond the first four pages to another page, the operating system is invoked to retrieve the needed page(s) from swap space and load it (them) into memory.

In loading a new page into the memory, if all memory frames are being used, then the operating system has to select one or more pages to discard from the memory. It does so by moving these pages back to swap space. The situation whereby a request is made to a page that is not in the memory is called a *page fault*. The process of moving pages between swap space and memory is known as *paging* or *swapping*.

In addition to performing paging, the operating system must also maintain a listing of which pages are currently stored in memory and where they are stored. This is done through a data structure called a *page table*. Pages are not necessarily loaded into consecutive frames. So, for instance, a reference to some application program's first page does not necessarily mean that the page is located in the memory's first frame,* nor would two

---

* In fact, this will never be the case because the early frames of memory are reserved for the operating system.

consecutive pages of a program necessarily be located in two consecutive frames of the memory.

The computer needs a mechanism to translate from the memory address as viewed from the program's point of view to the actual memory address. We refer to these addresses as the *virtual* or *logical address* and the *physical address*. A mapping process is used to convert from the virtual/logical address into the physical address. Operating systems maintain for each process a page table. The *page table* records the location of page in the memory (if it is currently in memory).

In Figure 8.15, we see an example of two processes that consist of five and seven pages, respectively, whereas memory consists of only four frames. Thus, to run both processes will require 12 total frames of memory, but there are only four. The operating system must decide which of the pages should be loaded at any one time. Imagine, for instance, that process A begins and the OS loads four of its five pages into the four frames. Later, if the processor switches from process A to process B, then some of A's pages must be swapped out in favor of B's pages. As time goes on, pages may be moved into and out of memory and then back into memory. Eventually, pages of a process may become more and more dispersed in terms of the frames that store them. In Figure 8.15, for instance, we see process A's first page (page 0 as we start our numbering at 0 instead of 1) is located in frame 1 while its fourth page (page 3) is located in frame 3.

The page table indicates in which frame the page is currently loaded, if any, so that the address can be modified appropriately. In Figure 8.15, we see that three of process A's pages are currently loaded into memory while only one page of process B is in memory. The figure also indicates whether pages had previously been loaded into memory and discarded. The page table for these two processes is provided in Table 8.2. In the table, N/A indicates that the given page is not currently in memory.

How many pages and frames might there be? This depends in part on the page/frame size, the size of a program, and the size of memory. Let us assume that we have a computer

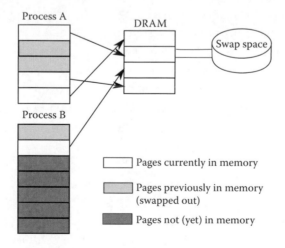

FIGURE 8.15   Paging example. (From Fox, R. *Information Technology: An Introduction for Today's Digital World*, FL: CRC Press, 2013.)

TABLE 8.2   Page Tables for Processes in Figure 8.15

| Process A | | Process B | |
|---|---|---|---|
| **Page** | **Frame** | **Page** | **Frame** |
| 0 | 1 | 0 | N/A |
| 1 | N/A | 1 | 2 |
| 2 | N/A | 2 | N/A |
| 3 | 3 | 3 | N/A |
| 4 | 0 | 4 | N/A |
| | | 5 | N/A |
| | | 6 | N/A |

whose page/frame size is 4096 bytes (4 Kbytes). We want to run a program that is 32 MBytes in size on a computer with 8 GBytes of memory. The number of pages is

32 MBytes/4 KBytes = 8 K pages or 8192 pages

The number of frames is

8 GBytes/4 KBytes = 2 M or roughly 2 million frames

So, this computer has over 2 million frames and this program requires 8192 of those frames to fit the entire program.

Will the operating system load an entire program or part of it? This depends on availability (how many free frames are there?) that itself is partially dependent on the number of programs currently in memory, the size of the program, and the need of which pages the program might need during its execution. It is likely that the full 32 MByte program would be loaded into memory because it is relatively small compared to the size of memory. If we instead had a large program, say 512 MBytes in size, it is unlikely that the entire program would be loaded into memory at one time. This idea of only loading pages into memory when needed is sometimes also referred to as *demand paging*.

Virtual memory is a great idea in that it allows us to run multiple programs whose total size is larger than that of the main memory. But there is a price to pay with virtual memory. The paging process itself is time consuming because of the disk access time required to copy a page from swap space to memory. Disk access is among the slowest activities of a computer. We want to minimize disk access as much as possible. Virtual memory provides a compromise between spending more money on memory and reducing the efficiency of our computer.

There are solutions to improve the performance of virtual memory. These include utilizing an intelligent strategy for page replacement (i.e., selecting a page to discard to free up a frame in a manner that decreases the total number of page faults) and formatting and organizing the swap space data differently from the rest of the file system. This is one of the many reasons why swap space is on a different partition from the rest of the operating system and data files.

With this brief introduction to virtual memory, we will reference virtual memory and swap space later in the book. We will also examine tools to gage the impact of virtual memory on the processor's performance.

## 8.7 SETTING UP NETWORK CONNECTIVITY AND A PRINTER

### 8.7.1 Establishing Network Connections

During your Linux installation, you should be able to establish your network connectivity. You can also perform this afterward. Once Linux is installed and you are logged in, you will be at the desktop. In CentOS (assuming Gnome), select the System menu and from there, the Preferences selection. From the list, select Network Connections. The Network Connections window will appear (see the left side of Figure 8.16). Here, you will find a list of all of your devices.

In our example, we have only a Wired Ethernet connection named System eth0. Selecting it allows us to select Edit… or Delete… In response to Edit…, the edit window appears (see the right side of Figure 8.16) allowing us to change the MAC address, add security, change IPv4, and IPv6 settings. We also can alter the device's startup behavior through two checkboxes. With `Connect automatically`, the device is started at boot time while `Available to all users` means that any user is able to access the network through the device. Click Apply… to close the Editing window (you will be asked to log in using the root password first) and then you can close the Network Connections window.

If, once you have installed Linux, you do not have network access, make sure a device is listed under Network Connections and that both of these checkboxes are selected. If you still do not have network access, it may be due to other issues such as the network service

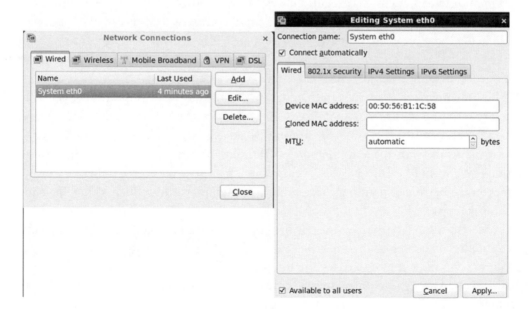

FIGURE 8.16 Network connectivity in CentOS.

not running, a firewall issue, or some physical problem with either your network card or your router. These issues are addressed in Chapters 12 and 14.

In Ubuntu, select System Settings to bring up the System Settings window (much like Windows' Control Panel). From here, select the Network button. A Network configuration window appears. Our choices are to select Wired and Network proxy (Wired is the only device to appear because we do not have any other device installed in this VM). From Wired, we can turn it on or off (it is currently on), view information about it, and click on Options... to bring up the Options window. The Options window presents identical information to the Editing window from CentOS. Compare this as shown in Figure 8.17 with that of Figure 8.16.

### 8.7.2 Establishing Printers

The process to connect to add a printer to your Linux system is very similar to adding a printer to a Windows computer. You will find the approach between CentOS and Ubuntu to be nearly identical; so, we only focus on CentOS. Under the System menu, select Administration and then Printing. This brings up a Printer configuration window that should have no printers listed. Click on New and before proceeding, you are asked to log in as root. After entering the root password, you are shown the New Printer window where you can select a printer from a parallel port, serial port, other, or a network printer.

Assuming that you want to connect to a network printer, you will have choices of Find Network Printer, AppSocket/HP JetDirect, Internet Printing Protocol (ipp), Internet Printing Protocol (https), LDP/LPR Host or Printer, and Windows Printer via SAMBA, see Figure 8.18. In selecting any of these, you will be asked to identify the host except for SAMBA where you are asked to provide an smb address. You can also try to find a printer if you select Find Network Printer. In our case, we are going to connect to an HP Laser Jet; so, we will fill in AppSocket/HP JetDirect. Under Host, we enter the IP address of the printer that then establishes the Forward button as available.

FIGURE 8.17   Network connectivity in Ubuntu.

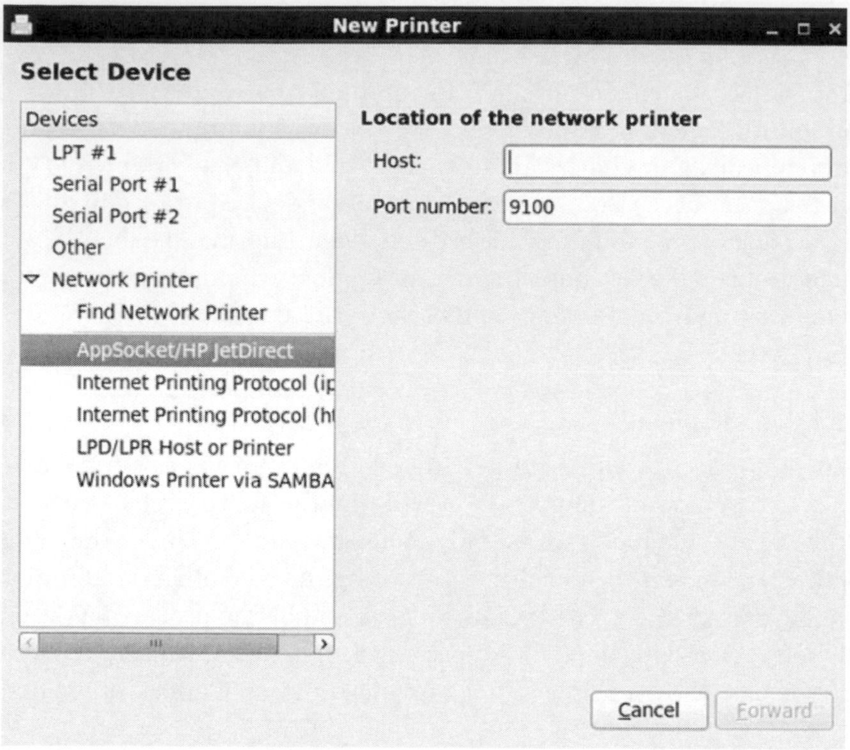

FIGURE 8.18    Configuring a printer.

Clicking Forward brings us to a page of installable options. We can, for instance, select trays, envelope feeder, amount of memory available on the printer, and so forth. Clicking Forward again brings us to a page where we describe the printer. We establish a "short name," a human readable name, and a location. Only the short name is required. We might use `lpr1` (laser printer 1) although in our case, the short name is already provided as `HP-Laserjet-4250`. Clicking on Apply finishes our installation of the printer. We are then asked to authenticate as root. With the printer installed, we are asked if we want to print a test page.

Our Printer configuration window now appears with a printer available. This printer is now the default printer. If we add another printer, we can change the default. This printer will be used by most applications software as it is established as the default printer.

We can also communicate to the printer through the command line through a variety of Linux printer commands. Table 8.3 lists some of the most common commands. For instance, to simply print a document, you would issue `lp` *filename* to print *filename* to the default printer, or `lp -d` *printername* *filename* to send *filename* to the printer *printername*. Both lp and lpr can be used to print multiple files with one command. Both printing commands, lp and lpr, are similar in usage and capability; lpr was developed for BSD Unix while lp was developed for Unix System V. You can also pipe the output of a program to lp or lpr as in `./program | lp`.

TABLE 8.3   Linux Printer Commands

| Command | Meaning | Options/Parameters |
|---|---|---|
| lp | Print file(s) | -d *destination* |
| | | -E (send file(s) encrypted to printer) |
| | | -h *hostname* |
| | | -n # (number of copies) |
| | | -q *priority* |
| lpr | Same as lp | Same as lp except –P *destination*, -H *server*, and -# for number of copies |
| lpq | Show printer queue | -E |
| | | -U username |
| | | -h hostname |
| | | -a (all printers) |
| lprm | Cancel print job | Job ID number (obtained through lpq) |
| cancel | Same as lprm | Specify printer name, add –a to cancel all print jobs |
| lpmove | Move print job from one printer to another | *Source-printer Destination-printer* |
| lpoptions | Select a default printer and/or modify options for a printer | -E enables encryption |
| | | -d *destination* |
| | | -o *options* |
| lpstat | Printer status information | -p *printer(s)* |

## 8.8 SELINUX

SELinux is *security-enhanced* Linux. This addition to Linux allows one to implement security options that go far beyond the simple rwx permissions available for data files, executable files, and directories. SELinux provides mandatory access control through a series of *policies*. Each policy defines access rules for the various users, their roles, and types onto the operating system objects. We will define these entities as they are used in SELinux below, but first, let us look at SELinux more generally.

SELinux became a standard part of Linux with version 2.6 and is available in many Linux distributions, including Red Hat and more specifically CentOS and Fedora, Debian and Ubuntu, and a basic form is available in OpenSUSE. SELinux implements a number of policies with names such as `targeted` (the default), `strict`, `mls` (multilevel security protection), and `minimum`. SELinux should be enabled upon system initialization after booting with the policy of targeted selection. You should find the SELinux configuration file (`/etc/selinux/config`) with these two directives

```
SELINUX = enforcing
SELINUXTYPE = targeted
```

The first directive can also take on values of `permissive` and `disabled`. The second can use any of the available policies. You can change the mode that SELinux is running through the `setenforce` command followed by `Enforcing`, `Permissive`, `1`, or `0`

(1 means Enforcing and 0 means Permissive). To see the current mode of SELinux, use `getenforce`.

## 8.8.1 SELinux Components

With these basics stated, let us now consider SELinux in some detail. First, we return to the types of entities that we deal with in SELinux: users, roles, types, and objects. We throw in another type called a context.

A user is someone who uses the system. However, the user is not equivalent to a specific user. Instead, there are types of users including, for instance, user, guest, and root. Another type of user is known as unconfined that provides a user with broader access rights than the user. User names are specified using *name*_u with already-established user names of `guest_u`, `root`, `staff_u`, `unconfined_u`, `user_u`, and `xguest_u`. Upon logging in, the normal setting is to establish the user as an unconfined user. While many users of the Linux system will map to one of these categories, it can also be the case that a user, during a Linux session, can change from one type to another (for instance through su or sudo).

The role allows SELinux to provide access rights (or place access restrictions) on users based on the role they are currently playing. Roles come from role-based access control. This idea comes from database applications. In large organizations, access to data is controlled by the *role* of the user. In such an organization, we might differentiate roles between manager, marketing, production, clerical, and research. Some files would be accessible to multiple roles (manager, for instance, might have access to all files) while others such as production would only have access to production-oriented files.

In SELinux, roles are generally assigned to the users such as `unconfined_r`, `guest_r`, `user_r`, and `system_r`. However, it is possible for a user to take on several roles at different times. For instance, the user unconfined_u may operate using the role unconfined_r at one time and system_r at another time.

The type specifies the level of enforcement. The type is tailored for the type of object being referenced whether it is a process, file, directory, or other. For instance, types available for a file include read and write. When a type is placed on a process, it is sometimes referred to as a domain. In this case, the domain dictates which processes the user is able to access. As with users and roles, SELinux contains predefined types of `auditadm_t`, `sysadm_t`, `guest_t`, `staff_t`, `unconfined_t`, and `user_t` among others.

We can now apply the users, roles, and types. We specify a context as a set of three or four values. These are at a minimum, a user, a role, and a type, separated by colons. For instance, we might have a context of `unconfined_u:object_r:user_home_t`. This is a context of the user unconfined_u on an object object_r with the type user_home_t. In fact, this context is the one defined for all users' home directories.

In this case, the context is incomplete. The fourth entry in a context is optional but specifies the security level. The security level itself is indicated by a sensitivity or a range of sensitivities, optionally followed by a category. For instance, we might find security levels of s0 to indicate sensitive level of s0, or s0-s3 if the sensitive level includes all of s0, s1, s2, and s3. The category can be a single-category value such as c0, a list of category values such as c0,c1,c2, or a range of categories denoted as c0.c3. The full security level might look like s0-s3:c0.c2. As an example, we might define c0 as meaning general data, c1 as being confidential data, c2 as being sensitive data, and c3 as being top secret data.

Let us look at some examples as preestablished in CentOS.

- User home directories: unconfined_u:object_r:user_home_t:s0

- User file: unconfined_u:object_r:user_home_t:s0

- Running user process: unconfined_u:unconfined_r:unconfined_t:s0-s0:c0.c1023

- /bin directory: system_u:object_r:bin_t:s0

- /bin/bash file: system_u:object_r:shell_exec_t:s0

- /etc/shadow file: system_u:object_r:shadow_t:s0

- /dev/sda1: system_u:object_r:fixed_disk_device_t:s0

- Running root process: unconfined_u:unconfined_r:unconfined_t:s0-s0:c0.c1023

As you can see in the above examples, contexts are already defined for most situations that we will face whether they are user processes, root processes, software-owned processes, user files and directories, or system files and directories. Also notice that most items use a security level of s0, which is the lowest level available.

## 8.8.2 Altering Contexts

You can obtain the contexts of a directory or file by using `ls  -Z` and the context of a running process through `ps  -Z` (or `id  -Z`). Many other Linux operations have a –Z option that can be used to see the context of the object in question. You can also find the predefined contexts in the directory `/etc/selinux/targeted/contexts` in which there are files for various types of processes as well as subdirectories of files for objects and user objects.

You can modify the context of an object in one of the two ways. First, if you issue a file command, you can include `–Z  context` to alter the default context. For instance, as we saw above, a user's file may have the context `unconfined_u:object_r:user_home_t:s0`. Let us assume we want to copy the file to /tmp and alter its context to `unconfined_u:object_r:user_tmp_t:s0`. We issue the command

```
cp -Z unconfined_u:object_r:user_tmp_t:s0 filename /tmp
```

The other means of modifying a context is with the `chcon` instruction. This instruction allows you to change the context of a file or directory. You can specify the new context, as we saw in the above cp example, or you can specify any of the user portion through –u *user*, the role portion through –r *role*, the type through –t *type*, or the security portion through –l *security*. The –R option for chcon operates on a directory recursively so that the change is applied to all files and subdirectories of the directory specified. You can also restrict the instruction to not follow symbolic links using –P (the default) or to impact symbolic links but not the files they link to (-h). If you have changed the context of an object, you can restore it using `restorecon` item as in `restorecon /tmp/file-name` from the above instruction.

### 8.8.3 Rules

With the contexts defined, this leads us to the rules. A rule is a mapping of a context to the allowable (or disallowable) actions of the user onto the object. For instance, we might have the following rule available for users to be able to access files within their home directory:

```
allow user_t user_home_t:file {create read write unlink};
```

This allows any user whose type is user_t to create, read (open), save or write to, or remove a link from a file.

Aside from *allow* rules, we will also define *type enforcement* rules. There are four types: type transition rules, type change rules, type member rules, and typebounds rules. Other types of rules include *role allow* rules that define whether a change in a role is allowed, and *access vector* rules (AV rules) that specify the access controls allowable for a process.

As an example, let us consider a type transition rule. We use this rule to specify that an object can be moved from a source type to a target type. The syntax of a type transition rule is

```
type_transition source_type target_type:class default_type
     [object_name];
```

The object_name is optional and might be included if we are dealing with a specified object (file).

Let us define as an example the transition and allow rules necessary for a user to create a file in their home directory. The user's process will be denoted as user_t, the directory will be denoted as user_home_t, and the new file will be denoted as user_home_t.

```
type_transition user_t user_home_t:file user_home_t;
```

We can then follow this rule with the permissions available for the creation action on both the directory (/var/log) and the file itself (wtmp).

```
allow user_t user_home_t:dir {read getattr lock search
     ioctl add_name remove_name write};
allow user_t user_home_t:file {create open getattr setattr
     read write append rename link unlink ioctl lock};
```

We see here that the directory needs permissions such as the ability to be read, get its attributes, lock it, search it, add a named entity, remove an item, or write to it. The second allow rule applies to files, permitting a file to be created, opened, read from, written to, appended, renamed, linked to, unlinked from, and locked. You might notice that the names of these access rights listed in the above rules are the names of system calls (refer back to Table 8.1).

Rules are placed together to make up a policy. Policies already exist for a number of scenarios. Policies are placed into policy packages and are found in the files under /etc/selinux/targeted/modules/active/modules. These are binary files and not editable directly. Instead, you can modify a policy package using the program audit2allow. You can also create your own policies using checkpolicy. The checkpolicy program will examine a given policy configuration and, if there are no errors, compile it into a binary file that can be loaded into the kernel for execution. We omit the details of creating or modifying policies here as they are both complex and dangerous. A warning from Red Hat's Customer Portal says that policy compilation could render your system inoperable!

## 8.9 CHAPTER REVIEW

Concepts and terms introduced in this chapter:

- Address space—the range of allowable memory addresses for a given process. Typically memory is divided into user address space and privileged or system address space. Within user address space, memory is further divided up among users and their individual processes.

- Device driver—a program that receives commands from the operating system and then translates these commands into program code that a given device (e.g., disk drive, printer) can understand.

- Dual booting—placing multiple operating systems on a computer's hard-disk drive so that the user can boot to either (any) of the operating systems.

- Frame—the division of main memory into fixed-sized storage units. One page fits precisely into one frame.

- Hybrid kernel—a type of kernel that combines features of the monolithic kernel and the microkernel.

- Installation type—in Linux, the decision of the installation type will dictate what software packages are initially installed. Installation types include desktop, minimal desktop, minimal (text based), server, database server, developer (programmer), and others.

- Kernel—the core component of the operating system responsible for most basic operating system functions such as user interface, memory management, resource management, and scheduling. Applications make requests of the kernel via system calls.

- Live boot—the ability to run a computer off of some media (e.g., optical disk, USB drive) rather than booting the operating system from hard disk. A live boot may have restrictions such as the inability to change system settings or install software.

- Logical volume manager—an approach to partitioning the file system whereby partitions are made logically through software rather than physically. This allows resizing and repartitioning without much risk of destroying data.

- Microkernel—a type of kernel whereby the kernel is kept to as small a size as possible by handing over a number of kernel duties to software servers. The microkernel approach to building operating systems has the potential for greater efficiency and fewer errors.

- Module—a means of keeping a monolithic kernel from becoming too complicated by breaking some kernel operations into separate components called modules. The modules loaded into a kernel can be altered at run time through insmod and rmmod, and the currently loaded modules can also be altered at run time through modprobe.

- Monolithic kernel—the opposite extreme of the microkernel where the kernel is a single, large, stand-alone unit. While the monolithic kernel requires fewer system calls than the microkernel, thus improving run-time efficiency, the monolithic kernel is challenging to program and modify.

- Operating system installation—the process of installing a new operating system onto a computer.

- Page—a fixed-size piece of a program. Programs are divided into pages so that pages are loaded into memory only as needed. Since a page is equal to a frame in size, the operating system must locate available frames when loading new pages. Pages not currently in memory are saved in virtual memory on the swap space.

- Page table—a data structure stored in memory and maintained by the operating system so that the logical memory addresses can be translated into physical memory addresses as pages may be distributed throughout memory in a seemingly random way.

- Partition—a logical division of the disk storage space to hold one portion of the operating system such as the kernel and core programs or the user directories or the swap space.

- Privileged mode—one of the two modes that most computers have; the privileged mode permits access to all instructions, memory, and resources. Only the operating system can operate in privileged mode.

- SELinux—security-enhanced Linux is a complex addition to recent Linux distributions to permit the use of mandatory access control policies over the simpler permissions already available in Linux. SELinux allows you to define users, roles, types, contexts, and rules that map the permissible operations available for a given context.

- Swap space—an area of the file system reserved for program pages that are not currently being used. Also known as virtual memory.

- Swapping—the process of moving pages from swap space to memory (and memory to swap space if a frame needs to be freed up for a new page).

- System call—a function call made between a running application or nonkernel portion of the operating system and the operating system kernel. Linux has between 300 and 400 system calls.

- User mode—when a computer is in user mode, commands issued of the processor are limited so that, for instance, access to devices and memory is restricted. Any such command is treated as a request to invoke the operating system, switch to privileged mode, and determine if the command can be carried out based on the user's access rights.

- Virtual machine—the simulation of a computer by means of a virtual machine software program and a virtual machine data file. The simulation mimics the capabilities of the hardware and operating system installed into the virtual machine data file.

- Virtual memory—the extension of main memory onto swap space so that the computer can run programs physically larger in size than that of memory.

## REVIEW PROBLEMS

1. What questions should you ask before installing an operating system?

2. Why is it safer to install an operating system into a virtual machine rather than onto a computer that already has an operating system?

3. When establishing a dual-boot computer, why should you back up your data files before installing the second operating system?

4. How does user mode differ from privileged mode?

5. What is address space? Can users access the operating system's address space?

6. What is a system call?

7. The monolithic kernel requires fewer system calls than the microkernel. In what way does this make the monolithic kernel more efficient than the microkernel?

8. One complaint with monolithic kernels is that they are very complex that can lead to surprising errors. What solution is used in Linux to reduce the complexity and also the inefficiency of a monolithic kernel?

9. What are the advantages and disadvantages of using a logical volume manager rather than creating physical disk partitions?

10. What might happen in creating physical disk partitions if you make a partition too small?

11. What might happen in creating physical disk partitions if you make a partition too large?

12. Research CentOS 6.0 and find out the minimum size needed for the /(root) partition and the /var partition. What did you find?

13. When installing Linux, you are asked to select the installation type (e.g., Desktop, Minimal, and Server). If you select Desktop but later wish to use your computer as a Server, would you have to reinstall the operating system? Explain.

14. In your CentOS installation, you specified a root password. In Ubuntu you did not. How then, in Ubuntu, are you able to perform system administrative tasks?

15. Compare the installation process of CentOS to Ubuntu in terms of which steps you had to perform and the order that they were performed. For instance, were any steps required by one and not used in the other?

16. Why is page swapping inefficient?

17. Assume that the operating system needs to swap in a page but needs to free a frame for this. It has selected one of the two possible pages to discard. Page 1 has been modified (i.e., the values stored in memory are different from what is stored on disk) while Page 2 has not been modified. Which page should the operating system discard and why?

18. What is a page fault?

19. Determine the memory size of your computer. Now, research some software and identify one or two titles that are larger in size than that of your computer's memory. Without virtual memory, could you run this program? With virtual memory, could you run this program?

20. Assume we have the following page table and answer the questions below. Assume a page/frame size is 4096 bytes.

| Page | Frame Location |
| --- | --- |
| 0 | 31 |
| 1 | 16 |
| 2 | 9 |
| 3 | N/A |
| 4 | N/A |
| 5 | 8 |
| 6 | 45 |
| 7 | N/A |

a. How many pages does this program have?

b. What is the size of the program in bytes?

c. How many of the program's pages are currently in memory?

d. How many of the program's pages are currently in swap space?

e. Are any of the program's pages not in swap space?

f. From the page table, can you determine the size of memory? If not, can you determine the minimal size of memory?

21. Assume that a computer uses page/frame sizes of 1024 bytes.

    a. A program is 64 MBytes. How many pages does it have?

    b. A computer has 8 GBytes of memory. How many frames does it have?

22. When multitasking, a computer is running multiple programs (processes) at one time. To handle this, the computer loads these programs (or portions of them) into memory. One strategy is to divide the memory equally among all programs. Another strategy is to divide memory so that each of the users have the same amount of memory, but for any one user, that space is further subdivided based on the number of processes. For instance, let us assume a computer has 1024 frames and has three users plus the operating system using the computer. Two of the users are running 1 process apiece and the other user is running 5 processes. We could divide the 1024 frames in at least two ways:

    a. 256 frames for each user and the operating system so that the third user must split the 256 frames among the 5 processes

    b. 128 frames for each process so that the third user actually gets 640 frames (5*128)

    Which of the two approaches seems fairest? Does your answer change if the computer typically only has a single user? Explain your answers.

23. For SELinux, explain the notation `user_u:user_r:user_t`.

24. For SELinux, what is the difference between enforcing and disabled?

25. Provide an allow rule for SELinux that would allow user_t to open, read, and close the /etc directory. NOTE: /etc is denoted by etc_t.

# CHAPTER 9

# User Accounts

T HIS CHAPTER'S LEARNING OBJECTIVES are

- To understand the role of the user account and the group account

- To know how to create, modify, and delete user and group accounts

- To understand strong passwords and the various tools available for password management

- To understand the role of the passwd, group, and shadow files

- To know how to use the various Linux facilities to establish user resources

- To know how to set up sudo access

- To understand issues involved in user account policies

## 9.1 INTRODUCTION

The user account is the mechanism by which the Linux operating system is able to handle the task of *protection*. Protection is needed to ensure that users do not maliciously or through accident destroy (delete), manipulate, or inspect resources that they should not have access to. In Linux, there are three forms of user accounts: root, user (human) accounts, and software accounts.

The root account has access to all system resources. The root account is automatically created with the Linux installation. Most software and configuration files are owned by root. No matter what permissions a file has, root can access it.

As root lies at one end of the spectrum of access rights, software accounts typically are at the other end. Most software does not require its own account. However, if the software has its own files and directory space that the user should not directly access, then the software is often given its own account. Software, unlike users, usually has no login shell. Thus, if hackers attempt to log into a Linux system under a software account, the hackers would find themselves unable to issue commands. You might recall from Chapter 3 that we can assign the execution permission to be 's' rather than 'x' so that the software runs under the file owner's permissions rather than the user's permissions. The software account is another approach to running processes whereby the software has its own directory space and access rights.

The user account lies in between these extremes. With each user account comes several different attributes (unless overridden):

- A username, user ID number (UID), and password (although the password may initially not have a value)

- An entry in both /etc/passwd and /etc/shadow indicating user account and password information

- A private group with a group ID number (GID), entered in /etc/group

- An initial home directory, by default under the directory /home, with default files

- A login shell, by default, Bash

In this chapter, we look at the programs available to create, modify, and delete users and groups. We look at the mechanisms available to the system administrator to automatically establish users with initial files. We also look at controlling passwords (i.e., using Linux tools to require that users to update their passwords in a timely fashion). We also discuss user account policies.

## 9.2 CREATING ACCOUNTS AND GROUPS

There are two approaches to creating user accounts and groups. First, there is a GUI tool, the User Manager, and second, there are command line programs. While the GUI tool is very simple to use, it is not necessarily preferred because, to create an account for many users, it requires a good deal of interaction. Instead, the command line program can be used in a shell script, which we will examine later in the chapter.

### 9.2.1 Creating User and Group Accounts through the GUI

The GUI User Manager program is launched either from the menu selecting System> Administration > Users and Groups, or from the command line issuing /usr/bin/ system-config-users. This brings up the tool shown in Figure 9.1.

In this figure, there are three (human) users already created: Student, foxr, and zappaf. The user information for these users is shown in the tool: username, UID, primary (or private) group, full name (none has been established for these users), login shell, and home directory. The system accounts (e.g., root, adm, bin) and software accounts are not shown in this tool.

From the User Manager GUI, you can add a user, delete a user, modify a user, add a group, delete a group, or modify a group. Here, we will concentrate on users. Later, we will look at groups.

Creating a new user is accomplished by clicking on the Add User button. This causes the Add User popup window to appear (see Figure 9.2). In this window, you enter the new

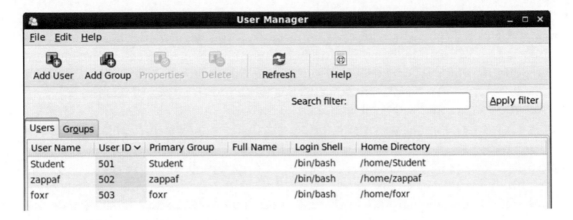

FIGURE 9.1    User Manager tool.

FIGURE 9.2    Adding users.

username, that user's full name, and an initial password (both of these are optional but recommended). You can also alter the login shell if desired from the default (bash) to any of the other available shells. Typically, Linux comes with sh (the original bourne shell), csh, and tcsh. If you add other shells (e.g., korn, ash, zoidberg), they should also appear. One last choice is /sbin/nologin. This is the choice for any software account as you do not want anyone to be able to log in as software and issue commands via a shell.

As you enter the username, the home directory is automatically filled out as /home/*username*. You can of course override this. For instance, if you segment user home directories based on type of user into /home/faculty, /home/staff, and /home/student, you would want to modify the entry. Alternatively, software is often given either no home directory or a home directory under /var. Unchecking the box for `Create a home directory` will create the user account without a home directory.

The last three checkboxes default to the selections as shown in the figure. `Create a primary group` is selected and `Specify user ID manually` and `Specify group ID manually` are unselected. The primary group is the user's private group, which is, by default, given the same name and ID as the username. There are few occasions where you would not want to give a user account its own private group. The default UID and GID are one greater than the last UID and GID issued. Unless you are creating a software account or have a specific policy regarding numbering, you will leave these as the default.

Based on the demands of your organization, you might come up with your own UID and GID numbering scheme. The general pattern in Linux is to assign software an account number between 1 and 99 while all users are given account numbers of 100 or greater (root being given the number 0). However, you might come up with a different numbering policy. For instance, a system that has administration, faculty, staff, and students might be numbered as follows:

- 100–199: Administration

- 200–500: Faculty

- 501–999: Staff

- 1000–2999: Graduate students

- 3000–9999: Undergraduate students

If that were the case, you would alter the ID from the default based on the type of account you were creating.

Group creation through the GUI is even simpler. Clicking on the Add Group button brings up the add group popup window. Here, you specify the group name and optionally alter the GID. The Group tab in the User Manager GUI provides a listing of all of the groups, their GIDs, and the users in each group. An example is shown in Figure 9.3. In this example, we see three private groups, Student, zappaf, and foxr. The group cool contains members zappaf and foxr but not Student. We might establish a group if we want directory or file permissions to be accessible for group members but not the rest of the world.

## 9.2.2 Creating User and Group Accounts from the Command Line

As stated earlier, the text-based approach to creating users (and groups) is often preferred because it can be used to create numerous accounts more easily than through

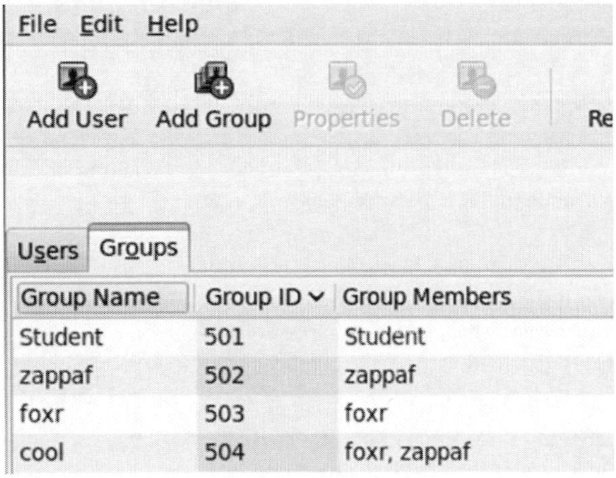

FIGURE 9.3 Adding groups.

the GUI. Imagine that you have to create 50 new users. You could issue the first use-radd command, and then use command line editing to quickly alter the command 49 times.

At a minimum, your useradd command will look like this:

```
useradd -m username
```

After issuing the command for the first new user, you type control+p, escape+b, control+k, and then type in a new name <enter>. You do this repeatedly for each new account. Using the GUI requires clicking on Add User, filling out the form (at a minimum, the user's name), and clicking the OK button. With the command line, you do not have to remove your hands from the keyboard, and you have just about the same to type since you have to specify the username in each version. Additionally, you can easily write a shell script that will use useradd to create new user accounts when it is passed a file of usernames (we explore such a script later in this section).

The useradd command allows a number of options that we would use to alter the default behavior. Some of these options are available via the GUI, but not all of them. See Table 9.1 to see the more useful options.

When a user is created, an entry is added to the /etc/passwd file. In earlier days in Unix, the passwd file stored passwords in an encrypted form. However, the passwd file has permissions of -rw-r-- r-- , meaning that it is readable by the world, this was felt to be a security flaw. In spite of the passwords being in an encrypted format, one could change their own password over and over in an attempt to compare it to other user-encrypted passwords in an attempt to guess those passwords. Now, passwords are kept in /etc/shadow, which is not accessible by anyone except root. So, whenever a new user is created, an entry is also added to /etc/shadow.

TABLE 9.1   Common useradd Options

| Option | Meaning | Example |
|--------|---------|---------|
| -c *comment* | Fills comment field, used to specify user's full name | -c "Richard Fox"—quote marks are necessary if the value has a blank space |
| -d *directory* | Used to alter the user's home directory from /home/username to directory specified | -d /home/faculty/foxr |
| -D | Print default values to see what defaults are currently set as, including directory, expiration value, default shell, default skeleton directory (see below), default shell, and whether to create an email storage location | |
| -e *date* | Set expiration date to date | -e 2014-05-31 |
| -g *GID* | Alter private group ID to this value, otherwise it defaults to 1 greater than the last issued GID | -g 999 |
| -G *groups* | Add user to the listed groups; groups are listed by name or GID and separated by commas with no spaces in between | -G faculty,staff,admin |
| -k directory | Change the default skeleton directory (this is explained in Section 9.6) | -k /etc/students/skel |
| -l | Do not add this user to the lastlog or faillog log files; this permits an account to go "unnoticed" by authentication logging mechanisms, which constitutes a breach in security | |
| -m | Create a home directory for this user | |
| -M | Do not create a home directory for this user (the default case so can be omitted) | |
| -N | Do not create a private group for this user | |
| -o | Used in conjunction with -u so that the UID does not have to be unique, see -u | -u 999 –o |
| -p *passwd* | Set the user's initial password; passwd must be encrypted for this to work | |
| -r | Create a system account for this user | |
| -s *shell* | Provide this user the specified shell rather than the default shell; for software, you will often use this to establish the shell as /sbin/nologin | -s /bin/csh |
| -u *UID* | Give the user specified UID rather than the default (one greater than the last UID); can be used with -o so that two users share a UID | -u 999 |

Here are some example useradd instructions with explanations:

- `useradd foo1`—create new user account foo1 with all of the default values except for a home directory (because –m was not used).

- `useradd -m foo2`—create new user account foo2 with all of the default values including a home directory at /home/foo2.

- `useradd -m -d /home/students/foo3`—create new user account foo3 with a home directory of /home/students/foo3

- `useradd -m -s /bin/csh foo4`—create new user account foo4 with a login shell of /bin/csh

- `useradd -m -u 1001 foo5`—create new user account foo5 with UID of 1001

- `useradd -m -o -u 1001 foo5jr`—create new user account foo5jr who will have the same UID as foo5

- `useradd -m -e 2015-12-31 -l -r backdoor`—interested in creating a backdoor account?

- `useradd -l -M -N -s /sbin/nologin softwaretitle`—create an account for softwaretitle that has no group, no login, and no home directory, and is not logged in lastlog or faillog log files.

You can examine the defaults set up for useradd with the instruction `useradd -D`. For instance, you might obtain output like

```
GROUP=100
HOME=/home
INACTIVE=-1
EXPIRE=
SHELL=/bin/bash
SKEL=/etc/skel
CREATE_MAIL_SPOOL=yes
```

In this case, we see that the default directory and shell are /home and /bin/bash, respectively, that the user's home directory will be populated by the items found in /etc/skel (see Section 9.6), and that a mail spool file will automatically be created. INACTIVE specifies the number of days that must elapse after a user's password has expired when their account becomes inactive. When the INACTIVE value is -1, there is no inactivity date. EXPIRE indicates the default by which new user accounts will expire. This might be set to say 12-31-2015 if you know that all of the accounts you are about to create should expire on that date. Finally, GROUP=100 indicates that if a user is not given a private group (which will happen if you use the -N option in useradd), then the user is added to the group `users`, which has a GID of 100.

You can also use the -D option to alter these defaults. For instance, to change the default shell to c-shell, use `useradd -D -s /bin/csh` or to change the default home directory to /home/faculty, use `useradd -D -d /home/faculty`. You can always reset these defaults. If you are creating two hundred new users and these users will all have the same shell, but not /bin/bash, alter the default to the new shell, create the two hundred new accounts, and then reset the default back to /bin/bash.

TABLE 9.2    Common groupadd Options

| Option | Meaning |
| --- | --- |
| -f | Force groupadd to exit without error if the specified groupname is already in use, in which case groupadd does not create a new group |
| -g *GID* | Use the specified *GID* in place of the default, if used with -f and the GID already exists, it will cause groupadd to generate a unique GID in place of the specified GID |
| -o | Used with -g so that two groups can share a *GID* |
| -p *passwd* | Assign the group to have the specified passwd |
| -r | Create a system group |

The groupadd instruction is far simpler than useradd. The basic format is

```
groupadd groupname
```

The options available for groupadd are limited. Table 9.2 describes the most common ones. The primary option you might use is -g to override the default GID. You might, for instance, group GIDs based on roles such as classification of group: students, faculty, staff having one set of GIDs (e.g., 101, 102, 103) and type of group: programmer, tester, designer, marketing using a different set of GIDs (e.g., 201, 202, 203, 204).

Let us consider an example scenario. You want to place any user who is not issued a private group account into a group other than users (GID 100). We will call this group others, which we will assign a GID of 205. First, we create the new group:

```
groupadd -g 205 others
```

Next, reset the GROUP default from 100 to 205 using

```
useradd -D -g 205
```

Note that for the -D -g combination to work, the GID must be an already-existing group ID. Now, when we use useradd -N *username*, *username* is automatically placed in the group 205 instead of 100. If we decide later to reset this so that such a user is placed in 100, we use useradd -D -g 100.

Whenever a new group is created, an entry is added to /etc/group. Since most user accounts generate a private group, a new user account will also add an entry to /etc/group. And since groups can have passwords, their passwords are placed in the file /etc/gshadow. So, any new group or user with private account causes a new entry to be placed into the /etc/gshadow file even though most groups do not have passwords.

## 9.2.3 Creating a Large Number of User Accounts

Both the GUI and the useradd command are simple to use if you can use the default values. If you are making numerous changes from the default, the GUI is certainly easier to use. So why then would anyone use the useradd command? Frankly, it will be less typing and less hassle in the long run when you have numerous, possibly dozens or hundreds, of accounts to generate. The command line instruction allows you to use command line editing for convenient and quick entry. Or, you can use a shell script, calling upon the useradd instruction.

Consider that you want to add three new accounts, Mike Keneally, George Duke, and Ruth Underwood. You want to use the defaults for all three. You might specify the first instruction:

```
useradd -c "Mike Keneally" -m keneallym <enter>
```

Now, use command line editing to alter this instruction and modify it for the other two new users. You could do this as follows:

- control+p—recall the instruction
- escape+b—move to beginning of username
- control+k (or escape+d)—delete username
- dukeg—enter new username
- control+a, escape+f, escape+f, control+f, control+f—move to the "M" in Mike Keneally
- escape+d, escape+d—delete Mike Keneally (if there were more than two names in quotes, do additional escape+d's)
- George Duke—type the new name
- <enter>

And repeat for Ruth Underwood. This saves a little mouse usage and has the same amount of typing because you would have to enter George Duke (name), dukeg (username), Ruth Underwood, underwoodr in the GUI as well as the command line.

Alternatively, a nice little shell script would also simplify matters. Place the three names in a text file and then use the following script. For this script, we will make the assumption that each name is fully lowercased so that the username is also fully lowercased.

```
#!/bin/bash
while read first last; do
    name="$first $last"
    username="$last${first:0:1}"
    useradd -c "$name" -m $username
done
```

This script should be easy to understand. First, we iterate over each row of the file, reading in two values and storing them in the variables first and last, respectively. We form the value for the name to be used in the comment field as "$first $last". We then form the username as last name and first initial, as in underwoodr. Now, we use $name and $username in useradd to create the account. Assuming this script is called create_users.sh and our list of usernames (just first and last names) is stored in new_users.txt, we run the script as ./create_users.sh < new_users.txt.

What happens if there is already a user in the system with the same username? If we already have a user Tom Fowler with a username of fowlert, then when we try to add Tim Fowler, we receive an error because fowlert already exists.

Let us use a slightly different naming scheme. Our usernames will be last name, first initial, number, where the number will be one greater than the previous user of the same name. This will require that we first inspect the usernames to see how many share the same last name and first initial.

```
#!/bin/bash
while read first last; do
    name="$first $last"
    username="$last${first:0:1}"
    n=`egrep -c $username /etc/passwd`
    n=$((n+1))
    username=$username$n
    useradd -c "$name" -m $username
done
```

Here, we have added two instructions to handle the numbering of each username. First, we use egrep to count the number of occurrences of the entries in /etc/passwd that have $username where username is last name and first initial. This is stored in the variable n. We add 1 to n so that the new user has a value one greater than the last user whose name matches.

Notice that if there are no current users of this name, n will store 0 before we add 1, so it becomes 1 so that the new user's number is a 1. Thus, the first time we have a fowlert, his username becomes fowlert1. The next time we look for a fowlert, we find one in the /etc/passwd file so that n becomes 2 and the next user is given the name fowlert2.

Another option to the mass creation of user accounts is to use the program newusers, stored in /usr/sbin. This program is similar in nature to the shell script developed above but requires a more complex text file of user data.

The format in the text file is a row of information describing each new user. The information is separated by colons and consists of the user's username, an initial user password, the user's UID and private group GID, comment, the user's home directory location, and the user's shell. If the username supplied already exists, then newusers will modify the name before generating the new account, for instance, by adding a number to the end of the name. An entry in the data file will look like this:

```
username:passwd:uid:gid:comment:dir:shell
```

The comment field should be the user's full name. The UID and GID are optional and can appear as ::, that is, nothing between two colons. The default for UID and GID are one greater than the previously used value. If the UID or GID are names that are already in use, then the UID or GID use the existing value, making it nonunique.

The newusers command has some modest error checking, for instance, by allowing duplicate UIDs and GIDs. If the path specified in the directory is erroneous, newusers does not terminate but continues without generating the directory. Thus, the new user will not have a home directory. However, newusers sends an error message to STDERR so that the system administrator can resolve the problem and create a home directory by hand. On the other hand, no checking is performed on the shell specified. If the shell is erroneously listed, then the entry in /etc/passwd will also be erroneous.

As newusers will automatically take the text-based password specified for a user and encrypt it, you are able to specify which encryption algorithm to use. This is accomplished using the -c *method* option, where method is one of DES, MD5, NONE, SHA256, or SHA512. The latter two are only available if you have the proper library available. An additional option is -r to create a system account.

Although this program is executable by the world, only root can actually run it because only root has write access to the /etc/passwd file. Further, the input file needs to be protected with proper permissions because it stores unencrypted passwords. As /root is typically not accessible to anyone other than root, it is best to store the newusers' input text files underneath /root.

## 9.3 MANAGING USERS AND GROUPS

With users and groups created, we must manage them. Management will include making modifications to users' accounts (e.g., changing shells or home directories) and groups (e.g., adding users to a group). A useful program to inspect a user's information is id. The id program returns the given user's UID, GID of the given user's private group, other groups that the user is a member of, and the user's SELinux context. For instance, id foxr might result in the output

```
uid=503(foxr) gid=503(foxr) groups=503(foxr),504(cool)
context=unconfined_u:unconfined_r:unconfined_t:
s0-s0:c0.c1023
```

Without the username, id returns the current user's information. The security context will not display for other users unless you issue the command as root.

### 9.3.1 GUI User Manager Tool

There are two ways to modify a user or a group. First, you can use the GUI User Manager tool. For either a user or software account, highlight the specific user from the list in the GUI and select Properties. This brings up a window that allows you to make changes to that user. Figure 9.4 illustrates the user property window for a user account.

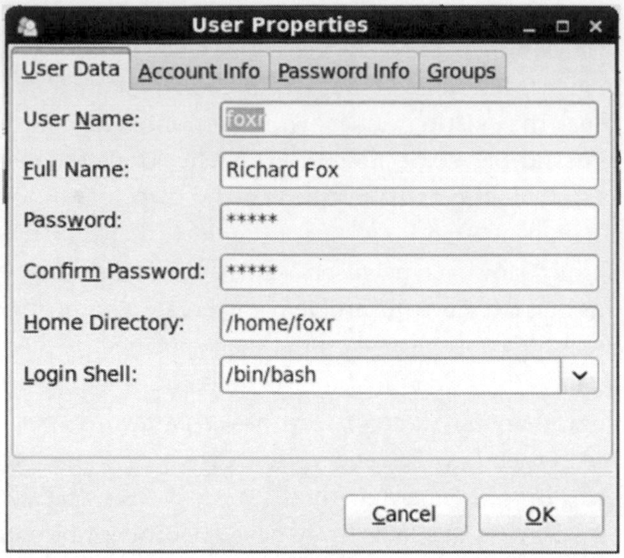

FIGURE 9.4   User properties.

The tabs along the top of the property window are User Data, Account Info, Password Info, and Groups. The User Data is the same as the information specified via the GUI Add User window: User Name, Full Name (comment field), Home Directory, Login Shell, as well as the password. To change the password, you would specify the password and then confirm it. The Account Info tab allows you to specify an expiration date. At this date, the account will expire, meaning that the user is no longer able to log into it. The account still exists. The Account Info tab also lets you lock the password (this disallows the user from changing passwords). The Password Info tab allows you to specify password expiration information (covered in Section 9.4). The Groups tab lists all groups in the system and allows you to change group membership for this user by checking or unchecking any group entry.

The Group Properties window contains only two tabs, Group Data and Group Users. Group Data allows you to change the group's name while Group Users lists all usernames and lets you add or remove users from the group.

The GUI also allows you to select a user or group and click on Delete. If you are deleting a user, you are asked whether to also delete the user's home directory, mail file, and temporary files. You will also be warned if that user has processes running (which usually means that the user is still logged in). For group deletion, if you select a user's private group will be told that you are not allowed to delete that group.

### 9.3.2  Command Line User and Group Management

The command line instructions equivalent to the GUI for user and group management are `usermod`, `userdel`, `groupmod`, and `groupdel`. The `usermod` operation has similar options to useradd. In addition, you can specify $-l$ *username2* to change the user's username to the *username2*. Other options include -L to lock this user's account, $-m$ *directory* to move the user's home directory to the new directory, and -U to unlock an account that was previously locked. The groupmod instruction is the same as groupadd except that it also has a $-n$ *newgroup* option to change the group name to *newgroup*, similar to useradd's -l option.

The userdel command, like the GUI, is used to delete user accounts. This can but does not have to delete the associated files (the user's home directory, temporary files, and mail file). The command is userdel *username* where *username* must be an existing user. If the user is logged in, you will be given an error and the user will not be deleted, unless you force the deletion (see the options below). The user deletion removes the user's entry in /etc/passwd, /etc/shadow, /etc/group (if the user has a private group) and /etc/gshadow (if the user has a private group). The user's name is removed from all groups.

The two important options are:

- -f—force deletion even if the user is currently logged in or has running processes. This option also deletes the user's home directory and mail file even if the home directory is shared between users. This option is drastic and should probably not be used.

- -r—deletes user files (home directory, mail). Files owned by the user in other locations are not deleted.

The userdel command exits with a success value. These are as follows:

- 0—success

- 1—cannot update passwd file

- 2—invalid syntax

- 6—user does not exist

- 8—user is currently logged in

- 10—cannot update group file

- 12—cannot remove home directory

In the case of all but 0 above, the command fails and you will have to reissue it after resolving the problem.

The userdel will not delete a user if he or she has running processes (even if not logged in) unless you use -f. As using -f can be overkill (for instance, by deleting a directory shared among users), it is best to manually kill the running processes of the user and then delete the user. Alternatively, you could always lock the user out of his or her account and wait for the processes to finish on their own. Note that even with -r, any files owned by this user outside of the user's home directory will not be deleted. Therefore, you would have to manually search for any files owned by this user to either change ownership or delete them.

The groupdel instruction is perhaps one of the easiest in Linux. There are no options; instead it is simply groupdel *groupname*. The group is deleted from the /etc/group and /etc/gshadow files, and the group is removed from any user's list of groups as stored in /etc/passwd. There may however be files owned by this group and so you would have to manually search for those files and change their ownership. As with userdel, groupdel returns an exit value as follows:

- 0—success

- 2—invalid command

- 6—group does not exist

- 8—group is an existing user's private group

- 10—cannot update group file

Care should be taken when deleting a user or a group. You should have policies established that determine when a user or group can be deleted. Section 9.8 discusses user account policies. Before deletion, ensure that the user is not logged in or that the group is not a private group. Finally, after performing the deletion, check the system to make sure that no remnant files owned by the user or group exist.

Just as we wrote a script to create user accounts, we might similarly want to create a script to delete user accounts. Below is an example of such a script. We will assume that the user account names are stored in a text file and we will input this file, deleting each account. Some questions to answer before writing this script are whether we should force deletion in case any of these users are currently logged in, and whether we want to remove the users' home directories and email space. Let us assume that we do not want to force deletion in any case and we will delete the user's directory/email only if there is an entry of "yes" in the file after the username. This requires that we input two values from each row of the file. We will call the variables username and deletion. In order to decide whether to delete the directory/email, we will use the condition [ -z $deletion ]. The -z option tests to see if the variable is null (has no value). So, if true, it means that there was no "yes" after the username and so we should not delete the directory/email. The else clause then will delete the directory/email using -r. Our script is shown below.

```
#!/bin/bash
while read username deletion; do
    if [ -z $deletion ]; then userdel $username
    else userdel -r $username
    fi
done
```

As we did not force deletions, this script fails to delete users who are logged in and/or have processes currently running. Thus, if the system administrator were to run this script, it may not complete the task fully. We could enhance the script by keeping a record of any user who, when we attempted to delete them, we received an error code of 8 (user logged in). We can do this by examining the value of $? after we execute the userdel command (recall $? stores the exit or return code of a function or command). Alternatively, we might want to see if the return code is anything other than a 0 as there are other reasons why a user may not be deleted (e.g., could not update the /etc/passwd file because it is currently open or otherwise busy, user does not exist). If we receive any error, we will store $user-name in a file of usernames that still require deletion. We can add this instruction between the fi and done statements.

```
if [ $? -ne 0 ]; then echo "$username" >>
    /root/not_yet_deleted.txt; fi
```

Or, we could use -f in our userdel commands to force deletion. However, the above if-then statement is still useful to keep track of any user not deleted no matter what reason caused it.

## 9.4 PASSWORDS

In this section, we examine user passwords. *Password management* involves three general operations. First is the initial generation of passwords. Second is establishing and enforcing a policy requiring strong passwords. Third is establishing and enforcing a policy whereby passwords must be modified in some timely fashion. We examine these three topics in this section, starting with means of automatically generating initial user passwords.

### 9.4.1 Ways to Automatically Generate Passwords

The apg program is an easy way to generate random passwords. It provides 6 passwords of 8 characters each, using randomly generated characters from /dev/random. The apg program can check the quality of a password by comparing it to a dictionary file to ensure that no random passwords come close to matching a dictionary entry. The program has many options, the most useful of which are listed in Table 9.3.

As apg may not be part of your initial Linux installation, you might have to install it. If you prefer, you can write your own command line instruction to generate passwords by using the output from /dev/urandom. The /dev/urandom device is a random number generator program serving as a device. It generates any character. If we use it to create passwords, we may find many of those characters generated should not be used in a password as they are control characters or nonvisible characters. We would want to delete any such characters from the password.

In order to remove specific characters from those generated by /dev/urandom, we can utilize the tr program (translate or delete characters). For tr, we add the −cd option to specify that we want to delete all but the characters provided in the included set. The set would be either one of [:alpha:] or [:alnum:], or a range of characters as in a-z, a-zA-Z

TABLE 9.3    Useful apg Options

| Option | Meaning |
| --- | --- |
| -a 0 or -a 1 | Select between the two embedded random number generation algorithms, defaults to 0 |
| -n *num* | Change the number of passwords produced from the default of 6 to *num* |
| -m *min* | Change the minimum length of the passwords produced to be a minimum of *min* characters (defaults to 8) |
| -x *max* | Change the maximum length of the passwords produced to be a maximum of *max* characters (defaults to 8) |
| -M *mode* | Change the mode, that is, the types of characters generated; mode can be S (include at least one nonalphanumeric character in every password), N (include at least one digit in every password), C (include at least one capital letter in every password) |
| -E *string* | Do not include any characters specified in the given *string*, not available for algorithm 0 |
| -y | Generate passwords and then encrypt them for output |

or perhaps bcdfghjklmnpqrstvwxyz if, for instance, we wanted to generate a string with no vowels. We would pipe the result to an instruction that could truncate the output. For this, head or tail would be useful.

We could use the following instruction to generate a random password of eight characters:

```
tr -cd '[:alpha:]' </dev/urandom | head -c8
```

This instruction would generate a randomly generated password of 8 letters (the use of -c8 causes head to return just the first 8 bytes, which is equivalent to the first 8 characters). Notice that we are not piping the result of /dev/random to tr but instead using /dev/urandom as input into tr because /dev/urandon is treated like a file rather than a program. Also, we could use [:alnum:] if we wanted passwords that included both letters and digits.

These are only two of the ways to generate passwords. There are many others, including using the date command combined with an encryption algorithm to take the characters in the date and generate a hash version of it. The hash version will look like a random series of characters. Alternatively, the encryption program openssl can generate random characters for us. Two commands are shown below using these approaches:

- `date%s | sha256sum | head -c8`

- `openssl rand -base64 6`

Obviously, a problem with the approaches shown above is that the passwords will truly look random and therefore will be very difficult to memorize. A user, when confronted with such a password, will surely change it to something easier to remember. However, this could result in a weak password. As a system administrator, you could also write a script that would generate a few random entries from the Linux dictionary and piece together parts of the words. Thus, while the password does not consist of a dictionary entry, it might be easier to remember. For instance, by obtaining the *com* from computer and attach it to the *got* in forgotten and the *ment* in government, giving comgotment. A script could be written to produce such a password.

If you use any of the above approaches to generate initial passwords, then you will have strong passwords. However, users will change their passwords (below, we discuss mechanisms to ensure that they change their passwords frequently). How can we ensure that they change their passwords to strong passwords? We will want to ensure three things. First, that the new passwords are strong. Second, that the new passwords are not old passwords. And third, that the new passwords do not have patterns of repeated characters from old passwords. For instance, if the user's current password is a1b2c3d4, we need to prevent the user from using a similar password such as a1b2c3d5 or z1b2c3d4.

### 9.4.2 Managing Passwords

Password management requires that passwords are modified in a timely fashion. For this duty, we turn to two standard Linux programs: chage and passwd. The chage program allows the system administrator to change user password expiration dates of a user. Through chage, the system administrator can force users to change passwords by specific dates or be locked out of their accounts. The format is of the instruction is

```
chage [options] username
```

We specify many of these options with a day or date. This value must be either the number of days that have elapsed since January 1, 1970 (known as the *epoch*), or the actual date using the format YYYY-MM-DD as in 2013-05-31 for May 31, 2013.

The chage program modifies entries in /etc/shadow because the shadow file stores not only encrypted passwords but also password expiration information. The format of the shadow file for each entry is shown below. The letters in parentheses are the options in chage that allow you to change those entries.

- Username
- Encrypted password
- Days since January 1, 1970 that the password was last changed (-d)
- Days before the password may change again (-m)
- Days before the password must be changed (-M)
- Days before warning is issued (-W)
- Days after the password expires that the account is disabled (-I)
- Days since January 1, 1970 that the account will become disabled (-E)

Two entries from the /etc/shadow file are shown below (with the encrypted passwords omitted and replaced with ...). For zappaf and foxr, they are allowed to change passwords

TABLE 9.4    Common chage Options

| Option | Meaning |
|---|---|
| -d *day* | Set the number of days of when the password was last changed. This is automatically set when a user changes password, but we can alter the stored value. If never changed, this date is the number of days since the epoch. |
| -E *day* | Set the day on which the user's account will become inactive (will expire), specified as a date (YYYY-MM-DD) or the number of days since the epoch. This option can be used to provide all accounts with a lifetime. Using $-E$ $-1$ removes any previously established expiration date. |
| -I *day* | Set the number of days of inactivity after a password has expired before the account becomes locked. Using $-I$ $-1$ removes any previously established inactivity date. |
| -l (lower case L) | Show this user's password date information. |
| -M *days* | The number of days remaining before the user must change their password. Should be used with -W. |
| -m *days* | Minimum number of days between which a user is allowed to change passwords. A value of 0 means that the user is free to change password at any time. If the value is greater than 0, this limits the user in terms of how often the password can be changed. |
| -W *days* | The number of days prior to when a password must be changed that a warning is issued to the user to remind them to change passwords. For example, if we use -W 7 -M 28 then the user is contacted in 21 days to remind them that their password needs changing within the next 7 days. |

in 1 day and are required to change them within 35 and 28 days, respectively, with warnings issued in 25 and 21 days, respectively. If they fail to change their passwords, their accounts become inactive within 20 and 10 days, respectively.

```
zappaf:...:15558:1:35:25:20:365:
foxr:...:15558:1:28:21:10::
```

The user zappaf is set to have his account expire in 365 days while foxr has no expiration date set. For foxr's entry, the :: at the end indicates that this last field (disable date) is not set. Finally, there could be a value at the very end of the line (after the last colon) but this field while currently reserved is not used.

The chage command has eight options, one is -h to obtain help. The other seven are described in Table 9.4. If no options are provided to chage, then chage operates in an interactive mode, prompting the user for each field.

The passwd program can also modify some of these entries. Many of the same options in chage are available in passwd, although they are specified differently. Table 9.5 illustrates the passwd options (refer to Table 9.4 as needed).

### 9.4.3 Generating Passwords in Our Script

You might have noticed that the User Manager GUI required that we specify an initial password for our users whereas useradd did not. We can enhance our previous script to create user accounts by also generating passwords. We will call upon apg to generate a password and then use passwd to alter the user's password from initially having none to

TABLE 9.5    Options for passwd

| Option | Meaning |
|--------|---------|
| -d | Disable the password (make the account passwordless) |
| -i *day* | Same as chage -I *day* |
| -k | Only modify the password if it has expired |
| -l | Lock the account (user cannot login until unlocked) |
| -n *day* | Same as chage -m *day* |
| -S | Output status of password for the given account, similar to chage -l |
| -u | Unlock the locked account |
| -x *day* | Same as chage -M *day* |
| -w *day* | Same as chage -W *day* |

having the one provided by the apg program. We need to inform the user of their initial password. We will therefore store each randomly generated password in a text file in /root's file space.

Given the user account creation script from Section 9.2.3, we would add these three lines prior to the done statement.

```
password='apg -n 1'
echo $password | passwd  --stdin $username
echo "$username $password" >> /root/tempPasswords
```

The first instruction uses apg to generate a single password (the -n option tells apg how many passwords to generate), storing it in the variable `password`. The passwd command is interactive. If we were to place `passwd $username` in our script, then executing the passwd command would pause the script to wait for user input. This negates the ability to automate the account generation process via a shell script. So, instead, we force passwd to accept its input from another source. In this case, we use a pipe. First, we `echo $password` to output the value stored in the variable. We pipe this into the passwd command. Now, we have to override the default input (keyboard) to instead come from the pipe (STDIN) by adding the option --stdin.

Finally, we have to make a record of this password so that we can later tell $username what his or her initial password is. We need to know both the username and the password, so we output both of these values from the variables storing them ($username and $password, respectively) and redirect them to the file /root/tempPasswords. We use >> to append to this file as we will most likely be creating numerous user accounts.

If you do not want to install apg, you can generate passwords through other mechanisms. The following uses /dev/urandom:

```
password='tr -cd '[:alpha:]' < /dev/urandom | head -c8'
echo $password | passwd --stdin $username
echo "$username $password" >> /root/tempPasswords
```

Now, we have a script that can generate numerous user accounts and initial passwords. In the exercises at the end of this chapter, you will be asked to enhance this script so that you can further tailor the user accounts. For instance, we might want to give some users the default shell but other users may want other shells.

## 9.5 PAM

PAM (or pam) is short for pluggable authentication module. Programs that deal with authentication in any form (chage, passwd, su, sudo, login, ssh, etc.) need to handle activities such as obtaining the password from the user, authenticate the username and password with a password file, maintain a session, and log events. Rather than requiring that a programmer define these operations in every program, we can make a series of library modules available, each module is capable of handling some aspect of authentication. We then need a mechanism to specify which functions of which modules a given application will call upon.

### 9.5.1 What Does PAM Do?

The idea behind PAM is that it is an *extensible* piece of software. As a system administrator, you configure how a given authentication program will perform its authentication duties through a configuration file. The configuration file lists for the various applications' duties which modules to call upon. You can add your own modules or third-party modules to PAM.

Let us consider as an example the passwd program. This program requires that the user authenticate before changing their password. This is a two-step process of obtaining a password from the user and authenticating the password. It then inputs a new password from the user and uses this to modify the /etc/shadow file.

Another program that requires user authentication is login which is called whenever a user attempts to log in from either the GUI or a text-based session. The login program requires more of PAM than passwd by first authenticating and checking that the account is appropriate in this setting as with passwd, but then maintaining a session. The session starts once the account is tested and terminates only when the user logs out or when a situation arises requiring that the user log out. In addition, the session maintains the user's namespace within the user's context (i.e., the session is run under the PID of the user with the user's access rights).

The strength of PAM is in the modules that implement the functions responsible for authentication operations. One function is called pam_open_session() which can be invoked through the module pam_unix_session.so. The collection of modules provides the proper support for most programs that require authentication. It is up to the system administrator to tie together those modules that will be used for the given program. Additionally, programmers can define their own modules for inclusion. Thus, a system that needs a specific module not available in PAM can be added (once written) and then used as the system administrator sees fit.

### 9.5.2 Configuring PAM for Applications

In order to configure PAM, the system administrator will create a configuration file for each program that will utilize some part of PAM. These files are placed in the directory /etc/pam.d. If you take a look in this directory, you will see already-existing configuration files for a number of programs such as atd and crond, sshd, login and passwd, reboot, halt and poweroff, and su (we will cover some of these programs in upcoming chapters). These configuration files consist of directives that specify the steps that the program needs to take with respect to using the various modules of PAM. These directives will consist of three parts: the module's type, a control flag, and the module's filename.

There are four module types. The `auth` type invokes the appropriate module to obtain the user's password and match it against whatever stored password file exists. Most modules are capable of using the /etc/shadow file while other modules might call upon other password files if authentication is being performed for accounts other than the login account (such as a software-specific account that might be found with an Oracle database).

The `account` module type handles account restrictions such as whether access is allowed based on a time of day or the current status of the system. The `session` type maintains a user session, handling tasks such as logging the start and end of session, confirming that the session is still open (so that a repeat login is not required) or even maintaining a remotely mounted file system. The `password` type deals with changing passwords, as with the passwd command.

The configuration file might include several directives with the same module type. These are known as a *stack*. Within a stack, we have the possibility that all of the modules must run and succeed or that only some of them have to succeed. The control flag is used to indicate what pattern of modules needs to succeed.

There are again four types or values for control flags. With `requisite`, the given module must succeed for access to be permitted. If the module fails, control is returned to the application to deal with the failure and decide how to proceed. The next control flag type is `required`. If multiple directives within a stack are listed as required, then all of the corresponding modules must succeed or else control is returned to the application to determine how to proceed with the failure. If there is only one directive in a stack, you could use requisite or required. The third control flag type is `sufficient`. Here, if any single module succeeds within the stack of sufficient directives, then success is passed on to the application. Thus, a stack of required directives acts as a logical AND operation while a stack of sufficient directives acts as a logical OR operation. The last control flag type is `optional` which does not impact the application but can be applied to perform other tasks. For instance, an optional directive whose module succeeds might be used to perform logging. Should that module fail, it does not return a failure to the application but logging would not take place.

As configurations are available for a number of applications, you can "piggy-back" on prior configurations. This is accomplished using the `include` statement in place of a control flag. The configuration file `system-auth` is utilized in a number of different configuration files, including `chfn`, `chsh`, `gdm`, `gdm-autologin`, `gnome-screensaver`, and `login`. This prevents a system administrator from having to recreate the same strategy over and over again.

The last *required* entry in the directive is the module itself. All of the modules are stored in /lib64/security and their names are of the form /pam_*xxx*.so where *xxx* is the remainder of their name. If you are using third-party modules or modules that exist in other libraries, you must specify the full path name. Otherwise, the path (/lib64/security) may be omitted. The module name is replaced by a configuration file name if the control flag is included.

Some modules utilize arguments as further specifications in the configuration statements. After the module name, it is permissible to include parameters such as tests passed on to the module so that the module has additional values to work with. For instance, you might have a condition such as uid < 500 or user != root. Parameters can themselves be followed by an option. Options include terms such as quiet, revoke, force, open, close, and auto_start.

### 9.5.3 An Example Configuration File

We wrap up this section by looking at one of the default /etc/pam.d configuration files. Here, we look at the one provided for su (to change user).

```
auth        sufficient      pam_rootok.so
auth        include         system-auth
account     sufficient      pam_succeed_if.so  uid=0 use_uid quiet
account     include         system-auth
password    include         system-auth
session     include         system-auth
session     optional        pam_xauth.so
```

This configuration file authenticates the user using pam_rootok.so. If this module succeeds, then the sufficient flag is met and no further authentication attempts are necessary. If this fails, then the su program moves on to the next step in authentication, which is to utilize the auth components found in the system-auth configuration file. This is shown below.

```
auth        required        pam_env.so
auth        sufficient      pam_fprintd.so
auth        sufficient      pam_unix.so nullok try_first_pass
auth        requisite       pam_succeed_if.so uid >= 500 quiet
auth        required        pam_deny.so
```

If we succeed in making it through system-auth's authentication phase, the next step is su's account stack. Here, we have a sufficient directive. If pam_succeed_if.so is successful, then we move on to the password directive; otherwise, we invoke system-auth's account stack. We see that the password directive calls upon system-auth's password stack. Finally, there are two session directives. The first, again, calls upon system-auth followed by an optional directive. In this case, the module, pam_xauth.so, executes but its success or failure status is immaterial.

Older versions of PAM used a single configuration file, /etc/pam.conf. In this file, the directives had an additional field at the beginning that listed the application such as su, login, or poweroff. If your version of PAM allows for separate configuration files (as we assumed here), then the pam.conf file will be ignored. For more information on PAM, you should explore the PAM documentation that comes with your particular dialect of Linux. It is best not to alter any configuration files unless you have a clear understanding not only of PAM configuration but also of each of the modules involved.

## 9.6 ESTABLISHING COMMON USER RESOURCES

For a system administrator, creation of accounts can be a tedious task. For instance, in a large organization, the system administrator may find that he or she is creating dozens or hundreds of accounts at a time and perhaps has to do this weekly or monthly. A shell script to automate the process is only one feature that the system administrator will want to employ. There are other mechanisms to help establish user accounts, including the automatic generation of user files and the establishment of environment variables and aliases. These items are all controlled by the system administrator.

### 9.6.1 Populating User Home Directories with Initial Files

The /etc/skel directory is, by default, the directory that the system administrator will manipulate to provide initial files for the user. Anything placed in this directory is copied into the new user's home directory when the directory is generated. The only change made to these items is the owner and group, which are altered from root to the user's username and private group name. Typically, the files in /etc/skel will be limited to user shell startup files (e.g., .bashrc) and common software configuration files and subdirectories (e.g., .mozilla/). Your installation of CentOS will probably have two or three files of .bashrc, .bash_profile, and maybe .bash_logout, along with two subdirectories with initial contents of .gnome2 and .mozilla.

The system administrator is encouraged to keep the startup files small and place system-oriented startup instructions in the startup files in /etc such as /etc/bashrc and /etc/profile. This makes it far easier for the administrator to update startup values like environment variables.

Consider defining an environment variable for the default text editor. This is called EDITOR. As a system administrator, you can define EDITOR either in /etc/bashrc (or /etc/profile) or in /etc/skel/.bashrc, or some other similar file that is automatically generated for each user. Let us assume you place this in /etc/skel/.bashrc as EDITOR=/bin/vi. After hundreds of user accounts have been generated, you decide to replace vi with vim. Now, not only do you have to modify the /etc/skel/.bashrc file to be EDITOR=/bin/vim, you also have to modify the hundreds of users' .bashrc files. Instead, placing this command in /etc/bashrc allows you to make just one modification and have it affect all users (or at least all users who run Bash). Alternatively, you could create a symbolic link to map vi to vim, but this solution would not be effective if you decided to change from vi to emacs.

The .bash_profile file will usually consist of just a few items. The file begins with an if-then statement testing if the user's .bashrc file exists and if so, executes it. Thus, the if-then statement orders the execution of the two files. While .bash_profile begins first, .bashrc is then executed before the remainder of .bash_profile executes.

The last two instructions in the .bash_profile file add $HOME/bin to the user's PATH variable and then exports the new version of PATH. The added directory is the user's home path followed by /bin, as in /home/foxr/bin. This is added so that users who wish to create their own executable programs (or who downloaded and installed executables in their own home directory space) can reference those programs without using a path.

The .bash_profile file typically consists of the following instructions:

```
if [ -f ~/.bashrc ]; then
        . ~./bashrc
fi

PATH=$PATH:$HOME/bin
export PATH
```

The .bashrc program may initially have only the following statement. This instruction tests for the availability of the file /etc/bashrc and if it exists, it is executed.

```
if [ -f/etc/bashrc ]; then
        . /etc/bashrc
fi
```

The user is free to modify and add to either or both of .bash_profile and .bashrc. It is the system administrator who would modify /etc/bashrc (as well as /etc/profile). The user might for instance add further environment variables to either .bash_profile or .bashrc. The user can define aliases in either of these files, unalias system-wide aliases (although this is not a particularly good idea), and place commands to execute their own scripts.

The /etc startup script files are far more involved than the .bash_profile and .bashrc files. These include /etc/bashrc, /etc/profile, /etc/csh.cshrc, and /etc/csh.login. The /etc/profile script is executed whenever a user logs in, irrelevant of whether the log in is GUI or command-line based and irrelevant of any shell utilized. This script first creates a PATH variable. This variable is established based on the type of user (system administrator or other). Included in the path will be /sbin, /usr/sbin, /usr/local/sbin for root, or /bin, /usr/bin, /usr/local/bin for normal users, among other directories. The variables USER, LOGNAME, MAIL, HISTCONTROL, HISTSIZE, and HOSTNAME are established and exported. Next, umask is set (discussed in the next subsection). Finally, the .sh scripts in /etc/profile.d are executed. Most of these files establish further environment variables related to the type of terminal (or GUI) being used, the language of the user, and specific software (e.g., vi). The system administrator can add further variables and aliases to the /etc/profile startup script.

The /etc/bashrc script will be executed by the user's own .bashrc script. The /etc/bashrc script is the primary mechanism whereby the system administrator will enforce startup features, including bash-specific variables and aliases.

The order that these files execute is based on the situation. Upon logging in and starting a bash shell, the behavior is

/etc/profile → ~/.profile (if it exists) → ~/.bash_profile → ~/.bashrc →/etc/bashrc

If the user opens a new bash shell, then the process is: ~./bashrc →/etc/bashrc. If the user logs in to a non-Bash session, /etc/profile and ~/.profile execute, but then either no further scripts execute, or scripts specific to that shell might execute (e.g., ~/.cshrc,/etc/csh.cshrc).

## 9.6.2 Initial User Settings and Defaults

One of the steps that /etc/profile performs is the establishment of the umask value. The umask instruction sets the default file permissions. Whenever a file or directory is created, its initial permissions are dictated by the umask value. The umask instruction is set for each user. In /etc/profile, we want to establish a umask value dependent on whether the user is root or anyone else. This allows us to have one default for root and one for any other user. The format for umask is umask *value* where *value* is a three-digit number. This is not the actual permissions (e.g., 755) but instead a number that is subtracted from 777 or 666.

As root, we might want to establish 644 for all files and 755 for all directories that we create. We obtain the umask value by subtracting 644 from 666 and 755 from 777. That is, 777–755 = 022 and 666–644 = 022. So, our umask for root is 022. For an ordinary user, we might use the default permissions of 664 for files and 775 for directories. Since 777–775 = 002 and 666–664 = 002, we use 002 for a user's umask. For software, we will probably also use 022.

An examination of the /etc/profile file shows us the following instruction to set the umask:

```
if [ $UID -gt 199 ] && [ "`id -gn`" = "`id -un`" ];
      then umask 002
else
      umask 022
fi
```

This instruction tests the user's UID to see if it is greater than 199 and tests to see if the user's ID name and group name match. The operation id -gn provides the user's group name and the operation id -un provides the user's username. We expect these two values to be equal for any normal user. If both the $UID is greater than 199 and the username and groupname are equal, then we are most likely dealing with a normal user as opposed to root or a software account. The normal user is given a umask value of 002 (to create permissions of 775 and 664). Root and software are given a umask value of 022 (to create permissions of 755 and 644).

In addition to the population of files and subdirectories of the new user's home directory through /etc/skel, the creation of a new user causes the creation of an email "spool" file. The spool file is a text file storing all of the received email messages (that have not been deleted). The location of the spool file can be specified through a default file called /etc/login.defs. This file in fact stores a number of default items that are utilized by different Linux commands.

Table 9.6 lists many of the directives available that the system administrator can use to specify desired system defaults. Directives such as CREATE_HOME, UID_MIN, UID_MAX, GID_MIN, and GID_MAX will impact the useradd instruction. Other entries impact other operations such as usermod, userdel, groupadd, groupmod, groupdel, login, passwd, and su. These defaults can also impact the contents of files such as /etc/passwd, /etc/group, /etc/shadow, and the mail spool files.

The system administrator can also place default restrictions on users in their shells. The ulimit gives the system administrator the tool to enforce such limitations. This command, when used, does not impact currently running shells or anything launched via the GUI but will impact future shells.

TABLE 9.6    Directives Available for /etc/login.defs File

| Directive | Usage | Possible Values |
|---|---|---|
| CREATE_HOME | When creating a user account, does useradd default to automatically creating a home directory or not creating a home directory? | yes, no |
| DEFAULT_HOME | If user's home directory is not available (e.g., not mounted), is login still allowed? | yes, no |
| ENCRYPT_METHOD | Encryption algorithm used to store encrypted passwords | SHA512, DES (default), MD5 |
| ENV_PATH, ENV_SUPATH | To establish the initial PATH variable for logged in users, for root | PATH=/bin:/usr/bin |
| FAIL_DELAY | Number of seconds after a failed login that the user must wait before retrying | 0, 5, etc. |
| MAIL_DIR, MAIL_FILE | Default directory, filename of user mail spool files | /var/spool/mail .mail or *username* |
| MAX_MEMBERS_PER_ GROUP | Maximum number of users allowable in a group, once reached, a new group of the same name is created, this creates a new line in /etc/group that shares the same name and GID (and password) | 0 (unlimited) |
| PASS_ALWAYS_WARN | Warn about weak passwords (even though they are still allowed) | yes, no |
| PASS_CHANGE_TRIES | Maximum number of attempts to change a password if password is rejected (too weak) | 0 (unlimited), 3 |
| PASS_MAX_DAYS, PASS_MIN_DAYS, PASS_MIN_LEN, PASS_WARN_AGE | Maximum, minimum number of days a password may be used, minimum password length, default warning date (as with chage -W) | Numeric value, 99999 for max, 0 for min are common defaults |
| UID_MIN, UID_MAX, GID_MIN, GID_MAX | Range of UID, GID available for useradd, groupadd | 500, 60000 |
| UMASK | Default umask value if none is specified (as in the /etc/profile) | 022 |

TABLE 9.7    Useful ulimit Options

| Option | Meaning |
|--------|---------|
| -c | Maximum core file size, in blocks |
| -e | Scheduling priority for new processes |
| -f | Maximum size of any newly created file, in blocks |
| -m | Maximum memory size useable by a new process, in kilobytes |
| -p | Maximum depth of a pipe (pipe size) |
| -r | Real-time priority for new processes |
| -T | Maximum number of threads that can be run at one time |
| -v | Maximum amount of virtual memory usage, in kilobytes |
| -x | Maximum number of file locks (number of open files allowable for the shell) |

Through ulimit, for instance, the system administrator can limit the size of a file created or the amount of memory usage permissible. To view the current limitations, use `ulimit -a`. To alter a limit, use

```
ulimit option value
```

where *option* is the proper option for the limit. Table 9.7 illustrates some of the more useful options.

## 9.7 THE SUDO COMMAND

The `sudo` command allows a user to execute commands as another user. Let us assume that zappaf has created an executable program that reads data files from zappaf's home directory. Because these data files contain secure information, zappaf has set their permissions so that only he has access. However, the program is one that he wants dukeg to be able to use. Instead of creating a group account of which zappaf and dukeg are members, zappaf decides to provide dukeg access via the sudo command. Now, dukeg can run this program through sudo.

The format of the sudo command is

```
sudo [-u username|uid] [-g groupname|gid] command
```

The *username/UID* or *groupname/GID* is that of owner of the program to be executed, not of the user wishing to run it. So, for instance, dukeg could issue `sudo -u zappaf program` where *program* is the program that zappaf wants to share with dukeg. You will notice that the user and group are optional. If not provided, then sudo runs *command* under root instead. Thus, sudo gives us the means of allowing ordinary users to run programs that are normally restricted to root.

Why would we want to give access to root-level programs to ordinary users? Because we are going to restrict sudo usage. In order for sudo to work, we must first establish the commands available to the various users of the system. This is done through a file called `/etc/sudoers`.

As an example of establishing access to a system administration program, let us consider the groupadd instruction. The groupadd instruction has permissions of `rwxr-x---` so that it is not accessible by the world. The user and group owner of groupadd are both root. Therefore, root is the only user that has access. However, groupadd is a relatively harmless instruction. That is, it will not interfere with user accounts or existing groups. So, we might allow users to use this instruction.

The process is twofold. First, the system administrator must edit the /etc/sudoers file. Entries in this file have the format:

```
username(s) host=command
```

*Username* is the username of the user who we will allow access to sudo. If you are not listed under the username, any attempt to use *command* via sudo will result in an error message and the event being logged. To specify multiple users, list their usernames separated by commas but with no spaces. Alternatively, you could specify `%users` to indicate all users of the system, or some other group using `%groupname`.

The value for host indicates for which machine this sudo command is permitted. You can use ALL to indicate that this sudo command is valid on all computers for a networked system or localhost for this computer. Alternatively, you could use this computer's HOSTNAME.

Finally, the command(s) should include full paths and options and parameters as needed. If you have multiple commands, separate them by spaces. Our entry for groupadd could be

```
%users      localhost=/usr/sbin/groupadd
```

Once edited, you save the sudoers file.

The second step is for a user to issue the command. The command is simply the command preceded by sudo, as in `sudo groupadd my_new_group`. The sudo program will require that the user submit his or her password to ensure that this user is authorized to execute the command. This password is timestamped so that future usage of sudo will not require the password. We see later that the timestamp can be reset.

Now that users can create their own groups, can they start populating those groups with users? The groupadd command does not allow us to insert users into a group. For that, we would need to provide access to usermod. We would want to only provide access to `usermod -G`.

So, now we modify the sudoers file again. Our entry in the sudoers file could be

```
%users      localhost=/usr/sbin/usermod -G
```

A user can now issue the command

```
sudo /usr/sbin/usermod -G group1,group2 newuser
```

to add *newuser* to *group1* and *group2*. As the sudoers file specifies that usermod includes the -G option, a user would not be able to issue a usermod command with other options.

We do not really want to provide sudo access to either of these two commands because they constitute a security breach. Imagine that we let zappaf use both of these instructions. The user zappaf could create a new group, hackers, using the command.

```
sudo /usr/sbin/groupadd -g 0 -o hackers
```

followed by

```
sudo /usr/sbin/usermod -G hackers zappaf
```

What has zappaf done? He has created a new group that shares the same GID as root and then added himself to this group, so zappaf is in a group shared with root. With such a privilege, it is possible that zappaf could exploit this to damage the system.

What follows are some examples that we might want to use sudo for. The first example allows users to view some directory that is normally inaccessible. The second example permits users to shut down and reboot the computer from the GUI. The third example allows users to mount the CD. There would be a similar capability for unmounting the CD.

- `sudo ls /usr/local/protected`
- `sudo shutdown -r +15 "quick reboot"`
- `sudo /sbin/mount/cdrom`

You might find other sudo options useful depending on circumstances. The -b option runs the command in the background. This could also be accomplished by appending & to the instruction. With -H, sudo uses the current user's HOME environment variable rather than the program's user. From earlier, if dukeg ran zappaf's executable program without -H, HOME stores /home/zappaf while with -H, HOME stores /home/dukeg. This could be useful if the program were to utilize or store files for the given user rather than the owner.

In addition to the described options, the option -K is used without a command (i.e., `sudo -K`). The result of -K is that the user's timestamp reverts to the epoch. In essence, this wipes out the fact that the user had successfully run sudo on this command in the past and will now require that the user enter their password the next time they use sudo. Similarly, the -k option will reset the user's timestamp. The -k option, unlike -K, can be used without or with a command.

The `visudo` program should be used to open and edit the /etc/sudoers file whenever you, as a system administrator, want to modify /etc/sudoers. visudo opens /etc/sudoers in vi but there is a difference between using visudo and opening the file yourself in vi, which is that visudo will check the file for syntax errors before closing the file. This will inform you of potential errors so that you can fix them at the time you are editing the file rather than waiting to find out there are errors when users attempt to use sudo at a later point in time.

## 9.8 ESTABLISHING USER AND GROUP POLICIES

To wrap up this chapter, we briefly consider user account policies. These policies might be generated by the system administrator, or by management, or both. Once developed, it will be up to the system administrator to implement the policies. Policies will impact who gets accounts, what access rights they have, whether they can access their accounts from off-site, or whether they can use on-site computers to access off-site material (e.g., personal email), and how long the accounts remain active, to name a few.

### 9.8.1 We Should Ask Some Questions before Generating Policies

We have to ask: what is the nature of our users?

- Will we have different levels (types) of users?

- Will users have different types of software that they will need to access and different files that go along with them?

- Will different levels of users require different resources, for instance, different amounts of hard disk space, web server space, and different processing needs?

The answers to these questions help inform us how to handle the users. If we have different levels of users, then we might have different types of accounts. A university might, for instance, provide different disk quotas and resource access to administrators, staff, faculty, and students. In an organization that clearly delineates duties to different software, we might want to establish software accounts so that only certain users have access. We might need to limit Oracle database client software because of a site license so that only those users who require access will be given an account whose role matches the files that they would need to access.

We have to look at our resources and ask whether we have enough resources so that restrictions are not necessary.

- Will we need to enforce disk quotas? If so, on all partitions or on specific partitions? Can different users have different quota limits? (See Chapter 10 for a discussion on disk quotas.)

- Will we need to establish priority levels for user processes?

- Will users have sole access to their workstations; will workstations be networked so that users could potentially log into other workstations; will resources be sharable?

If users have sole access to their workstation, it sounds like this would simplify the task of the system administrator. However, if network resources are being made available, the system administrator would still need to maintain networks accounts. Additionally, giving users sole access to their workstation could potentially lead the users to take actions that might harm the workstation. Giving the user the ability to download software onto their workstation could open up the workstation to attack by viruses, Trojan horses, and spyware.

We also want to ask questions about the accounts themselves.

- Will accounts exist for a limited amount of time or be unlimited? For instance, in a university setting, do we delete student accounts once the student graduates? If so, how long after graduation will we eliminate them? For a company, how long should an account remain in existence after the employee has left the company?

- What password policy will we enact? No organization should ever use anything other than strong passwords. How often should passwords be changed? Can passwords be changed to something similar to previous passwords?

The policies that we establish will be in part based on the type and size of the organization. Larger organizations will have more resources, but also greater demand on those resources. Smaller organizations may not be able to afford the more costly file servers and so may have greater restrictions on file space usage.

Policies will also be driven by management through some form of risk assessment and management. Risk assessment and management are based on the identification of organization assets and those assets' vulnerabilities and threats. Assets are not limited to the physical hardware such as the computers, file servers, and printers.

Probably even more critical as assets are the data owned by the organization. A company with clients would not want the client data to be accessible from outside. Any such access would constitute a breach in the privacy of the individuals' data that the organization has retained. Some of this data would be confidential, such as credit card numbers, and thus a breach in privacy would impact not only the clients but also their trust in the organization.

Protection of confidential and sensitive data might provide us with the most important of our policies. We might, for instance, limit access to the data to a select group of employees. We would require that these employees log in to the database system before access. We would require that the passwords be strong and we would enforce timely password changes. We might also require that database passwords differ from the passwords used to log into their computers.

Other policies might be less critical from a security perspective but equally important for management. For instance, should employees be allowed to use work email for personal business? If so, this could result in a security hole and a use of organizational resources that seems wasteful. On the other hand, employees may resent not being able to send personal emails from work. Should we limit website access? We might, through a proxy server or firewall, prohibit access to such websites as those of Facebook, Twitter, and even ESPN. Finally, should system administrators have the rights to (or responsibility to) examine users' file spaces such as to see if users are using email for personal use or to examine web browser caches for inappropriate website access?

### 9.8.2 Four Categories of Computer Usage Policies

Below, we break policy issues into four categories: user accounts, passwords, disk space, and miscellany. For each of these, we discuss possible options.

For user accounts, the questions are

- Does every user get an account? This would be the most likely case.

- Should users share accounts? This can be a security violation and is generally not encouraged.

- How long do user accounts remain active after the user is no longer with the organization? Companies may disable such accounts immediately. Other organizations might wait a few weeks. Universities often keep student accounts active for some time after graduation, possibly even permanently (or alternatively, the accounts remain active but are shifted to other servers). If accounts are disabled or deleted, do the users get some notification? What happens to any files owned by deleted accounts?

- What resources come with an account? File space? Web server space? Remote access? Access to one workstation or all workstations? Access to one printer, multiple printers, or all printers?

Password policies revolve around what sort of password management the organization wants to enforce

- Will users be given an initial password?

- Will the organization enforce strong passwords?

- How often will passwords require changing?

- Can passwords be reused? If so, at what frequency? If not, can password variations be permitted?

Disk space utilization policies concern quotas and types of file space access:

- Will files be placed locally or on a file server? If the latter, is file sharing allowed?

- Will users have disk quotas?

- Should files be encrypted?

- Does file space also exist for the web server?

- Are the users allowed to store anything in their file space?

- Are there policies giving system administrators permission to search user file space for files that should not be there (e.g., illegal downloads)?

Miscellaneous topics include how to implement and enforce protection and security, backups and disaster planning, and replacement of resources. As most of these issues are not related to users and accounts, we will hold off on discussing these until later in the textbook.

## 9.9  CHAPTER REVIEW

Concepts introduced in this chapter:

- Epoch—the date January 1, 1970, used in some Linux commands and files to count the number of dates until an action should take place (e.g., modifying a password).

- GID—the ID number assigned to a group. This is an integer used for bookkeeping.

- Group—an account generated to permit a third level of access rights so that resource access can be established for a specific collection of users. In Linux, each user account is typically accompanied by a private group populated by only one user.

- PAM—pluggable authentication module allows a system administrator to tailor how a program or service will achieve authentication by calling upon any number of modules.

- Password—a means of implementing access control by pairing a username with a password known only to that user. Passwords, for security purposes, are stored in an encrypted manner.

- Private group—a group account generated for most user accounts whose name matches that of the user. The private group should contain only a single user, that of the owner. Files and directories created by a user default to being owned by this private group.

- Strong password—a set of restrictions placed on passwords to make them hard to crack. Typically, a strong password should combine letters with at least one nonalphabetic character, be at least eight characters long, and be changed no less often than every three months.

- UID—the ID number assigned to a user. This is an integer used for bookkeeping.

- User—a person who uses the computer to run processes.

- User account—the account generated so that users can log into the computer and so the operating system can differentiate access to resources. In Linux, there are three types of accounts: root, normal users, and software.

- User Manager—the Linux GUI application used to create, modify, and delete user and group accounts. See also the Linux commands groupadd, groupdel, groupmod, useradd, userdel, and usermod.

- User policies—usage policies established by management in conjunction with the system administrator(s) and implemented by the system administrator to dictate such aspects of user capabilities such as software access, download capabilities, file space quotas, website and email usage, and password management.

Linux commands covered in this chapter:

- apg—third-party software package to automatically generate random passwords.

- chage—control password expiration information.

- groupadd—create new group.

- groupdel—delete existing group.

- groupmod—modify existing group (use usermod to add users to a group).

- newusers—program to generate new user accounts given a text file of user data. This is an alternative to developing your own script as we did in this chapter.

- pam—password authentication module to handle authentication responsibilities for most or all applications that require user authentication.

- passwd—used to modify user passwords but can also be used to control password expiration information similar to chage.

- sudo—allows a user to execute a program as another user. Most commonly used so that the system administrator can give access to some root programs to other users.

- tr—used to translate characters from one format to another. When coupled with /dev/urandom, we can take the randomly generated characters and convert them into readable ASCII characters to generate random passwords.

- ulimit—used to establish limits on resources in the shell session.

- umask—used to set default permissions when new files and directories are created.

- useradd—create a new user.

- userdel—delete an existing user.

- usermod—modify attributes of an existing user.

- visudo—open the /etc/sudoers file in vi for editing and syntax checking.

Linux files covered in this chapter:

- .bash_profile—script file placed in user home directories, executed whenever a user opens a new Bash session. Users can modify this file to add environment variables, aliases, and script code.

- .bashrc—script file placed in user home directories, executed by .bash_profile. Users can modify this file to add environment variables, aliases, and script code.

- /dev/urandom—software serving as a device that provides random number generation. Can be used to generate random passwords.

- /etc/bashrc—script file executed whenever a user starts a new Bash session. Controlled by the system administrator.

- /etc/group—file storing all of the groups defined for the system and the groups' members. This file is readable by the world.

- /etc/login.defs—default values used by a number of different programs such as useradd.

- /etc/pam.d—directory of configuration files used by PAM.

- /etc/passwd—file storing all user account information. This file does not include passwords (as the name implies). The file is readable by the world, so any user can view account information about users (e.g., user's username, home directory, login shell, UID, full name).

- /etc/profile—script file executed whenever a user logs into the system. Controlled by the system administrator.

- /etc/shadow—file storing password information for all users and groups. All passwords are encrypted. Other information such as password expirations are listed here. This file is accessible only by root.

- /etc/skel—directory controlled by the system administrator containing initial files and directories to duplicate when a new user is added to the system. Anything stored here is copied into the new user's home directory upon user account creation.

- /etc/sudoers—file storing sudo access rights. See the sudo instruction. This file should only be opened using visudo.

## REVIEW QUESTIONS

*Note*: In questions dealing with usernames, assume usernames will be of the form last-name followed by first initial like foxr, zappaf, or dukeg.

1. Of the various items to enter when adding a new user using the GUI (see Figure 9.2), which fields are required and which are not? Which can you leave unmodified?

2. Using the GUI (see Figure 9.1), how would you delete a user? A group?

3. Using the GUI (see Figures 9.1 and 9.2), how would you add a user using all of the defaults except that the user would have a home directory of /home/faculty/*username* and would be given csh as their login shell?

4. Using useradd, provide the instruction needed to create a new account for Chad Wackerman using all of the defaults.

5. Using useradd, provide the instruction needed to create a new account for Chester Thompson using all of the defaults except giving his account a login shell of /bin/csh.

6. Using useradd, provide the instruction needed to create a new account for Terry Bozzio using all of the defaults except giving his account a home directory of /home/musicians/*username* and adding him to the group musicians.

7. Using useradd, provide the instruction needed to create a new account for Adrian Belew whose starting home directory will obtain its contents not from /etc/skel but from /etc/skel2.

8. Under what circumstances might you choose to use the -N option for useradd?

9. What is wrong with the following useradd instruction?

```
useradd -c Frank Zappa -m zappaf
```

10. What is wrong with the following useradd instruction?

```
useradd -m -o underwoodi
```

11. Using useradd, provide the instruction needed to create a new account for the software audacity, which will have its own home directory of /var/media/audacity, a login shell of /sbin/nologin, and no private group account.

12. By default, the account expiration date is set to −1 indicating that there should not be an initial expiration date. You want to change this so that the expiration date is automatically set for new users to be December 31, 2018. How can you do this?

13. Modify the account creation script from Section 9.2.3 of this chapter as follows:

    • All entries of the text file of users will be listed as Lastname Firstname Major

    • If the Major is CSC, the student is assigned the shell tcsh instead of bash.

14. Modify the account creation script from Section 9.2.3 of this chapter as follows:

    • All entries of the text file of users will be listed as Lastname Firstname Type

    • Type is either Faculty or Student. All home directories are placed in either /home/faculty/*username* or /home/student/*username* based on the value of Type.

15. Modify the account creation script from Section 9.2.3 of this chapter as follows:

    • All entries of the text file of users will be listed as Lastname Firstname Role

    • Role will be Administrator, Database, or Network. If Administrator, add this user to the group management. If database or network, add the user to the group technical. If database, also add the user to the group oracle.

16. In the account creation script from Section 9.2.3 comes the following instruction. Explain what this instruction does and how it works.

```
n = `egrep -c $username /etc/passwd`
```

17. You forget to specify an expiration date for a new user, martinb. You want his account to expire on June 1, 2020. How can you do this using usermod?

18. Repeat #17 using chage.

19. Examine /etc/passwd. What are the home directories established for these users?

    a.  root

    b.  bin

    c.  lp

    d.  mail

    e.  halt

20. Examine /etc/passwd. What login shell is specified for the following users?

    a.  root

    b.  bin

    c.  sync

    d.  shutdown

    e.  mail

21. Examine /etc/group. What users are members of each of these groups?

    a.  root

    b.  bin

    c.  adm

    d.  tty

    e.  lp

22. With root access, examine /etc/shadow. You will find large numbers in the first numeric position in many of the lists. What does this number represent?

23. With root access, examine /etc/shadow. Many of the entries end with the entries 99999:7:::. What does 99999 represent?

24. Provide a chage command for user keneallym to expire his password in 30 days, sending a warning 5 days in advance.

25. Provide a chage command for user underwoodr to expire her password in 90 days, sending a warning 10 days in advance.

26. Provide a chage command to lock the account for dukeg.

27. Provide a chage command to change the minimum time between password changes to 7 days for user marst.

28. Redo question 24 using the passwd command.

29. Redo question 25 using the passwd command.

30. What does it mean to lock an account using the passwd command?

31. Which script creates the user's PATH variable?

32. You want to add the directories /opt/bin and ~/myscripts to your PATH variable. How would you do this and in which script file would you want to place the command?

33. Which script file begins executing first, .bashrc or .bash_profile? Which of these files finishes executing first?

34. What permissions would a file have if you created the file after doing umask 034?

35. What permissions would a directory have if you created the file after doing umask 015?

36. What would the setting DEFAULT_HOME=no in the file /etc/login.defs do?

37. How would you, as a system administrator, control the number of seconds that would elapse before a user can attempt to login again after a failed attempt?

38. What instruction would you use to establish the initial PATH for root? For nonroot users?

39. What is the difference between the module types auth and account in a PAM configuration file?

40. Assume there are four session directives for a given PAM configuration file. Three are listed as required and the fourth as optional. What happens if the third directive's call to the given module fails? What happens if the fourth directive's call to the given module fails?

41. What does the include statement do in a PAM configuration file directive? In what way does the include statement help the system administrator?

42. Notice that /etc/shadow is not readable by anyone but root. As root, you want to give permission to read this file to user zappaf on the local computer. What entry would you add to the /etc/sudoers file to establish this?

43. Anyone in the group *faculty* has sudo access to the killall command. You, a faculty member, want to kill all copies of the process *someprocess*. What command would you issue using sudo?

44. Why should you use visudo to modify the /etc/sudoers file rather than directly editing it in vi?

45. Aside from the examples covered in Section 9.7, name three additional examples of why you might set up sudo privileges for one or more users.

Assume you are the sole system administrator for a small company of 10–20 users. Your organization has a single file server that primarily stores software and group documents. User documents can either be stored on the file server or, more often, user workstations. The company currently has no restrictions placed on users in terms of personal use of company computers. Answer questions 46–49 regarding user account policies for this organization. Offer a brief explanation to each answer.

46. Should there be a policy prohibiting users from downloading software on their workstations?

47. Should users have quotas placed on their file server space should they decide to store files there?

48. Should the company install a proxy server to store commonly accessed web pages and also use it to restrict access to websites deemed unrelated to company business?

49. Should users be able to access the file server from off-site computers?

You have been asked to propose user account policies for a large company of 250 employees. The company utilizes file servers to store all user files. In addition, the company has its own web server and proxy server. Employees are permitted to have their own websites if they desire them. The company has a policy restricting computer usage to "professional business" only. Answer questions 50–53. Offer a brief explanation to each answer.

50. Should users have disk quotas placed on them for either their file server space or web space?

51. Should you establish rules that prohibit certain types of content from employee websites?

52. Should the proxy server be set up to restrict access to websites such as Facebook and personal email servers (e.g., Gmail, Yahoo mail)?

53. Should the system administrator be allowed to examine individual employee storage to see if they are storing information that is not related to the business?

54. As a system administrator, describe a policy that you would suggest for enforcing strong passwords (e.g., minimum length, restriction on types of characters used, and duration until passwords need to be changed).

55. If you were the system administrator for a university, would you recommend that students, once graduated, can keep their accounts? If so, for how long? 6 months? 1 year? Forever?

56. Repeat #55 assuming you are the system administrator for a corporation of thousands of employees.

57. Repeat #55 assuming it is a small company of 50 employees.

# CHAPTER 10

# The Linux File System

T HIS CHAPTER'S LEARNING OBJECTIVES are

- To understand the types of objects that Linux treats as files and how they differ

- To understand the role of the inode

- To be able to utilize Linux file system commands of badblocks, cpio, df, du, dump, mkfifo, mount, stat, umount, and tar

- To understand common partitions in Linux, the role of the files /etc/fstab and /etc/mtab and the meaning of the term "file system"

- To be able to establish disk quotas

- To understand the role of the top-level Linux directories

## 10.1 INTRODUCTION

In Chapter 3, we viewed the Linux file system from a user's perspective. Here, we examine the Linux file system from a system administration point of view. For a single user of a standalone workstation, there are few duties that the system administrator will be required to perform. However, it is still important to understand the concepts of the file system. These concepts include partitions, file layout, and the role of the top-level directories. The system administrator for a larger Linux system or a network of workstations will have to monitor file system usage and performance, perform backups, and handle other duties.

This chapter breaks the Linux file system into four basic topics. First, we examine a file. For this, we consider what types of entities make up files in the file system. We will find that in Linux many types of entities are treated as files. This lets us apply file operations to any or all of them. By examining the storage structure of a file through the inode, we can see how files are stored. Second, we examine the partition. We discussed the role of partitioning during the Linux installation in Chapter 8. Now we look at what a partition is, why we partition our file space, how to handle a partition, and how to repartition the file system. Third, we look at the top-level directories. Although we briefly examined this layout in Chapter 3, here we focus on them in more detail, spotlighting the more important directories, subdirectories, and files of note. Finally, we look at some administrative duties to support a secure and reliable file system. Before we examine the specifics of the Linux file system, we will first consider a generic file system.

## 10.2 STORAGE ACCESS

A collection of storage devices present a file space. This file space exists at two levels: a logical level defined by partitions, directories, and files, and a physical level defined by file systems, disk blocks, and pointers. The users and system administrators primarily view the file space at the logical level. The physical level is one that the operating system handles for us based on our commands (requests).

The devices that make up the file space include hard disk drives, optical disk drives and optical disks, solid-state drives, including USB drives, and in some cases magnetic tape drives and magnetic tape media. In some cases, storage is shifted into memory in the form of ramdisks or other mechanisms whereby memory mimics file storage. Ramdisks are discussed in Section 10.5 when we examine some of the devices found under /dev. Collectively, these devices provide us with storage access, where we store executable programs and data files. We generally perform one of two operations on this storage space: we read from a file (load the file into memory, or input from file) and we write to a file (store/save information to a file, or output to file).

### 10.2.1 Disk Storage and Blocks

As hard disk storage is the most common implementation of a file system, we will concentrate on it although some of the concepts apply to other forms of storage as well. To store a file on disk, the file is decomposed into fixed-sized units called *blocks*. Figure 10.1 illustrates a small file (six blocks) and the physical locations of those blocks. Notice that the last block may not fill up the entire disk block space, so it leaves behind a small *fragment*.

The operating system must be able to manage this distribution of files to blocks in three ways. First, given a file and block, the operating system must map that block number into a physical location on some disk surface. Second, the operating system must be able to direct the disk drive to access that particular block through a movement of both the disk and the drive's read/write head. Third, the operating system must be able to maintain free file space (available blocks), including the return of file blocks once a file has been deleted. All of these operations are hidden from the user and system administrator.

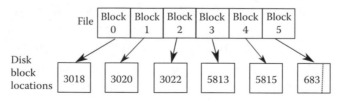

FIGURE 10.1 File decomposed into blocks. (Adapted from Fox, R. *Information Technology: An Introduction for Today's Digital World*, FL: CRC Press, 2013.)

Let us consider how a disk file might be broken into blocks. We might assume that our file system already has hundreds or thousands of files stored in it. The files are distributed across all of the disk's surfaces (recall from Chapter 3 that a hard disk drive will contain multiple disk platters and each platter has two surfaces, a top and bottom). Given a new file to store, the file is broken into blocks. The first block is placed at the first available free block on disk.

Where should the next disk block be placed? If the next block after the first is available, we could place the block there, giving us two blocks of contiguous storage. This may or may not be desirable. The disk drive spins the disks very rapidly. If we want to read two blocks, we read the first block and transfer it into a buffer in the disk drive. Then, that data are transferred to memory. However, during that transfer, the disk continues to spin. When we are ready to read the second disk block, it is likely that this block has spun past the read/write head and now the disk drive must wait to finish a full disk revolution before reading again. Distributing disk blocks so that they are not contiguous will get around this problem. If you examine Figure 10.1, you will see that the first three disk blocks are located near each other but not in a contiguous block. Instead, the first block lies at location 3018, the second at 3020, and the third at 3022.

Whether initial blocks are contiguous or distributed, we will find that further disk blocks may have to be placed elsewhere because we have reached the end of the available disk blocks in this locality. With the deletion of files and saving of other files, we will eventually find disk blocks of one file scattered around the disk surfaces. This may lead to some inefficiency in access in that we have to move from one location of the disk to another to read consecutive blocks and so the seek time and rotational latency are lengthened.

Again, referring back to Figure 10.1, we might assume that the next available block after 3022 is at 5813 and so as the file continues to grow, its next block lies at 5813 followed by 5815. As the next block, according to the figure, lies at 683, we might surmise that 683 was picked up as free space because a file was deleted.

## 10.2.2 Block Indexing Using a File Allocation Table

How do we locate a particular block of a disk? File systems use an *indexing* scheme. MS DOS and earlier Windows systems used a *file allocation table* (FAT). For every disk block in the file system, the next block's location is stored in the table under the current block number. That is, block i's successor location is stored at location i.

The FAT is loaded from disk into memory at the time the file system is mounted (e.g., at system initialization time). Now, a search of memory would track down a particular disk block's location. For instance, if the file started at block 1500, we would look at FAT location 1500 to find the file's second block location. If it were 1505, we would look at 1505 to find the file's third location, and so forth.

In Figure 10.2, a partial listing of a FAT is provided. Here, assume a file starts at block 151. Its next block is 153 followed by 156, which is the end of the file (denoted by "EOF"). To find the file's third block, the operating system will examine the FAT starting at location 151 to find 153 (the file's second block) and then look at location 153 to find 156, the file's third block. Another file might start at block 154. Its third block is at location 732. The entry "Bad" indicates a bad sector that should not be used.

File allocation table (portion)

| Block | 150 | 151 | 152 | 153 | 154 | 155 | 156 |
|---|---|---|---|---|---|---|---|
| Next location | 381 | 153 | Bad | 156 | 155 | 732 | EOF |

FIGURE 10.2 Excerpt of a file allocation table. (Adapted from Fox, R. *Information Technology: An Introduction for Today's Digital World*, FL: CRC Press, 2013.)

More recent versions of Windows operating systems use the NTFS file system (new technology file system), which utilizes a data structure called a B+ tree for indexing. In Linux, indexing is handled through the inode data structure. We explore this in Section 10.3 where you will see that every inode contains pointers to disk blocks that either store data or additional pointers.

### 10.2.3 Other Disk File Details

Aside from determining the indexing strategy and the use/reuse of blocks, the file system must also specify a number of other details. These will include naming schemes for file entries (files, directories, links). It is common today for names to permit just about any character, including blank spaces; however, older file systems had limitations such as eight-character names and names consisting only of letters and digits (and perhaps a few types of punctuation marks such as the hyphen, underscore, and period). Some file systems do not differentiate between uppercase and lowercase characters while others do. Most file systems permit but do not require file name extensions.

File systems will also maintain information about the entries, often called metadata. This will include the creation date/time, last modification date/time and last access date/time, owner (and group in many cases), and permissions or access control list. The access control list enumerates for each user of the system the permissions granted to that user so that there can be several levels of permissions over the Linux user/group/other approach.

Many different file system types have been developed over the years. Many early mainframes had their own, unique file systems. Today, operating systems tend to share file systems or provide compatibility so that a different file system can still be accessed by many different types of operating systems. Aside from the previously mentioned FAT and NTFS file systems, some of the more common file systems are the extended file system family (ext, ext2, ext3, ext4, derived originally from the Minix OS file system) used in Linux. NFS (the network file system) is also available in Linux. Files-11, which is a descendant of the file system developed for DEC PDP mainframes, and the Vax VMS operating system (and itself a precursor of NTFS) are also available. While these multitudes of file systems are available, most Linux systems primarily use the ext family as the default file system type.

## 10.3 FILES

In the Linux operating system, everything is treated as a file except for the process. What does this mean? Among other things, Linux file commands can be issued on entities that are not traditional files. The entities treated like files include directories, physical devices, named pipes, and file system links. Aside from physical devices, there are also some special-purpose programs that are treated like files (for instance, a random number generator). Let us consider each of these in turn.

## 10.3.1 Files versus Directories

The directory should be familiar to the reader by now. It is a named entity that contains files and subdirectories (or devices, links, etc.). The directory offers the user the ability to organize their files in some reasonable manner, giving the file space a hierarchical structure. Directories can be created just about anywhere in the file system and can contain just about anything from empty directories to directories that themselves contain directories.

The directory differs from the file in a few significant ways. First, we expect directories to be executable. Without that permission, no one (including the owner) can cd into the directory. Second, the directory does not store content like a file; instead it merely stores other items. That is, whereas the file ultimately is a collection of blocks of data, the directory contains a list of pointers to objects. Third, there are some commands that operate on directories and not files (e.g., cd, pwd, mkdir) and some commands that operate on files but not directories (e.g., wc, diff, less, more). We do find that most Linux file commands will operate on directories themselves, including for instance cp, mv, rm (using the recursive version), and wildcards apply to both files and directories.

## 10.3.2 Nonfile File Types

Many devices are treated as files in Linux. These devices are listed under the /dev directory. We categorize these devices into two subcategories: character devices and block devices. Character devices are those that input or output streams of characters. These will include the keyboard, the mouse, a terminal (as in terminal window), and serial devices such as older MODEMs and printers. Block devices communicate via blocks of data. The term "block" is traditionally applied to disk drives where the files are broken into fixed-sized blocks. However, here, block is applied to any device that communicates by transmitting chunks of data at a time (as opposed to the previously mentioned character type). Aside from hard disk, block devices include optical disk and flash drive.

Aside from the quantity of data movement, another differentiating characteristic between character and block devices is how input and output are handled. For a character device, a program executing a file command must wait until the character is transferred before resuming. For a block device, blocks are buffered in memory so that the program can continue once the instruction has been issued. Further, as blocks are only portions of entire files, it is typically the case that a file command can request one portion of a file. This is often known as *random access*. The idea is that we do not have to request block 1 before obtaining block 2. Having to read blocks in order is known as *sequential access*. But in random access, we can obtain any block desired and it should take no longer to access block j than block i.

Another type of file construct is the *domain socket*, also referred to as a local socket. This is not to be confused with a network socket. The domain socket is used to open communication between two local *processes*. This permits interprocess communication (IPC) so that the two processes can share data. We might, for instance, want to use IPC when one process is producing data that another process is to consume. This would be the case when some application software is going to print a file. The application software produces the data to be printed, and the printer's device driver consumes the data. The IPC is also used to create a rendezvous between two processes where process B must wait for some event from process A.

There are several distinctions between a network and domain socket. The network socket is not treated as a file (although the network itself is a device that can interact via file system commands) while the domain socket is. The network socket is created by the operating system to maintain communication with a remote computer while domain sockets are created by users or running software. Network sockets provide communication lines between *computers* rather than between processes.

Yet another type of file entity is the *named pipe*. You have already explored the pipe in earlier chapters. The named pipe differs in that it exists beyond the single usage that occurs when we place a pipe between two Linux commands.

To create a named pipe, you define it through the `mkfifo` operation. The expression FIFO is short for "first-in-first-out." FIFO is often used to describe a queue (waiting line) as queues are generally serviced in a first-in, first-out manner. In this case, mkfifo creates a FIFO, or a named pipe. Once the pipe exists, you can assign it to be used between any two processes. Unlike an ordinary pipe that must be used between two Linux processes in a single command, the named pipe can be used in separate instructions.

Let us examine the usage of a named pipe. First, we define our pipe:

```
mkfifo a_pipe
```

This creates a file entity called a_pipe. As with any file or directory, a_pipe has permissions, user and group owner, creation/modification date, and a size (of 0). Now that the pipe exists, we might use the pipe in some operation:

```
ps aux > a_pipe
```

Unlike performing `ps aux`, or even `ps aux | more`, this instruction does not seem to do anything when executed. In fact, our terminal window seems to hang as there is no output but neither is the cursor returned to us. What we have done is opened one end of the pipe. But until the other end of the pipe is open, there is nowhere for the `ps aux` instruction's output to "flow."

To open the other end of the pipe, we might apply an operation (in a different terminal window since we do not have a prompt in the original window) like:

```
cat a_pipe
```

Now, the contents "flow" from the ps aux command through the pipe to the cat command. The output appears in the second terminal window and when done, the command line prompt returns in the original window.

You might ask why use a named pipe? In fact, the pipe is being used much like an ordinary pipe. Additionally, the named pipe does roughly the same thing as a domain socket—it is a go between for IPC. There are differences between the named pipe and pipe and between the named pipe and the domain socket. The named pipe remains in existence. We can call upon the named pipe numerous times. Notice here that the source program is immaterial. We can use a_pipe no matter what the source program is.

Additionally, the mkfifo instruction allows us to fine tune the pipe's performance. Specifically, we can assign permissions to the result of the pipe. This is done using the option –M *mode* where *mode* is a set of permissions such as –M 600 or –M u=rwx,g=r,o=r. The difference between the named pipe and the domain socket is a little more obscure. The named pipe always transfers one byte (character) at a time. The domain socket is not limited to byte transfers but could conceivably transfer more data at a time.

## 10.3.3 Links as File Types

The final file type is the link. There are two forms of links: hard links and soft (or symbolic) links. A hard link is stored in a directory to represent a file. It stores the file's name and the inode number. When creating a new hard link, it duplicates the original hard link, storing the new link in a different directory. The symbolic link instead merely creates a pointer to point at the original hard link.

The difference between the two types of links is subtle but important. If you were to create a symbolic link and then attempt to access a file through the symbolic link rather than the original link, you are causing an extra level of indirect access. The operating system must first access the symbolic link, which is a pointer. The pointer then provides access to the original file link. This file link then provides access to the file's inode, which then provides access to the file's disk blocks (we discuss inodes in the next subsection).

It may sound like symbolic links are a drawback and we should only use hard links. This is not true though for several reasons. First, hard links cannot link files together that exist on separate partitions. Additionally, hard links can only link together files whereas symbolic links can link directories and other file system entities together. On the positive side for hard links, they are always up to date. If you move the original object, all of the hard links are modified at the same time. If you delete or move a file that is linked by a symbolic link, the file's (hard) link is modified but not the symbolic link; thus you may have an out-of-date symbolic link. This can lead to errors at a later time.

In either case, a link is used so that you can refer to a file that is stored in some other location than the current directory. This can be useful when you do not want to add the file's location to your PATH variable. For instance, imagine that user zappaf has created a program called my_program, which is stored in ~zappaf. You want to run the program (and zappaf was nice enough to set its permissions to 755). Rather than adding /home/zappaf to your PATH, you create a symbolic link from your home directory to ~zappaf/my_program. Now you can issue the my_program command from your home directory.

*One last comment*: You can determine the number of hard links that exist for a single file when you perform an ls –l. The integer value after the permissions is the number of hard links. This number will never be less than 1 because with no hard links, the file will not exist. However, the number could be far larger than 1. Deleting any of the hard links will reduce this number. If the number becomes 0, then the file's inode is returned to the file system for reuse, and thus access to the file is lost with its disk space available for reuse.

If you have a symbolic link in a directory, you will be able to note this by its type and name when viewing the results of an `ls -l` command. First, the type is indicated by an 'l' (for link) and the name will contain the symbolic link's name, an arrow (->) and the location of the file being linked. Unfortunately, unlike the hard link usage, if you were to use `ls -l` on the original file, you will not see any indication that it is linked to by symbolic links.

### 10.3.4 File Types

Collectively, all of these special types of entities are treated like files in the following ways:

- Each item is listed when you do an ls.

- Each item can be operated upon by file commands such as mv, cp, rm and we can apply redirection operators on them.

- Each item is represented in the directory by means of an inode.

You can determine a file's type by using ls -l (long listing). The first character of the 10-character permissions is the file's type. In Linux, the seven types are denoted by the characters in Table 10.1.

To illustrate the types of files, see Figure 10.3. Here, we see excerpts from three directories. At the top is a subdirectory of /tmp. The /tmp directory stores files created by running software. In many (most) cases, the types of files created in /tmp are domain sockets, whose type is denoted by an 's'. The next group of items listed come from the /dev directory where you can find character-type devices (c), block-type devices (b), directories (d), symbolic links (l), and regular files (-), although no regular files are shown in the figure. Finally, the last six entries come from a user's file space where we see a named pipe (p), several regular files with a variety of permissions, and a directory.

Every file (no matter the type, e.g., regular file, character type, block type, named pipe) is stored in a directory. The directory maintains the entities stored in it through a list. The listing is a collection of hard and soft links. A hard link of a file stores the file's name and

TABLE 10.1    Characters for Each Type of File

| Character | Type |
| --- | --- |
| - | Regular file |
| d | Directory |
| b | Block device |
| c | Character device |
| l | Symbolic link |
| p | Named pipe |
| s | Domain socket |

```
srwxr-xr-x. 1 Student Student 0 Aug 12 08:34 socket
srwxr-xr-x. 1 Student Student 0 Aug 12 08:34 socket.pkcs11
srwxr-xr-x. 1 Student Student 0 Aug 12 08:34 socket.ssh
crw-rw----. 1 root root     10,  61 Aug 12 08:33 cpu_dma_latency
crw-rw----. 1 root root     10,  62 Aug 12 08:33 crash
drwxr-xr-x. 5 root root         100 Aug 12 08:33 disk
brw-rw----. 1 root disk    253,   0 Aug 12 08:33 dm-0
brw-rw----. 1 root disk    253,   1 Aug 12 08:33 dm-1
lrwxrwxrwx. 1 root root           3 Aug 12 08:33 dvd -> sr0
lrwxrwxrwx. 1 root root           3 Aug 12 08:33 dvdrw -> sr0
prw-rw--w-. 1 Student Student     0 Aug 20 08:37 mypipe
-rw-rw-r--. 1 Student Student    24 Aug 14 14:02 num.txt
-rwxr-xr-x. 1 Student Student   219 Jul  9 11:06 q10
-rwxr-xr-x. 1 Student Student   179 Jun 28 14:38 s5
-rwxr--r-x. 1 Student Student   116 Jul  9 11:04 sub
drwxrwxr-x. 2 Student Student  4096 Aug 19 09:30 temp
```

FIGURE 10.3   Long listings illustrating file types.

the inode number dedicated to that file. The symbolic link is a pointer to a hard link stored elsewhere. As the user modifies the contents of the directory, this list is modified.

New files require new hard links pointing to newly allocated inodes. The deletion of a file causes the hard link to be removed and the numeric entry of hard links to a file to be decremented (e.g., in Figure 10.3, deleting the item disk in /dev would result in the hard link count being reduced to 4). The inode itself remains allocated to the given file unless the hard link count becomes 0.

### 10.3.5  inode

We now turn to the inode. When the file system is first established, it comes with a set number of inodes. The inode is a data structure used to store file information. The information that every inode will store consists of

- The file type

- The file's permissions

- The file's owner and group

- The file's size

- The inode number

- A timestamp indicating when the inode was last modified, when the file was created, and when the file was last accessed

- A link count (the number of hard links that point at this file)

- The location of the file (i.e., the device storing the file) and pointers to the individual file blocks (if this is a regular file), these pointers break down into

  - A set number of pointers that point directly to blocks

- A set number of pointers that point to indirect blocks; each indirect block contains pointers that point directly to blocks

- A set number of pointers that point to doubly indirect blocks, which are blocks that have pointers that point to additional indirect blocks

- A set number of pointers that point to triply indirect blocks, which are blocks that have pointers that point to additional doubly indirect blocks

Typically, an inode will contain 15 pointers broken down as follows:

- 12 direct pointers

- 1 indirect pointer

- 1 double indirect pointer

- 1 triply indirect pointer

An inode is illustrated in Figure 10.4 (which contains 1 indirect pointer and 1 doubly indirect pointer but no triply indirect pointer because of a lack of space).

Let us take a look at how to access a Linux file through the inode. We will make a few assumptions. First, our Linux inode will store 12 direct pointers, 1 indirect pointer, 1 doubly indirect pointer, and 1 triply indirect pointer. Blocks of pointers will store 12 pointers no matter whether they are indirect, doubly indirect, or triply indirect blocks. We will assume that our file consists of 500 blocks, numbered 0 to 499, each block storing 8 KB (the

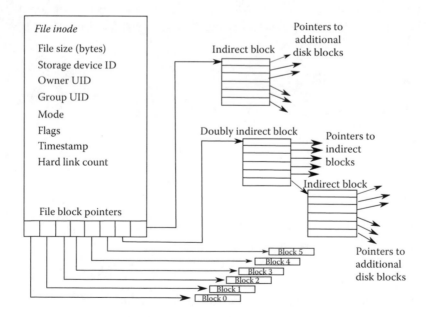

FIGURE 10.4 inode structure with pointers to disk blocks. (Adapted from Fox, R. *Information Technology: An Introduction for Today's Digital World*, FL: CRC Press, 2013.)

typical disk block stores between 1 KB and 8 KB depending on the file system utilized). Our example file then stores 500 * 8 KB = 4000 KB or 4 MB. Here is the breakdown of how we access the various blocks.

- Blocks 0–11: direct pointers from the inode.

- Blocks 12–23: pointers from an indirect block pointed to by the inode's indirect pointer.

- For the rest of the file, access is more complicated.

  - We follow the inode's doubly indirect pointer to a doubly indirect block. This block contains 12 pointers to indirect blocks. Each indirect block contains 12 pointers to disk blocks.

    - The doubly indirect block's first pointer points to an indirect block of 12 pointers, which point to blocks 24–35.

    - The doubly indirect block's second pointer points to another indirect block of 12 pointers, which point to blocks 36–47.

    - …

    - The doubly indirect block's last pointer points to an indirect block of 12 pointers, which point to blocks 156–167.

  - We follow the inode's triply indirect pointer to a triply indirect block. This block contains 12 pointers to doubly indirect blocks, each of which contains 12 pointers to indirect blocks, each of which contain 12 pointers to disk blocks. From the triply indirect block, we can reach blocks 168 through 499 (with room to increase the file to block 1895).

Earlier, we noted that the disk drive supports random access. The idea is that to track down a block, block i, we have a mechanism to locate it. This is done through the inode pointers as just described.

The above example is far from accurate. A disk block used to store an indirect, doubly indirect, or triply indirect block of pointers would be 8 KB in size. Such a sized block would store far more than 12 pointers. A pointer is usually 32 or 64 bits long (4 or 8 bytes). If we assume an 8 byte pointer and an 8 KB disk block, then an indirect, doubly indirect, or triply indirect block would store 8 KB/8 B pointers = 1 K or 1024 pointers rather than 12.

When a file system is created, it comes with a set number of inodes. The actual number depends on the size of the file system and the size of a disk block. Typically, there is 1 inode for every 2–8 KB of file system space. If we have a 1 TB file system (a common size for a hard disk today), we might have as many as 128 K (approximately 128 thousand) inodes. The remainder of the file system is made up of disk blocks dedicated to file storage and pointers. Unless nearly all of the files in the file system are very small, the number of inodes should be more than sufficient for any file system usage.

The system administrator is in charge of administering the file system. Rarely, if ever, will the system administrator have to worry about inodes or pointers. The Linux file system commands instead operate at a higher level of abstraction, allowing administrators and users alike to operate on the file entities (files, directories, links, etc.). It is the device drivers, implemented by system programmers, which must deal with inodes and the location of disk blocks.

Let us consider, as an example, file creation. First, a new inode must be allocated from the available inodes. The next inode in order is used. The inode's information is filled in, consisting of the file type, initial permissions (defaulting from the user's umask value), owner and group of the user, the file system's device number, and a timestamp for file creation. If the file is to store some initial contents, disk blocks are allocated and the direct pointers are modified in the inode to point at these blocks. Finally, the content can be saved to those blocks. Additionally, the directory is modified to store the new hard link.

If we were instead to move a file within the same partition, then all we have to do is transfer the hard link from one directory to another. Copying a file requires a new inode and disk blocks with the original copy's blocks copied into the new copy's blocks. Deleting a file requires the removal of the hard link from the directory along with a decrement to the hard link count. If this reaches zero, the inode is returned to the collection of free inodes. Disk blocks themselves are not modified (erased) but instead are returned to free blocks to be reused in the future. If a new file is created, the returned inode might be reused at that time.

## 10.3.6 Linux Commands to Inspect inodes and Files

There are several tools available to inspect inodes. The following commands provide such information:

- `stat`—provides details on specific file usage, the option `-c  %i` displays the file's inode number

- `ls`—the option –i displays the inodes of all entries in the directory

- `df  -i`—we explore df in more detail in the next section but this command provides information on the utilization of the file system, partition by partition. The -i option includes details on the number of inodes used.

We wrap up this section by examining the information obtainable from stat. The stat command itself will respond with the name of the file, the size of the file, the blocks used to store the file, the device storing the file (specified as a device number), the inode of the file, the number of hard links to the file, the file's permissions, UID, GID in both name and number, and the last access, modification, and change date and time for the file. The stat command has many options. The most significant are listed here:

- -L—follow links to obtain inode information of files stored in other directories (without this, symbolic links in the given directory are ignored)

- -f—used to obtain statistics on an entire file system rather than a file

TABLE 10.2    Formatting Characters for -c, Bottom Half for -c -f

| Formatting Character | Meaning |
| --- | --- |
| %b, %B | Number of blocks (file size in blocks), size of blocks |
| %d, %D | Device number storing file in decimal, in hexadecimal |
| %f | File type (see Table 10.1) |
| %g, %G, %u, %U | GID, group name, UID, user name |
| %h | Number of hard links |
| %i | inode number |
| %n | File name |
| %s | Size in bytes |
| %x, %y, %z | Time of last access, modification, change |
| %a | Free blocks available to ordinary users |
| %b | Total blocks in file system |
| %c | Total file inodes in use |
| %d | Free inodes |
| %f | Total free blocks |
| %l | Maximum allowable file name length |
| %s | Block size |
| %T | Type of file system |

- -c *FORMAT*—output only the requested information where *FORMAT* uses the characters listed in Table 10.2, additionally when used with -f (file system stats) there are other formatting characters available (see second half of Table 10.2)

Let us take a closer look at stat with two examples. First, we view information of a series of regular files. Specifically, we use stat to provide for us the size of each file in blocks and bytes, the file name, the inode number of the file, and the time of last access. This command is given as stat -c "%b %s %n %i %x". A partial output is shown below. Each file consists of only 8 blocks with sizes that vary from 48 bytes to 413 bytes. The inodes are all in the 53x,xxx range. Finally, the last access time and date are given.

```
8 361 courses.txt 530970 2013-03-06 07:59:03.458188951 -0500
8 117 mykey.pub 530991 2013-03-06 07:59:03.436319789 -0500
8 413 names.txt 530974 2013-03-06 07:59:03.426421446 -0500
8 80 s1 531019 2013-03-06 07:59:03.440076584 -0500
8 48 s2 531021 2013-03-06 07:59:03.426421446 -0500
```

Second, we inspect devices from /dev where we look at the file's type, described in English using %F, among other things. The command is stat -c "%d %u %h %i %n %F". We also view the device number, UID of the owner, number of hard links, inode number, file name, and file type. Below are some excerpts from this output.

```
5 0 1 12556 autofs character special file
5 0 1 6106 dm-0 block special file
5 0 1 10109 dvdrw symbolic link
```

```
5 0 3 5341 input directory
5 0 1 2455364 log socket
11 0 2 1 pts directory
5 69 1 5185 vcs character special file
```

In this second listing, we see the devices autofs, dm-0, dvdrw, input, log, pts, and vcs. All except pts are located on device number 5. All except vcs are owned by user 0 (root); vcs is owned by vcsa, the virtual console memory owner account. Most of the items have only one hard link, found in /dev. Both input and pts are directories and have more than one hard link. The inode numbers vary from 1 to 2,455,364. The file type demonstrates that "files" can make up a wide variety of entities from block or character files (devices) to symbolic links to directories to domain sockets. This last field varies in length from one word (directory) to three words (character special file, block special file).

You might want to explore the inode numbers in your file system to see how diverse they are. Each new file is given the next inode available. As your file system is used, you will find newer files have higher inode numbers although deleted files return their inodes. The following script will output the largest and smallest inode numbers of a list of files passed in as parameters.

```
#!/bin/bash
largest=`stat -c "%i" $1`
largestFile=$1
smallest=`stat -c "%i" $1`
smallestFile=$1
shift
for item in $@; do
      number=`stat -c "%i" $item`
      if [ $number -gt $largest ]; then
            largest=$number; largestFile=$item;
      fi
      if [ $number -lt $smallest ]; then
            smallest=$number; smallestFile=$item;
      fi
done
echo The largest inode from the files provided is
echo $largest of file $largestFile. The smallest
echo inode from the files provided is $smallest
echo of file $smallestFile
```

Notice in this script the use of the shift command so that all of the parameters after the first are shifted down. Since we had already processed the first parameter ($1), we no longer need it.

Whenever any file is used in Linux, it must first be opened. The opening of a file requires a special designator known as the *file descriptor*. The file descriptor is an integer assigned to the file while it is open. In Linux, three file descriptors are always made available:

- 0 – stdin

- 1 – stdout

- 2 – stderr

Any remaining files that are utilized during Linux command or program execution need to be opened and have a file descriptor given to that file.

When a file is to be opened, the operating system kernel gets involved. First, it determines if the user has adequate access rights to the file. If so, it then generates a file descriptor. It then creates an entry in the system's file table, a data structure that stores file pointers for every open file. The location of this pointer in the file table is equal to the file descriptor generated. For instance, if the file is given the descriptor 185, then the file's pointer will be the 185th entry in the file table. The pointer itself will point to an inode for the given file. As devices are treated as files, file descriptors will also exist for every device, entities such as the keyboard, terminal windows, the monitor, the network interface(s), the disk drives, as well as the open files. You can view the file descriptors of a given process by looking at the fd subdirectory of the process' entry in the /proc directory (e.g., /proc/16531/fd). There will always be entries labeled 0, 1, and 2 for STDIN, STDOUT, and STDERR, respectively. Other devices and files in use will require additional entries. Alternatively, the lsof command will list any open files.

## 10.4 PARTITIONS

The disk drive(s) making up your Linux storage space is(are) divided into partitions. Each partition will contain an independent *file system* unto itself. Multiple partitions may be placed on one physical device or a single partition could be distributed across multiple devices. The distribution of partitions to physical device(s) should be transparent to the user and, unless a device fails, it should be transparent to the system administrator as well.

### 10.4.1 File System Redefined

The term "file system" requires a more precise definition before we continue. To this point, we have referenced a file system as the collection of all storage devices available. That is, the file system contains all of the files and directories of our computer. Now we see that a specific device (e.g., an internal hard disk) is divided into logical sections called partitions. Each partition stores a file system. Thus, the term "file system," as applied thus far is inappropriate because in fact there are (most likely) multiple file systems making up your storage space.

From here on, we will refer to the collection of storage devices as the storage (or file) space and use the term "file system" to define what a partition stores. The file system now includes with it information such as a type, mount options, and mount point. Since partitions can take on different types of file systems, a file space might comprise multiple file systems.

There are many reasons to partition the file space. First, because each partition is independent of the others, we can perform a disk operation on one partition, which would not impact the availability of the others. For instance, if we wanted to back up one partition, we would want to make it inaccessible during the backup process. But the other partitions would remain accessible (as long as the partition we were backing up was not the root partition). Second, we can establish different access options on each partition. One partition might be read-only, another might have disk quotas placed on the individual directories, and another might involve journaling. Third, partitions will also allow us to specify disk space utilization more carefully. Some partitions would be expected to grow in size while others would remain static in size.

The segmentation of the storage devices into partitions is done at a logical level, not necessarily at a physical level. For instance, if we have a single hard disk for our system, it would be logically divided into partitions, not physically. When we view the storage devices in /dev, we will see one specific disk device per file system. If our computer uses a SATA hard disk, we will see under /dev devices such as /dev/sda1, /dev/sda2, and /dev/sda3. This should not be taken as having three SATA hard disks but instead that we have partitioned the one SATA hard disk into three file systems.

## 10.4.2 Viewing the Available File Systems

How do you know which device maps to which partition? There are several different ways to determine this. One way is to examine the contents of the /etc/fstab file. This file is the file system table, which specifies which partitions should be mounted at system initialization time. An example of an /etc/fstab file is shown in Figure 10.5. The table stores for each partition to be mounted the device's name, the device's mount point, the file system type, and the mount options. Let us explore these in more detail.

In Figure 10.5, the file system's device name is given in the first column. Most of the entries are stored on a SATA hard disk, which is indicated as /dev/sda#. The individual numbers denote different logical devices even though they may be placed on the same physical hard disk. In some installations of Linux, the device name is given using a UUID specifier, which can look extremely cryptic as it is presented as an address using hexadecimal values.

For a remotely mounted file system, the device name is the name of the remote file system's host and the mount point within that host. In Figure 10.5, we see one example of a

| | | | | | |
|---|---|---|---|---|---|
| /dev/sda1 | / | ext4 | defaults | | 1 1 |
| /dev/sda5 | /home | ext4 | defaults | | 1 2 |
| /dev/sda3 | swap | swap | pri=2000 | | 0 0 |
| /dev/sda2 | /var | ext4 | defaults | | 1 3 |
| proc | /proc | proc | defaults | | 0 0 |
| /dev/cdrom | /media/cdrom | auto | ro, noauto, user, exec | | 0 0 |
| tmpfs | /dev/shm | tmpfs | defaults | | 0 0 |
| www.someserver.com: /home/stuff | /home/coolstuff | nfs | rw, sync | | 0 0 |

FIGURE 10.5   /etc/fstab entries.

partition to be mounted remotely, www.someserver.com: /home/stuff. There may also be special file systems that are not placed on physical devices. In Figure 10.5, we see two such file systems, one for proc and one for tmpfs. tmpfs is the device name for shared memory (which is stored in memory as a ramdisk rather than hard disk; see Section 10.5 for details on ramdisks). Another file system which might appear is devpts to communicate with terminal windows. Not shown here is sysfs, which stores the /sys directory, used for plug-and-play devices.

The second column in fstab is the mount point. This is the logical location in the file system where the file system is made accessible. This will usually be at the top level of the Linux file system, that is, a directory under /. There are some occasions where a mount point will be placed within a subdirectory. For instance, we might find mount points under /usr or under /home. The remotely mounted partition listed in Figure 10.5 will be mounted under /home/coolstuff while the cdrom will be mounted under /media/cdrom.

Not all file systems have explicit mount points. The swap file system is one such example. The consequences of not having an explicit mount point is that the file system in question could not be mounted or unmounted using the typical mount and umount commands (covered later in this section). We discussed swap space in Chapter 8 and will refer to it in more detail later in this chapter.

The third column in the fstab specifies the file system's type. These include ext, ext2, ext3, ext4, and others. The ext file system is appropriate for either internal or removable storage. Today, we do not see ext in use but instead most systems are based on ext2, which is far more efficient; ext3, which provides journaling; or ext4, which can accommodate very large files (terabytes in size). The term *journaling* means that the file system tracks changes so that the hard disk contents can be rolled back. Other alternatives that are often used in a Linux operating system include NFS (the networked file system), JFS (a journaling file system), umsdos (a version of MS DOS for Linux), iso9660 (used for optical disk drives), and proc, which is a virtual file system. The idea behind a virtual file system is that it is stored in memory instead of on a storage device. The proc file system contains data on all running processes. In Figure 10.5, we see the remote file system is of type nfs (networked file system).

The fourth field in fstab for each file system consists of the mount options specified for that partition. The defaults option is common and the obvious choice if you do not want to make changes. Aside from defaults, you can specify any of a number of options. You can specify whether the file system should be mounted at boot time or not using auto and noauto, respectively. The option user/nouser specifies whether ordinary users can mount the given partition (user) or whether only root can mount it (nouser). The user option is commonly used for devices that a user should be able to mount after system initialization takes place such as with an optical disk (cdrom) or a USB flash drive.

The exec/noexec option indicates whether binary programs can be executed from the partition or not. You might apply noexec if you do not want users to be able to run programs stored in a particular partition. As an example, we may want to place noexec on the /home partition. This would prevent a user from writing a program, storing it in their home directory, and executing it from there. Instead, any such program would have to be moved to another partition, for instance, /usr. This restriction may be overly cautious if we expect our users to write and execute their own programs. But if we do not expect that of our users, the precaution could prevent users from downloading executable programs into their home directories and executing them. This is a reasonable security measure in that we are preventing users from possibly downloading programs that are some form of malware.

Two other options are ro versus rw (read-only, read/write) and sync/async. In the former case, the ro option means that data files on this partition can only be read. We might use this option if the files are all executable programs, as found in /bin and /sbin. Partitions with data files such as /var and /home would be rw. The latter option indicates whether files have to be accessed in a synchronized way or not. With sync, any read or write operation must be completed before the process moves on to the next step. This is sometimes referred to as *blocking I/O* because the process is blocked from continuing until I/O has completed. In asynchronous I/O, the process issues the I/O command and then continues on without stopping.

The defaults option consists of rw, exec, auto, nouser, and async. Other options are available. The swap partition has an added option of pri = 2000. This is known as a configuration priority. When multiple partitions have a priority, mounting of those partitions is handled based on the order of the priority (the higher the number, the higher the priority).

The final column in the fstab file is a sequence of two numbers. This pair of values indicates the order by which all partitions will be examined by fsck and the order that the partitions will be archived by dump, respectively. The most common entry is 0 0.

You can explore the currently mounted partitions through the df command. This command displays file system usage. Specifically, it lists the amount of disk space that is being used and the amount that is free for each mounted partition. An example of output from df is shown in Figure 10.6. Notice how the information somewhat overlaps what we find in the /etc/fstab file although df does not display all of the file systems (for instance, missing are swap and proc).

The first column of the df output is the mounted partition's device name while the last column is the partition's mount point. In between, we see usage and free information. In this case, we are given these values in terms of used and available disk blocks. We can alter this behavior to illustrate the number of used and free inodes or used and free disk space by capacity (e.g., 31G).

| Filesystem | 1 K-Blocks | Used | Available | Use% | Mounted on |
|---|---|---|---|---|---|
| /dev/sda1 | 4,031,680 | 2,955,480 | 871,400 | 78 | / |
| /dev/sda5 | 6,349,763 | 359,761 | 5,667,443 | 6 | /home |
| /dev/sda2 | 1,007,896 | 178,220 | 778,476 | 19 | /var |
| www.someserver.com: /home/stuff | 635,008 | 35,968 | 566,784 | 6 | /home/coolstuff |
| tmpfs | 510,324 | 284 | 510,040 | 1 | /dev/shm |

FIGURE 10.6   Results from df.

Options to alter the output of df are shown below:

- -a—output all file systems, including "dummy" file systems
- -B size—output number of blocks in size increments, for instance, -B 1K would output file system capacity and availability in 1K block sizes while -B 8K would use 8K block sizes
- -h—use human-readable format for sizes using units in K, M, G, T, and so on
- -i—output inodes rather than disk blocks

The df command can be used on individual file systems, as in `df /home`. The parameter(s) can also be individual files, in which case df outputs the usage information for the file's file system. For instance, `df /home/foxr/somefile` would output the same information as `df /home`.

Finally, the /etc/mtab file contains the up-to-date mount information. That is, while fstab is a static description of the file space's partitions, mtab shows you what is currently mounted, including file systems that are mounted remotely. If you have not mounted or unmounted any partitions, then you should have the same partitions listed in both mtab and fstab. However, mtab is interactive; that is, it is modified as partitions are mounted and unmounted. If you have unmounted some partitions and want to reestablish the mounting as you expect to see it upon a fresh boot of the OS, you can issue the mount all command, which is `mount -a`. We explore mounting later.

### 10.4.3 Creating Partitions

When do you create your initial partitions? You do so when you first install your operating system, as we saw in Chapter 8. How do you decide upon the partitions for your system? What size do we reserve for our partitions and what types should these partitions be (what types of file systems?) Where will we make the mount points for the partitions? These are decisions that we will make based on the type and usage of our computer.

Is it a standalone workstation for a single user? If so, fine-grained partitions are not really needed and most of the space can be reserved for /home. Is the workstation one that multiple users might be using at a time? This may lead you to a greater degree of partitioning such as having a separate /usr partition for software. Is the computer a server that will handle dozens, hundreds, or even millions of requests a day? If so, then /var will store a greater amount of information and so might require extra space.

Fortunately, these partitioning decisions are fairly uniform in that we do not have to consider too many other questions. This simplifies the partitioning process and lets us have confidence in our decisions. Below is a description of the different partitions that you might choose and why.

/—every Linux installation requires a root partition for the Linux OS. This will include the directories /bin, /usr/bin, /sbin, and /usr/sbin unless we specify a finer-grained distribution. In addition, / will store the root account's home directory (/root).

If we do not separately create a /boot partition, then /boot (the OS kernel and boot loader programs) will be stored under the root partition as well.

/home—we separate the user space from the root partition for security purposes. The /home partition should be the partition in which all user directories and files are stored. While / is not expected to grow, /home will grow as you add users to the system and those users create files. The /home directory is a potential target for placing disk quotas whereas / would not need this.

/var—this top-level directory houses many of the system data files. The term "var" indicates "variable" in that these files will change over time. Not only does this require that the partition be rw (readable, writable) but also that we will need to ensure room for these files to grow over time.

Among the files that you would find here are mail directories (one for each user account), log files, which are generated by various software and operating system events, print spooler files, and possibly even server files such as the web space for a website. Unlike /home, there would be little point to placing disk quotas on this file space because the files will primarily be owned by various software (or root).

swap—The swap space is essential so that the Linux operating system can expect to run efficiently no matter how many processes are launched. Processes are copied into memory to be executed. However, when memory becomes full, unused portions of those processes are held in swap space. This includes both executable code and data space. The items moved between swap space and memory are known as pages and the pages are only loaded into memory on demand (as needed). So memory is in reality only storing portions of what is running while their entire backing content is stored in swap space.

The swap space for Linux should be at least the size of the computer's RAM. This is different from Windows systems where swap space is often recommended to be at least twice the size of the main memory. If you examine the swap partition entry in /etc/fstab (see Figure 10.5), you will see that this file system does not have a mount point like the other partitions. You will also see that the swap partition does not even appear when you use the df command (see Figure 10.6).

Linux comes with `swapon` and `swapoff` instructions to mount and unmount the swap space. Normally, you would never unmount the swap space. But if you wanted to switch swap spaces, for instance, if your original swap partition was deemed too small and you have created a new swap partition on another device, you can swapoff and then swapon using the new device. When using these commands, the option -a will turn *all* swap space off or on.

Other partitions that you might make include /usr to store all executable user software, /tmp, which is used by running software to store temporary files, /boot to store the boot program itself, /opt and /usr/bin. The /opt directory is for third-party software and /usr/bin allows you to separate the binary portions of the /usr directory from other files that support application software (such as man pages, configuration files, shared library files, and source code).

You might expect files in /usr and /opt to be read-only, so you would mount them using the ro option instead of the default rw. The /tmp directory is writable by anyone and everyone so it might be mounted using a umask of 000 (recall the umask from Chapter 9 is used to establish default permissions by subtracting the umask from 777 for directories, thus /tmp would have permissions of 777).

## 10.4.4 Altering Partitions

What happens if you decide, after OS installation, that you want to increase the number of partitions or change the size of a partition? This can be a challenging problem.

Consider, for instance, that you divided your file space into partitions where root is 16 GB, swap is 32 GB, /usr is 100 GB, /var is 20 GB, and /home has the remainder of the disk space (say 830 GB to give you a total of approximately 1 TB of disk space). You now decide that you want to expand /usr from 100 GB to 200 GB and you want to add a partition for /usr/bin separate from /usr.

In making these changes, you will have to impinge on another partition's space. You would not want to take this from / or /var as those are fairly small. The natural option is to take space from /home. However, in doing so, you must make sure that reducing the /home partition does not destroy data. For instance, if 700 GB of space has already been taken up in /home, reducing its size would necessarily result in a loss of data.

Before attempting to repartition your Linux file space, you should make sure that you have backed up the data partitions (/home, /var). This ensures that if you are encroaching on another partition's disk space and the repartitioning deletes files, you have them available to be restored.

The parted program can be used to handle repartitioning. This is a command-line program used to view, alter, add, and delete partitions. The program can run with or without user interaction. Typing parted drops you into the parted interpreter with a prompt of (parted). Now you enter partitioning commands.

There are numerous commands available. The more useful ones are listed in Table 10.3. The value NUMBER refers to the numeric value associated with the partition, which can be obtained through the print command. With certain options, parted runs without interaction, for instance, -l (or --list)does the same as the list command. You are unable to perform tasks such as mkpart, mkfs, and resize without the interactive interpreter.

You can also create a file system through the command line instructions `mkfs` and `mke2fs`. By using either of these two commands, you are able to specify file system options that are not provided in parted. For instance, mke2fs allows you to provide a stripe size if you are configuring the file system for use with RAID (see Chapter 14 for a discussion of RAID), or journaling options for an ext3- or ext4-type file system. For either command, you must specify the partition's device name such as /dev/sda6.

Other options are available. If the specific type is not provided, mkfs defaults to ext2 while mke2fs defaults to the type listed in the `/etc/mke2fs.conf` file.

## 10.4.5 Using a Logical Volume Manager to Partition

Instead of risking the need for repartitioning in the future, another option is to use a logical volume manager (LVM). LVM is an approach to disk partitioning where partitions' sizes are not physically imposed at the time of initial disk partitioning. Instead,

TABLE 10.3    Parted Commands

| Command/Syntax | Meaning |
|---|---|
| check NUMBER | Perform a simple check on the specified partition |
| cp FROM TO | Copy the file system on the partition numbered FROM to the partition numbered TO; this can permit you to copy a file system to another device |
| mkfs NUMBER TYPE | Create a file system of type TYPE on the partition NUMBER |
| mkpart TYPE START END | Create a partition of the specified TYPE (one of primary, logical, extended) where START and END are disk locations indicated by byte position (e.g., 512 GB, 1024 GB), percentage of total capacity or sector number |
| mkpartfs TYPE FS-TYPE START END | Create a partition of the specified TYPE and assign it a file system of type FS-TYPE |
| move NUMBER START END | Move the partition NUMBER to the specified location |
| name NUMBER NAME | Assign NAME to the partition NUMBER |
| print [devices\|free\|list\|all\|NUMBER] | List the partitions based on the parameter provided (e.g., all partitions, the partition numbered NUMBER), list and all do the same thing |
| rescue NUMBER START END | Rescue a lost partition near START |
| resize NUMBER START END | Alter partition NUMBER's size as specified |
| rm NUMBER | Delete the partition NUMBER |

partitioning is done in somewhat of a virtual way. Or alternatively, you might think of LVM as a software layer residing over the physical disk where partitioning is handled through this software layer.

The way LVM works is to set up sets of partitions, called volumes. First, there are *physical volumes*. The physical volumes are grouped together to form volume groups. The volume groups represent the sum total of the storage available for partitioning.

For instance, if you have two hard disk drives, /dev/sda1 and /dev/sdb1, these two can be united for a volume group. Now, partitioning can be done so that a partition is placed anywhere on either (or both) of the physical disk drives.

Note that the hard disk drives can still be physically partitioned but there is no longer a one-to-one mapping of a file system to a partition. Instead, the physical partitions become portions of a volume group.

A physical volume is itself made up of *physical extents* (PE). The PEs are fixed in size,* much like the page/frame size of virtual memory. A physical volume then is a collection of PEs. The number of PEs for the physical volume is based on the PE size (dictated by the operating system) and the size of the physical volume itself. From our previous paragraph, imagine that /dev/sda1 is 1 TB and /dev/sdb1 is 512 GB in size. These two physical volumes combine for a single volume group of 1.5 TB. The PE sizes, for our example, are 4 MB. This gives us a total of over 393,000 PEs.

We now divide the volume group into partitions, known as *logical volumes*. The logical volume is established with a certain minimum size and an estimated total size.

---

* To be more accurate, the PEs of Linux are fixed in size. Other operating systems may implement variable-sized PEs.

This minimum size dictates the number of initial PEs allocated to that logical volume. The estimated total size is only a suggestion of how many PEs the volume might eventually use.

As the logical volume grows in size, it is allocated more PEs. In this way, a poorly estimated total size should not require repartitioning. As long as PEs are available, any of the logical volumes (partitions) can grow. In addition, if disk space is freed up from a partition (for instance, by deleting unnecessary files), those PEs are handed back to the pool for reuse for any of the logical volumes.

Figure 10.7 provides an illustration of the concepts described here. We see all of the available physical volumes (two disks in this example) united together as a single volume group. Each physical volume has PEs. A logical volume is a collection of PEs. The PEs are allocated on an as-needed basis. Because of this, a logical volume's PEs are not all located within one section of a disk drive but instead can range to include space from all of the physical devices.

There are many reasons to use LVM over physical disk partitioning (recall that disk partitioning does not physically alter the disk but portions out the disk space for different uses). The biggest advantage of LVM is that the system administrator will not have to worry about creating too small a partition, which later might need resizing. The LVM greatly simplifies any disk maintenance that the administrator may be involved with. Another benefit of LVM is in creating a backup of a partition. This can be done using LVM without interfering with the accessibility of that partition. LVMs are also easily capable of incorporating more hard disk space as hard disk drives are added to the system without having to repartition or partition anew the added space. In essence, a new hard disk is added to the single volume group providing additional PEs to the available pool. You can also establish redundancy with LVM, simulating RAID technology.

There are some reasons for not using LVMs. First, the LVM introduces an added layer of indirection in that access to a partition now requires identifying the PEs of the partition. This can lead to challenges in disaster recovery efforts as well as complications to the boot process. Additionally, as a partition may no longer occupy disk blocks in close

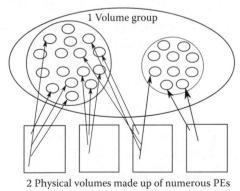

FIGURE 10.7   Implementing logical volumes out of physical volumes using LVM.

TABLE 10.4    Logical Volume Manager Commands

| Command | Meaning |
| --- | --- |
| pvchange, vgchange, lvchange | Change attributes of a physical volume, volume group, logical volume |
| pvck, vgck | Check physical volume, volume group |
| pvcreate, vgcreate, lvcreate | Create a new physical volume, volume group, logical volume |
| pvmove | Move the PEs of a physical volume |
| pvresize | Resize a partition |
| vgextend, lvextend | Add a physical volume to a volume group, extend the size of a logical volume |
| vgmerge | Merge two volume groups |
| vgreduce | Remove a physical volume from a volume group |
| lvreduce, lvremove, lvresize | Reduce the size of a logical volume, delete a logical volume, resize a logical volume |

proximity to each other but instead may be widely distributed because the PEs are provided on an as-needed basis, the efficiency of accessing large portions of the partition at one time decreases. In Figure 10.7, we see that one of the logical volumes (the third from the left) exists on two separate devices.

As booting takes place prior to the mounting of the file systems, the boot sector needs to reside in an expected location. If the /boot partition were handled by the LVM, then its placement is not predictable. Therefore, we separate the /boot partition from LVM, creating (at least) two physical partitions, one containing /boot and the other containing our volume group(s). If we have one volume group, then we have two physical partitions. The volume group partition(s) then contain all of our logical partitions. If you reexamine the first example of partitioning that we covered in Section 8.3.2 of Chapter 8 when we used the LVM, you will see that we did exactly this: two partitions of /boot and a volume group, the latter of which was partitioned using an LVM.

You establish your disk partitioning using LVM at the time you initially partition your file space. However, you can modify the physical and logical volumes later through a suite of commands. A few of the more useful commands are provided in Table 10.4. These commands are also available through the interactive, text-based lvm2 program (which is usually called lvm in Linux systems). The lvm2 program is a front-end or interface for the programs listed in Table 10.4. Note that Table 10.4 is only a partial listing of the LVM programs available for creating and maintaining LVM-based partitions.

## 10.4.6 Mounting and Unmounting File Systems

Let us focus again on file systems with a discussion of mounting and unmounting them. We use mount and umount to handle these tasks. Unfortunately, the mount command can be complicated. We will focus on its more basic options.

At its simplest form, the mount command expects two arguments: the file system to mount and the mount point. The mount point must already exist. For instance, if we want to mount a new file system to /home2, we would first have to create /home2. After issuing the command mkdir   /home2, we would have an empty directory. This is the mount point for our new partition. The file system being mounted will use a device name, for instance, we might issue the command

```
mount /dev/sda3 /home2
```

assuming that the physical device sda3 stores the partition for the /home2 file space.

If we want to use nondefault options when mounting a device, we will use `-o option`. Mount options are varied. We discussed many of them earlier in this section (auto, noauto, sync, async, ro, rw, etc.). There are a large number of other options available. A thorough look at the mount man page will convince you that learning all that mount can do can be challenging!

In addition to the typical format for mount, we can issue mount with just one of the following options:

- -a—mount all partitions as specified in the /etc/fstab directory
- -l—list all mounted partitions
- -l -t *type*—list all mounted partitions of the *type* specified (e.g., ext4, nfs)

Although most partitions will be mounted at the time the Linux kernel is initialized, some partitions will need to be mounted later, or unmounted and remounted. For instance, adding an external storage device such as an external hard disk or a USB drive will require mounting. When done with the device, for safety's sake, you would want to unmount it. Or, you may wish to make a particular partition inaccessible for some time, for instance, if you were going to move the physical device or perform a backup. The partition would be unmounted temporarily and then remounted.

And then you might want to mount a *remote file system*. A remote file system is a device storing a file system that is made available by network. The remote file system might be stored on a file server shared among multiple computers. Alternatively, the remote file system might be some computer that you have access to over the Internet. In any event, mounting the file system is not very difficult.

Let us take as an example that the remote computer www.someserver.com has a directory /home/stuff. We want to mount this directory under our own /home/cool-stuff directory. To accomplish this, we must make sure that the service netfs is running (see Chapter 11). We also must create the directory /home/coolstuff. Then we perform the mount operation.

```
mount www.someserver.com:/home/stuff /home/coolstuff
```

Notice that our mount command is very similar to the previous example, the only difference is that our device's name is now of the form *remote_name:remote_mount_point* (as opposed to local_device_name).

Because we are manually mounting this file system, we would be best served by adding detail to the mount command that we might otherwise find in the /etc/fstab file. These details would include the file system type and options that specify how we will access the remote partition. To specify the type, use -t *type* as in -t nfs. To specify options, use -o *options*. We might use -o noauto,ro to indicate that this partition will not be automatically mounted and the file system will be read-only. We would also most likely want to mount a remote file system as asynchronous (async) because we do not want local processes to block while waiting for confirmation from the remote file server. However, async is part of the defaults so we do not have to explicitly list it.

Once mounted, users can cd into and access files in this remote computer's /home/stuff directory by specifying /home/coolstuff on the local machine. In this case, users will only be able to read from this file system.

Ordinarily, you are not able to mount remote file systems just by using a mount command. Instead, the administrator of the remote machine must set this up. The steps are listed here. We will assume that we are the system administrators for www.someserver.com and want to let others mount /home/stuff.

First, we have to specify that we want to export this file system. This is done using the /etc/exports file. We would add the following line to this file:

```
/home/stuff *(rw,noexec)
```

If we had other file systems to export, we would have an entry (line) for each one (or we could combine file systems on single lines if they share the same export options).

The first entry in the line is the file system's mount point. The second item, *, indicates the remote machine(s) that we want to permit access to. The * is used to mean any remote computer. We might wish to restrict this, for instance, to some IP address(es) as in 10.2.0.0/16, which means any computer whose IP address starts with 10.2, or 10.2.56.30 to indicate that only the machine 10.2.56.30 can access it.

The items in parentheses are the options that we are placing on this file system. These options then apply to how the remote computer(s) interact with this file system. For instance, we might specify ro which would mean that any machine that mounted the file system would only be able to read from it (not write to it) even though locally, the file system might be accessible as rw. In the example above, we are assigning rw and noexec options. Recall noexec means that users are not allowed to execute software from this directory.

Next, we need to restart the nfs service. This can be done graphically through the services tool or from the command line using either /sbin/service nfs restart or /etc/init.d/nfs restart. We will explore services in Chapter 11. Finally, we have to issue the exportfs command. Generally, the command is exportfs *filesystem*, but we can use exportfs -a to export all of the file systems in the /etc/exports file. Now, the file system should be mountable from remote locations.

The umount command unmounts a partition. It is far simpler than mount. The common application is to state umount *mount-point* as in umount /home/coolstuff. The umount command comes with options including –n to prevent the unmounting from being written to mtab (which would make mtab out of date), -r so that if the unmounting fails, the partition is moved from its current status to read-only, -l to unmount the file system now but to actually clean up the records of its mounting/unmounting later, -a to unmount all file systems currently mounted (the root file system is an exception) and −O *options* to specify that the unmounting should take place on all file systems that match the specified options, such as ro or noexec.

When unmounting a file system, you might receive the following error message:

```
umount: /dir: device is busy.
```

The value */dir* is the mount point you were attempting to unmount (e.g., /home).

You are unable to unmount *busy* file systems. A busy file system does not mean that it is being used exactly at that moment but instead that its status is currently *open*. For instance, if you have logged in through the GUI as a normal user and then used su to change to root, you would be unable to umount /home. This is because at least one user is logged in and so /home has a status of "busy." Even umount −f (force) will not work on a busy partition.

In order to unmount either /home or /var, you would have to log in through the text-based interface and log in as root. You could more easily unmount a remote file system (e.g., /home/coolstuff from above) or /cdrom or one of the file systems that may not be in use such as /opt or /usr/bin.

## 10.4.7 Establishing Quotas on a File System

A disk quota limits the amount of space available to a user or a group. You establish quotas on a per-partition basis. The only partition that makes any sense to establish quotas on is /home as we would only want to place quotas on the user's space. Here, we look at steps to establish disk quotas.

The first step is to modify the partition's options. The best way to do this is to edit the /etc/fstab file. We want to add the option usrquota to any other options listed. Most likely, the only option is defaults, so our entry will now appear as defaults,usrquota. We could use grpquota in addition to or in place of usrquota if we wanted to establish quotas on groups owning files. With this modification made, before we can attempt to place quotas on the users (or groups), we would have to remount the /home partition. We could unmount and then mount the partition although we could more easily issue the command mount −o remount /home. Here, -o indicates an option to the mount instruction. Usually, these options are placed on the partition (e.g., sync, ro, or rw) but here, the option indicates how to mount the partition.

In order to place quotas on users (or groups), we must create a quota database, which we accomplish by running quotacheck. We supply the options -c to create the database file and -u (or -g) to indicate that quotas are to be established on users (groups). The -u option is assigned by default, so it can be omitted. The quotacheck program places the database file, aquota.user, at the root level of the given file system; in our case, it creates /home/aquota.user. Note that this is not a text-based database file and so cannot be read or written to through a text editor.

In order to create the database, quotacheck performs a scan of the file system in question. During this scan, any writes performed to the file system could cause corruption of the file system and so it is best to run this program with the given file system unmounted. Otherwise, quotacheck will attempt to remount the file system upon completion of the scan. There are several things you could do in order to run quotacheck successfully. You could run quotacheck during system initialization before the file system has been mounted. You could use `telinit 1` to change run levels, unmount the file system, run the program, and then remount the file system. Or, you could issue the -m option. If you do none of these, quotacheck will yield an error as it is unable to remount the file system. Our command becomes `quotacheck -cmu /home`.

At this point, you can place quotas on individual users through the command `edquota`. The only required parameter is the username (groupname) of the user you wish to place (or alter) a quota on. The username (groupname) can be either a name or a number (UID/ GID). The edquota command drops you into a vi session with the user's quota information displayed. You can now make changes, save, and exit vi. You might find an entry such as the following:

```
Disk quotas for user foxr (uid 500):
Filesystem          blocks   soft   hard   inodes   soft   hard
/dev/sda5           1538     0      0      91       0      0
```

Soft and hard refer to *soft limits* and *hard limits*. Hard limits cannot be exceeded and if an attempt is made to save a file whose size causes a user to exceed a hard limit, then an error will arise. Users are allowed to exceed the soft limits for an amount of time indicated by a grace period. Once that grace period has elapsed, the soft limit becomes a hard limit until the user reduces his or her storage to again be below the soft limit.

As you can see in the above entry, you are able to establish quotas on both (or either) file blocks and inodes. The value 0 indicates no limit.

The grace period can be established on all file systems, a single file system or on individual users/groups. For a file system, use `edquota [-f filesystem] -t time` where time is a value followed by one of seconds, minutes, hours, or days. For instance, `edquota -t 10 hours` would set a grace period of 10 hours for all file systems (or in our case, only /home since /home is the only file system with quotas implemented on it).

If you want to issue a different grace period for a specific user or group, use `edquota -T username/groupname time`. You can use the option `unset` in place of a time when issuing a grace period on an individual user to indicate that while all other users have the grace period as specified via -t, this user has an unlimited grace period.

The edquota command is interactive in that you are dropped into an editor. This may create a very onerous task if you had to establish quotas for dozens or hundreds of users. Instead, you might prefer to issue quotas from the command line (using command line editing) or even more preferable, through a shell script. To accomplish this, you would instead use the `setquota` instruction. The basic form of this instruction is

```
setquota [-u|-g] name bsoft bhard isoft ihard filesystem
```

where name is the *user* (or *group*) name or *UID* (*GID*), *bsoft*, *bhard*, *isoft*, and *ihard* are the numeric values for the soft limits and hard limits for blocks and inodes, and *filesystem* is the file system that the quota should be implemented on. Use one of -u and -g to indicate whether this is being applied to a user (the default) or a group.

What follows is a simple shell script that establishes quotas on all users in /home using default values. Notice that user will store the full path of the user's directory, such as /home/foxr. The basename command prunes away the directory portion, leaving just the user's name, so that for instance user will store foxr. As /home may contain content other than user directories, we use the condition -d to ensure that $user stores a directory name. We will also want to avoid any directory that is not of a user such as lost+found.

```
#!/bin/bash
for user in/home/*; do
     user=`basename $user`
     if [[ -d $user && $user != lost+found ]]; then
          setquota -u $user 1024 1200 100 150 /home
     fi
done
```

This script assigns the same quota values to all users; you could instead create a file of user names with the four values to be applied for each user. Another option is to create two or more categories of user and establish the common quotas based on the category. Implementing a script like this is left as an exercise in the review section.

Two last instructions of note are quota and repquota. These report on the established quotas set on users. The instruction quota *username* will output the information as shown previously for foxr. The instruction repquota *filesystem* will output all quotas for the users/groups of the given file system.

## 10.5 LINUX TOP-LEVEL DIRECTORIES REVISITED

We visited the top-level Linux directory structure in Chapter 3. Here, we focus our attention on the roles of the directories as they pertain to system administration. The top-level directory structure is shown in Figure 10.8. User home directories are located under /home but may be further segregated, as shown here as faculty and student subdirectories. Other directories such as /usr, /etc, and /var contain standardized subdirectories.

### 10.5.1 Root Partition Directories

We start out our examination with /, which is the topmost point of the Linux file space. All directories are attached to root by mounting the file systems here. The root directory should only contain subdirectories, not files. All files would be placed further down the file space hierarchy. Notice that while we refer to this directory as root, there is also a /root directory. The /root directory is the system administrator's home directory. The /root directory is explored below. In the meantime, we will refer to the topmost directory either as / or the root directory and the system administrator home directory as /root.

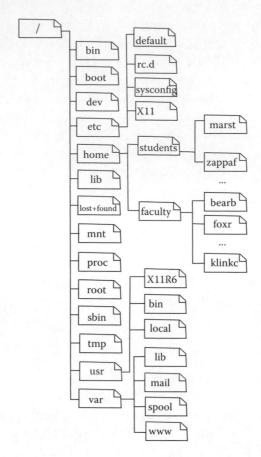

FIGURE 10.8   Top-level Linux directory structure. (Adapted from Fox, R. *Information Technology: An Introduction for Today's Digital World,* FL: CRC Press, 2013.)

The root partition directories will include /bin, /sbin, /etc, and /root. Depending on how you partitioned your storage space, you might also find /boot and /usr here. We will make the assumption that /boot is on the root partition while /usr is in a separate partition. Also, part of the root partition will be directories for /lib, /lost+found, /mnt, /opt, and /tmp. Although /etc should be part of the root partition, we will examine /etc in a separate subsection because of its importance.

The /bin directory contains binary files (executable programs). It is the primary repository of file system commands. For instance, you would find ls, cp, mv, rm, pwd, ln, touch, chown, chmod, vi, cat, more, mkdir, rmdir, and tar in this directory. You will also find executables for other common Linux operations in this directory such as echo, sort, date, ping, traceroute, df, bash, gzip, gunzip, and grep. Interestingly, not all file or basic commands are found in /bin. Those that are not are commonly found in /usr/bin.

The /sbin directory contains other binary files. The main difference between /bin and /sbin is that /sbin contains programs that are intended for system administration usage. While many of these programs are accessible by normal users, most users would have no need to run them. The programs vary greatly in their roles but include disk operation programs (e.g., badblock, fsck, fdisk, mkfs, quotaon), network programs (arp, ifconfig, ip, iptables), troubleshooting commands (kdump, killall5, mkdumprd), and programs dealing with system initialization and shutdown such as init, run level, halt, reboot, and shutdown. Additional system administration programs are found in /usr/sbin.

The /boot directory contains files needed for boot loading the operating system. One of the most common Linux boot loaders is GRUB (grand unified bootloader). This program runs shortly after the computer hardware is tested. GRUB lists the available operating systems and waits for the user to select which operating system to boot to. GRUB then finds and loads that operating system kernel and the boot process completes by initializing and running that operating system. GRUB is capable of booting to Linux, BSD Unix, MacOS X, DOS, and is capable of calling the Windows Boot Manager to boot to Windows NT, Vista, 7 or 8. The GRUB program and support code are all stored in the grub subdirectory of /boot. In this directory, one of the files is called device.map. This file, edited by GRUB (and also editable by root), contains the mapping information of GRUB device names to Linux device names.

The /boot directory also contains a program called vmlinuz. This is the Linux kernel in what is known as object format. This program is not directly executable but can be used by other software to generate executable code. The vmlinuz program actually has a lengthier name such as vmlinuz-2.6-32-279.el6.x86_64 based on the specific version of Linux installed. There will also be config and System.map files that match the extension to vmlinuz (e.g., config-2.6.32-279.el6.x86_64). If you happen to have multiple kernels available, they will all be stored in this directory with different extended names. We will explore /boot's contents in more detail in Chapter 11 when we examine the Linux boot and initialization process.

In newer Linux distributions, you will also find an efi subdirectory of /boot. EFI stands for extensible firmware interface. This is a new approach to booting that does not rely on BIOS. One advantage of EFI booting is that through EFI the file space can be made accessible. In order to use EFI, the boot process bypasses BIOS to execute right from the EFI firmware.

The /root directory is the home directory for root. If you look at the contents of /root, you will see that it looks much like any user's home directory, including, for instance, its own .bashrc and .bash_profile files. So why is it placed in a separate directory? The main reasons are security and convenience. The root directory should not be accessible by other users. Its permissions default to 550 (readable/executable by owner and group only). The executable access is there so that it can be cd'ed into only by root. We place this restriction on /root because the system administrator(s) may wish to store confidential information here, such as passwords generated for new user accounts. For convenience, /root is located under the root (/) partition. This allows root to log in, unmount /home, perform operations on /home while still having access to a home directory. If root's directory was under /home, /home would always be busy.

The /lib directory is in a way an extension to the /usr directory. /lib will store shared library files. These are files (usually executable code) that are to be used by multiple programs. Shared files are of particular use in open source software. The idea is that members of the open source community utilize other code as much as possible when writing new applications. This simplifies everyone's efforts. In newer Linux systems, there are two lib directories, one for older, 32-bit software (/lib) and one for 64-bit software (/lib64).

The /lost+found directory will hopefully always be empty. Its role is to store files that were found to be damaged in a system that did not go through the proper shutdown process. This might be the case when there was a sudden system failure or power outage. Upon rebooting the system, Linux runs a program called `fsck` which checks the integrity of the file systems. Any corrupted files are moved into /lost+found. So, if /lost+found is empty, then the file systems are currently corruption free. If files are found here, it does not necessarily mean that files are inaccessible or that data have been lost, although that certainly could be the case.

The /tmp directory is a directory for temporary files created by running software. As software may not have access to its own file space, the /tmp directory is made available. Interestingly, this directory is writable by world. As such, a system administrator may wish to place /tmp in its own partition so that it does not constitute a security problem with the rest of the operating system. The /tmp directory is divided up into subdirectories for different processes and different users. Largely what you find in these directories are domain sockets. Upon a reboot, anything previously in /tmp will be gone as a new image is created.

The /mnt directory is a mount point that system administrators can use to temporarily mount partitions. This would not be a permanent mount point but instead often used for quick mount and umount operations, such as to test the availability of a remote file system, or to perform some copy or move operation and then unmount the partition.

The directory /opt is available to install third-party software. In essence, if software is not part of the distribution, it is intended to be installed in /opt. This does not necessarily mean that installation programs will put it in /opt or that a system administrator must use /opt. It is also not necessarily straightforward identifying if a software package is part of the distribution or third party. When software is not installed in /opt, it will instead usually be placed under /usr/local.

### 10.5.2 /dev, /proc, and /sys

We group these three directories together because they are both *virtual* file systems. By virtual, we mean that they do not exist in storage space but instead are located in memory. You will find these three devices in the /etc/fstab file to be mounted at system initialization time but they do not appear when you use the df command. You would see the file systems of proc and sysfs when you use `df -a`.

The /dev directory stores the devices connected to the computer. These are mostly input, output, and storage devices, although there are exceptions to that statement. We already explored many of the devices found in /dev (Chapter 3 and earlier in this chapter). Now that you have read through the earlier parts of this chapter, hopefully you have a better understanding of concepts such as character and block devices and the role of sda1, sda2, sda3 (for SATA hard drives), hda1, hda2, hda3 (for IDE hard drives), and so on. You will also find symbolic links to other locations in the file space or to devices in /dev. For instance, the device `core` is a link to /proc/kcore and the device `fd` is a link to /proc/self/fd while devices such as `dvd` and `dvdrw` are links to `sr0`.

The lp devices are printers. The loop devices are used to connect the file space to specific files as stored on a block device. This becomes necessary for special types of block devices such as optical drives and floppy disks.

Another collection of devices go by the name ram0, ram1, and so on. These are used as *ramdisks*. The ramdisk is an older idea whereby RAM is used to simulate disk. That is, items from disk are loaded into RAM so that access to those items can be performed far more efficiently. As the ramdisks are limited in size compared to the file space available, you have to load the ramdisks wisely. Commonly, files that are receiving multiple disk accesses are loaded into available ramdisks until they are no longer needed (or until newer file accesses arise).

The ramdisk dates back to around 1980. Early personal computers had a limited address space for memory, in part caused by the limited number of bits that could be stored in registers or carried across the address bus. However, some PC owners were willing to buy additional memory. Early PC operating systems would load select files into ramdisk in anticipation of being used in the near future. The downside to this technology was that RAM memory was still very expensive and so a ramdisk might not be practical for many home computer users.

Linux systems offer ramdisks under two formats. First, there are specific ramdisk devices stored under /dev/ram. Second is the tmpfs partition that provides shared memory in the form of a ramdisk.

Finally, there are a number of tty devices. These are available terminal windows. You might find 64 tty's and 4 ttyS's, the latter of which are serial consoles.

The /proc directory contains information generated by the Linux kernel about the running processes. Each running process is indicated by its own subdirectory whose name matches the PID for that process. An example of this directory is shown in Figure 10.9. The directories with names such as 1, 10, 1041, through 2225 are all of processes. There are

```
1      14     1655   1952   2038   254   8            iomem          partitions
10     1403   1670   1963   2040   26    9            ioports        sched_debug
1041   1404   1672   1965   2041   27    936          irq            schedstat
11     1444   1687   1974   2043   28    937          kallsyms       scsi
1189   1445   17     1975   2044   29    938          kcore          self
12     1484   1757   1977   2045   3     939          keys           slabinfo
1214   15     1763   1979   2047   30    acpi         key-users      softirqs
1260   1566   18     1981   2050   31    buddyinfo    kmsg           stat
1277   1575   1804   1982   2054   32    bus          kpagecount     swaps
1288   1590   1815   1984   2057   348   cgroups      kpageflags     sys
1293   1598   1820   1986   2060   349   cmdline      loadavg        sysrq-trigger
13     16     1830   1995   2086   358   cpuinfo      locks          sysvipc
1300   1606   1839   1997   21     37    crypto       mdstat         timer_list
1301   1617   1847   1998   22     38    devices      meminfo        timer_stats
1318   1630   1848   1999   2225   4     diskstats    misc           tty
13202  1637   19     2      23     40    dma          modules        uptime
13210  1642   1922   20     24     41    driver       mounts         version
13325  1644   1928   2014   243    435   execdomains  mpt            vmallocinfo
13326  1646   1937   2019   245    5     fb           mtd            vmstat
1356   1651   1941   2021   25     6     filesystems  mtrr           zoneinfo
1367   1653   1943   2022   252    7     fs           net
1394   1654   1949   2023   253    71    interrupts   pagetypeinfo
```

FIGURE 10.9   Contents of the /proc directory.

other directories containing kernel generation information about the system itself such as acpi, bus, driver, fs, and irq. The remainder of the content of this directory are files about system processes.

The files stored both within /proc and in the subdirectories are not true files but instead information about the processes. You will find that all of these have a size of 0. Some of the files seen in the /proc directory also occur within each subdirectory such as cgroups, cmdline, schedstat, and stat. Others are unique for running processes (e.g., environ, maps, mounts, pagemap, and status) and others are unique for the kernel.

Here we examine some of the contents for the processes.

- cmdline—the command line instruction that launched this process (if launched from the command line, empty otherwise)

- environ—this process's environment variables (if any)

- fd—file descriptors (open files)

- io—this process' I/O utilization

- limits—the limits under which this process was established (e.g., maximum amount of CPU time, maximum file size, and maximum memory utilization)

- mounts—mount information

- cwd—a link to the current working directory of the process

- exe—a link to the process' executable file

- root—a link to the root directory of the process

For the system files, the cmdline file stores the kernel's startup instruction, including all parameters supplied. Other files stored in /proc include information on CPU usage, memory usage, average CPU load, file system usage, mounted partitions and their usage, and virtual memory usage.

Whereas /proc stores information about running processes, /sys stores information about devices and device drivers along with device configurations. The /sys directory's file system is called sysfs. In /sys, you will find subdirectories for block devices, the bus, firmware, file systems, and power management. You will also find subdirectories storing crash/core information about the kernel. A quick look at this directory's long listing will show you that, like /proc, the "files" here have a size of 0.

### 10.5.3 The /etc Directory

Although part of the root partition, we consider the /etc directory separately because it is a directory that you, as a system administrator, will be involved with most often. This directory stores *configuration files* for tuning the operating system to your needs as well as other forms of data files such as the /etc/passwd file storing user account information. Most of the files in /etc are text files, editable with vi. The /etc directory also has several subdirectories that contain further groups of configuration files. You have already seen a few of these files such as passwd, group, and shadow, and you will explore many more of these files in later chapters as we examine services, system initialization, networking, scheduling, and other topics.

Below we spotlight some of the other important files and subdirectories. Note that the contents of /etc will vary by distribution and even by version of each distribution. The following list comes from what you can find in CentOS 6.

- abrt/—a directory containing configuration files for dealing with software crashes.
- aliases—a file of user aliases (i.e., user names that are aliased to other users); typically you will find several software usernames are aliases for root such as bin, daemon, and adm. You are able to add further aliases such as www to webmaster and webmaster to a particular user.
- anacrontab—configuration file for the anacron program.
- at.allow, at.deny—files that list users allowed to or not allowed to use at.
- audit/—directory that contains configuration files for the auditd service.
- auto.master, auto.misc, auto.net, auto.smb—configuration and map files for automatic mounting.
- avahi/—contains configuration files for the avahi daemon used to register local IP addresses and DNS-related services.
- bashrc—the file that executes whenever a new bash shell opens; edited by the system administrator.
- cron.d/, cron.daily/, cron.hourly/, cron.monthly/, cron.allow, cron.deny, crontab—directories and configuration files that support the crontab daemon for scheduling processes.
- cups/—the directory containing cups printing configuration files.
- dnsmasq—configuration file for dnsmasq.d daemon that performs DNS forwarding for a NAT network server.
- fonts/—directory of font configuration files for system fonts.
- gdm/—directory of configuration files for the gnome desktop manager.
- hosts, hosts.allow, hosts.deny—the host tables, files that store IP alias to IP address mapping, and lists of hosts allowed to or not allowed to connect to this computer.
- httpd/—directory containing the Apache web server configuration, data, and executable files (if not installed elsewhere).
- init, inittab, init.d/—the startup service, a configuration file storing the default run level, and a directory containing all system service control scripts.

- logrotate.conf, logrotate.d—configuration file and daemon for log rotation.

- mime.types—mapping of mime type to file extension.

- nfsmount.conf—nfs mount daemon configuration file.

- nsswitch.conf—configuration file for the network service switch daemon.

- oddjob—configuration file for oddjobd daemon.

- pam.d/—directory containing pluggable authentication module program, data, and configuration files.

- rc, rc.local, rc.sysinit—startup scripts in charge of starting system services.

- rc0.d/, rc1.d/, …, rc6.d/—directories containing symbolic links to services that are to be started for the given run level (e.g., rc5.d stores the services to be started for run level 5).

- selinux/—directory containing security enhanced Linux configuration files.

- sysconfig/—directory containing scripts and configuration files for network operations including the Linux firewall (iptables, ip6tables).

- yum, yum.conf, yum.repos.d—yum configuration files.

### 10.5.4 /home

The /home directory contains all of the users' home directory space. This is where users by default will store their personal data files, scripts, multimedia files, and so forth. Group directories can also be established under /home.

Being on a separate partition allows us to treat this file system differently from others. We might for instance implement a disk quota on /home to help control the usage of this partition. If this file system were implemented with the option noexec, users would not be able to run software directly from their own file space, perhaps discouraging them from downloading and installing software. By keeping /home separate, it also helps facilitate performing timely backups because we can unmount the file system without impacting other file systems.

### 10.5.5 /usr

The /usr directory stores application software and supporting files. Like the root level, /usr comes with its own top-level directories: bin, etc, games, include, lib, lib64, libexec, local, sbin, share, src, X11R6. Aside from the binary files (executables), the directories contain documentation files (under/usr/local/man), library files, C header and object files, and programs that support the X windows system. The src subdirectory stores Linux kernel source files (as well as header files and documentation).

Many of these subdirectories are themselves broken into subdirectories. For instance, /usr/local might contain its own bin, etc, games, include, lib, man, sbin, share, and src directories. The difference between /usr and /usr/local is that /usr/local, historically, was to store the portions of the software that were local to that computer whereas /usr would store software that was expected to be found on every computer in the network. Today, this is no longer honored and third-party software is often found in /usr/local rather than in /opt or /usr.

### 10.5.6 /var

The /var directory contains system and software data. These data files will differ over time, thus the "var" name (variable). As with /usr, /var will most likely have a set structure of top-level directories. These will include account, cache, db, games, lock, log, mail, opt, run, spool, tmp, www, and yp. Some of these are worth commenting on.

The `account` directory stores process account information as generated by the program `accton`. The `/var/cache` directory is broken into several subdirectories. The `man` subdirectory stores recently accessed man pages. These are cached as man requires reformatting. A cached man page will take less time to appear. Examine the man pages shortly thereafter and it takes less time because the cache is storing the formatted version. Other cache subdirectories are available for recent access by cups (the printer service), yum, and others.

The /var/games subdirectory stores temporary data created from a program in /usr/games. The /var/lock subdirectory stores the files that software are currently accessing. These are known as *lock* files because when one process accesses a file, it becomes locked for any other process.

The /var/log subdirectory stores log files that are automatically generated by the kernel and running software. The /var/mail and /var/spool directories store the mail files and print files, respectively.

Finally, the /var/www directory is often used if your computer is running a web server. This directory becomes the web server's webspace. That is, all (or most) web pages are stored here along with other website files such as scripts, icon files (small images), and error pages.

### 10.5.7 Directories and Changes

If you examine the usage of these various directories presented in this section, you find that /home and /var are the two that tend to change over time. Because of this, it is important to backup these two partitions in a timely manner. Thus, you would need to unmount the partition to be backed up, perform the backup, and then make it available again. While the partition is unavailable, you might not permit user logins so you have to be wise with the backup process. We will discuss backups in the next section as well as in Chapter 14. The /usr directory will also change when you install or uninstall software. However, as changes to /usr are less frequent, backing up this partition will be a lesser priority.

If you are an active system administrator, you might also change /etc and /root frequently. You would generally find that other system directories, /bin, /sbin, /lib, /boot, and so forth, would not change over time unless you were to upgrade your operating system itself.

The /home and /var partitions not only change but grow over time. When you partition your storage space, it would be wise to allow plenty of room for growth. Alternatively, you could use an LVM so that you would not have to worry about repartitioning in the future.

Whether you use physical partitions or LVM, unexpected growth in these file systems may require that you add storage devices to expand their capacities. As a last resort, you might have to look for files to delete to free up space.

## 10.6 OTHER SYSTEM ADMINISTRATION DUTIES

Accessibility of the file systems is a primary concern of the system administrator. This amounts to ensuring that the needed file systems can be accessed by users when they log in. It also requires protecting the contents of the file systems. Protection takes on two different forms: ensuring that files have adequate permissions and backing up of files so that they can be restored in the case of corruption, deletion, or improper modification. In addition, the system administrator will have other file system-related duties. We have already discussed accessibility in terms of mounting and unmounting. In this section, we focus on the other duties.

### 10.6.1 Monitoring Disk Usage

The system administrator should occasionally monitor the status of the file systems, or at least the more critical ones. There are numerous Linux commands that provide a picture of file space usage and available capacity. We have already seen the df command, which provides a summary of the mounted partitions' capacities. For more specific information, the du (disk usage) command will provide the (estimated) size of individual files, directories, and entire partitions. The du command generally is of the form

```
du [options] file(s)
```

Common options for du are listed in Table 10.5.

You can obtain an entire directory's disk usage by using du -s /dir as in du -s /home/foxr, or du -sh /home/foxr if you prefer. What about obtaining a disk usage summary of all regular files? For that, we might want write a script that iterates through files, testing each with the -f condition.

What follows is a script that will search the directory specified as a parameter for the size of all regular files. The du -s command will respond with both the file's size and the file's name. By using awk '{print $1}', we only obtain the file size.

TABLE 10.5    Common du Options

| Option | Meaning |
|---|---|
| -a | Provide details on files and directories |
| -B *size* | Provide sizes in blocks of *size*, for example, -B 4K |
| -b | Provide sizes in bytes |
| -c | Provide a total at the end of the listing |
| -h | Provide human-readable output (summarizes in terms of K, M, G, etc.) |
| -L/-P | Follow/do not follow symbolic links |
| -s | Summary only, do not list specific file sizes |
| --time | Add last modification time/date for each item |

```
#!/bin/bash
for item in `ls $1/*`; do
    if [ -f $item ]; then
        size=`du -s $item | awk '{print $1}'`
        sum=$((sum+size))
    fi
done
echo $1 has regular files of size $sum
```

Both du and df can give us statistics on file space and file utilization. The stat command, covered earlier, provides specific file details on individual files or file systems, including, for instance, size, number of blocks, inode, device number, permissions, last access, and modification times and dates. We examine many other monitoring commands, such as vmstat for virtual memory usage, in Chapter 14.

### 10.6.2 Identifying Bad Blocks and Disk Errors

Linux offers several instructions to examine the file systems for errors. The `badblocks` instruction calls upon other programs (such as fcsk) to do much of the work. For bad-blocks, specify a device to scan. This device will be one of the file systems (partitions) whose name might look something like /dev/hda1 or /dev/sda5. Table 10.6 lists some common options for badblocks.

If bad blocks are found, you would next use the `fsck` program to attempt to repair those bad blocks. Or, you could use fsck to find errors itself. Note that you should never use fsck on a mounted drive as it can (and probably will) cause damage to that file system.

The fsck command receives any of a mount point, file system name or file system node (UUID) as a parameter. If no file system is specified, by default, fsck will examine every file system, in order, as listed in /etc/fstab. You can also specify the file system type to be tested using -t *type*. For instance, `-t ext4` would search all ext4 type file systems as listed in /etc/fstab. With this option, you can also accompany it with further options that must match that file system to be searched. The list of options `-t ext4 opts = ro,nosync`

TABLE 10.6  Common Options for badblocks

| Option | Meaning |
|---|---|
| -c *num* | The number of blocks to scan at a time, the default is 64 |
| -e *num* | Force badblocks to exit after *num* bad blocks have been found (if your file system is truly damaged, you may not care to find all of the bad blocks, so -e 100 for instance would abort badblocks after 100 bad blocks are found) |
| -i *file* | Scan for bad blocks but skip any block numbers that are stored in the file *file*; this is useful if you already know of bad blocks and you want badblocks to avoid scanning those particular blocks |
| -p *num* | Force badblocks to make *num* repeated passes; the default is 0 |
| -s | Show the progress as a percentage of the file system completed |
| -w | Use write-mode to test for badblocks; badblocks writes some pattern to each block and then reads the block to see if it can obtain what was written |

would check file systems of type ext4, which are read-only and not synchronous. Additional options include

- -C—display progress.

- -M—do not check mounted file systems.

- -N—do not attempt repair, just report what should be done, -n does much the same thing but used in different file system checkers (see below).

- -R—skip the root file system if it is already mounted (this can be used in conjunction with -A, which performs all file system checks in a single run).

- -a—automatically repair the file system with no interaction with the user.

There are actually several different fsck programs available in Linux. These include e2fsck, fsck.ext2, fsck.ext3, and fsck.ext4. With multiple file system types now available in Linux, the fsck program, which was set up to use the ext style file system, will call upon these other file system checking programs to complete its task. Some of the above parameters will not work with various checking programs (thus, -N and -n are both available).

## 10.6.3 Protecting File Systems

The system administrator must make sure that the file system is protected so that it is available as needed. Timely backups are one approach to protecting the contents of the file systems. A backup, perhaps stored on tape or to a separate file system (e.g., a file server/remote hard disk) allows the system administrator to replace damaged data files by a simple restoration process. An alternative is to provide redundancy of the files stored in the file space so that files and error correction information are both stored. This can be done using *RAID* (redundant array of inexpensive disks).

To support backups and redundancy, Linux has a number of useful programs. Two common programs are dump and restore. These two programs perform backups and restoration of entire file systems (as opposed to individual files and directories). By default, dump compares the previous backup to the current state of the file system to determine which files need to be backed up. This looks at the last modification date of each file comparing it to the date of the last backup. Any file that has not been previously backed up is new (since the last dump) and so is automatically backed up. The restore program is able to restore a full backup, an incremental backup (e.g., files backed up since a particular modification date), a specific file system, directory, or even single file.

You are able to control the performance of dump by providing a backup level (0–9) where 0 indicates a full backup. With any other number, dump will back up all files that have not been backed up since the last time dump was run on this file system with a smaller level. For instance, if you run dump with 6 in one week and then run dump with 7 in the next week, then only those files modified in the past week are backed up. If however you run dump with 5, then all files modified since your last dump run at levels 0–4 are backed up.

The dump program produces backup files in 10 KB block sizes. You can override this by specifying –B *blocksize*. The maximum block size is 64 KB. You can also specify that dump compress the backup files as they are being saved. You are also able to alter dump's behavior based on the target location of the backup (e.g., remote hard drive, mounted tape drive).

Prior to performing a full or even incremental backup, you should switch to a text-based single user mode and unmount the file system(s) you intend to back up. It would also be wise to run fsck on the file system(s) before the backup is attempted.

Older programs for performing backups of files are `tar` and `cpio`. The tar program, tape archive, was originally intended to copy files into an archive to be stored on tape for backup purposes. Today, tar is just as commonly used to bundle programs together for convenient transmission over the Internet. The tar program expects either `-cf` or `-xf` to specify that a new archive should be created (c) or a current archive should be extracted (x). The -f option indicates that the target or source is a file. When extracted from an archive, you only specify the archive name but when creating an archive, you specify both the file to be created and the files/directories to be archived.

Here are a couple of examples of creating archives. In the first example, we are tarring a collection of files. In the second example, we are tarring all items under the user foxr's home directory, including the directory itself.

- `tar -cf archive.tar file1.txt file2.txt file3.txt *.dat`
- `tar -cf archive2.tar /home/foxr`

If we untar archive2.tar using `tar -xf archive2.tar`, we not only get all of the files that were placed in the tar file, but also the directory foxr. We would only untar this in some other directory, not /home because the target foxr would already exist. The tar command has numerous options. Table 10.7 describes some of the more common ones.

In order to archive to some externally mounted device such as a tape (or a floppy disk), you would specify the device's name in place of the destination file. For instance

- `tar -cf /dev/tape0 /home/foxr`
- `tar -cfM /dev/df0 /home/foxr`

In the second example, the M option specifies that there will be multiple floppy disks used so that, when one disk is full, tar will pause for the user to replace the disk with a new floppy.

TABLE 10.7  Common Options for tar

| Option | Meaning |
| --- | --- |
| -A | Concatenate tar file(s) onto an existing tar file |
| -d | Use −diff to compare two archives |
| -r | Append files onto an existing tar file |
| -t | Output just the list of content (file names, directory names) of a tar file, not the content itself |
| -u | Append only files newer than the tar file; used for incremental backup; note that this only adds new files, it does not alter or add modified files |
| -j, -J, -z | Compress the tar file using bzip2, xz, gzip |
| -I *file* | Compress using the executable compression program *file* |

The cpio program is newer than tar and can access either cpio or tar-based archives. Cpio runs in one of three modes: copy-in, copy-out, and copy-pass. The copy-in and copy-out modes are similar to extraction and creation operations in tar, respectively. This may seem counterintuitive in that "in" usually denotes input into the program, but here "in" references movement of data with respect to the file system (from archive into file system) and "out" references movement out of the file system (from file system out to archive). The options for copy-in and copy-out mode are -i and -o, respectively. Keep in mind then that -i extracts from the archive while -o creates the archive.

The copy-pass mode is like using copy-out and copy-in combined and then discarding the archive. That is, copy-pass takes files from one location and copies them to another as if you were creating and then extracting from an archive but without retaining the archive itself.

Cpio in copy-out mode, unlike tar, is interactive. If using copy-out mode, you are placed at a prompt where you type in file names. The instruction for copy-out mode is

```
cpio -o -F filename
```

where *filename* is the archive to create. As you type each file name, that file's contents are copied into an archive.

A session to create or add to an archive might look like this.

```
$ cpio -o -F new_archive
  file1.txt
  file2.txt
  file3.txt <control+d>
  183 blocks
$
```

After typing control+d, the cpio program archives the files whose names were entered. These are placed into the new file new_archive. Once done, cpio responds with the archive's size in blocks and returns you to your Bash prompt. If the archive already exists, it is over-written. You can append to an existing archive by adding the option --append.

The copy-in mode can be used to either output stored content (restore the content) or output the names of the items in the archive. To extract the files from the archive, use

```
cpio -i -F archive
```

and to list the contents, use

```
cpio -i -F archive -list
```

Another common backup tool is rsync. This program, while capable of performing backups, is available as a network-based file copying tool. As such, it is similar to Linux programs rcp (remote copy) and scp (secure copy). The command that you would issue varies depending on whether you are doing a remote or local copy and in which direction. There are three possibilities:

- rsync *source destination*—local copy

- rsync *username@host:source destination*—copy from remote to local computer

- rsync *source username@host:destination*—copy from local to remote computer

*Source* is an individual file, files indicated via wildcards, or a directory. *Destination* is a directory. The username can be omitted if your user account is the same on both remote and local computers. You are asked to provide *username's* password. If you omit the destination, rsync merely outputs the list of course files and does no copying.

Here are a few simple examples.

- rsync /home/foxr/* /home/foxr/backup—same as cp /home/foxr/* /home/foxr/backup

- rsync foxr@10.11.12.13:/home/foxr/* .—copy remote files to current directory

- rsync foxr@10.11.12.13:/home/foxr—display all files of that remote directory

The rsync program's power comes from an algorithm it utilizes to compare source files to any at the destination location. Rsync can perform three types of comparisons: last modification date, size, and checksums. If the source file has either been modified since the last copy or is larger, it is copied over. To use checksums instead of modification time and size, add the option -c. Other options control whether links are followed, whether a copy is done recursively, and whether permissions, ownership, and so on are preserved.

The rsync command is built to perform copies across the network. Neither tar nor cpio does this unless you are copying from or to a remotely mounted file system. If you are only creating a backup on your local machine, you are not affording yourself much protection. The best way to perform a backup is to store the archived material elsewhere.

As specified earlier, one alternative to backing up a file system is to use redundancy in the form of RAID. Aside from needing specialized hardware (multiple disks), you need RAID software. In Linux, the program for interacting with RAID hardware is called md, multiple device driver, also known as Linux Software RAID. It is located under /dev/md. The md program can handle any of RAID levels 0, 1, 4, 5, 6, or 10 (it does not handle RAID 2 or 3).

RAID level 4 is generally not used because all redundancy information is placed on a single drive. If two or more disk accesses are attempted across the RAID device, only one can be accommodated because the redundancy drive becomes a bottleneck. RAID levels 0, 1, 5, 6, and 10 have the potential for handling two or more accesses at a time because data and redundancy information are divided across multiple disks. We will not go into any more detail on md because it is a complex process that would take a chapter unto itself to explore adequately. The interested system administrator should read md's man page as well as mdadm and dmraid. RAID technology is discussed in more detail in Chapter 14.

Another form of data protection is through permissions. As a system administrator, you may not be responsible for examining user directories and files to see if they are using adequate permissions. However, if you feel that users have information that should remain secure, or if the organization has a policy that requires that files have secure permissions, you might explore this. There are many ways to search user directories and files for bad permissions. We explored the find command in Chapter 3 and saw that it could search for files of a given permission. For instance, we might issue a command like the following:

```
find / -perm 666 -or -perm 646 -or -perm 446
```

In the above instruction, we are seeking any files whose permissions give others (world) write access. Similarly, there may be files in specific directories that should not be readable by anyone but the user or group. We could further elaborate upon the find command by executing a chmod command on those files found. This might look like

```
find / perm ... -exec chmod 660 {} \;
```

On the other hand, we might want to write a script that similarly examines file permissions and catalogs those files whose permissions do not seem suitable. Below is an example of such a script. Note that stat -c "%a" will return the permission of the given file as a three-digit number.

```
#!/bin/bash
for file in $(ls -R /); do
     if [ -f $file ]; then
          number=`stat -c "%a" $file`
          if [[ number -eq 666 || number -eq 646 ||
               number -eq 446 || number -eq 466 ]]; then
                    echo $file $number >>
                    /root/badfilepermissions.txt
          fi
     fi
done
```

The system administrator can inspect the badfilepermissions.txt file to see which files should be altered. Another script can be written to easily alter file permissions. The script below uses the while read statement to iterate through every entry in the file and alter the permissions of each file to the value given as a parameter. This script, call it changepermissions.sh, could then be executed from the command line as

```
./changepermissions.sh 660 < /root/badfilepermissions.txt
```

```
#!/bin/bash
while read file number; do
     chmod $1 $file
done
```

Ensuring that data are available is only one part of the duties of protecting disk data. Another side to this is the use of encryption. We explored the idea of encryption in Chapter 5 when we introduced the open source encryption tool openssl. Although openssl is primarily intended on encrypted messages to be sent over network, it can also be used to encrypt files in a local file system. You can also specify that a partition be encrypted when you create the partition. Alternatively, you can apply encryption programs later, including Loop-AES, DM-crypt, PGP, and TrueCrypt.

## 10.6.4 Isolating a Directory within a File System

We end this section with one last tool, `chroot`. This program is utilized alongside of another process. What chroot does is isolate a process at the time the process is launched so that it is limited to the file space specified. Inside this file space, the process operates as if there were no other file system available. Thus, the process is unable to breach the root level of this file space and affect other files.

Consider for instance a web server that operates on scripts, password files, log files, error files, and the web documents. Let us assume the entire collection of web server files (including its own binaries) is located under /usr/local/apache2. The web server has no need to access files in /etc, /boot, /dev, /home, or /var. By launching the webserver with a chroot of /usr/local/apache2, it is unable to access anything above this directory. This protects your system in that inadvertent or erroneous code cannot damage your system, nor can a hacker using the web server to attack your system and damage any part of the system outside of the web server.

The chroot command has other useful applications aside from creating a secure or isolated space. You can use it to create an isolated file system to test code that you are developing. This is sometimes known as a *sandbox*. You do not deploy the software for testing on a normal system but instead isolate it within a sandbox. If you are running software that invokes services, files, or programs whose names conflict with system names already installed, using chroot allows the isolated file space to use the same names without the system confusing which specific files/programs are being requested.

The chroot command is used as follows:

```
chroot [options] directory [command]
```

The *command(s)* listed are executed with *directory* as the root level of the file system that the command(s) is able to access. We might start our Apache web server (whose controlling service is called apachectl) as

```
chroot /usr/local/apache2 apachectl start
```

In the above instruction, `apachectl` is the command with a parameter of `start`, and `/usr/local/apache2` is the root of file system to be isolated. There are only a few options available for chroot, including userspec and groups to indicate the user and/or groups to use for the root of the isolated file system.

## 10.7 CHAPTER REVIEW

Concepts and terms introduced in this chapter:

- Block—fixed-sized unit of storage in the file space. Typically, files are broken into blocks and distributed across the hard disk surfaces.

- Block device—type of device, denoted by type 'b' in a long listing, that performs input/ output on blocks (rather than characters); most storage devices are block devices.

- Character device—type of device, denoted by type 'c' in a long listing, that performs input/output on characters (rather than blocks); keyboard and mouse are examples of character devices.

- Directory—organizational unit to house files and subdirectories; denoted by 'd' in a long listing.

- Domain socket—a mechanism to support interprocess communication; denoted by 's' in a long listing.

- ext (extended file system type)—family of file systems supported by Linux; ext is not used but ext2, ext3, and ext4 are all common (moreso ext4 today).

- FAT (file allocation table)—used in older Windows operating systems to store the disk block layout so that obtaining the ith block of a file can be easily determined without having to perform i-1 disk accesses.

- FIFO—first-in-first-out, an expression used to describe how elements waiting in a queue are serviced; in Linux, a fifo is a named pipe.

- File space—the collection of devices used for storage; typically consisting of an internal hard disk, optical disks, and USB drives mounted as needed and possibly externally connected hard disk drives or hard disk drives accessed remotely by network.

- File system—the storage structure of a partition, including a specific type.

- File type—Linux denotes file types to differentiate between regular files, directories, symbolic links, block devices, character devices, named pipes, and domain sockets; the file type is indicated as the first letter of the permissions in a long listing and can also be obtained using the stat command.

- Hard link—the name of the file and its inode number. Two files that are hard linked together permit access to the file via either link. Deleting one "file" deletes a hard link but not the file. Only if no other hard links exist will the deletion of the last remaining hard link cause the file to be deleted, returning the inode to the file system for reuse.

- Index—a means of indicating where a disk block is to be found; a mapping process is required to convert from a file's disk block i to the location on disk of that block.

- Indirect block—inodes come with several direct pointers to the first group of disk blocks for the file; the remainder of the disk blocks are pointed to by pointers in indirect blocks; the inode has pointers to indirect blocks, doubly indirect blocks and triply indirect blocks.

- inode—a data structure storing information about a specific file including pointers to its blocks or indirect blocks, creation/modification/access information, permissions, ownership, file type, and device number; any Linux file system contains a set number of inodes.

- Link—either a hard link or a symbolic link.

- Logical volume manager (LVM)—a software means of partition management so that partition sizes can be changed without requiring direct changes to the file system itself; this makes partition management safer and easier.

- Mounting—making a partition available.

- Mount options—control access to the partition such as making it readonly (ro) or read/write (rw), synchronous (sync) or asynchronous (nosync) and permitting anyone to mount the partition (user) or not (nouser), among others.

- Mount point—the logical location of a mounted partition, this will be some directory such as /opt, /mnt, or /usr/local/mountpoint.

- Named pipe—a mechanism to link the output of one process with the input to another, like a Bash pipe, but in this case the named pipe persists as a file-like object.

- Network file system (nfs)—a form of file system that permits mounting of partitions over the network.

- Partition—a logical division of the file system to protect the contents from other partitions.

- Physical extent (PE)—a fixed-sized unit of storage allocated to a logical volume on demand through LVMs.

- Pointer—an indicator of where a disk block is located.

- Quota—a limit established by the system administrator on the number of blocks (or inodes) that a given user or group is permitted to use.

- Remote file system—a partition that is made available over the network.

- Symbolic (soft) link—a pointer to a hard link. The symbolic link takes up less space in a directory than a hard link as it just stores a pointer and not the file's name. Deleting the original file will leave the symbolic link pointing to an inode of a nonexistent file. Soft links are indicated in a long listing with the letter 'l' as the file's type, its name listed differently, pointing the name and the actual location of the file, such as an entry like `link- >/usr/local/bin/someprogram`.

- Top-level Linux directories—standardized directories that you would find in any Linux operating system.

- Unmounting—removing a partition from being accessible; you would do this if you had to work on the partition, for instance, to perform a backup or repair bad blocks.

Linux commands covered in this chapter:

- badblocks—locate bad blocks within a particular device or partition.

- chroot—run the given application(s) within the specified file system as if the file system were the root level so that the application(s) cannot access outside of the file system.

- cpio—backup utility.

- df—report on file system usage (amount available, amount used) for all or given partitions.

- du—report on disk usage for given file(s) or directory(ies).

- dump—backup utility that can perform incremental backups, used in place of cpio in most cases.

- edquota—to establish quota values for a user or group using the vi editor.

- exportfs—permit specified file system to be mounted remotely.

- fsck—file system check; can locate bad blocks and repair files damaged by remaining open at the last system shutdown.

- lvm2—a program to handle maintenance on partitions using an LVM.

- mount—mount specified partition at the specified mount point.

- nfs—service that permits mounting of remote partitions.

- parted—a utility to handle partitions for instance by resizing, renaming, and moving them.

- quotacheck—used to generate a database of users/groups for the given file system; this is the first step in establishing quotas.

- repquota—used to display all quotas established for the users/groups of the given file system.

- setquota—used to establish quota values for a user or group from the command line.

- stat—display file statistics.

- tar—tape archive, historically used to perform backup to tape but today is most commonly used to create archives of files and directories.

- umount—unmount a partition.

Linux directories and files of note discussed in this chapter:

- aquota.user—the database of users/groups of a file system generated by the quotacheck instruction, stored at the root level of the given file system (e.g., /home/aquota.user).

- /bin—location of common binary files (Linux commands and programs).

- /boot—location of boot loader program (e.g., GRUB) and Linux kernel, required for booting Linux.

- /dev—directory storing interfaces to most of the available devices (both physical like hard disk, optical disk, modem and logical like terminal windows (tty), programs like random and zero, and ramdisks).

- /etc—stores system configuration files; system administrators will often use the files in this directory.

- /etc/mtab—the currently mounted partitions; kept up-to-date.

- /etc/fstab—the file system table, specifies mount operations at system initialization time.

- /home—the users' home directory space.

- /proc—stored in memory rather than on the file system, this directory stores information about all running processes.

- /root—the system administrator's home directory.

- /sbin—system administration binary files (commands, programs).

- /usr—application software and other common programs that are not found under /bin and /sbin.

- /var—system data files that grow over time such as log files, email files, and print spooler files.

## REVIEW PROBLEMS

1. What is the difference between how early Windows and Linux index disk blocks?

2. What is the most recent version of the extended file system as used in Linux?

3. Which of the following file systems can Linux utilize: NFS, FAT, NTFS, Files-11?

4. What is a B+ tree (you might have to research this)?

5. Match the character descriptor (as reported by `ls -l`) with the type of file

   a.  -                         i.   directory

   b.  b                        ii.  symbolic link

|  |  |  |  |
|---|---|---|---|
| c. | c | iii. | domain socket |
| d. | d | iv. | regular file |
| e. | l | v. | block device |
| f. | p | vi. | character device |
| g. | s | vii. | named pipe |

6. What type of device is a USB drive?

7. What type of device is a modem?

8. What does the command `mkfifo foo` do?

9. You have created a named pipe, mypipe, and performed `wc* > mypipe`. What happens? How do you complete the pipe?

10. Assume an inode stores 10 direct pointers and 2 indirect pointers (no doubly or triply indirect pointers). Assume that an indirect block stores 10 pointers. How many disk blocks could the largest file contain if this was the case?

11. Assume an inode stores 12 direct pointers, 2 indirect pointers, 2 doubly indirect pointers, and 1 triply indirect pointer. Assume that an indirect block stores 24 pointers. How many disk blocks could the largest file contain in this case?

12. What is an i-list?

13. Provide two ways to obtain a file's inode number.

14. You issue the instruction `stat -c "%h" somefile` and the response is 3. What does this mean?

15. How does the -f option alter how stat works?

16. Why do you need to partition a Linux file system?

17. If you were to install Linux, which partitions would you create and why?

18. An optical drive will often have partition options of: ro, noauto, user, exec. Explain what each means and why specifically we would find each of these for the optical drive.

19. Under what circumstance would you specify the ro option for a remotely mounted partition that contains user home directories?

20. The /proc directory will reside on its own partition of type proc. Why does this partition not have a similar type as most of the rest of the file system (e.g., ext4)?

21. You want to find out how much of a particular partition has been filled, that is, how much of its capacity still remains. Which instruction might you use: stat, ls, du, df, or fsck?

22. Examine the file /etc/fstab. You will find that most entries have in their last column 0 0. What partitions have numbers other than 0 0? What do the numbers mean?

23. What is the difference between the contents of /etc/fstab and /etc/mtab?

24. Assume that /home has been unmounted. /home is stored physically on the device /dev/sda3. Write the appropriate command to remount /home.

25. What happens if you issue the command mount -a?

26. What service do you need to start in order to mount a remote partition over the network?

27. You want to mount a remote partition over the network. Would you need to run exportfs?

28. The following entry is found in the /etc/exports file. What does it mean?

    ```
    /usr/bin/somestuff 10.11.0.0/16(ro,sync)
    ```

29. As the system administrator you want to unmount /home so that you can perform a backup. After issuing umount /home, you are told that the device is busy. Why might this happen?

30. What is the role of the parted instruction?

31. What advantages would an LVM have over specifying partitions during Linux installation?

32. If we were to use an LVM to provide our file system partitioning, why would /boot remain as a physical partition?

33. Assume that you have two disk drives each containing 1 TB. From these two drives, you create two physical partitions, /boot and volumegroup1. The /boot partition is 128 MB. Assume that a physical extent is 8 MB. How many PEs are available for your LVM?

34. In order to establish a disk quota for users, you must specify that the _____ should have an option of usrquota. Should the blank be filled in with device name, file system name, directory name, or user name?

35. Before issuing the instruction quotacheck, you should make sure that the file system is unmounted. If it is not, what option might you use on quotacheck?

36. What is meant by a soft limit versus a hard limit?

37. How can you establish a soft limit's grace period?

38. What is the role of the file aquota.user?

39. What is the drawback of the edquota instruction that the setquota instruction might be preferred?

40. Imagine that we have three categories of user. Users under category 1 have no quota restrictions. Users under category 2 have no inode restrictions but block restrictions of 4096 for a soft limit and 5400 for a hard limit. Users under category 3 have restrictions of 4096/5400 for blocks and 200/250 for inodes (soft/hard limits). Write a shell script that will read from a text file all of the users and their category, 1, 2, or 3 and establish the proper quota for each user. The username and category are the only two items found on a line, and each user is on a separate line of the text file.

41. Write a script, similar to the one presented near the end of Section 10.4.7, which will input from a text file a list of user names along with a soft limit and a hard limit for blocks. Then set for this user a disk quota based on those limits. There will be no limit for inodes. The file system will be /home.

42. Alter your script from question #41 so that you input a user name, a string, and two numbers. The string will be either block or inode and the two numbers will then reflect the soft and hard limit on blocks or inodes for the given user.

43. Explore the contents of /bin and /usr/bin. Under which of these directories would you find all of the common file system commands (ls, cp, rm, etc.)?

44. Examine /usr/bin. List five commands (programs) that you have learned to this point of the textbook that are found in this directory.

45. Explore the contents of /sbin and /usr/sbin. In general, which of these directories contains more programs that you are familiar with?

46. Why should /root have a home directory separate from /home?

47. Examine your /proc directory and you will find subdirectories whose names are numbers. Each of these is a running process. The number is that process' PID. You can count the number of running processes using ps aux | wc -l. Come up with a command to count the number of numbered subdirectories (do not count subdirectories that are not numbered as these are not processes). You might find the count differs by one as the ps command itself is listed and therefore was counted in the wc command even though the ps command terminates before the output appears. What command did you come up with to count the number of numbered /proc subdirectories?

48. Under /proc, you will find subdirectories for all running processes. Each one of these subdirectories contains a file called cmdline. What does this file store? Examine some of these files of processes with small PIDs. Why do they not have values? Examine some of these files of processes with large PIDs. Do they have values? What can you infer about whether a process will have a value for cmdline or not?

49. The file /dev/null is like a trash can. You can dump items to it and those items disappear. Perform the following two operations as root:

```
cp /etc/passwd /dev/null
echo Hello >> /dev/null
```

Does /dev/null change?

50. What is the purpose of the /lost+found directory?

51. As a system administrator, you plan on temporarily mounting a remote partition so that you can copy material into /usr. Where would you mount this partition and why?

52. Provide an argument for installing third-party software in /opt. Provide an argument for installing third-party software in /usr/bin.

53. Examine the shell script in Section 10.6.1. You will find this instruction:

```
size=`du -s $item | awk '{print $1}'`
```

Why is the awk statement needed? Write this script in vi and try to run it with only the instruction

```
size=`du -s $item`
```

What happens? Why?

54. Write a tar command to create an archive of all of the contents in /home. Call the file home.tar. What command did you come up with?

55. Provide two examples of how you might use tar aside from performing a tape backup.

# CHAPTER 11

# System Initialization and Services

THIS CHAPTER'S LEARNING OBJECTIVES are

- To understand the boot process and why it is necessary

- To understand how the Linux kernel initializes

- To understand the init process and the Linux startup scripts

- To understand the role of the runlevel in Linux and the startup of services given the runlevel

- To understand the roles and types of the Linux services

- To be able to control services through the GUI and command line

- To know how to configure services through GUI tools and configuration files

## 11.1 INTRODUCTION

The Linux boot and initialization process is well established. It is largely automated requiring little to no system administrator interaction. However, there are reasons for learning the process. If something goes wrong, understanding the process will help you troubleshoot and resolve any issues. There are some aspects of system initialization that the system administrator may wish to tailor to the needs of the organization. Additionally, if the system is a dual booting one, understanding the boot process becomes critical.

Services are programs run by the operating system in the background to handle requests of various agents on demand. Services are configured by the administrator and the administrator is able to start or stop services as needed. Services cover a full range of activities, including scheduling, logging, network communication, file system communication, device interaction, and power management. Although many services are initially configured, you might find it necessary to refine the configuration files.

It may seem odd to couple these two topics into one chapter. However, the two are related in that during system initialization one of the startup scripts is responsible for starting services. And so in this chapter, we first examine the boot and system initialization processes, ending with the starting of system services. We then examine many of the services in detail to see what they do and how to configure them.

## 11.2 BOOT PROCESS

The boot process is necessary because any computer relies on the operating system to present an interface to the user. The operating system is responsible for interpreting user commands and when those commands involve running programs, it is the operating system that must locate the executable program in the file system and load it into memory so that it can begin execution. So the operating system must itself be loaded into memory and running before a user can use the computer to load and run programs.

### 11.2.1 Volatile and Nonvolatile Memory

The operating system is loaded into random access memory (RAM) memory. RAM is *volatile memory* meaning that it retains its contents only while power is being supplied to it. Shut down the computer and RAM loses its contents, resulting in RAM being empty. Turn on the computer and RAM is initially empty.

To run a program, you need the operating system loaded into memory and running. This leads to a paradoxical situation: how do we get the operating system loaded into memory and running if the operating system is needed to be loaded in memory and running in order to load and run programs? If memory were not volatile, we could keep the operating system in memory permanently; we could install the operating system the first time and leave it there. But this is not the case.

The reason that RAM memory is volatile is because of the technology we use. There are two forms of RAM memory: SRAM and DRAM. SRAM, or static RAM, is built out of transistors where several transistors make up each cell (storage location). One cell stores one bit. DRAM is made up of one transistor and one capacitor per storage location (bit). DRAM is smaller and cheaper, which results in a greater amount of DRAM storage available than SRAM. SRAM is used to build cache memories and registers while DRAM makes up what we often refer to as "main memory." See Table 11.1 that differentiates between SRAM, DRAM, and ROM.

For SRAM and DRAM to maintain their storage, a constant supply of electrical current is needed from a voltage source. For SRAM, without that current, any charge is immediately lost. For DRAM, the capacitor, once charged, retains that charge for a very short amount of time before the charge dissipates. DRAM chips are set up to recharge the cells that have a charge. If power is no longer supplying that chip, they cannot recharge the cells and so very quickly, any current is lost.

For us using the computer, all of memory's contents would disappear once we shut the power off. Whatever had been stored there is lost. Shutting down the computer (or just

TABLE 11.1  Differences between Memory Types

| Type | Volatility | Typical Amount | Relative Expense | Usage |
| --- | --- | --- | --- | --- |
| DRAM | Volatile | 4–16 GByte | Very cheap | Main memory: stores running program code and data, graphics |
| SRAM | Volatile | 1–2 MByte | Moderately expensive | Stores recently and currently used portions of program code and data |
| ROM | Nonvolatile | 4K or less | Very expensive | Stores unchanging information: the boot program, basic I/O device drivers, microcode |

unplugging it from the power supply) causes memory to become empty. Upon rebooting, memory remains empty and thus we need to first locate and load the operating system.

An alternative form of memory from SRAM and DRAM is called ROM, read-only memory. This form of memory is *nonvolatile* meaning that when the power is shut off, ROM does not lose its contents. With the power off, ROM is not accessible but once power has been restored, ROM retains the contents that it had prior to losing power. ROM is set up with the information permanently stored in it. Thus, it cannot change or be written to. This is why we call this form of memory "read-only." We will use ROM to help us solve the paradox of restoring DRAM with the operating system upon booting/rebooting the computer.

## 11.2.2  Boot Process

We need some initialization program that can, upon starting the computer, locate and load the operating system into DRAM. This initialization process will be called *booting* (taken from the term "bootstrapping"). The boot process begins whenever a computer is cold booted (turned on) or soft booted (rebooted from software). We will store permanently some portions of the boot process (a program) in ROM chips.

For any computer, the first step in the boot process is to access the ROM BIOS. BIOS stands for basic I/O system. In Linux, the boot program starts at ROM BIOS address 0xFFFF0. This notation is a hexadecimal address. The first task for the BIOS is to perform a Power-On Self-Test (POST), which examines various pieces of hardware connected to the computer to ensure that they are working properly.

Specifically, the POST tests the CPU registers, main memory, hardware devices such as the interrupt controller, disk controllers, and timer, and then identifies all devices currently connected via the system bus (namely, the keyboard, mouse, monitor). Additionally, it assembles a list of all devices that can be booted from (those devices that could store the operating system). These include hard disk(s), floppy disk, optical disk, flash drive, and network. The POST step may be skipped during a reboot (warm boot) as these devices are already on and functioning.

Loading the operating system may start automatically. Or, if instructed to by the user, the BIOS can present the list of bootable devices and await the user's selection of a boot location. Most commonly, the list of bootable locations is preenumerated and prioritized. The user can later alter this prioritized list. Usually the network is the last on the list and the hard disk is second to last. In this way, a user can override booting from what might be stored on hard disk if they insert a bootable USB flash drive, optical disk, or floppy disk.

As BIOS is not alterable (because it is stored in ROM), computers come with another form of memory, CMOS. This stands for complementary metal-oxide semiconductor, which runs off of a battery placed on the motherboard. This memory retains the prioritized boot ordering, which can be altered by the user if desired. The CMOS may also store other alterable boot information and include a clock of the current date and time.

## 11.3 BOOT LOADING IN LINUX

### 11.3.1 Boot Loaders

The typical situation in Linux is to boot the operating system from the internal hard disk. We will assume this is the case here. The very first sector of the Linux disk space is set aside for the *master boot record* (MBR). The MBR contains the *boot loader* program (or a portion of it) as well as a partition table and a magic number. The magic number is used for a validation check only. The partition table contains the location on disk of active partitions where bootable operating systems can be found. The MBR is 512 bytes in length. The first part of the MBR consists of 446 bytes that contain part of the boot loader program.

For the Linux operating system, there are two commonly used boot loaders: GRUB (grand unified bootloader), which can be used to load multiple types of operating systems (e.g., Windows and Linux or two versions of Linux) and LILO, which can only be used to load multiple types of Linux operating systems (e.g., CentOS and Ubuntu). These two boot loader programs are too large to fit in the 446 bytes available, so the boot loaders are divided into two parts, generally thought of as the installer and the kernel loader, also referred to as stage 1 and stage 2 of the boot loading process. Stage 1 merely finds and loads stage 2 into memory. Stage 2 is responsible for finding and loading the operating system kernel.

LILO (Linux Loader) is the older of the two boot loaders. It can boot an operating system from either floppy or hard disk and it can select any of up to 16 different boot images. LILO is file system independent, which is both a weakness and a strength (in that LILO is simpler and does not have a reliance on accessing the file system). It can also be parameterized based on the boot image selected. You can alter LILO's behavior (once Linux has booted) by editing the file /etc/lilo.conf. This will change how LILO works in future boots. This configuration file stores global options, such as the boot location, the type of installation (e.g., menu- or text-based) and the location of file system mapping information. The remainder of the file contains specific configuration information for each of the boot images.

GRUB can operate in two or three stages (in which case, there is a middle stage called stage 1.5). Stage 1.5 contains device drivers so that GRUB can access different types of file systems. Because of this, GRUB can actually interact with the Linux file system prior to the kernel being run. LILO on the other hand must perform the kernel loading operation without any access to information that might be stored in the file system. With this capability, GRUB is able to load stage 2 directly from the file system rather than a reserved location on the hard disk. This also permits access to GRUB stage 2 by system administrators to edit its functionality.

GRUB stage 2 loads a default configuration file in order to present to the user a selection of kernels to select between. Through this GUI interface, the user can also drop into a GRUB command line to perform boot loader operations such as changing kernel parameters, enabling or disabling kernel modules, and editing or replacing boot configuration files. Once a kernel is selected, GRUB uses a mapper to locate the kernel and load it into memory.

A third boot loader is called loadlin. This boot loader runs under either DOS or Windows. This means that you are able to boot to Linux from DOS/Windows as opposed to the more traditional boot loader, which runs during the boot process. What loadlin does is replace the images in memory of DOS/Windows with the Linux kernel and configuration parameters while it executes. As such, loadlin is more useful if you desire booting to Linux occasionally from a fully booted Windows machine.

With access to the MBR and boot loader program, the boot process has shifted from running program code from ROM to running code from hard disk. ROM chips are expensive and so an economic strategy is to store as much of the boot process on disk (or other storage device) as possible. From this point forward, we will see that the boot process involves accessing storage and using volatile memory.

## 11.3.2 Loading the Linux Kernel

The Linux kernel is stored as a partially compressed image. What is loaded into memory is not entirely executable. The Linux kernel is stored on hard disk as the file `vmlinuz` (the full file name includes the version number and architecture type, for instance, vmlinuz-2.6.32-279.el6.x86_64). The first portion of the kernel image is executable and it contains two parts, a 512-byte boot sector and a kernel setup program. As the kernel setup portion runs, it performs some basic initial hardware setup. It then uncompresses the latter portion of the vmlinuz file into `vmlinux`, the rest of the kernel (See Figure 11.1). It should be noted that the Linux kernel can be stored on a USB drive or optical disk and inserted prior to (or during) the boot process as vmlinuz is bootable from these sources.

With the kernel, vmlinux, uncompressed, kernel initialization begins. Its first step is to continue initialization of hardware. Among other things, the power system tests the various components, searches for and loads ramdisks*, accesses the buses and through the buses the various components connected to the computer (e.g., monitor, keyboard, memory, disk controller(s), timer, plug-and-play devices) as well as setting up interrupt handling mechanisms (IRQs).

One of the ramdisks is loaded with `initramfs`, which is the initial Linux root file system. This file system is transferred into RAM for use during kernel initialization for

| Boot sector | Setup sector | Compressed kernel image (vmlinux) |
|---|---|---|

FIGURE 11.1   vmlinuz file.

---

* Refer to Section 10.5 for a discussion on ramdisks.

efficient access. The initramfs file system allows the kernel to access a file system without having to mount any part of the normal Linux file system, thus permitting the kernel to operate using file operations. This file system is only stored temporarily and removed from RAMdisk midway through kernel initialization.

The initramfs contains directories that mirror the real (or permanent) Linux file system: bin, dev, etc, lib, loopfs, proc, sbin, sys, and sysroot. Prior to Linux 2.6.13, this initial file system was called initrd. One difference between initramfs and initrd is the initrd had to be stored via a block storage device whereas initramfs is more flexible.

Unlike the matching directories in the permanent Linux file system, the initramfs directories are minimal, containing only those executables and configuration files needed to finalize the kernel initialization process. The /dev directory for instance permits the kernel to communicate with hardware devices during hardware initialization while /bin and /sbin contain the executable programs necessary for the kernel to start virtual memory and mount the root partition.

At this point in time, the kernel executes a command called `pivot_root`, which alters the root partition from initrd to /. This dismisses initramfs (removes it from memory) and establishes the permanent file system. At this point, only the root file system is mounted. Recall that the root partition will contain /bin, /sbin, and /etc. These directories will all be needed during the remainder of operating system initialization. Later, the other file systems in /etc/fstab (e.g., /var, /home) will be mounted.

To complete kernel execution, the kernel will find and execute the init process (usually located at /sbin/init). It is the init process that is responsible for initializing and establishing the user space. In essence, init is responsible for bringing the system up to a usable state, doing everything that the kernel initialization did not. Now, the kernel moves to the background, waiting to be called by other processes.

## 11.4 INITIALIZATION OF THE LINUX OPERATING SYSTEM*

The init program is the first process to execute in any Linux session. This is a permanent process in that it will run during the entire session that Linux is running, but it will largely sit in the background. It is only stopped when Linux is shut down. The init process is also stopped if the telinit command is used (which causes Linux to switch to a different runlevel, and a new init process is started). Staying in the background allows Linux to retain init as an active process without it requiring resources. It is moved into the foreground as needed. For instance, init moves to the foreground to adopt an orphaned process, and then it moves back into the background. You will always find init to be process 1 (PID of 1).

### 11.4.1 inittab File and Runlevels

The init process handles the remainder of the operating system's initialization through startup scripts. In older versions of Linux, the init process had been synchronous in that each step would happen in order. This unfortunately could lead to a situation where, if

---

* See www.nku.edu/~foxr/linux/ for a discussion of changes made to the initialization process in Red Hat version 7 and CentOS version 2.7.

one step of the process were to stall, the entire process would hang. This could happen, for instance, if init attempted to communicate with a device that was not responding. In order to improve the startup process, the init process has been replaced with `Upstart`, an event-based version of init. Thus, the init process is now asynchronous. This allows init to invoke the various startup scripts, tasks, and services in a scheduled order but without having to wait for a previous step to respond before moving on.

Additionally, Upstart can handle plug-and-play devices by discovering new devices connected at any time. Upstart has been made available for Fedora, Ubuntu, CentOS, Google's Chrome and Chromium OS, and openSUSE. It is available as an option for Debian.

The first visible step in init is to invoke the script `inittab`. This script has at least one statement to establish the *runlevel* (see below). The file may include other statements that establish actions to perform based on various conditions. The format of inittab instructions is *name:#:action:process*. The *name* an identifier of up to four letters. The # is a number in the range from 0 to 6 indicating the runlevel. The action specifies what action should take place on the named process.

To establish the default runlevel, inittab will include a statement with the action `init-default` and with no process. The initdefault action sets the runlevel, to be used later in the operating system initialization process. The following statement is used to establish a runlevel of 5.

```
id:5:initdefault:
```

Editing this file and altering the number from 5 to another runlevel will cause the system to enter another runlevel the next time you boot the computer.

Other statements can be added to inittab to control the actions for differing conditions. These instructions can contain actions such as `respawn`, `wait`, `boot`, `sysinit`, `powerfail`, or `ctrlaltdel`. Several examples follow to explore these actions and the syntax of the statements:

- `rc::bootwait:/etc/rc`—this command executes the /etc/rc script during the init process but does not establish a runlevel. The `bootwait` action specifies that the init process must wait until /etc/rc completes execution.

- `2:1:respawn:/etc/getty 9600 tty2`—the action `respawn` indicates that the given process should run when a current tty terminates. The process is /etc/getty, which opens a new tty (terminal window with a login screen). Thus, if someone logs off, a new login window will appear. This command sets the runlevel to 1.

- `ca::ctrlaltdel:/sbin/shutdown -t90 120 "shutting down now"`—when the user presses ctrl + alt + del, /sbin/shutdown will run with parameters −t90 120 "shutting down now". The −t90 indicates that the message "shutting down now" is sent out followed by a 90-seconds delay. The shutdown occurs in 120 seconds. Therefore, after pressing ctrl + alt + del, there is a 30-seconds delay before the message followed by a 90-seconds delay before shutdown takes place.

- `umnt:1:once:umount -a`—the action `once` indicates that upon changing to this runlevel, perform the umount operation, unmounting all mounted partitions leaving access only to / (the root partition).

- `si::sysinit:/etc/rc.d/rc.sysinit`—the action `sysinit` indicates that the process should run during system initialization and before any boot or bootwait entries. In this case, upon the inittab executing, the script rc.sysinit executes. This entry is no longer needed because Upstart takes care of executing the initial shell scripts for us.

- `pw::powerwait:/usr/sbin/saveall`—the action `powerwait` indicates that the process `saveall` should run if an uninterruptible power supply (UPS) has kicked in because the computer has lost its power supply. Typically, a UPS will only have battery power for 20–30 min and so it is important in the interim that the system be shut down properly. The fictitious process /usr/sbin/saveall presumably will be a program added to the system that ensures that all files are saved and closed properly so that a sudden loss of power does not cause damage to the file system.

You might notice in several cases, no runlevel was specified. The inittab allows us to define a number of situations whereby processes are invoked without changing the default runlevel.

Linux has seven different runlevels. These levels dictate which services should start and which should not. Through the selection of services, Linux either boots to a GUI or a text-based platform, provides network access or not, and permits multiple users or only single user (root) access. The runlevels are given in Table 11.2. The most common runlevel is either 5 for most workstations or 3 for some servers (which are not expected to be used as a

TABLE 11.2    Linux Runlevels

| Runlevel | Name | Common Usage |
|---|---|---|
| 0 | Halt | Shuts down the system; not used in inittab as it would immediately shut down on initialization. |
| 1 | Single-user mode | Useful for administrative tasks including unmounting partitions and reinstalling portions of the OS; when used, only root access is available. |
| 2 | Multiuser mode | In multiuser mode, Linux allows users other than root to log in. In this case, network services are not started so that the user is limited to access via the console only. |
| 3 | Multiuser mode with Networking | Commonly used mode for servers or systems that do not require graphical interface. |
| 4 | Not used | For special/undefined purposes. |
| 5 | Multiuser mode with Networking and GUI | Most common mode for a Linux workstation. |
| 6 | Reboot | Reboots the system; not used in inittab because it would reboot repeatedly. |

regular workstation but instead logged into remotely by system and server administrators). The actual runlevel definitions differ from distribution to distribution but this is the most common usage.

### 11.4.2 Executing rcS.conf and rc.sysinit

After inittab executes, the init process continues to invoke other startup scripts. These scripts are located in /etc/init (not to be confused with the init process, called /sbin/init). The next script up is rcS.conf. It contains the following for loop:

```
for t in $(cat /proc/cmdline); do
    case $t in
            emergency)
                    start rcS-emergency
                    break
            ;;
    esac
done
```

The command $(cat/proc/cmdline) iterates through every word found in /proc/cmdline. This file stores the kernel's startup command. If any word is emergency, then the script invokes the rcS-emergency script. The break statement forces the for loop to exit as there is no need to continue searching through the command line once emergency is found. The rcS-emergency.conf invokes the script /etc/sysconfig/init and then executes two processes as necessary, plymouth and /sbin/sulogin. Plymouth is a *bootsplash* program; it displays graphically information about the boot process. It is interactive in that the user can provide input in spite of Linux not yet enabling things such as terminal windows. The sulogin program provides the user with the following prompting message:

```
Give root password for system maintenance
(or type Control-D for normal startup):
```

This allows the user to enter single user system administrator mode (runlevel 1) or continue with the boot process normally.

If rcS-emergency does not execute, or if the user continues with the normal boot process, rcS.conf then executes the /etc/rc.d/rc.sysinit script. This script is in charge of initializing hardware, loading kernel modules, mounting special file systems (e.g., /proc, /sys), and establishing numerous environment variables. It is also responsible for determining the status of SELinux. The rc.sysinit script contains a series of if-then statements to start up processes to establish further environment variables. For instance, you will find

```
if [ -f/etc/sysconfig/network ]; then
    ./etc/sysconfig/network
fi
```

This statement tests to see if the network script exists and if so, executes it. The network script will establish additional environment variables (e.g., NETWORKING= yes, HOSTNAME).

Other if-then statements determine whether to mount some other file systems. For instance, you will find the following code:

```
if [ ! -e /proc/mount ]; then
     mount -n -t proc/proc/proc
     mount -n -t sysfs/sys/sys >/dev/null 2 > &1
fi
```

This statement tests to see if /proc/mount does not yet exist and if not, it performs the two mount operations mounting the /proc directory and the /sys directory. In the case of /sys, the notation 2 > &1 refers to redirecting standard error (2) to the same location that standard output (1) is pointing to. In essence, this discards (sends to /dev/null) error messages that arise from the mount command so that they are not reported to the console.

Next, the script /etc/init.d/functions is run. In this script, the system's PATH variable is established and exported. The primarily role of this script though is to define functions that can be invoked from other script files yet to be run.

There are a variety of functions defined here, including functions to check the PID of a process, output the PID of a process that is already running, set the PID of a new process, kill a running process, start a new process, and a group of functions that will be used to interpret the information in /etc/fstab so that they can be initially mounted.

Returning to rc.sysinit, it next deals with either establishing or disabling SELinux. At this point, rc.sysinit is in charge of initializing hardware and identifying plug-and-play devices. User-defined and system-defined modules can be loaded using modprobe and kernel parameters can be altered using sysctl. The file system is mounted, calling upon functions defined in the functions script, and establishing the file /etc/mtab.

During mounting, disk quotas are checked and the quotaon process is started if any of the file systems have established disk quotas. Additionally, mounted file systems are established as read-only or readable and writable, encryption is established (if called for), and if a file system is determined to have corruption, fsck is suggested to the user. A number of other bookkeeping activities are performed, including cleaning up log files under /var/log, creating new log files to report on the most recent system initialization, and dmesg is called to report on the latest kernel initialization (dmesg is described at the end of this section).

### 11.4.3 rc.conf and rc Scripts

The script /etc/init/rc.conf is now executed, which then executes the script /etc/rc.d/rc passing it the established runlevel. The rc script is responsible for establishing the services needed for the given runlevel. This script contains the following two loops (excerpted here):

```
for i in /etc/rc$runlevel.d/K*; do
     $i stop
```

and

```
for i in /etc/rc$runlevel.d/S*; do
     $i start
```

The variable `runlevel.d` stores the runlevel as provided by rc.conf, which was established back in inittab. If, for instance, the runlevel is 5, then $runlevel is 5 and we have the for loop

```
for i in/etc/rc5.d/K*; do
```

An examination of /etc shows that there are directories named rc0.d, rc1.d, rc2.d, through rc6.d These seven directories contain symbolic links to service control scripts, all of which are stored in /etc/init.d.

These symbolic links use the following form for their names where *name* is the service name.

```
K##name  or
S##name
```

The K or S is used to indicate whether the particular service should be stopped (killed) or started, respectively. The ## are two digits that indicate the order that the services should be stopped or started. The order is controlled by globbing when K* or S* is filename expanded. Since all of the services to be killed will be listed by K*, the filenames are expanded based on the two-digit number rather than the name that follows. Otherwise, services would be killed in alphabetical order.

The first for loop iterates through all of the `K##name` services, stopping them in the order that the services are numbered. The second for loop iterates through all of the `S##name` services, starting them in the order that the services are numbered.

You might ask why services need to be stopped if you are just now initializing the system. The `telinit` instruction allows you to change runlevels from the command line. So, if you are currently in runlevel 5 and wish to switch to runlevel 1, `telinit 1` will accomplish this. However, in switching from 5 to 1, you must kill any service that is started by runlevel 5 but which should not be running for runlevel 1. Similarly, exiting from runlevel 1 back to runlevel 5 will require restarting those services.

Figure 11.2 shows the contents of the rc5.d subdirectory. Recall that runlevel 5 is multiuser, networked, and GUI. Thus, we would expect services that deal with multiple users, the network, and the GUI to all be started. Other services that are not needed by this runlevel are stopped. The system administrator can decide whether any unstarted services should be started at a later time.

```
K01smartd         K80kdump            S13cpuspeed         S28autofs
K02oddjobd        K84wpa_supplicant   S13irqbalance       S30nfs
K05wdaemon        K87restorecond      S13rpcbind          S50bluetooth
K10psacct         K88sssd             S15mdmonitor        S55sshd
K10saslauthd      K89rdisc            S22messagebus       S70spice-vdagentd
K15httpd          K95firstboot        S23NetworkManager   S80postfix
K50dnsmasq        K99rngd             S24avahi-daemon     S82abrt-ccpp
K50netconsole     S01sysstat          S24nfslock          S82abrtd
K50snmpd          S02lvm2-monitor     S24rpcgssd          S82abrt-oops
K50snmptrapd      S08ip6tables        S24rpcidmapd        S90crond
K69rpcsvcgssd     S08iptables         S25cups             S95atd
K73ypbind         S10network          S25netfs            S99certmonger
K74ntpd           S11auditd           S26acpid            S99local
K75ntpdate        S11portreserve      S26haldaemon
K75quota_nld      S12rsyslog          S26udev-post
```

FIGURE 11.2   Symbolic links to services for runlevel 5.

What we see in Figure 11.2 are 22 stopped services and 37 started services. For instance, httpd is killed because this service by default is not needed (this is the Apache web server and most Linux users will not be running a web server). Similarly, smartd, a service to monitor the reliability of your hard disk drive(s) is stopped because, unless there is a reason to suspect drive failures, this service would be resource-heavy and somewhat irrelevant. On the other hand, we see the Linux firewall (iptables, ip6tables), the ssh daemon, the nfs daemon, the cron and at daemons, and the printer service (cups) are all started as these are commonly used network or multiuser services.

You can modify which services are automatically started and stopped by changing the symbolic links under the appropriate rc#.d directory, altering the K and S labels. For instance, if you wanted httpd to start under runlevel 5, then under /etc/rc5.d, change K15httpd to S85httpd (we explore later where we can find the two-digit values 15 and 85).

Alternatively, there is a program called `chkconfig`. Without options, it displays for each service, which runlevels it is started and stopped for. For instance, the sshd service contains the following entry:

```
sshd 0:off 1:off 2:on 3:on 4:on 5:on 6:off
```

You can alter any service's behavior using

```
chkconfig --level levelnumber servicename status
```

where *levelnumber* is the runlevel, *servicename* is the name of the service, and *status* is on or off.

## 11.4.4 Finalizing System Initialization

Now that services are started, Linux is ready for use. The script /etc/rc.d/rc.local is the last script executed. This script is available for system administrators to perform additional automated startup operations. If you, as a system administrator, wish to have some external file system mounted and made available for all users, you could include the mount command in this script. If you want to have scheduled processes run, you could add scheduling commands to this script.

Other scripts may be started by Upstart prior to the rc scripts. Again, these are found in /etc/init and include scripts to define actions for ctrl + alt + del (control-alt-delete.conf), scripts to start terminal windows (start-ttys.conf, tty.conf), and scripts to start up a GUI display manager (prefdm.conf).

As a system administrator, you are able to adjust the operating system initialization process by adjusting some of these start-up scripts (although there would probably be little need to do so). Any such changes should be done cautiously with the current version of the scripts saved under different names so that you can roll the system back as necessary.

You can also explore the kernel initialization process. After the system has been brought up, the command dmesg will respond with the kernel ring buffer. These are the messages that the kernel produces when initializing. Although much of the information that dmesg will display is cryptic, you can examine it for possible boot errors (as well as to learn more about the Linux boot process).

## 11.5 LINUX SERVICES

In this section, we briefly explore the various services in Linux. In Linux, a service is generally referred to as a daemon (pronounced "demon"). We will consider several services in more detail in the next two sections when we look at how those services can be configured. These sections are incomplete as there are far too many services to cover.

A service is a piece of operating system code used to handle some type of request. The service has several distinctive features. First, it runs in the background so that it does not take up processor time unless called upon. Second, services can handle requests that come from many different sources: users, applications software, hardware, other operating system services, messages from the network. Third, services are *configurable*. Configuration is usually handled through configuration files, which are often stored in the /etc directory (or a subdirectory). Finally, services can be running or stopped, and you (the system administrator) can control which services are running or stopped and which services are automatically started at system initialization time based on the runlevel (see the discussion of the rc script from Section 11.4.3).

There are several files related to every service. These are the services themselves (executable programs, typically stored in /usr/sbin, /sbin, or in some cases, /usr/bin), configuration and data files found in or under /etc, and the service controlling scripts. The service controlling scripts are found in /etc/init.d. It is these scripts that the symbolic links from the /etc/rc#.d directories reference (recall Figure 11.2). We examine below how to use these controlling scripts and we examine some of the configuration files in Section 11.7.

### 11.5.1 Categories of Services

Each Linux service provides a different form of support to software, the user, the file system, the network, or some other aspect of the operating system. The Linux services can be broken down into several different categories. Some services support a range of requests while other services must combine to perform their task. Here, we will break the types of services into the following categories. It should be noted that other Linux administrators and authors may select a different categorization of services.

- Boot

- File system

- Hardware

- Language support

- Logging

- Network, web/Internet

- Power management

- Scheduling

- System maintenance

Notice the separation between network and web/Internet services. While network services must be running to communicate with the Internet or with websites, the various web/Internet services do not have to be running to communicate within the local area network. For instance, the network service itself is necessary to communicate via network and rdisc is used during system initialization to discover your computer's local subnet router. The service certmonger maintains website security certificates, keeping them up-to-date. Without this service, the user is still able to access the network and the Internet although not necessarily in a secure fashion. The iptables/ip6tables services are the Linux firewall. As with certmonger, if these services are not running, network and Internet access are still available but not securely. Other service categories above could be broken into a finer grain such as with hardware services as there are different services to support different devices such as a Bluetooth service, services to support USB devices, services to support printing, and so forth. Table 11.3 illustrates many of the services, including their name, their role, and their type.

## 11.5.2 Examination of Significant Linux Services

In CentOS 6.0, there are over 60 services available. Different Linux distributions will use different services. There are nearly 80 in Ubuntu version 12. Many overlap and some go by slightly different names (e.g., cron vs. crond). Here, we briefly examine several services.

- abrtd—this is the automated bug report daemon. When a piece of applications software crashes, this daemon will collect information about the software and the crash such as the core dump and the command line instruction, and perform some action. The file /etc/abrt/abrt.conf contains the configuration information for this daemon. Specifically, it includes rules that dictate which software should be reported and what action(s) taken by abrtd for that software. Actions are primarily report

TABLE 11.3  Linux Services

| Name | Type | Description |
|------|------|-------------|
| acpi | Power management | Laptop battery fan monitor |
| acpid | Event handling | Handles acpi events |
| anacron | Scheduling | For scheduling startup tasks at initialization time |
| apmd | Power management | Laptop power management |
| arpwatch | Web/Internet | Logs remote IP addresses with hostnames |
| atd | Scheduling | Executes at jobs based on a scheduled time and batch jobs-based CPU load |
| auditd | Logging | The Linux auditing system daemon that logs system, software, and user-generated events |
| autofs | File system | Automatically mounts file systems at initialization |
| bluetooth | Hardware | Bluetooth service |
| certmonger | Web/Internet | Maintain up-to-date security certificates |
| cpufreq, cpufreqd | Hardware | Configures and scales CPU frequency to reduce possible CPU overheating |
| crond | Scheduling | The daemon for handling cronttab jobs |
| cups | Hardware | Service for printing |
| cvs | System | Managing multiuser documents |
| dhcpd | Web/Internet | Configure DHCP access |
| dnsmasq | Web/Internet | Starts/stops DNS caching |
| gpm | Hardware | Mouse driver |
| haldaemon | Hardware | Monitors for new or removed hardware |
| httpd | Web/Internet | The Apache web server |
| iptables, ip6tables | Web/Internet | The Linux firewalls |
| mdadm | File system | Manages software for RAID |
| named | Web/Internet | Starts/stops the BIND program (DNS) |
| netfs | File system | Allows remote mounting |
| netplugd | Network | Monitors network interface |
| network | Network | Starts and stops network access |
| nfs | File system | Enables network file system sharing |
| nscd | Network | Password and group lookup service |
| oddjobd | System | Fields requests from software that otherwise do not have access to needed Linux operations |
| postfix | Network | Mail service |
| prelude | Network | Intrusion detection system service |
| rdisc | Network | Discovers routers on local subnet |
| rsync | File system | Allows remote mounting of file systems |
| smartd | Hardware | Monitors SMART devices, particularly hard drives |
| snmpd | Network | Network management protocol for small networks |
| sshd | Network | Service to permit ssh access |
| syslog | Logging | System logging |
| ypbind | Network | Name server for NIS/YP networks |

logging operations by placing information in a database or log file. In addition, plugins can be used to provide more specific types of actions such as invoking a debugger or sending a crash report by email to a prespecified email address. By default, logging information is filed in the /var/spool/abrt directory. The abrtd daemon calls upon subservices for added functionality such as abrt-gui which displays logged information graphically and abrt-cli which provides similar information via the command line, or abrt-applet which, as its name implies, is an applet that can provide real-time feedback of software crashes.

- anacron—this service is a scheduler, responsible for running tasks (processes, scripts) based on a preestablished schedule. Anacron will run tasks that were scheduled at times that the system was not available. Anacron runs such tasks at the first opportunity after the system has been restarted/resumed. The configuration file for anacron is /etc/anacrontab. Of the scheduled activities that might be preset for anacrontab are the entries in the directories /etc/cron.daily, /etc/cron.weekly, and /etc/cron.monthly based on those intervals (daily, weekly, monthly). See crontab below for more detail.

- atd—the at daemon is a one-time scheduler. It runs processes that were scheduled through either the at or batch commands. In at, the task is scheduled for a specific time and date. These can be absolute values such as today 13:50 (1:50 PM), tomorrow 9:00 AM, or 12/31/15 12:00 PM, or a time relative to now such as now + 5 minutes, now + 2 hours, now + 1 day. You can also use terms such as noon, midnight, and teatime. The batch program executes scheduled processes when the CPU's load drops below 80%. The at daemon also permits a user to inspect and remove waiting at and batch jobs using the commands atq and atrm, respectively. We examine at and batch in Chapter 14.

- auditd—this is the Linux Auditing System daemon, responsible for creating log entries of specified activities no matter who performs those actions. There are two programs and three files related to auditd. First is the auditd program itself. Second is auditctl, which the system administrator will use to start and stop the audit daemon. It is auditctl, which, when started, will read in the configuration rules.

This leads us to the three files. The first is the auditd.conf file, which is a standard configuration file. The configuration file specifies how auditd will operate such as the location of the log file, the action of whether log files will be rotated, and the number of log files to rotate between, but it does not specify the logging activities. Instead, logging activities are stored as rules in a second data file, audit.rules. Rules vary from generic to specific.

In Table 11.4, we see some of the types of rules that you can specify. One such rule could be used to log attempts to write to or alter the attributes of the file /etc/foobar. txt using

```
-w /etc/foobar.txt -p wa
```

TABLE 11.4    audit.rules Rule Examples

| Syntax | Meaning |
|---|---|
| -D | Delete any previously defined rules |
| -b # | # is a number, establish # buffers, for example, -b 1024 |
| -f # | Set failure flag to # (0 is silent, 1 is print failure messages, 2 is panic or halt the system) |
| -w *directory* | Log attempts to access the *directory* |
| -w *filename* | Log attempts to access the *file* |
| -w *filename* –p [rwxa]* | Log attempts to read *file* (r), write to *file* (w), execute *file* (x), or change *file*'s attributes (a). The * indicates that any combination of the options r, w, x, and a can be listed. |
| -a *action,list* –S *syscall* –F *field = value* | Log system calls; *action* is either always or never, *list* is one of task, entry, exit, user or exclude. The –S option allows you to specify a Linux operation such as chmod, mkdir, or mount. The –F option allow you to fine-tune the match by testing some system or user parameters such as EUID |

and another rule could be used to log all normal users (UID > 199) who issue rmdir or mkdir operations:

```
-a entry,always -S mkdir -S rmdir -F UID>199
```

The third file is another configuration file, stored under /etc/sysconfig/ auditd. This file contains options for auditd rather than configuration rules. We will briefly examine the two .conf files in Section 11.7 of this chapter.

- crond—this is the daemon for handling cron jobs, which, unlike at and batch jobs, are scheduled to recur based on some pattern such as hourly or weekly. To run a cron job, you use the program crontab. The crontab program is somewhat more complicated to use than the at program. For crontab, you specify a file that contains the recurrence information and the process(es) to be executed. You specify each process and recurrence in one line of the file. The file can contain any number of process/recurrence entries. Since the entry must fit on one line, lengthy entries of processes can be placed into a script so that the process listed is merely the script to execute. The cron daemon, crond, examines all scheduled cron jobs against the current time to see if any should be executed. This can happen as frequently as every minute.

Each user is able to submit a single crontab job. Thus, a user who wishes to add other recurring events will have to modify the current crontab job (if one has been submitted). The crontab instruction permits options of -l to inspect the current crontab jobs, -r to remove the current crontab jobs (as they are all bundled in one file, this removes the entire file) and -e to edit the current crontab jobs (if the user chooses to handle the scheduled jobs via an editor rather than through a file).

The recurrence information is broken into five parts: the minute, the hour (0–23), the day of the month (1–31), the month (1–12, or jan, feb, mar), and the day of the week (0–6, where 0 is Sunday and 6 is Saturday). Only relevant portions are specified; otherwise, * is used to indicate a wildcard. You can also specify multiple recurrences by separating the different recurrences with a slash (/). Below are some examples of crontab entries. We cover crontab in more detail in Chapter 14.

- `0 12 * 15 *  ./foobar`—run the script foobar at 12 noon on the 15th of every month

- `0 0 1 1 *  ./foobar`—run the script foobar on January 1 at midnight (once per year)

- `30 15 * * 6  ./foobar`—run the script foobar every Saturday at 3:30 PM (15 = 3 PM in military time)

- `0/15/30/45 * 1 1 *  ./foobar`—run the script foobar on January 1 every 15 min of every hour

As stated earlier, a scheduled job missed because of downtime does not get executed when the system is brought back up. Instead, if the job is critical enough, it should be scheduled with anacron.

- cups—the common Unix printing system is used to control print jobs on the printers connected to a Linux computer. The printers may be directly connected to the computer or available via a network even if the network consists of computers running different versions of Linux or Unix. In Chapter 8, we briefly looked at how to install a printer on a Linux computer, the result of which is a configuration file storing information about the installed printer. In addition, installing a printer requires that the proper device driver for the printer is installed. When the user issues a printer command (`lp`, `lpr`, `lpq`, `lprm`), cups takes over to handle the command by accessing the appropriate printer. An additional command, `lpc`, allows you to alter a printer's properties (this is also available via the Printer Configuration GUI).

- dnsmasq—this service is something like a mini-DNS server for Linux (DNS was introduced in Chapter 5 and is also explored in Chapters 12 and 15). The role of dnsmasq is threefold, all of which revolve around caching DNS information. First, it caches responses from DNS server access so that further repeated accesses can be handled locally while the entries remain stored in cache. Second, it uses the /etc/ hosts file, which contains specific IP alias to address mapping as set up by the system administrator to bypass DNS server access. Third, it responds to requests from DHCP servers regarding IP alias to IP address mappings. With dnsmasq, we reduce the amount of access to the DNS server and thus reduce waiting time in Internet-based communication. The service uses `/etc/dnsmasq.conf` as a configuration file to specify details such as a DNS server to query in case the local cache does not contain the relevant information, what network interface(s) to use (e.g., eth1 for any DHCP requests), and the amount of time that items should remain cached.

- iptables and ip6tables—these are the Linux firewall for IPv4 and IPv6 messages, respectively. The names of the services are also the names of a file of firewall rules, which are located in /etc/sysconfig. In addition to the rules files of `iptables` and ip6tables, the configuration files are iptables-config and ip6tables-config. We explore specific firewall rules in Chapter 12.

- logrotate—this service performs operations on log files, including rotating logs files, compressing log files, and emailing log files. Log rotation renames the current log file (usually stored under /var/log) and creates a new log file so that no log file grows too large. Typically, log files are rotated automatically based on some rotation rate such as weekly or monthly. Often, log files' names are affixed with either a date or a number indicated their age, for instance, mylog (the current log file), mylog.1, mylog.2, and mylog.3 (older log files of one, two, and three rotations in the past). The subdirectory /etc/logrotate.d contains numerous configuration files, one per type of log file. Each of these configuration files lists specific log files (e.g., /var/log/messages) followed by a series of directives for handling that group of log files such as the maximum size of a log file, the rotation rate, the number of log files to retain, and so forth.

- nfs—permits the exportation of file systems for remote mounting. The system administrator places file system information in the /etc/exports file. For each file system that can be remotely mounted, the entry must include the file system's local mount point, the location(s) where the file system can be remotely mounted (this might be * to indicate anywhere, or specific or partial URLs or IP addresses as in 10.11.12.0/24 meaning any computer in the network 10.11.12) and any options for the file system such as ro (read only) and sync. We examined NFS in Chapter 10.

- syslogd—there are two other logging services of note, syslogd, and klogd. The klogd service logs kernel messages and is not configurable. The syslogd service logs nonkernel operating system messages. It is configurable and can also be set up to log kernel messages and application software messages. The syslogd daemon has a configuration file, /etc/syslog. We explore how to configure syslogd in Section 11.7. It should be noted that the most recent versions of Linux have renamed the daemon to rsyslogd and the configuration file to /etc/rsyslog.

Another service of note is called oddjobd. Service requests are made by the user or system administrator, the operating system, external messages coming in from the network, or other software. In some cases, software will not have permissions to execute a given service. The oddjobd service provides a mapping so that specific software can make requests of specific services.

The above list describes only some of the many services available. Missing from the above discussion, among others, are services dealing with the computer network, port mapping, and discovering the network router. Many of these network services are explored in Chapter 12.

## 11.5.3 Starting and Stopping Services

As a system administrator, it will be your task to handle services. Your options are to start or stop services, and to reconfigure them. Decisions include whether a service should be running at all times or on demand, whether a service should start at system initialization time (recall the rc.conf script from Section 11.4 that started and stopped services based on runlevel), or whether a service should not be run. And then, you will have to determine whether to use the default configuration or alter the configuration of a service. We examine how to start and stop services here, saving configuration for the next two sections.

There are usually four different things you can do to a service: start, stop, restart, or obtain the service's status. You can control the service from either a GUI program or from the command line. The service configuration tool allows you to select any service and view its status, start, stop, or restart it. Although it is called a *configuration* tool, in fact, you cannot configure the service from it. This GUI is shown in Figure 11.3 where some of the many services can be seen listed.

For each service listed, you will see two symbols to the left of its name. The first is whether the service is current enabled (green), disabled (red), or customizable (three slider controls). These three states indicate whether the service is set up to automatically start (enabled), automatically stop (disabled), or start in specific but not all runlevels. It should be noted that "customizable" only permits changes in levels 2, 3, 4, and 5. Services that start or stop in levels 0, 1, and 6 are unalterable through this tool. You might notice in the GUI that there are additional buttons for Enable, Disable, and Customize so that you can alter the service's startup behavior.

The second symbol, immediately to the left of the name, indicates whether the service is currently running (a plugged in symbol) or stopped (a plug not plugged in symbol). You can start, stop, and restart a service through buttons near the top of the GUI. Finally, when you click on any service, a description of that service is provided for you in the Description box. In Figure 11.3, we see a brief description for the atd service.

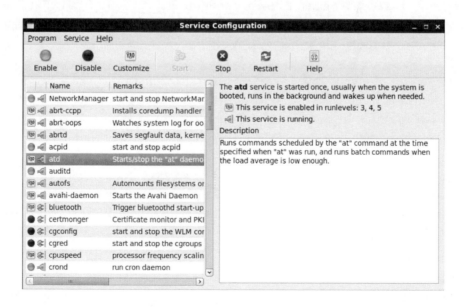

FIGURE 11.3    Linux service configuration GUI.

You can also control the services through the command line in one of two ways. First, you can use the `service` program, located in /sbin. The format of this command is

```
/sbin/service service command
```

where *service* is the service in question, and command is one of `start`, `stop`, `restart`, or `status`. The start and stop commands should be obvious, starting or stopping a given service. The restart command first stops the service and then starts it. You would use restart after you have modified the service's configuration file. Status provides you the status of the service, whether it is running or stopped.

You can also control the service through the service control script. The service control scripts are all stored in /etc/init.d, for instance, /etc/init.d/nfs. You can directly provide the command to the service as in

```
/etc/init.d/nfs command
```

where the command again is one of `start`, `stop`, `restart`, or `status`. If your current working directory is /etc/init.d, then you can issue the command as

```
./nfs command
```

## 11.5.4 Examining the atd Service Control Script

The start/stop commands are not actually issued to the service itself but instead to the script that controls the service. These scripts are stored in /etc/init.d with symbolic links from the various /etc/rc#.d directories to them. To get a better understanding of these scripts, we examine the /etc/init.d/atd script,* which controls atd, the at daemon. Below, we see this script file. The script's code is interspersed among comments that describe what the script is doing.

```
#!/bin/sh
#
# atd Starts/stop the "at" daemon
#
# chkconfig: 345 95 5
# description: Runs commands scheduled by the "at" command at the time \
# specified when "at" was run, and runs batch commands when the load \
# average is low enough.

### BEGIN INIT INFO
# Provides: atd at batch
# Required-Start: $local_fs
```

---

* The atd script as well as the atd service, the at program, and all of Linux is protected under the GNU General Public License, v3.0. For more information, see http://www.gnu.org/copyleft/gpl.html. The script is being reprinted here without permission as per part 4 of the GPL, which guarantees the right to provide "verbatim copies" of course code as long as the copyright notice is conspicuously and appropriately published.

```
# Required-Stop: $local_fs
# Default-Start: 345
# Default-Stop: 95
# Short-Description: Starts/stop the "at" daemon
# Description: Runs commands scheduled by the "at" command at the time
# specified when "at" was run, and runs batch commands when the load
# average is low enough.
### END INIT INFO
```

The above lines of the script are all comments explaining information about the script. The line chkconfig provides the default enable information, that is, the levels in which atd is automatically started: 3, 4, and 5. The next number, 95, is atd's start priority. This number corresponds to the two-digit value in any symbolic link name starting with an S that points to atd. For instance, in /etc/rc5.d, the symbolic link is named S95atd. The last digit, 5, is the stop priority that corresponds to the two-digit value in any symbolic link name starting with a K that points to atd. So, for instance, /etc/rc1.d has the symbolic link K05atd. The provided line indicates which programs rely on this service. As we see, atd, at, and batch all rely on this atd script to start the atd service. Required-Start and Required-Stop list services that rely on this service. In this case, $local_fs is a variable. Default-Start and Default-Stop are as explained earlier in the paragraph.

```
# Source function library.
./etc/rc.d/init.d/functions
```

The first executable line of atd executes the script /etc/rc.d/init.d/functions. This file contains a variety of commands and function definitions that may be used by the scripts in the /etc/init.d. Stepping through the functions file, we see the following actions:

```
TEXTDOMAIN=initscripts
umask 022
PATH="/sbin:/usr/sbin:/bin:/usr/bin"
export PATH
```

In these first four instructions, we see the assignment of two environment variables and a umask statement. The environment variable PATH is set up so that the various scripts in this directory can reference instructions without full paths. Similarly, if a file has to be created, it will be given initial permissions of 644 (666–022). We will see in atd that it creates a file known as lockfile, used to determine whether the atd service is running or stopped. Both the umask statement and the creation of PATH are used so that the atd service has values assigned to it in spite of root already having a umask value and PATH variable.

As the script file function continues to execute, it defines a number of additional environment variables, including CONSOLETYPE, COLUMNS, and SETCOLOR_NORMAL, all used to properly display in a terminal window or on console. Several if-then and if-then-else statements are used to define further environment variables. The script then defines a number of functions that will be used by the various script files such as atd. For instance, the function checkpid will be used to determine if a PID is already assigned to a process, which is useful to see if the process is currently running or not. The function pidofproc obtains the PID of a running process. The function killproc is used to stop a running process.

Another function of note is pids_var_run. This function is used by status (explained below) to test to see if a given process has both a PID and a pid file. The pid file is stored under /var/run to indicate that a particular process has started running. The user must have sufficient access rights to view the pid file or else this function terminates with a return value of 4 to indicate insufficient privilege to access the file. If there is a pid file but no PID for this program, then the function returns 1 indicating that the program is dead even though a pid file exists. If there is a PID and pid file, the function returns 0 indicating the program is running; otherwise, the function returns 3 indicating that the program is not running. The function daemon is used to start an actual daemon service.

Let us now focus on the `status` function from the functions file. We will refer to the status function every time we issue the command /sbin/service *servicename* status. This function begins by defining four local variables and assigning lock_file and pid_file to null. Next, it tests the number of parameters passed to this function.

If it received no parameters, it outputs a usage statement and returns error code 1. Otherwise, the function expects to receive one or more parameters. If there is one parameter, the parameter is the name of the daemon's executable program, such as atd. The other possibility is that the function call includes one or two options: -l *lock_filename* and –p *pid_filename*. If the –p option is used, then status calls upon the function pids_var_run. Given the response from this function, if the *pid_filename* exists and there is a PID associated with this process then status outputs

*name* (pid *PID*) is running . . .

where *name* is the service name and *PID* is the process ID assigned to this service, and returns a value of 0. Otherwise, based on the return value from the pids_var_run function, status will output one of

```
name dead but pid file exists
```

or

```
name status unknown due to insufficient privileges
```

If the –p option is not used, status then tests for the existence of the lock_file. If it exists, status responds with the output

*name* dead but subsys locked

meaning that the process is not running but the *lock_file* still exists and needs to be deleted. Finally, if there is no lock_file, then status responds with

*name* is stopped

Other functions found in the function script are called `success`, `failure`, `passed`, `warning`, and `action` among other supporting functions utilized by the various functions mentioned above. With functions now loaded, the atd script continues.

```
exec=/usr/sbin/atd
prog="atd"
config=/etc/sysconfig/atd
```

These three lines of code establish variables used within the atd script for convenience. Notice that `prog` is the name of the service program to be executed and `exec` is the full path to execute the program.

```
[ -e /etc/sysconfig/$prog ] && ./etc/sysconfig/$prog
lockfile=/var/lock/subsys/$prog
```

The above two lines first test to make sure the executable program for the service exists and then executes it. Then, `lockfile` is set to the name of a file that atd will create to place the lockfile in its appropriate location, /var/lock/subsys. At this point, the atd script file defines a number of functions called from a case statement at the bottom of the file.

```
start() {
        [ -x $exec ] || exit 5
        [ -f $config ] || exit 6
        echo -n $"Starting $prog: "
        daemon $exec $OPTS
        retval=$?
        echo
        [ $retval -eq 0 ] && touch $lockfile
}
```

The start function will attempt to start the atd service. It first tests to see if the program /usr/sbin/atd is executable. If not, start immediately exits with an error code of 5. It then tests to see if the configuration file /etc/sysconfig/atd exists. If not, start exits with an error code of 6. Otherwise, it outputs a message that it is starting atd, calls the daemon function to execute /usr/sbin/atd with options of $OPTS. As OPTS was not defined in this script, it will only have a value if it was established and exported by another script or set at the command line prompt. The daemon script will return 0 if successful or an error code if unsuccessful. This value is placed into the variable `retval`. Finally, if retval is equal to 0 (success), then the `lockfile` is created to indicate that atd was successfully started.

```
stop () {
        echo -n $"Stopping $prog: "
        if [ -n "`pidfileofproc $exec`" ]; then
```

```
        killproc $exec
        RETVAL=3
    else
        failure $"Stopping $prog"
    fi
    retval=$?
    echo
    [ $retval -eq 0 ] && rm -f $lockfile
}
```

The stop function operates somewhat like the opposite of the start function. First, it outputs a message indicating that atd is being stopped. It then calls upon the function pidfileofproc to see if atd has a PID and therefore is running. If so, the function killproc is called to kill atd and the variable RETVAL is set to 3; otherwise, the function failure is called indicating that stop failed while trying to stop atd. The variable retval (which differs from RETVAL) is set to the value of the most recent function call, which will either be that of killproc or failure. Killproc will return 0 upon success and so if retval is 0, it means that atd was successfully stopped. The lockfile file is then removed to indicate that the process is no longer running.

```
restart() {
    stop
    start
}

reload() {
    restart
}

force_reload() {
    restart
}
```

Restart simply calls the two functions stop and start, respectively. You would use restart if you have modified atd's configuration file or you find atd to not be functioning correctly and wish to restart it. You can also separately call stop and start. Also shown above are reload and force_reload, both of which call restart to give you the same result. This allows the system administrator to restart atd using any of restart, reload, or force_reload.

```
rh_status() {
    status $prog
}
```

The rh_status function appears to do nothing useful; it merely calls status passing it the name of the program, atd. In fact, rh_status is a type of wrapper function. We use rh_status because we may want to invoke it directly or indirectly through another function, rh_status_q (shown below). As there is no –l or –p option provided in the call to status, status will test for a PID and if one exists, output the "running" message; otherwise, it will respond with the "dead," "insufficient privileges," or "stopped" message.

```
rh_status_q() {
     rh_status >/dev/null 2>&1
}
```

This function is used to invoke the rh_status function from above. The notation `> /dev/null  2>&1` is common in system administration scripts indicating that STDERR messages should be redirected to the same output location as STDOUT. In the clause `2>&1`, the 2 refers to STDERR, the > redirects any output of STDERR to &1, and the &1 indicates "point to STDOUT's destination." Prior to this, the code `> /dev/null` redirects STDOUT to /dev/null. This instruction ensures that any output or error from rh_status is not output to the user's terminal window but instead sent to /dev/null.

Why would we want to prevent output from being seen in the terminal window when we are calling upon status to find out the service's status? The reason for this is that calling upon rh_status_q to obtain the status of the service is being done to obtain not text output but a return code. We hope to receive a 0 to indicate that the service is running as expected. We will see below that rh_status and not rh_status_q is used when we want to obtain the service's status while rh_status_q is used to start or stop the service. That is, we use rh_status to output the status of the service while we use rh_status_q to obtain the service's status in a nonoutput situation.

```
case "$1" in
     start)
            rh_status_q || exit 0
            $1
            ;;
     stop)
            rh_status_q || exit 0
            $1
            ;;
     restart)
            $1
            ;;
     reload)
            rh_status_q || exit 7
            $1
            ;;
     force-reload)
            force-reload
            ;;
     status)
```

```
                rh_status
                ;;
        condrestart|try-restart)
                rh_status_q || exit 0
                restart
                ;;
        *)
                echo $"Usage: $0 {start|stop|status|restart|
                        condrestart|try-restart|reload|
                        force-reload}"
                exit 2
esac
exit $?
```

The script ends with the above case statement. This statement tests the parameter passed to the script from the command line. Expected parameters are start, stop, restart, reload, force-reload, status, condrestart, and try-restart. If start is provided, then the case statement calls rh_status_q to determine if the service is already running. If rh_status_q exits with a value other than 0, then the exit 0 statement is executed and the script terminates. This indicates that there is nothing to do if we wish to start an already-running service. Otherwise, rh_status_q will return some other code (hopefully 3 indicating that the service is stopped, but possibly 1, 2, or 4 indicating some kind of error with the service). In any of these cases, $1, which is the word "start," is invoked. This calls the start function, which starts the atd service.

Notice that we are using rh_status_q rather than rh_status. The former function does not output status information but instead returns the status code (0, 1, 2, 3, 4). The same strategy is used for stop.

If the parameter, $1, is restart, then the case statement merely calls $1, which itself calls the restart function. For reload, rh_status_q is invoked as with start and stop. In this case, an error code of 7 is returned if the status is 0 (running); otherwise, the reload function is called. The reload function, as shown earlier, calls the restart function.

For force_reload, it merely calls force_reload. For status, the rh_status function is called. This function does output to the terminal window unlike rh_status_q. If condrestart or try-restart are specified as the parameter, then rh_status_q is called and restart is only called if rh_status_q returns a nonzero value.

Finally, the default clause is invoked if there was no parameter supplied, or the parameter is not one of the legal parameters. The default clause causes the script to then output a usage statement indicating how to use this atd script. In this echo statement, the $0 parameter is the script name itself (atd). The script exits with the error code of the last function called, which will be a 0 if everything worked correctly.

## 11.6 CONFIGURING SERVICES THROUGH GUI TOOLS

There are two ways to configure a service. In some cases, there are GUI tools (or wizards) available. This is true of the Linux firewall, Bluetooth, kdump (kernel crash dumps), cups (printing service), along with a few others. Or, you can alter the service's configuration file(s) and restart the service. In this section, we briefly explore a few GUI tools.

## 11.6.1 Firewall Configuration Tool

You can bring up the Firewall Configuration tool from the Administration submenu of the System menu (in Gnome), or by typing /usr/bin/system-config-firewall (see Figure 11.4). Making changes to either the GUI or the firewall's configuration file requires root access.

First, you can select to enable or disable the firewall. In the figure, the firewall is currently enabled so you can only choose to disable it. The reload button causes any changes that you made but did not apply to be discarded. The wizard button allows you to make simple changes to the firewall such as whether to use a desktop-based configuration or a server-based configuration. For any other, more specific changes, you will have to actually enter information in the GUI. After making any changes to the GUI, you have to click on Apply for the change to take effect.

Aside from enabling/disabling the firewall or using the wizard to change settings, any other types of changes are made through a menu of rule types provided along the left side of the GUI. These selections offer different options for how you can specify firewall rules. You can specify firewall rules based on service type or port number, or through custom rules.

In Figure 11.4, Trusted Services has been selected and you can see some of the network programs listed. None of these programs have been selected so that the firewall currently will not permit messages from these types of programs. For instance, any attempt to send an FTP or DNS request will be rejected. If you were to scroll down, you would see by default only one program enabled, SSH. We would change our firewall if we wanted to

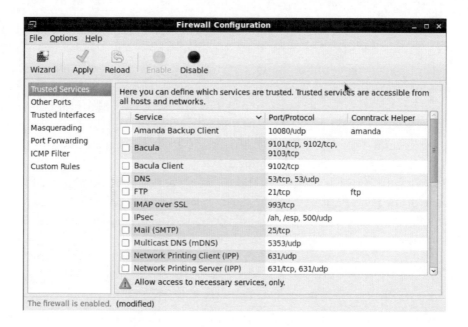

FIGURE 11.4  Firewall Configuration Tool.

permit other types of messages to be allowed through. For instance, if you were going to implement a web server on this Linux machine, you would have to allow HTTP and probably HTTPS messages (not seen in the figure). Or, if you wanted to establish a DNS server, you would enable DNS.

For each service in the list, we also see the port(s) and protocol(s) associated with it. These details help flesh out the firewall rules. For instance, SSH uses TCP over port 22. If you select DNS and apply the change, a rule is then generated for the firewall that permits TCP and UDP messages over port 53. Conntrack Helper is short for Connection Tracker Helper. These programs help track connections of various types. For instance, amanda is a program used to track Amanda Backup Client messages.

The Other Ports selection allows you to add ports that are not listed under Trusted Services. For instance, telnet is available over port 23. You would have to select this through Other Ports rather than Trusted Services. Additionally, you could open up ports that are not the typical port such as adding port 8080 to HTTP. HTTP requests default to port 80. Aside from the port, or range of ports, you also specify the acceptable protocol for the given port.

Trusted Interfaces include eth+ devices (e.g., eth0, eth1, eth2, etc.), ippp+ devices, isdn+ devices, ppp+ devices, tun+ devices, and wlan+ devices. Each network interface is listed separately. The default is for none of these to be selected. You can also add your own interface to the list if you have interface devices not listed.

Masquerading contains the same interfaces as trusted interfaces. The difference is that masquerading hides your local network so that your host appears as a single Internet address. If masquerading is turned on, then port forwarding is enabled.

Port Forwarding allows you to specify forwarding rules. You would use forwarding if your device were not an end-user of the network but a device that routed messages (e.g., router, gateway). Forwarding permits messages to be forwarded from your device to another. Port forwarding are the rules that specify which messages should be passed along and which should not. These rules must list the interface(s), port(s), and protocol(s) of incoming messages and the destination address(es) and port(s) permitted. For instance, you might specify that TCP messages received from eth+ over port 80 can be forwarded to IP address 10.11.1.2. If any other local machine's IP address is present in the destination address of a received message, it is not forwarded.

ICMP Filter permits messages that are sent by troubleshooting programs such as ping. Selections include destination unreachable, parameter problem, redirect, router advertisement, and time exceeded. Finally, custom rules allow you to enter your own firewall rules. These rules are specified using the iptables format, which we will explore in Chapter 12.

## 11.6.2 kdump Configuration Tool

The Kernel Dump Configuration Tool can be used in place of the kdump configuration file kdump.conf. The kdump service is called upon when the kernel crashes so that a record of why the kernel crashed can be saved as a core dump. As with the Firewall, this GUI offers convenient ways to alter kdump's behavior. Figure 11.5 illustrates the main window for the Kernel Dump Configuration Tool. From the main panel, your choices are to enable/

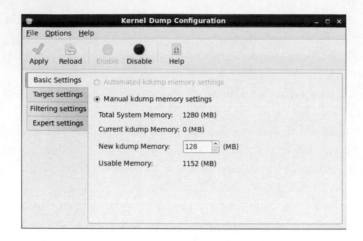

FIGURE 11.5    Kernel dump configuration GUI tool.

disable the kdump service and establish the size of a kernel dump. There are three other settings to select, Target settings, Filtering settings, and Expert settings.

Through the other tabs, you are able to specify other kdump information. In Target settings, you specify the location and filename for a kernel dump, which can be somewhere in the local filesystem, for example, /var/crash, a raw device, or a network location accessed either through NFS or SSH. The Filtering settings tab gives you the ability to establish how much data to dump by specifying a filtering level. Expert settings allow you to change the kernel (default or a custom kernel) and the command line specification for starting the kernel. Figure 11.6 combines these three tabs' windows.

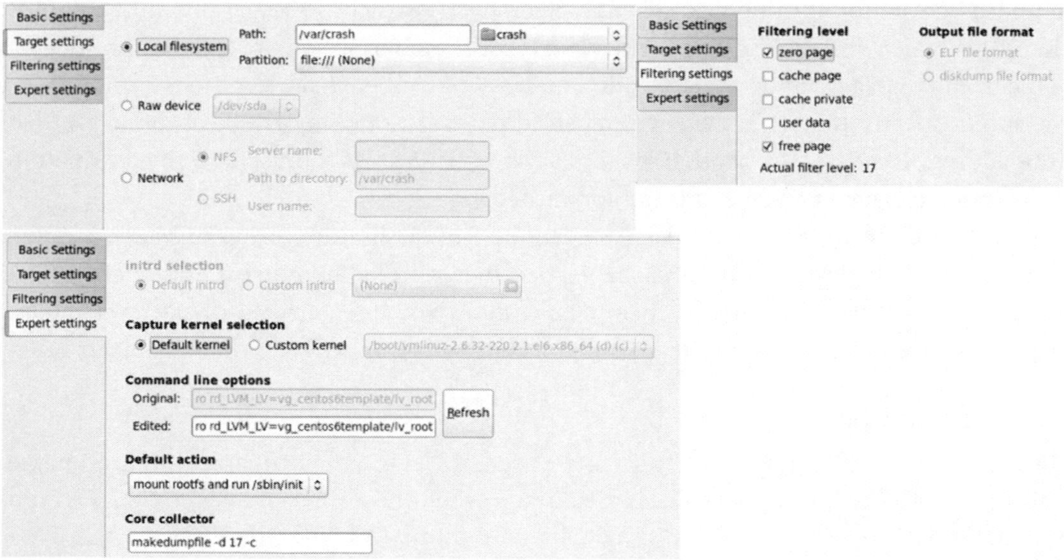

FIGURE 11.6    Other Kdump GUI Settings.

## 11.7 CONFIGURING SERVICES THROUGH CONFIGURATION FILES

We now turn to using configuration files to configure services. This approach requires that you edit the appropriate service configuration file(s). In many cases, there is a single configuration file for the given service. In some cases though, there might be several files such as a configuration file and a rule file. Most of these files are located under /etc, although in many cases, they are in subdirectories such as /etc/sysconfig, or they are located under a directory specific to the service.

The entries in these files differ from service to service but primarily the format is to list directives with values. Directives are options using reserved terms and may appear like assignment statements, such as

```
AUTOCREATE_SERVER_KEYS=YES
```

as found in the sshd configuration file /etc/sysconfig/sshd, or they may appear as directives and arguments (parameters), as in

```
path /var/crash
```

from the kdump configuration file /etc/kdump.conf.

In other cases, the configuration file stores lists of entries such as a list of usernames or a list of IP addresses and IP aliases. The /etc/at.allow file lists usernames of all users who would be allowed to issue at commands. The /etc/hosts file lists IP address and IP alias pairs for IP aliases that can bypass the DNS process by converting the IP alias into its IP address locally.

### 11.7.1 Configuring Syslog

Our examination of configuration files will start with a look at the logging service syslogd. The configuration file is /etc/syslog.conf. In newer versions of Linux, the service is called rsyslogd and the file is rsyslog.conf. The rsyslog.conf file begins with a number of $ModLoad statements to load needed modules for logging actions, followed by any global directives. These statements do not occur in the older syslog.conf file.

The file then consists of logging rules. Rules take on the following format:

```
source.priority    action
```

A *source* is the type of software that might generate an activity to be logged. For instance, mail represents the system mailer, cron is the crontab service, and authpriv represents any authentication software. You can also assign specific application software to logging rules using the reserved sources of local0, local1, local2, ..., local7. An * can be used in place of the source indicating any source. For instance, *.*priority* would indicate any event of the given *priority* level as generated by any system software.

The *priority* dictates which type of event should be logged. There are nine levels of priority: none, debug, info, notice, warning, err, crit, alert, and emerg. The meanings of these

TABLE 11.5   Priority Levels for syslog.conf/rsyslog.conf

| Priority Level | Meaning |
| --- | --- |
| none | No priority |
| debug | Log debugging messages; used by programmers and software testers |
| info | Log informational messages generated by the program to specify what it is doing |
| notice | Log events worth noting such as opening files, writing to disk, mounting attempts |
| warning | Log detected potential problems |
| err | Log errors that arise that do not cause the program to terminate |
| crit | Log errors that arise that will cause the program to terminate |
| alert | Log errors that not only cause the program to terminate but may also cause problems with other running programs |
| emerg | Log errors that could cause the entire OS to crash |

priorities is given in Table 11.5. Each level is progressively more urgent. Using an * for priority indicates that any message sent by the source no matter what level the priority is to be logged.

The *action* is the location of where the log event should be sent. This is typically a file. The log files are text files and in most cases, they will be stored under /var/log. You can replace the log file with /dev/console to direct log messages to the main console window. Alternatively, the use of * for action causes the messages to be redirected to all open terminal windows. This might be useful if you want a specific type of message to be viewed by all users.

The initial entries in the rsyslog.conf/syslog.conf may look like the following list.

```
#kern.*                   /dev/console
*.info;mail.none;authpriv.none;cron.none
                          /var/log/messages
authpriv.*                /var/log/secure
mail.*                    -/var/log/maillog
cron.*                    /var/log/cron
*.emerg                   *
uucp,news.crit            /var/log/spooler
local7.*                  /var/log/boot.log
```

These rules work like a nested if-then-else statement. If any message is informational or if mail, authpriv, or cron have a message whose priority is "none," log those messages to /var/log/messages. Otherwise, if any authpriv message (other than "none") is generated, log those messages to /var/log/secure, and so forth. The hyphen appearing before the log location for mail.* indicates that writing to this file does not need to be synchronized (messages are not necessarily written to the file in the order that they are received). In most cases, we want file access to be synchronized but in the case of this log file, removing the necessity for syncing the file may improve performance. This is an issue only because the maillog may be a very large file. The * used as an action for any emergency message is sent to all users. You should be able to interpret the rest of the rules.

You might notice also that the very first rule above is commented out. The klogd dae-mon is already logging kernel messages for us. By uncommenting this rule, we would also see all such messages sent to the administrator's console.

Now let us imagine as a system administrator, we wanted to add our own log file. In this case, we want to log all messages that originate from any of the services. The source would be daemon. We could add the rule

```
daemon.*                /var/log/daemons
```

Or, if we prefer to only view important messages from our services, we might use daemon.warn so that only warnings and higher-level messages are logged. Now, with our conf file modified, we need to save this file and restart the syslog/rsyslog service. We would issue one of the two instructions:

```
/sbin/service syslog restart
/etc/init.d/syslog restart
```

You would use rsyslog for a newer Linux system.

## 11.7.2 Configuring nfs

The nfs service supports network file sharing by permitting file systems to be remotely accessible over network. Although originally intended for local area network file sharing, file systems can be remotely accessed over any network. We explored this in Chapter 10 when we demonstrated how to set up a file system to be remotely accessible.

The nfs service accesses three separate files. First is /etc/fstab. All file systems listed in /etc/fstab are automatically mounted when the system is booted, or the mount  -a com-mand is issued. To mount a remote file system, the system administrator can either issue the mount command from the command line, or preferably, add the remote file system to the /etc/fstab file.

Second is /etc/exports. Any file system that is to be exported (made available remotely) is listed here. The format here is

```
local_mount_point    network_address(es)(options)
```

as in

```
/home/coolstuff    10.11.0.0/16(ro,sync)
```

In this case, /home/coolstuff can be mounted remotely by any machine whose IP address starts with 10.11. The file system is synchronized and read-only (the file system itself may be writable on the local host but read only for those remotely mounting it).

Third is the nfs configuration file, /etc/nfsmount.conf. This file contains three sec-tions: mount point options, server options, and global options. The mount point options are specific for each mount point. The server options are specific to a single server. The global options are true of all file systems mounted. Options include whether mounting is performed in the background or foreground, the protocol used for accessing the file system (e.g., tcp, udp), block sizes for reads and writes, timeout amount, cache amounts, and default mounting options such as version of nfs.

Altering any of these three files will require restarting the nfs service. One additional instruction is exportfs. This instruction maintains a table of exported file systems via the nfs protocol. This can be useful if you are permitting remote mounting of a file system so that you can reference which file systems need to be exported. By default, this table is stored in `/var/lib/nfs/etab`.

### 11.7.3 Other Service Configuration Examples

Unlike the complexity of either syslog or nfs, the configuration file for atd is simple. Stored as /etc/sysconfig/atd, this file will contain any options that you wish to force atd to use. These are the same options that you can specify when atd is executed. For instance, -l specifies a limiting load factor. By default, atd will only run a waiting batch job when the CPU load drops below 80%. With –l, you are able to alter this behavior. The option –b allows you to specify the minimum amount of time between atd running any two waiting batch jobs. By default, this value is 60 seconds.

As mentioned in Section 11.5.2, the auditd software logging service has no less than three configuration files. The primary configuration file is located under /etc/audit/ `auditd.conf`. This file stores the configuration for how auditd will run. For instance, it specifies the location and name of the log file generated, the format of the log file, the number of log files to retain, the maximum size of a log file before log file rotation should kick in (or some other action, including suspending the daemon until some other action takes place, or send a message to syslog to log a warning), and flushing operations to delete entries from log files. The /etc/audit directory also contains a rules file, `audit.rules`. See Table 11.4 for a description of the rule format. The file /etc/sysconfig/auditd.conf contains directives for extra auditd options. Specifically, this file is used to control how auditd starts and stops as opposed to the configuration of how auditd runs. This file controls the language that auditd uses (defaults to U.S. English) and whether the audit system, including system calls, should be shut down when auditd is not running.

The cups configuration file, /etc/cups/cupsd.conf, contains directives much like those found in the Apache conf file (we explore this in Chapter 15). An examination of this file shows us values such as

- `LogLevel warn`—level for message logging

- `SystemGroup sys root`—user name and group name that this process runs under

- `Listen ...`—IP address(es) and port(s) to listen for messages under

- `BrowseOrder allow,deny`—establish order to read BrowseAll and BrowseDeny

- `BrowseAllow all`—specify who is allowed to share this printer

- `BrowseDeny IPaddr`—specify those locations who are not allowed to access the printer

Many of the directives are placed inside of containers such as <Location>, <Policy>, and <Limit>. We will withhold any further explanation of containers until we examine Apache.

The kdump service can be controlled through the GUI that we explored in Section 11.6. There is also a configuration file, `/etc/kdump.conf`, which can be used to control kdump. The directives found here include the options as found in the GUI. For instance, you can specify the values in the local file system for the path and partition of the generated dump file, or the raw device or network destination for the dump, as found under the Target settings tab. This file though contains additional directives such as the specification of a script or executable program to run immediately prior to and after the dump.

Not all configuration files are coupled with a specific service. Instead, some files store configuration data that may be used by multiple services or the kernel. In these cases, altering the configuration file does not require restarting a service. We see a list below of some of the more prominent files and their roles. We have already explored many of these and will see more of these in Chapter 12.

- /etc/fstab—list of file systems to mount upon system initialization or execution of the command `mount -a`.

- /etc/group, /etc/passwd, /etc/shadow—group account, user account, and user password files. These files are accessed by groupadd, groupdel, groupmod, useradd, userdel, usermod, chage, passwd, and authentication programs.

- /etc/hosts—lists IP alias to IP address mapping to bypass a DNS request.

- /etc/hosts.allow, /etc/hosts.deny—list of IP addresses of machines that are permitted access or not permitted access to this computer.

- /etc/nologin—a text message that, if it exists, prevents nonroot users from logging in. In such a case, this text message is displayed to users. This allows the system administrator to prevent all user logins during maintenance.

- /etc/resolv.conf—lists the IP addresses of the computer's DNS server(s)

There are also many notable user configuration files. These files, stored in the user's home directory (or a subdirectory), are all dot files (start with a dot). These include login and shell starting scripts such as .bashrc, .bash_profile, .cshrc, and .profile. There are also logout scripts such as .bash_logout to specify what happens as a shell exits. Another Bash file is .bash_history, which contains the history list. Other files are software specific such as .emacs and .virc (or .vimrc) to store startup information for emacs and vi, respectively. Similarly, .mail.rc is the user initialization file for the mail program. There are also a number of dot directories storing various software-related initialization and data files such as

.gnome2 for the Gnome display, .mozilla for the Firefox browser, and .openoffice.org for
OpenOffice user preferences.

## 11.8 CHAPTER REVIEW

Concepts and terms introduced in this chapter:

- Booting—the process of starting a computer. The boot process includes running a
  power on self-test, locating bootable devices, loading the operating system kernel,
  and initializing it.

- Boot loader—a program that performs the portion of booting that locates and loads
  the operating system kernel.

- Configuration file—a file of directives or options that define how a service will execute. Changing the configuration file will alter the service's behavior.

- CUPS—common Unix printer system, which is a service controlling access to system
  printers, including the ability to print, track print jobs, cancel print jobs, and alter
  printer configuration information.

- init—the first process run in Linux, its role is to initialize the operating system so that
  it is ready for user interaction.

- Master boot record—a reserved location on the hard disk storing a portion of the
  boot loader so that the boot loader can be found and begin execution during the boot
  process.

- Nonvolatile memory—a form of memory whose contents are retained even without
  power. ROM is a form of nonvolatile memory.

- Ramdisk—using memory to mimic the file system so that the operating system can
  access contents using file commands without the slower interaction with disk files.
  Linux uses ramdisks extensively.

- ROM—read-only memory, made up of nonvolatile memory. This type of memory has
  its contents permanently fixed in place so it can be read from but not written to. The
  primary use of ROM is to store the boot program (or a portion of it).

- Rules file—some services use multiple configuration files, separating directives/
  options from rules that specify the types of tasks the service should handle.

- Runlevel—a number, 0 to 6, that indicates the services that are available (started/
  stopped). The common runlevels are 5 (GUI, multiuser, network) and 3 (text-based,
  multiuser, network). The runlevel is established early in the init process execution
  using the file /etc/inittab.

- Service—an operating system program that responds to service requests from any
  number of sources. Services are background processes, which only execute when called
  upon.

- Upstart—recent versions of Linux have modified the init process so that it now runs in an event-based way so that devices that take longer to respond (or are unresponsive) have no impact on the init process continuing through its tasks.

- Volatile memory—a form of memory that requires a constant power input to retain its contents. Both SRAM (cache, registers) and DRAM (main memory) are forms of volatile memory.

Linux commands covered in this chapter:

- chkconfig—view or alter the runlevels that each service is started or stopped in

- dmesg—display the kernel ring (messages generated during kernel initialization)

- /sbin/init—first process run by the Linux kernel, responsible for bringing the rest of the operating system up to usage after kernel initialization

- /sbin/service—command used to start/stop services

- telinit—command to switch runlevels after init is running

Linux files, scripts, and directories covered in this chapter:

- /etc/inittab—file storing the startup runlevel (along with other behaviors)

- /etc/init.d/—directory storing scripts used to start and stop services

- /etc/init.d/cups—subdirectory storing cups service configuration files and specific printer configuration files

- /etc/init/rcS.conf—script that tests for an emergency situation upon startup and then executes other initialization scripts

- /etc/init/rc.conf—script to provide the runlevel to the rc script

- /etc/rc.d/rc—based on the runlevel, starts and stops services

- /etc/rc.d/rc.sysinit—startup script that initializes hardware, loads modules, mounts partitions, defines environment variables, and other tasks

## REVIEW PROBLEMS

1. What is the difference between RAM and ROM?

2. What is the difference between DRAM and SRAM?

3. Why do we store the boot program in ROM instead of RAM?

4. List several bootable devices (devices from which the operating system can be booted from).

5. Why might a user want to change the order that bootable devices are tested?

6. You want to set up a dual boot computer between Windows 7 and Red Hat Linux. Which boot loader program should you use?

7. You want to set up a dual boot computer between Ubuntu and Red Hat Linux. Which boot loader program could you use?

8. What is the role of initramfs?

9. What does pivot_root do?

10. What does the number 2 mean in the following instruction? What does initdefault mean?

```
id:2:initdefault:
```

11. What is the difference between runlevel 2 and runlevel 3?

12. Why would you never use runlevel 6 in your initdefault statement in inittab?

13. Examine the directory /etc/rc5.d. What types of files are stored here?

14. In the directory /etc/rc2.d, what does the 60 refer to in the file K60nfs?

15. What files in /etc/rc0.d start with the letter S? Why are there so few?

16. Compare the S files in /etc/rc0.d to those in /etc/rc6.d. What are the differences? Explain them.

17. For runlevel 5, you want the services dnsmasq and nfs started and abrt-ccpp stopped. What would you do to accomplish this?

18. When using the program chkconfig, you see the following output for kdump:

```
0: off 1:off 2:off 3:off 4:off 5:off 6:off
```

What can you conclude by this?

19. Match the following activities with the service that handles it:

| | | | |
|---|---|---|---|
| a. | monitors laptop battery fan | i. | postifx |
| b. | maintains website security certificates | ii. | cvs |
| c. | manages multiuser documents | iii. | named |
| d. | monitors for added or removed hardware | iv. | acpi |
| e. | manages software for RAID | v. | certmonger |
| f. | runs the BIND DNS server | vi. | ypbind |
| g. | enables network file system sharing | vii. | mdadm |
| h. | mail service | viii. | nfs |
| i. | name server for NIS networks | ix. | haldaemon |

20. How do the three scheduling services (see Table 11.3) differ?

21. What is the difference between the auditd files auditd.conf and audit.rules?

22. For auditd, write a rule that will log all attempts to

    a. access/etc.

    b. execute the file/usr/bin/passwd.

23. Provide two ways to determine if the service cpuspeed is currently running.

Questions 24–30 pertain to the atd service script (from Section 11.5).

24. In comments under chkconfig, we see 345, 95, and 5. What do these three numbers mean?

25. The function script contains a umask instruction and the definition of a PATH variable. Since umask and PATH already exist for root, why are these needed?

26. When doing /sbin/service atd start, if the function call to start results in an error code of 5, what does this mean?

27. When doing /sbin/service atd start, if the function call to start results in an error code of 6, what does this mean?

28. When doing /sbin/service atd stop, if the function call to stop results in an code of 3, what does this mean?

29. What is the difference between rh_status_q and rh_status?

30. Assume the atd service (the executable) can receive a parameter, -d, to display all waiting at and batch jobs. We want to add display as a legal parameter to atd. Specifically, how would you modify the atd script to accomplish this? *Hint*: you need to modify the case statement and add the code, preferably in a function, to handle the display action.

31. Using the Firewall configuration GUI, how would you set it up so that your firewall accepts SMTP messages?

32. Using the Firewall configuration GUI, how could you add telnet, port 23?

33. Which priority level is more severe (with respect to logging messages), err or crit? Explain.

34. You have modified the file /etc/sysconfig/syslog.conf by adding

```
lpr .* /var/log/printer
```

You have sent several print jobs to the printer and even witnessed some errors, yet nothing has been logged to the log file. Why not?

35. Which files would you work with to configure nfs?

36. What type of information would you place into either of the files /etc/hosts.allow and /etc/hosts.deny?

37. What type of information would you place into either of the files /etc/at.allow and /etc/at.deny?

38. What is the /etc/nologin file used for?

# CHAPTER 12

# Network Configuration

THE LEARNING OUTCOMES OF this chapter are

- To understand the nature of a computer network

- To understand the TCP/IP protocol

- To be able to configure the Linux network through configuration files and services, including the Linux firewall

- To understand the difference between static and dynamic IP addresses and how to use DHCP

- To be able to use the network program ip

- To be able to write scripts that support network-monitoring activities

## 12.1  INTRODUCTION

Without network access, you are limited to the software that comes with your computer and are unable to access any network resources (printers, file servers, web servers, etc.). To establish network access, you need to do the following:

- Have a network interface that connects your computer to your network access point. In Linux, all computers have an interface called lo. lo is your *loopback device*, also known as your local network or local host. Software will sometimes communicate through this device as if the device was sending messages out onto the network, but lo does not actually reach a network. Therefore, you need at least one other interface which can take on many different forms. The most common interface is an Ethernet device. This might be an Ethernet card to connect your computer to a LAN, which is a wired connection, or a wireless card. The Ethernet device, in Linux, is denoted with the name eth. The actual name will be `eth#` where # is a digit, for instance eth0. It is possible that your computer will have multiple Ethernet devices, in which case, the numbers will differ, for instance, by having eth0 and eth1. Other types of interfaces include ppp (point-to-point interface), tr (token ring interface), slip or plip (serial line Internet protocol, parallel line Internet protocol), and sit (simple Internet transitions).

- Run your network services. There are numerous services related to the network. The most important is `network`. Others include the firewall service, interface device services, network authentication services, and network file system services. If you are uninterested in the services offered, these others do not need to be started (although it is important to always run the firewall, we will discuss the firewall in Section 12.6). The network service should automatically be started in run level 3 or 5; so, you should not have to worry about it. We discuss some of these services in Section 12.3.

- Have access to a network-broadcast device. In a network, this would be a physical connection from your computer's network interface device to a hub, switch, or router (most likely a network switch or router). From a home computer, you will probably connect via a wireless card to a MODEM in your household. This wireless connection is similar to your networked computer connecting to a router. In your case, the MODEM will serve as a router within your household. The modern MODEMs permit multiple devices to connect to them in case you want to develop your own home network. Your MODEM then connects to an Internet service provider via the telephone line, coaxial cable, or fiber-optic cable.

In this chapter, we will explore the services, configuration, and data files available in Linux to configure your computer to communicate with the network. We will assume that you will be communicating not just with a local area network but with the Internet through the TCP/IP protocol.

## 12.2 COMPUTER NETWORKS AND TCP/IP

### 12.2.1 Broadcast Devices

Before we begin our examination of Linux network configuration, let us first discuss computer networks. A computer network is a collection of computers and other devices connected together through some medium, for instance, twisted-wire pair or fiber-optic cable. Aside from computers, printers, servers, and so forth, we also need network-broadcast devices. These devices fall into four categories: hubs, switches, routers, and gateways. The intention of these devices is to permit computers to communicate with each other without having to directly connect each computer to every other computer.

In the hub, the devices within some small network (a subnet) are connected to one hub. Any message that reaches the hub is then broadcast to all connected devices. The switch is a more capable device. Like the hub, computers in a subnet connect to it. But the subnet retains device addresses so that any incoming message is directed to the device that matches the message's destination address. These addresses are not IP addresses but instead addresses for the lowest layer of the network (the link layer), for instance MAC (media access control) addresses. The router connects together networks; so, it is sort of like a switch for switches. Messages that arrive at a router are then directed to the proper network or subnet. At this level, addresses are network addresses, for instance, IPv4 addresses. Finally, gateways are used to connect one network to another type of network. You may find a gateway to be used as an organization's point of presence onto the Internet.

Figure 12.1 illustrates the layout of a local area network. In this network, we find two subnetworks, each with their own broadcast switch. Connected to one switch are four desktop computers and a printer. Connected to the other switch are three desktop computers and a printer. The two switches are connected to a router. Each of the resources connects to its broadcast device via a twisted-wire pair cable. Not shown here is the connection of the router to other networks.

### 12.2.2 The TCP/IP Protocol Stack

For Internet communication, all devices must utilize the TCP/IP protocol. This four-level protocol is actually a suite of protocols in which each layer can be implemented by one of the many different specific protocols. What TCP/IP provides is the rules for how communication can take place between resources on the Internet. These rules include how messages are broken into packets, how addressing and error-handling information is added to the packets, how the packets are treated as they move from location to location, how two-way communication is established, and how received packets are pieced together to make a message. Figure 12.2 demonstrates the four layers of the TCP/IP protocol. We briefly examine each layer below.

First, application software takes your message and produces the initial, application-neutral message to be transmitted. The application software utilizes one of the many protocols

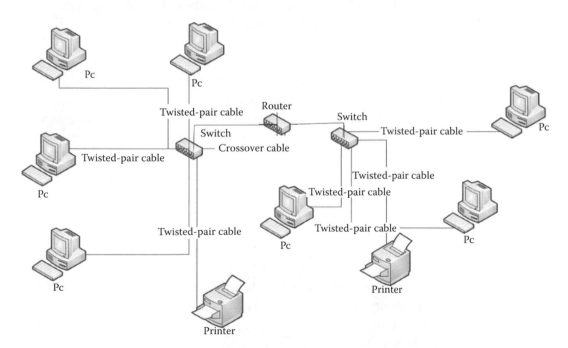

FIGURE 12.1   A local area network. (Adapted from Fox, R. *Information Technology: An Introduction for Today's Digital World*, FL: CRC Press, 2013.)

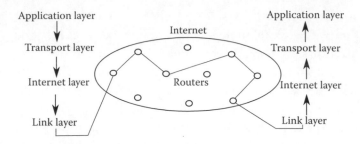

FIGURE 12.2   The TCP/IP protocol stack.

available to implement the application layer. These include DNS, FTP, IMAP, LDAP, MIME, NFS, POP, SSH, SMTP, SNMP, and Telnet. The idea is that no matter what platform of computer is used to create and transmit the message, the recipient need not be using the same platform or even the same application software. Thus, the application layer presents to us a process-to-process communication.

The transport layer provides a uniform interface between two resources whether they are on the same type of network or different type of network. In effect, this layer is able to hide the underlying implementation of the network. The transport layer supports host-to-host communication. This layer is responsible for pulling apart the message as generated by the application layer into distinct packets and supplying information such as error correction details (e.g., a checksum) and a count of the number of packets (e.g., packet 6 of 9) so that the transport layer of the recipient can determine if an error arose and if the message is complete. This layer also has the responsibility of identifying the proper application to receive the message through the message's port number used. We discuss ports in more detail later.

There are several different protocols available in the transport layer. The primary protocol is TCP, which is confusing because both the application and transport layers make up the TCP portion of TCP/IP. Another protocol used at this level is UDP. The primary distinction between these two protocols is TCP's necessity for packet reception. A computer that receives TCP packets is either erroneous or incomplete in that not all packets arrive based on the count (e.g., 8 of the 9 packets arrive) requests to have packets resent. In UDP, safe delivery is not guaranteed. The result is that TCP is slower but safer. UDP facilitates real-time communication (e.g., streaming video) because dropped packets are deemed less important than time while TCP guarantees full delivery. Other protocols available for the transport layer include SCTP, DCCP, and UDP Lite.

The remainder of the TCP/IP protocol stack comprises the IP portion (Internet protocol). It is at these layers that we see the rules of Internet addressing. Its top layer is the Internet layer. At this layer, IP is responsible for addressing packets and sending them off across the Internet. Routers operate at this layer to receive a packet and decide which network or subnet to pass the packet onto.

The IP layer utilizes IP addresses. Internet addressing was originally devised to use single 32-bit binary numbers, each of which is divided into four "octets" of 8 bits each. We are used to seeing this address written in decimal, for instance, as 10.11.12.13. Each octet

TABLE 12.1    Number of IP Addresses Based on Bit Size

| Format | IPv4 | IPv6 |
| --- | --- | --- |
| Bits used for address | 32 | 128 |
| Number of unique addresses | $2^{32} = 4,294,967,296$ | $2^{128}$ that is roughly 340,000,000,000,000,000,000,000,0 00,000,000,000,000 $(3.403*10^{38})$ |

can store a number between 0 and 255. This address is known as the IP version 4 address, or IPv4.

With 32 bits, we have something like 4 billion unique IP addresses. The original IP addressing scheme was selected decades ago when the number of computers that made up the then-Internet was numbered in the hundreds. Shortly thereafter, the number of Internet resources increased to tens of thousands. It was not until recently that the number of resources that need IP addresses have grown to outnumber the 4 billion addresses available through IPv4. To rectify this situation, an addition has been added to IP called IPv6. IPv6 addresses are 128 bits in length, usually stored using 32 hexadecimal numbers. For instance, an IPv6 address might be 518f:a487:aa01:1566:85b3:143a:abcd:8750. Table 12.1 illustrates the difference in the number of available addresses between IPv4 and IPv6. Clearly, the number of IPv6 addresses should last us a very long time!

Aside from IPv4 and IPv6, other protocols exist to implement this layer. These include ICMP and ICMPv6, both of which are used to handle network configuration and network error-checking messages. Both the ping and traceroute programs use ICMP (we discuss these in Section 12.5). IGMP is a protocol for handling multicast broadcasts.

The link layer is the lowest layer in the TCP/IP protocol stack and handles the physical communication across the network media. This layer is responsible for converting messages, stored in binary, into the type of signal that the media will handle (e.g., sound, electrical current, and pulses of light) and converting the received signals back into binary. This may include network tasks such as modulation and multiplexing. This layer may also add start and stop bits to the packet.

This layer handles message addressing not with IP addressing but with MAC addressing. Switches operate at this layer, unlike routers that operate at the Internet layer. This means that the link layer can differentiate between computers on its local subnet but not between resources located outside the subnet, and must rely on a local router to move the message beyond the subnet. This layer also handles transmission error detection, bit synchronization, and dealing with variance in signal strength. Link layer protocols include ARP, Tunnels, PPP, ISDN, DSL, and Ethernet.

To transmit a message, the message is processed going down the TCP/IP protocol stack from application layer to transport layer to Internet layer to link layer. The message is then sent across the network as individual packets. When a packet arrives at an intermediate location, a router operating at the Internet layer, examines the destination address to continue sending the packet on its way. Once it arrives at the destination, the packets are assembled back into a single message, passing the packets up from link layer to Internet layer to transport layer and finally, the message is presented to the application software.

## 12.2.3 Ports

The *port* is a 16-bit number that denotes a mapping of a message (or packet) to its protocol. The protocol in turn tells the computer how to interpret the message. In 16 bits, you can store a number between 0 and 65,535; thus, port numbers range between 0 and 65,535.

Most network applications have a port number designated to it although most of the 65,536 port numbers are not reserved. Table 12.2 provides a listing of some of the most common ports and their uses. These are known as *registered ports*. Notice in some cases, a protocol is mapped to more than one port (e.g., FTP, HTTP). Also listed in this table is whether the port can handle TCP packets, UDP packets, or both. The first 1024 port numbers are often referred to as the well-known ports. There are many registered ports with numbers of 1024 or higher but not all of these ports are currently in use as they have been reserved for future use or are not officially reserved. Port numbers that are not reserved allow future applications to eventually claim them.

TABLE 12.2   Registered Ports

| Port | Packet Type (TCP, UDP) | Usage |
| --- | --- | --- |
| 20 | Both | FTP data |
| 21 | TCP | FTP control |
| 22 | Both | ssh (also SCP, SFTP) |
| 23 | Both | Telnet |
| 25 | TCP | SMTP |
| 43 | TCP | WHOIS |
| 53 | Both | DNS |
| 57 | TCP | Mail transfer protocol |
| 67 | UDP | Bootstrap Protocol (used by DHCP) |
| 68 | UDP | Bootstrap Protocol (used by DHCP) |
| 70 | TCP | Gopher |
| 80 | TCP | HTTP |
| 109, 110 | TCP | POP2, POP3 |
| 118 | Both | SQL |
| 123 | UDP | Network time protocol |
| 161 | UDP | SNMP |
| 194 | Both | IRC (Internet relay chat) |
| 443 | TCP | HTTPS |
| 514 | UDP | Syslog (Linux system logging) |
| 530 | Both | RPC |
| 636 | Both | LDAP |
| 989 | Both | FTPS data (FTP over TLS/SSL) |
| 990 | Both | FTPS control (FTP over TLS/SSL) |
| 992 | Both | Telnet over TLS/SSL |
| 2049 | Both | NFS |
| 3128 | TCP | Squid proxy |
| 6660–6669 | TCP | IRC |
| 6888–6900 | Both | BitTorrent |
| 8008, 8080, and 8090 | TCP | Alternate for HTTP |

There are also many unofficially reserved ports, primarily used by specific software. For instance, 531 is unofficially reserved for the AOL (America Online) Instant Messenger and 843 is for Adobe Flash software communication. Other unofficially reserved ports exist for VMware, Oracle, Cisco, Novell, Symantec, multiplayer computer games, and so forth.

While the port itself is not an address, it is used with an IP address to form a more specific destination. Together, these two pieces of information can be used by a firewall to safeguard a computer from messages that should be discarded rather than processed. We explore the Linux firewall in Section 12.6.

## 12.2.4 IPv6

As described earlier and in Table 12.1, IPv4 addressing only affords about 4 billion unique addresses. With the great success of mobile devices, the number of unique IP addresses needed at any time now exceeds 4 billion. In fact, IPv4 addresses to be awarded by the Internet-Assigned Numbers Authority (IANA) to the Asia-Pacific Network Information Center (APNIC) were exhausted by 2011 and other areas of the world have similarly run out of allocatable IP addresses. Thus, we need to shift to a different form of addressing that provides a greater number of unique addresses. This is the primary motivation for the creation of IPv6.

The main distinction between IPv4 and IPv6 is the size of the address. Rather than a 32-bit number, the IPv6 address is 128 bits. It is also expressed in hexadecimal notation rather than decimal, as a sequence of eight groups of four hexadecimal digits. Each group is separated by a colon (:).

For shorthand, leading 0s of a group can be omitted. For example, an IPv6 address might be 1234:5678:90ab:cdef:0012:0034:0000:5678 that could also be expressed as 1234:5678:90ab:cdef:12:34::5678. The notation :: between the last two groups in the shorthand version denotes a group that is entirely composed of zeroes.

If multiple groups contain zeroes, we replace all of them with a single :: notation. For instance, 1234:5678:9a00:0000:0000:0000:00bc:def0 could appear as 1234:5678:9a00::bc:def0. Counting from the left, we find three groups followed by :: followed by two groups. We are missing three groups, which appear after the third group.

As we need to replace IPv4 with IPv6, we need to modify how the Internet works. This requires modifying both operating systems in their implementation of TCP/IP and hardware such as routers. The first part of this replacement process has been completed in that nearly all modern operating systems are capable of handling IPv6. But there are still gaps. Not all networks have been modified (e.g., software running on routers, physical devices) to handle IPv6.

Until all aspects of the Internet are capable, we have a mixed set of requirements. And because of additions made in the IPv6 protocol (such as a different header), the two protocols are not interoperable meaning that we need two distinct mechanisms available to handle the two protocols. Some of the current approaches include a dual-stack implementation of TCP/IP where there are two pathways that a message can take through the TCP/IP protocol depending on whether the particular packet in question uses IPv4 or IPv6, and tunneling where the IPv6 mechanisms are hidden from the network through the tunnel. Thus, it falls on the devices that create the tunnel to understand how to handle IPv6.

IPv6 is significant for more reasons than just (greatly) enlarging the address space. One area that IPv6 tackles is of security implemented within the network itself. TCP/IP with IPv4 lacks any built-in network-based security. Therefore, any use of encryption in our communication must be handled through additional protocols (see Chapter 5 for details on encryption). IPv6 implements the Internet protocol security (IPsec). Interestingly, the demand for such security was so great that it has been implemented into many IPv4 networks.

Another feature of IPv6 is the ability for a host to automatically configure itself with respect to addressing and locating its router/gateway. Also, IPv6 headers can have optional components. The required portion is simplified over the IPv4 header by discarding seldom used parts and requiring a fairly minimal-sized description of the packet. However, the optional portion, or extension, can be used to specify information on security or other options such as size. The size of packets was also altered for IPv6. In IPv4, packets are limited to 64 K octets. With IPv6, a packet can be expanded to include what is known as a *jumbogram*. The jumbogram can contain more than 4 billion octets.

All in all, the design of IPv6 is well thought out, having been engineered over a period of years starting in 1998. This is not the case with IPv4 that was designed for an incarnation of the Internet preceding its popularity and in fact preceding the popularity of personal computers. But until all networks on the Internet are compliant with IPv6, we will continue to see a large number of IPv4 users. According to www.worldipv6launch.org as of October 2013, only about 12% of Internet communication with popular websites use IPv6 while other estimates state that only about 16% of all Internet networks support IPv6.

### 12.2.5 Domains, the DNS, and IP Aliases

Now that we have covered some details of TCP/IP, let us examine how the Internet itself works. You want to send a message to another computer. This message might be an HTTP request generated by your web browser in response to clicking on a link, it might be an email message sent from your email program, it might be a ping request to see if a given network resource is responding, or it might be an ssh command so that you can remotely log into another computer. Most likely, all you know of the remote computer (the destination) is its IP alias. This alias must first be translated into an IP address through address resolution.

Your local site probably has a DNS server. The DNS is set up to perform alias to address translation for you. If the destination is not part of your local domain, it is likely that your DNS server does not have that information locally. If it does, it can look up the IP address of the requested resource and respond with that address. Otherwise, it kicks the request up to another DNS server located elsewhere on the Internet. Eventually, the request will be kicked up to a DNS server that knows the domain of the destination computer. Your computer will either receive an IP address of the domain or a message that the alias could not be resolved. Given the address, your computer can now add this to your message and send the message out onto the Internet. If no DNS server knows of the domain specified, you will be given an error.

What do we mean by a domain? The Internet is divided into top-level domains such as .com, .edu, .gov, and .net. Within each of these general domains, specific domains are established such as amazon.com or nku.edu.

DNS name servers generally come in two forms, *authorities* and *caches*. An authority is responsible for its own domain. This means that for each domain, there will be at least one authoritative DNS server that contains information about that domain. Other DNS servers can cache information about domains outside of their own. Thus, your local DNS server may know how to find a domain such as nku.edu. If not, then it will know of a DNS server to communicate with one that will either know of the domain or know of another DNS server to hand the request onto. Typically, a specific resource's address will only be stored in an authoritative DNS server's table. But the domain information itself should be known elsewhere.

As an example, you want to reach www.nku.edu. The domain is nku.edu. While your local DNS server may not know nku.edu, it knows of a DNS server that does. Your request goes to that DNS server that responds with the IP address for the DNS server for nku.edu. Now, your request for www.nku.edu goes to that DNS server that knows about the specific machine www. In its DNS table is a mapping of www to the IP address. This address is sent back to your computer and finally, you have the IP address needed to communicate with the computer www.nku.edu. While this seems complicated, the actual amount of time that this operation might take will typically be a few seconds or less, probably less than a second. Further, this entire interaction happens transparently for the user.

To establish proper communication, you will have to set up a table that indicates the location of your DNS server(s). In Linux, these IP addresses are placed in the file `/etc/resolv.conf`. We will look at setting up a DNS server in Linux using the bind program in Chapter 15 (available at http://www.crcpress.com/product/isbn/9781482235890).

If there are machines whose IP addresses are static and which you communicate with often, you may wish to bypass the entire DNS address-translation process. In Linux, you can set up your own mapping information in the file `/etc/hosts`. Before any name resolution is attempted, Linux first examines the hosts file to see if there any entries that match the request. Entries in this file are denoted as ip address followed by host name(s).

Let us consider a local server with a static IP address called `ourserver.internalnet.com`. The organization has aliased this machine to the name `internalserver.com`. As we might contact this server frequently and since it has an IP address, this server seems a useful target for inclusion in the /etc/hosts file to reduce the amount of traffic for our DNS servers. If the IP address is 10.11.12.13, we could add this entry to /etc/hosts.

```
10.11.12.13 ourserver.internalnet.com internalserver.com
```

What else do you have to do to establish how your computer communicates on a network? First, your computer needs an IP address. This is discussed in Section 12.4. Second, you need for your computer to respond to incoming messages. The network service needs to be running, and this is discussed in Section 12.3. You may also wish to protect your computer so that incoming messages are scrutinized first. We use a firewall for this. The Linux firewall is discussed in Section 12.6. Third, you need to establish your interface(s) to the network. We also discuss this in Section 12.3.

## 12.3 NETWORK SERVICES AND FILES

The primary network services are all controlled by script files stored under /etc/init.d as are most of the important services in Linux (refer back to Chapter 11). The services themselves are typically found under /usr/sbin. Many of the network services can be thought of as umbrella services in that other services call upon them to help fulfill their tasks. Other supporting scripts for network services are found under /etc/sysconfig and /etc/sysconfig/network-scripts.

### 12.3.1 The Network Service

The network service is used to start or stop network communication. If this service is not running, then you are unable to communicate over the network. Even lo (the loopback device over 127.0.0.1) is unreachable. Unless you wish your computer to serve as a stand-alone machine, you will want your network service running in all situations except when you are working on the network (changing interface devices or altering some network configuration). All other services related to the network rely on network running. Thus, if you stop network, services such as netconsole, httpd, iptables, autofs, ntpd, and dnsmasq are useless even if they are running.

When you start the network service through the controlling script (from /etc/init.d), it first executes the script `/etc/init.d/functions` which contains numerous script functions that support other scripts. It then executes the script `/etc/sysconfig/network` to establish two environment variables: `NETWORKING=yes` and `HOSTNAME=`*hostname*. This script might also establish values for variables such as GATEWAY, GATEWAYDEV, and NOZEROCONF that store the IP address of the network's gateway device, the type of device, and whether zero-configuration networking* is available or not. The network script makes sure the program /sbin/ip exists because, without this, the computer is unable to assign IP addresses to the interface device(s).

Next, the network service's script executes the script `/etc/sysconfig/network-scripts/network-functions`. The functions defined here perform three types of tasks. First, they query devices for status such as whether an interface is up or down, what the MAC address is of a given interface, or whether there is a wireless device available. Second, they establish values such as hostnames, IP addresses, and add to default routes. Third, they can locate the gateway of the local network.

The network service script concludes by establishing all available interface devices (other than lo). This includes establishing information about each device such as the device's name and manner by which an IP address is obtained. We describe this in more detail in the next section.

At this point, your interfaces should be up and available along with IP addresses as obtained by a DHCP server or established as static (see Section 12.4). Most of these activities are handled through scripts in /etc/sysconfig/network-scripts. These include ifup-eth,

---

* Zero-configuration networking permits computers on a network to establish communication automatically without the need for an administrator to set up the necessary services. For instance, in a network with zero configuration, computers should be able to automatically obtain an IP address and the location of their local DNS servers. Avahi-daemon, discussed later, is an example of a zero-configuration service.

ifup-ipv6, ifup-ppp, ifup-routes, and init.ipv6-global. The ifup scripts start the various device interfaces, obtain IPv4 and IPv6 addresses, and establish data to be placed into the ifcfg configuration files (e.g., ifcfg-eth0 and ifcfg-lo).

## 12.3.2 The /etc/sysconfig/network-scripts Directory's Contents

Let us explore the /etc/sysconfig/network-scripts directory in a little more detail. The listing is shown in Figure 12.3. There are four types of files here. First, there are two data files: `ifcfg-eth0` and `ifcfg-lo`. These files contain configuration information about this computer's two interface devices, eth0 and lo. Next, there are several executable scripts. These are all named `ifdown-device` and `ifup-device` as in `ifdown-eth` and `ifup-ippp`. There are two such scripts, `ifdown-isdn` and `ifup-isdn`, which are actually symbolic links to `ifdown-ippp` and `ifup-ippp`, respectively. The files `ifdown` and `ifup` are also symbolic links, but in this case to services located in /sbin: `/sbin/ifdown` and `/sbin/ifup`. Finally, the files `network-functions` and `network-functions-ipv6` are script files containing network script functions, as mentioned earlier.

Most of the scripts in this directory are of the form `ifdown` or `ifup`. As you might expect, `ifdown` is used to bring an interface down while `ifup` is used to start an interface. The scripts ifdown and ifup expect an interface as an argument as in `ifdown eth0`. The remainder of these scripts are named after specific interfaces, so that, for instance, `ifdown eth0` and `ifdown-eth` will accomplish the same thing. The interfaces covered by these scripts are bnep, eth, ippp, ipv6, isdn, post, ppp, routes, sit, and tunnel. The ifup version also includes scripts for aliases, plip, plusb, and wireless. We will not explore the content of these `ifup` and `ifdown` scripts as they are very involved. If you are interested in studying these devices, you should explore a text on Linux networking.

Let us instead focus on one of the two configuration files, `ifcfg-eth0`. You will have a configuration file for each network interface device. In the computer illustrated in Figure 12.3, there are only two interface devices, eth0 and lo. The ifcfg-eth0 file contains a listing of a number of environment variables as used by the networking services. These values are established at the time the network is brought up (or when ifup-eth0 is called upon to bring up the Ethernet device).

Some of the most important variables and their content are shown in Table 12.3 along with the meaning and range of values permissible. Of particular note are BOOTPROTO, BROADCAST, GATEWAY, HWADDR, and ONBOOT. These five values specify the mechanism by which this device will be given an IP address (statically assigned, assigned by DHCP, or none), the IP address of this device's broadcast device, the IP address of this

```
ifcfg-eth0     ifdown-isdn      ifup-aliases   ifup-plusb     init.ipv6-global
ifcfg-lo       ifdown-post      ifup-bnep      ifup-post      net.hotplug
ifdown         ifdown-ppp       ifup-eth       ifup-ppp       network-functions
ifdown-bnep    ifdown-routes    ifup-ippp      ifup-routes    network-functions-ipv6
ifdown-eth     ifdown-sit       ifup-ipv6      ifup-sit
ifdown-ippp    ifdown-tunnel    ifup-isdn      ifup-tunnel
ifdown-ipv6    ifup             ifup-plip      ifup-wireless
```

FIGURE 12.3  Contents of the network-scripts directory.

TABLE 12.3 Contents of eth0 Configuration File

| Variable | Range/Type of Value | Meaning |
|---|---|---|
| BOOTPROTO | "static," "dhcp," "none" | Source of the IP address (static or via DHCP server or none at all) |
| BROADCAST | IP address | Broadcast device's address (typically you will use this variable or GATEWAY but not both) |
| DEVICE | Alphanumeric | Device's name (e.g., eth0, ippp, and lo) |
| DHCP_HOSTNAME | IP alias | Name of DHCP server |
| DHCP_TIMEOUT | Integer | Number of seconds before timing out when waiting for DHCP server to respond |
| GATEWAY | IP address | IP address of subnet router/gateway |
| HWADDR | Hexadecimal address | MAC address of device |
| IPADDR | IP address | Set by system administrator for static IP |
| IPV6INIT | Yes, no | Initialize IPv6 address by default |
| NAME | Alphanumeric | Name of device, for example, Ethernet, loopback |
| NETMASK | Subnet mask | The mask used to obtain the local network portion of the IP address, for example, 255.248.0.0 |
| NETWORK | Network address | IP address of the local network |
| NM_CONTROLLED | Yes, no | Whether the device is controlled by a network manager program |
| ONBOOT | Yes, no | Whether to start this interface upon boot or have it manually started |
| TYPE | Alphanumeric | Type of device, for example, Ethernet, ppp |
| USERCTL | Yes, no | Is the user allowed to control this device? |
| UUID | Hexadecimal address | Address of physical device |

device's gateway device, the MAC address of this device, and whether the device should be started upon boot.

Notice that BROADCAST and GATEWAY are IP addresses of the network/subnetwork's broadcast device and gateway device, respectively. These variables are typically not assigned if BOOTPROTO is dhcp. If BOOTPROTO is static, you should have one but not necessarily both values assigned. Also, if BOOTPROTO is dhcp, you would not have an entry for IPADDR; instead, this is filled in only if you want to statically assign an IP address to your interface.

Let us consider the role of the value NETMASK. A broadcast device may need to determine a network address from an IP address. This is performed by ANDing the IP address to the network's netmask. Thus, we store this value in the variable NETMASK. Let us look at a couple of examples.

If the netmask is 255.255.240.0 and the IP address is 10.11.12.13, then we obtain the local network's address as 255.255.240.0 AND 10.11.12.13. First, we have to convert these two sets of octets from decimal to binary.

```
255.255.192.0 = 11111111.11111111.11110000.00000000
10.11.12.13 = 00001010.00001011.00001100.00001101
```

We now apply AND in a bitwise manner, that is, column by column. A binary AND operation results in 1 only if both bits are 1; otherwise, the result is 0. The first 20 bits of the netmask are 1. When ANDing any bit by 1, we get that bit. So, the first 20 bits of the network address will be the first 20 bits of the IP address. The last 12 bits of the netmask are 0; so, the result of the AND operation will also be 0 for the last 12 bits.

Our result is as follows:

```
        11111111.11111111.11110000.00000000
AND     00001010.00001011.00001100.00001101
        00001010.00001011.00000000.00000000
```

This gives us a network address of 00001010.00001011.00000000.00000000, or 10.11.0.0.

This particular example perhaps is not particularly illustrative of the application because it appears that we are just dropping the last two octets of the IP address, converting 10.11.12.13 into 10.11.0.0. This is not always going to be the case depending on the netmask and the IP address. Using the same netmask, what would be the result if our IP address was 128.58.221.39?

```
        11111111.11111111.11110000.00000000
AND     10000000.00111010.11011101.00100111
        10000000.00111010.11010000.00000000
```

In this case, the network address is 10000000.00111010.11010000.00000000, or 128.58.208.0.

### 12.3.3 Other Network Services

Now that we have explored the network service, let us turn our attention to some of the other network services. First, we look at snmpd. This is the SNM (simple network management) protocol daemon. The role of this service is to listen for SNMP messages and respond to them. The incoming packets are requests for information from a remote device and commands to alter internal settings. Snmpd is primarily used by system administrators to control other network devices, such as servers and routers, across the network. A related service is snmptrapd, used to start the SNMP trap daemon that performs logging and communication with the operating system for SNMP messages. These services use two configuration files: /etc/snmp/snmpd.conf and /etc/snmp/snmptrapd.conf.

By default, most network applications use specific ports. For instance, ssh is tied to port 22. The portreserve service does much as the title suggests; it reserves a port for a given application (protocol) while the application is communicating over the network. Additionally, portreserve prevents other programs from utilizing a port that should be reserved for a specific application. Once the application is done with the port, portrelease can be used to release the port so that other applications can use it. To control these services, there are controlling scripts /etc/init.d.

Avahi performs "service discovery" across a network. In essence, through Avahi, your computer can locate and utilize network services available to clients on your local area network. These include print services and the location of available printers and file services and location of file servers. Avahi is a zero-configuration service in that it can run without user intervention. As with many of the other services, `avahi-daemon` is stored under /usr/sbin with a controlling script in /etc/init.d. Avahi has a configuration file and other supporting files located under the `/etc/avahi` subdirectory.

The `rdisc` service locates your subnet's router. It does this using the ICMP router discovery protocol. Once the router has been identified, this service modifies your computer's router tables to indicate default routes.

The `dnsmasq` service starts a DNS caching server. Recall from Section 12.2 that most Internet communications use IP aliases instead of IP addresses. We need to perform an address translation from alias to address and we typically use a DNS server for this. The dnsmasq program can operate as a DNS server for a small network, or merely as a cache of previous requests of address translations. It can also serve as a DHCP server (see Section 12.4). The dnsmasq service is stored in /usr/sbin with the controlling script to start and stop the service in /etc/init.d. The configuration file for dnsmasq is `/etc/dnsmasq. conf`. A standard CentOS installation provides a version of this configuration file where all directives are commented out so that you would have to edit this file to alter dnsmasq's behavior. There are a number of other supporting files and directories under /var and /etc/sysconfig.

The `postfix` service controls email. Postfix itself is quite complex and beyond the scope of this chapter. Briefly, however, the `master` daemon runs the postfix daemon that itself calls upon the `sendmail` program. Postfix has two configuration files: `/etc/post-fix/main.cf` and `/etc/postfix/master.cf`, and several data files including `/etc/postfix/access`, `/etc/postfix/canonical`, and `/etc/postfix/transport`. Both the postfix and sendmail programs are located under /usr/sbin but while postfix has a controlling script in /etc/init.d, sendmail does not have a controlling script.

`httpd` is the Apache web server. If you have installed Apache through yum or the Add/Remove software, you will find it under /usr/sbin with the controlling script in /etc/init.d. This default installation places the configuration file in `/etc/httpd/conf/httpd.conf` and the web space along with other supporting files in directories under /var/www. If you were to perform your own installation from open source, you can control the placement of these files under one directory such as /usr/local/apache2. In such a case, you would find the controlling script to be called `apachectl` rather than `httpd`. We will explore Apache installation and configuration in detail in Chapter 15 (available at http://www.crcpress.com/product/isbn/9781482235890).

The script `/etc/init.d/sshd` starts and stops the ssh daemon, `sshd`. Without sshd running, access to your computer via ssh is not available. Ssh provides secure access by encrypting any communication over the network. For ssh to work, your system must be configured to handle encryption.

There are several different configuration files that ssh will utilize. These include

- /etc/ssh/ssh_random_seed
- /etc/ssh/sshd_config

- /etc/ssh/ssh_config

- /etc/ssh/ssh_hot_key

- /etc/ssh/ssh_host_key.pub

In addition, the firewall must be set up to permit ssh messages (this is the case by default).

We already discussed nfs in Chapter 10. This service provides the functionality to offer local file systems as targets to be mounted remotely across a network. In addition, to remotely mount a file system, you need netfs. An additional related service is nfslock. This service performs file locking on networked file systems. File locking ensures that if a file is open by one process (by one user), it is inaccessible to other users, unless the file is opened as a read-only file.

The next service is netconsole, a module to provide remote access via a terminal window. Its configuration file is /etc/sysconfig/netconsole that gives you the ability to define the port number for a network console, the interface device, an IP address for a remote syslog server, and a port to listen for the remote syslog daemon.

The service certmonger provides digital certificate monitoring. These certificates are used to both provide encrypted communication with web servers and ensure that the web servers are legitimate. The certmonger service monitors already-established and downloaded certificates to see if any certificate has expired and if so, attempt to obtain a newer certificate. The certmonger program is located under /usr/sbin and there may be configuration information under /etc/sysconfig/certmonger. Additionally, certmonger maintains data files under both /var/run and /var/lock. Figure 12.4 contains an expired certificate (look at the "Expires On" field and you will see that it expired in 2011). It would be certmonger's job to update this certificate the next time it is called upon.

Two related services are ntpd and ntpdate. These services obtain the time and date from a remote NTP server to modify the internal clock. While ntpdate will modify both time and date, ntpd will only modify time.

## 12.3.4 The xinetd Service

One last service of note is xinetd (or the older inetd). This service is known as a *super-server* in that it is capable of controlling multiple running services. In this case, the super-server listens to the various ports for incoming messages and then invokes the service that corresponds to the port. For a system that might expect to receive any number of network communications, having xinetd running may be preferable than keeping a number of network-oriented services running all the time. On the other hand, a dedicated server would run the appropriate service(s) instead. For instance, for a web server running Apache, we would expect the Apache web server program (httpd) to be running all the time rather than having it started and stopped by xinetd on demand.

The xinetd service will be invoked when an incoming message arrives. It will first utilize the file /etc/services, which provides a mapping of services to ports. For instance, if a message arrives over port 22, it will be mapped to ssh. With this information, xinetd will

FIGURE 12.4    An expired certificate.

examine its own configuration for how to respond to an ssh message (presumably resulting in the starting of the sshd service).

The xinetd service is configured in two ways. First, it has its own configuration file, /etc/ xinetd.conf. The configuration file will consist of default directives. An example for the defaults is shown below.

```
defaults {
        instances           = 50
        log_type            = RSYSLOG authpriv
        log_on_success      = PID HOST DURATION EXIT
        log_on_failure      = HOST
        cps                 = 25 10
        umask               = 002
}
includedir /etc/xinetd.d
```

Instances provide a maximum number of simultaneous requests that can be handled. The logging directives specify who is responsible for logging and what to log on a successful versus failed connection attempt. In this case, rsyslog is asked to perform logging using authpriv as the source where a successful access will log the PID, host IP address, duration, and exit status of the communication while the failed attempt will log the host IP address. The directive cps establishes the number of connections per second of any given service and the amount of time that a service must wait before it can be restarted, respectively. The directive umask establishes the umask value for the service.

The includedir directive establishes the directory that will store various services' configuration files. Each service configuration file contains the options for the given service. For instance, the rsync service configuration file might look like the following:

```
service rsync
{
        disable             = yes
        flags               = IPv6
        socket_type         = stream
        wait                = no
        user                = root
        server              = /usr/bin/rsync
        server_args         = --daemon
        log_on_failure     += USERID
}
```

In this example, we see that the rsync service, by default, will be disabled (not running). Its flags are only IPv6. The type of communication is a stream (as opposed to dgram, raw, or seqpacket). The directive wait indicates that this service should run as a single thread so that multiple requests cannot be handled in parallel but instead, any successive requests for rsync must wait until the current request has terminated. It runs under the user root and the executable of the program is at /usr/sbin/sync that runs with only one option, --daemon. Finally, the += notation for log_on_failure indicates that USERID should be added to any log_on_failure options provided in the xinetd.conf file.

In addition, you can add directives for protocol, group (like user), nice to specify a priority value for the service, only_from to provide a list of remote hosts from which an incoming message will invoke the service (any other incoming messages will not be handled by xinetd), no_access to provide a list of remote hosts that this service will not handle (the opposite of only_from, you would only use one of these two directives), and access_time to control when the service can be invoked. You can also override any of the defaults, for instance, by including instance, cps, log_type, and umask in the specific service's file. There are also about a dozen flags available aside from IPv6.

The xinetd service is preferred over inetd because inetd has greater limitations. For one, xinetd has security-oriented options such as the only_from, no_access, and time directives. There are also more logging options available through xinetd. Finally, in inetd, all the specific services that inetd could handle would be enumerated in its configuration file, line by line. The use of individual service configuration files in xinetd is considered cleaner and easier to maintain.

### 12.3.5 Two /etc Network Files

The /etc directory stores other files that impact the network that we need to explore. First, is the file /etc/hosts. Through this file, the system administrator can establish IP alias to IP address mappings that can bypass a DNS lookup. The advantage in doing this is to save time. However, given that IP addresses can change over time, the risk you face is that an IP address entered into the `/etc/hosts` file may become out of date and lead to errors in attempting to resolve a mapping request. Therefore, you should only place static IP addresses in this file and only those that you are sure will not change, or are static IP addresses that you control (so that if you change the static addresses, you can modify the hosts table).

The syntax for the host table is a series of rows, each specifying a mapping. The entries are in this form:

```
ip_address ip_alias1 ip_alias2 ...
```

As you can see, multiple IP aliases can share the same IP address. For instance, we might find a computer like www.nku.edu has a real (canonical) name of machine1.nku.edu and may also be known as ns1.nku.edu. If that machine's IP address was 10.11.12.13, the entry might read

```
10.11.12.13    www.nku.edu machine1.nku.edu    ns1.nku.edu
```

The order of the aliases is unimportant, but the entries are scanned in order; so, we would probably want to list them in decreasing order of usage so that the most common name came first.

The hosts table can also store other information. First, we can establish which should be examined first in an IP address resolution: `hosts` or `bind`. Hosts refer to accessing this file while bind is a DNS server program for Linux. We examine bind in detail in Chapter 15 (available at http://www.crcpress.com/product/isbn/9781482235890). If we want to make sure the `/etc/hosts` table is consulted before a name server, we would use

```
order hosts, bind
```

The order command can also include nis to indicate that name resolution takes place through NIS (the network information service).

The other file to explore is `/etc/resolv.conf`. This file is used to specify the location(s) of your IP alias name resolver, that is, your DNS server. If properly configured, this file will be automatically filled with the proper name server addresses upon starting your network service. This is accomplished via zero configuration. Otherwise, you might be responsible for setting up this file yourself.

The primary entry for this file is the list of IP addresses for the local DNS server(s). Any such entry is placed after the directive `nameserver`. Other common directives are `search` and `domain`. Only nameserver is required. The search and domain entries can be used to establish the search domain name and the domain name of your computer. The search domain name, by default, is the same as the domain name. However, you can add to the search entry other domain names to use when the current domain is one of several subdomains. Other directives include `sortlist` to establish an ordering behind IP addresses that are returned by the C function `gethostbyname`, and `options` to alter default values of variables used by the various C functions that implement name resolution (e.g., `res_init`, `res_query`, and `res_search`). As an example of a resolv.conf file, we might see the following entries:

```
domain somedomain.somecompany.com
search somedomain.somecompany.com
nameserver 10.11.12.13
nameserver 10.11.12.14
nameserver 172.15.183.1
```

Here, we see that this machine is part of somedomain.somecompany.com and has three local name servers to call upon.

## 12.4 OBTAINING IP ADDRESSES

IP addresses can be assigned in two ways: static IP addresses assigned once by an administrator (and seldom changed once established), and dynamic IP addresses typically assigned whenever the computer is booted/rebooted. Static IP addresses are permanent, or persistent. Changing one requires modifying DNS tables. Today, most IP addresses are generated dynamically. For this, you need access to a DHCP. The DHCP server will need to know the range of available IP addresses that it can hand out. This will vary from organization to organization as IP addresses are provided to each organization by an IP authority. In this section, we look at setting up static IP addresses, obtaining IP addresses from a DHCP server, and setting up a DHCP server in Linux. We start with the easy case, static IP addresses.

### 12.4.1 Static IP Addresses

To establish a static IP address, you must first have an address available. For an organization, you will be allotted some number of IP addresses for your resources. Someone then has the task of assigning the addresses to the individual machines that will have static addresses. Once assigned a static IP address for your computer, you would perform the following activities (for Red Hat Linux).

First, you would edit the file `ifcfg-eth0` under `/etc/sysconfig/network-scripts` (see Section 12.3). In this file, you will place the IP address in the variable `IPADDR` and assign `BOOTPROTO` to "static." The former establishes the IP address and the latter indicates that the computer's address is static and not dynamic. Additionally, you would assign `HOSTNAME` the value of the machine's hostname, which would then be used to form the machine's IP alias. For instance, if the hostname is Machine1 and it is located within the domain somecompany.com, the full IP alias becomes Machine1.somecompany.com. You would also supply the netmask in the variable `NETMASK`.

In addition to modifying the ifcfg file for your interface device(s), you must modify the file /etc/sysconfig/network. Specifically, you want to include both the host name and the IP address of the machine's *gateway*. The gateway is your local network's connection to other networks. The gateway is the destination for messages from your computer that are to reach other networks. You would assign these two values to the variables HOSTNAME and GATEWAY, respectively.

Now that your computer knows its own IP address and IP alias, we now must alert other computers of your address. Recall from Section 12.2 that DNS servers are used to perform IP alias to IP address mappings. Your own local organization or your Internet service provider will store its own DNS server. This server primarily maps IP aliases to internal computers. That is, your local DNS is an authority of the local domain, knowing for each specific internal host with a static IP address what that address is or where to go to retrieve that information. This DNS server's table must be modified to include an entry for your computer. Once both your computer and your DNS server have been provided the IP address, you are ready to go. Just restart the network service on your computer and you will find that your IP address has been assigned (e.g., use ip addr or ifconfig).

If you wish to change static IP addresses, you would modify the ifcfg file to update the IPADDR and possibly NETMASK entries. You would also have to notify your local DNS server(s) of the modification. And, of course, you would have to restart your network service to use the new IP address. If you do not modify the DNS server, you may find that responses to your outgoing messages do not make it back to your computer.

If you wish to establish a static IP address for a Ubuntu machine, the process is slightly different. Rather than editing the ifcfg file(s), you would instead modify /etc/network/interfaces by specifying the IP address, network address, netmask, broadcast address, and gateway address. What follows is an example:

```
iface eth0 inet static
        address 10.11.12.13
        network 10.11.0.0
        netmask 255.255.192.0
        broadcast 10.11.51.1
        gateway 172.83.11.253
```

## 12.4.2 Dynamic IP Addresses

DHCP was developed to replace the outdated Bootstrap Protocol from the early 1990s. The Bootstrap Protocol was set up to supply clients with IP addresses upon boot, but did not have the functionality that DHCP provides. With DHCP, not only can you obtain dynamic IP addresses but DHCP will also respond with other configuration information such as the addresses of the local network's router and the network's DNS server(s). Additionally, because dynamic addresses are only granted temporarily, with DHCP, IP addresses can be set to expire.

As the DHCP server must be able to inform local clients of network-specific information, the DHCP server's domain is limited to small networks. The DHCP server will only service computers that share the same network link (the bottommost layer in TCP/IP). Thus, any organization whose network consists of many subnets will require several DHCP servers. Each server should be provided its own range of IP addresses.

Upon booting, a client of the DHCP server will send a broadcast query on its local network, requesting an IP address and configuration from the DHCP server. If the local subnet does not have its own DHCP server, the router can forward the request to the DHCP server located on other subnetworks and await a response to return to the client.

The server stores a table of available and used IP addresses. It selects an unused address and sends to the client the address, the default gateway address, the domain name, the DNS server addresses, and possibly other information (time, date, and duration of validity for the IP address). Now, the client can set its own internal information such as in the file `/etc/sysconfig/network-scripts/ifcfg-eth0`. Once this information has been stored, the client can (re)start its network service. For convenience, a client can request from the DHCP server the last used IP address. If available, the DHCP server can permit this. By doing so, other tables that have cached your computer's IP address will not need to be modified.

Linux uses the program `/sbin/dhclient` to perform the operations needed to set up your client computer with dynamic IP addresses. In CentOS, all you have to do is modify the appropriate ifcfg file (e.g., ifcfg-eth0) to call upon `dhclient`. Specifically, you will modify the `BOOTPROTO` from `static` to dhcp. You would also remove the entries for both `IPADDR` and `NETMASK`, as these will be filled in by the DHCP server.

Another entry worth exploring is `PERSISTENT_DHCLIENT`. By setting this variable to 1, you are asking your computer to query the local router or DHCP until an IP address has been granted. Without this, if your first request does not reach a DHCP or if for some reason, you are not provided an IP address, then you have to try again. You would restart the network service and upon restarting this service (or the next boot), you will obtain your IP address and configuration information from the DHCP server.

## 12.4.3 Setting up a DHCP Server

To set up a DHCP server in Linux, you will need the `dhcp` package. You can obtain this via source code or an executable using rpm or yum. Using yum, issue the instruction `yum -y install dhcp`. Installation includes numerous files. First, the dhcpd service-controlling script is placed in `/etc/init.d/dhcpd`. The dhcpd service is stored as `/usr/sbin/dhcpd`. There are also configuration and data files created and stored as `/etc/sysconfig/dhcpd` and `/etc/dhcp/dhcpd.conf`.

To establish your DHCP server, you will have to modify the conf file. When DHCP is installed, a sample conf file is provided for you under `/usr/share/doc/dhcp*/dhcpd.conf.sample` (where * denotes the version, for instance `dhcp-4.1.1`). As with most conf files, you will have to modify the directives to fit your particular situation. These directives are listed in Table 12.4.

The conf file begins with global directives that apply to all subnets of the local area network. The `domain-name` and `domain-name-servers` directives might be defined here if they cover the entire network. The conf file will then contain one or more subnet entries. Each subnet entry includes a netmask and the range of IP addresses that the DHCP

TABLE 12.4   DHCP Directives

| Directive | Meaning |
|---|---|
| subnet | DHCP's network (or subnet) address |
| netmask | DHCP's local network (or subnet) netmask |
| range | Range of IP addresses available to assign to clients, ranges are indicated by separating IP addresses with a space as in 10.11.12.1 10.11.12.20 |
| routers | IP address (or alias) of router(s) that DHCP server will respond to |
| domain-name | the organization (or subnet) domain name |
| domain-name-servers | IP addresses (or aliases) of domain DNS servers |
| default-lease-time *value* | *value* is the amount of time (in seconds) that an IP address can be made available to a client; once exceeded, the IP address expires and the client must renew or request a new one |
| max-lease-time *value* | *value* is the maximum amount of time IP address is leased |
| authoritative | If listed, means this DHCP is the official server for network |
| log-facility *level* | Use level listed for syslogd logging (e.g., local7) |
| group | Specify parameters that apply to a group of subnets |

will be able to issue to the clients on that subnet. Options include the broadcast IP address and router IP address.

Your DHCP may have multiple sets of entries if the DHCP is responsible for multiple subnetworks. What follows is an example where the DHCP server hosts two subnets, 10.11.0.0 and 10.11.128.0.

```
option domain-name somecompany.com;
option domain-name-servers 10.11.1.1 10.11.1.2;

subnet 10.11.0.0 netmask 255.255.128.0 {
      range 10.11.12.0 10.11.12.20;
      option broadcast-address 10.11.12.22;
      option routers 10.11.12.21;
}

subnet 10.11.128.0 netmask 255.255.128.0 {
      range 10.11.128.129 10.11.128.253;
      broadcast-address 10.11.128.255;
      option routers 10.11.128.254;
}
```

The two entries shown above specify two subnets that the DHCP server handles. You will notice that both subnets use the same netmask (255.255.128.0 corresponds to the first 17 bits of the 32-bit address, that is, the network address is made up of the first 17 bits, and the machine address is the remaining 15 bits). This DHCP is allowed to hand out IP addresses within the range 10.11.12.0–10.11.12.20 for the first subnet and addresses of 10.11.128.129–10.11.128.253 for the second subnet. The two subnets each have their own broadcast device and their own router; so, there are two sets of addresses, one per entry.

Since both subnets are part of the same organization's network, they share the same domain-name and domain-name-servers, defined in a global section.

This example does not illustrate the use of *leases*. A lease means that the IP address, when issued to a client, is of limited duration and once it expires, the IP address is returned to the DHCP server's pool of addresses. A client with an expired IP address must ask for a new one. This might be set up whereby the client is able to reattain the same IP address (renew the lease) or it might be set up where the client loses that address and upon receiving a new IP address, it can very well differ.

We can also establish one or more entries for pools. A pool contains a set of IP addresses that are available to computers that are not covered by one of the subnet entries. This allows you to not only offer addresses outside the given subnets, but to also establish different options such as less or more restrictive leases, or access outside a firewall to one set of addresses while those in the pool are restricted from Internet communication.

With your configuration file now available, you must perform two or three more tasks. First, you must restart the service. Second, you must permit DHCP requests to come in through your firewall. Typically, DHCP requests come over port 67 under the UDP protocol. You could add a rule to your firewall (see Section 12.6) to permit protocol UDP port 67. Third, if necessary, you must modify your local routers to know where the DHCP server is located (the machine's IP address).

The DHCP software comes with IPv6 components (e.g., `/etc/sysconfig/dhcpd6` and `/etc/dhcp/dhcpd6.conf`). Additionally, dhcpd stores lease information under `/var/lib/dhcpd/dhcpd.leases` and `/var/lib/dhcpd/dhcpd6.leases`.

## 12.5 NETWORK PROGRAMS

The primary duties of the system administrator with respect to the network are to ensure that the network is accessible and that access to the network is secure. We consider security in Section 12.6. Here, we examine some of the available Linux programs that let the system administrator (or the user in many cases) query network status.

### 12.5.1 The ip Program

There have been a whole host of programs available for querying network information. Most of these are now obsolete with the same functionality woven into the program `ip`. It makes ip an umbrella program. It can be used to show routing tables, show device information, set routing information, set device information, or provide network tunnels.* As ip can do so many things, there are different specific commands available. The basic syntax for ip is

```
ip [options] object {command}
```

---

* A tunnel is a network connection between two devices, for example, two computers communicating by ssh. What distinguishes a tunnel from an ordinary connection is that the tunnel is long term and the computers at both ends of the tunnel have two different IP addresses, their normal address, and one dedicated for the tunnel while it exists. Additionally, the tunnel can be used to encapsulate another protocol; for instance, within a normal TCP tunnel, we might use the SSH protocol to ensure encrypted communication.

where *object* is one of `link`, `addr`, `addrlabel`, `route`, `rule`, `neigh`, `tunnel`, `maddr`, `mroute`, or `monitor`. These are described below. The *command* will vary depending on the object selected. Options are –V for version, -s for statistics, -r for resolve, -f for family that is followed by `inet`, `inet6`, `ipx`, `dnet`, or `link`, and –o to output on one line.

The object specified dictates just what ip will return or set.

- link—can have a command of `set` or `show`. For set, you are then required to specify the device such as lo or eth0. This should be followed by one of

  - up/down

  - arp on/off

  - promisc on/off

  - allmulticast on/off

  - txqueuelen *numpackets*

  - name/alias *newname*

  - address/broadcast new_link_layer_address

  - mtu *MTUvalue* (maximum transmission unit)

  - netps *PID*

  - vf *NUM* (virtual function device number)

- addr—can have commands of `add`, `del`, `show`, and `flush` to add or remove an IP address from the device, show the current IP address(es), or flush (remove) an address based on some specified criteria. With add, you can specify a new name for a device, a peer or broadcast address(es), label, or scope. The show command can be followed by one of dev *name*, scope *scope_val*, to *addr*, label *value*, dynamic, permanent, tentative, deprecated, primary, or secondary.

- addrlabel—to label an IPv6 address for later use, this object can be followed by one of `add`, `del`, or `list`.

- route—this object allows you to manipulate or inspect any of the routing tables. There are several types of routing tables maintained: `unicast`, `unreachable`, `black-hole`, `prohibit`, `local`, `broadcast`, `throw`, `nat`, `anycast`, and `multicast`. The command for route is one of `add`, `change`, `replace`, `del` (delete), `show`, `flush`, or `get`. For add, change, and replace, you may follow this with the word `to` and then both the type of table (e.g., `unicast`, `nat`) and IP or IPv6. Alternate options for add, change, and replace are to specify a new preference number, a table ID, a new device name, a new src or via address, and a new mtu value among other values. The usage of the arguments for del differ somewhat from add, and show has additional arguments available. The flush command flushes the given table(s) while the get command can be used to retrieve one specific value from a table. Here are a few examples of route commands.

- ip route add 10.11.12.0/24 dev eth0

- ip route add default via 10.11.12.13

- ip route add 10.11.1.2 via 10.11.30.1

- ip route add nat 192.51.185.16 via 10.11.12.1

- ip route change default via 10.11.30.6 dev eth0

- ip route show cache

- rule—change the route selection algorithm so that the routing table used in a particular setting changes.

- neigh—add, change, delete, replace, or show, where neighbors are objects specified as bindings between a protocol address and a link layer address.

- maddress—multicast address, commands are show, add, and delete.

- mroute—multicast-routing cache management, show is the only command here.

- tunnel—add, change, and delete a network tunnel.

- monitor and rtmon (state monitoring).

## 12.5.2 Other Network Commands

The ip instruction has replaced route that previously let you view and manipulate the IP routing table. This table would contain the gateway address and the gateway's router address along with netmasks for each interface device (excluding lo).

To view interface IP addresses, you can use ip addr. Alternatively, the ifconfig command will respond with the same information. The ifconfig statement, like ip, also allows you to configure your network interfaces. The command is followed by the interface in question (e.g., lo or eth0) and any options that you want to establish or the new address for that interface (if you omit the options/address, ifconfig responds with the already-established address). The command will look like one of these:

```
ifconfig interface options
```

or

```
ifconfig interface address
```

where options include interface to change the name of the interface (as in ifconfig eth0 interface foobar to change the name of eth0 to foobar), up/down to start or stop the interface, `arp/-arp` to enable or disable the use of the ARP proto-col on this device, `promisc/-promisc` to enable or disable promiscuous mode on this interface, or `allmulti/-allmulti` to enable or disable multicast mode on this device. Additionally, you can change the MTU parameter, destination IP address if the device is a ppp, change the netmask, and change the device's IP address.

Neither route nor ifconfig are necessary because ip can accomplish all these tasks. However, both route and ifconfig continue to be available in current versions of Linux. You might find them easier to use over the rather complicated ip, but eventually, these instruc-tions might be deprecated; so, it is best to learn ip now.

We have already discussed `ping` and `traceroute` in Chapter 5. As a system admin-istrator, you might find these commands useful in testing out your network access and the availability of other devices that you are in charge of. However, both commands can lead to security holes in your network as others might try to investigate your local area network through these programs. Through ping or traceroute, a clever hacker could accumulate legal IP addresses of your network. This is known as a *reconnaissance attack*. With such IP addresses, the hacker could exploit this information using other forms of attack such as denial of service, intrusion, and IP spoofing. You can configure your firewall to prevent incoming messages from ping or traceroute.

`Netstat`, like ifconfig and route, is an older program available to obtain network con-nection and routing table information as well as interface statistics and multicast member-ships. The netstat command will dump all available statistics if not provided any options. The options -t (or --tcp) and -u (or --udp) provide information on TCP and UDP commu-nication, respectively. The -r (or --route) option provides the same output as the route com-mand. The option -i (or --interfaces) followed by an interface name provides information about that interface.

The netstat command now has been superseded by the command `ss`. The ss command is a utility to investigate network sockets. In essence, this program dumps socket statistics. It permits many of the same options as netstat, for instance, -t (--tcp), -u (--udp), and -r. We also have `nstat` and `rtacct` to monitor the kernel's snmp communication and report statistics on network and interface usage.

## 12.6 THE LINUX FIREWALL

A *firewall* is a program that examines incoming and outgoing network messages and decides which messages are permitted to be passed through the firewall. The firewall itself uses a collection of rules that define attributes of messages that should or should not be allowed through. Rules can pertain to incoming messages only, outgoing messages only, or possibly both. The criteria tested by the rules can include matching any source or destina-tion IP addresses, ports, protocols, size, and/or interface to specific values.

A *stateful* firewall is able to make decisions on groups of messages that make up a single network connection. This is useful when a message is a part of a group of messages that make up an established session between two machines. The firewall is an essential tool in today's computing to prevent or reduce external attacks. However, as we will see, a firewall is only as good as the rules defined.

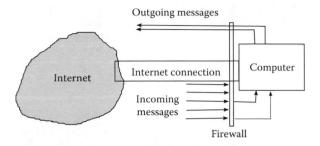

FIGURE 12.5    A computer's firewall.

A firewall can be set up to protect a single computer or an entire network. Typically, an organization will employ multiple firewalls, one at the Internet point of presence, perhaps as a proxy server (or in addition to a proxy server), and one for every computer. Figure 12.5 illustrates the concept of a firewall protecting a single computer. In the figure, we see two of five incoming messages are accepted and the other three are not while both outgoing messages are permitted to leave the firewall.

The Linux firewall can be configured through the GUI (as covered in Chapter 11) or by modifying the `iptables` (and `ip6tables`) file that contains the firewall rules, and the `iptables-config` (and ip6tables-config) file that contains the configuration directives. These files are stored in /etc/sysconfig. Here, we will concentrate on configuring the firewall through these files rather than the GUI. We will only consider using the IPv4 portion of the firewall (iptables and iptables-config) as it is more critical that you set up your firewall for IPv4 than IPv6. You would find the ip6tables/ip6tables-config to be similar.

### 12.6.1  The iptables-config File

The iptables-config file contains a handful of directives pertaining to your firewall. This file controls general aspects of the firewall. Table 12.5 provides the directives for this file along with their meaning and default value. Aside from the first directive (IPTABLES_ MODULES), the values are either "yes" or "no."

Many of the directives from Table 12.5 impact information displayed in response to issuing the status command to the iptables controlling script (i.e., in response to /sbin/ `service iptables status`). The iptables service's status information differs from the status information obtained from most other services that usually just respond with a message that the service is running or stopped. With iptables, you obtain detailed information of the firewall's rules. Figure 12.6 demonstrates the response from a status request. The iptables-config file, in this case, has directives of IPTABLES_STATUS_VERBOSE=yes and IPTABLES_STATUS_LINENUMBERS=yes.

### 12.6.2  Rules for the iptables File

The iptables file contains the rules for your firewall. The rules specify how incoming (INPUT) messages will be handled, how outgoing (OUTPUT) messages will be handed,

TABLE 12.5    Iptables-config Directives

| Directive | Meaning | Default |
|---|---|---|
| IPTABLES_MODULES | List of modules to load after firewall rules are applied | "" (none) |
| IPTABLES_MODULES_ UNLOAD | Unloads modules on firewall stop or restart | Yes |
| IPTABLES_SAVE_ON_STOP | Rules may be added to your firewall from the command line; if this directive is set to yes, then all current rules are saved to iptables upon stopping the firewall | No |
| IPTABLES_SAVE_ON_RESTART | If set to yes, saves all current rules to iptables upon restarting the firewall | No |
| IPTABLES_SAVE_COUNTER | Saves all chains of rules and counters for rules to iptables upon stop or restart of firewall, or the command `/sbin/service iptables save` | No |
| IPTABLES_STATUS_NUMERIC | Prints IP addresses and ports in numeric format when `status` of firewall is requested | Yes |
| IPTABLES_STATUS_VERBOSE | Prints statistics about packets and bytes when `status` of firewall is requested | Yes |
| IPTABLES_STATUS_ LINENUMBERS | Prints line numbers of rules when `status` of firewall is requested | Yes |

```
Table: filter
Chain INPUT (policy ACCEPT 0 packets, 0 bytes)
num   pkts bytes target     prot opt in     out     source               destination
1       30  1899 ACCEPT     all  --  *      *       0.0.0.0/0            0.0.0.0/0            state RELATED,ESTABLISHED
2        0     0 ACCEPT     icmp --  *      *       0.0.0.0/0            0.0.0.0/0
3        1    41 ACCEPT     all  --  lo     *       0.0.0.0/0            0.0.0.0/0
4        0     0 ACCEPT     tcp  --  *      *       0.0.0.0/0            0.0.0.0/0            state NEW tcp dpt:22
5      205 43144 REJECT     all  --  *      *       0.0.0.0/0            0.0.0.0/0            reject-with icmp-host-prohibite
d

Chain FORWARD (policy ACCEPT 0 packets, 0 bytes)
num   pkts bytes target     prot opt in     out     source               destination
1        0     0 REJECT     all  --  *      *       0.0.0.0/0            0.0.0.0/0            reject-with icmp-host-prohibite
d

Chain OUTPUT (policy ACCEPT 36 packets, 66782 bytes)
num   pkts bytes target     prot opt in     out     source               destination
```

FIGURE 12.6    Verbose status from iptables.

and how forwarded (FORWARD) messages will be handled. The most critical of the rules are the incoming message rules. With improper (or no) INPUT rules, your computer is open to receiving messages of any type. You want to ensure that only specific types of messages are permitted into your computer. Although OUTPUT rules are less common, you might wish to use some rules to control outgoing messages as well, for instance, to prevent users from accessing particular websites.

The iptables service works by using *chains* of rules. For any incoming packet, the packet is compared to a chain of input rules. The firewall continues to compare the message to the rules of the given chain until either it reaches the end of the chain, in which case, a default rule is applied, or a rule with a *target* value matches the message, in which case, the rule triggers and the target is followed. Rules have the following format:

```
-A chain [options] [-j target]
```

The value of *chain* will be one of INPUT, OUTPUT or FORWARD, or a chain that you have defined. The value of *target* will be one of ACCEPT, REJECT, DROP, or LOG that indicates what should be done with the message (these four values are described later). The –A option indicates that this rule should be *appended* to the given chain. Thus, we are adding to the chain so that if previous rules do not match, the firewall can continue to examine more rules of the chain.

The options in the rule specify the criteria by which the rule will match. You can think of these as the conditions of an if-then statement. Some of the most useful and common options available for the iptables rules are presented in Table 12.6. Notice `--dport` and `--dports` are subtly different. The former is used if you are comparing the message's port to a single port of interest while the latter compares the message's port against a list of ports. Similarly, there is a distinction between `sport` and `sports`.

Rules may have any number of options. For the rule to be true, *all* options specified must be true. If a rule is true, the action specified under the target will take place. If any target other than LOG is used, chaining will stop. For instance, if a rule causes a packet to be accepted, then no more chaining takes place. The four targets are listed below.

- ACCEPT—permit the packet entry to the system

- REJECT—reject the packet and notify the sender

- DROP—reject the packet without notifying the sender

- LOG—log the packet but continue chaining rules to reach one of the other targets

TABLE 12.6  iptable Rule Options

| Option | Meaning | Example |
|---|---|---|
| -m state --state *value(s)* | True if the message's state matches the listed value(s) (see also Table 12.7) | `-m state --state ESTABLISHED,RELATED` |
| -p *protocol* | True if the message is of the given protocol (e.g., udp, tcp, and icmp) | `-p tcp` |
| -i *interface* | True if the message is received by the given interface | `-i eth0` |
| -o *interface* | True if the message is being sent over the given interface | `-o lo` |
| -s *address* | True if the message originated from the given IP address | `-s 10.11.12.13` <br> `-s 10.11.0.0/16` |
| -d *address* | True if the message is being sent to the given IP address | `-d 172.19.31.141` |
| --dport *port* | True if the message is intended to be received at the given port | `--dport 431` |
| --sport *port* | True if the message originated from the given port | `--sport 22` |
| --dports *port1,port2,…* | True if the message is intended for any of the given ports | `--dports 80,8080,443` |
| --sports *port1,port2,…* | True if the message originated from any of the given ports | `--sports 67,68` |

Each option comes with at least one parameter. For instance, with –s or –d, you would indicate one (or more) IP address. With --dport, you list one port and with --dports, you list multiple ports, separated by a comma.

The –m option is used to employ an additional *matching module*. We saw in Table 12.6 that one useful module is called `state` that can be used to match the message to one or more possible states of interest such as NEW or ESTABLISHED. There are numerous other modules available as shown in Table 12.7. Each type of module has its own syntax to describe the criteria to match against. For instance, we might want to match this message to the address type of the source sending the message. This might be a BROADCAST device, a UNICAST device, and so on. Therefore, we use a rule like

```
-A INPUT -m addrtype BROADCAST -j ACCEPT
```

One additional option of note is –g *chain*. This option is not part of a rule's criteria but instead specifies that if the rule applies, then rather than stopping, chaining should continue through the chain *chain*.

TABLE 12.7    Modules for Further iptables Rule Criteria

| Module | Meaning | Extensions |
|---|---|---|
| addrtype | Match based on source address *type* or destination address *type* | --src-type *type*<br>--dst-type *type*<br>Types include BLACKHOLE, BROADCAST, MULTICAST, NAT, and UNICAST |
| conntrack | Match based on the connection's status | --ctstate *state*<br>States include INVALID, ESTABLISHED, NEW, RELATED, SNAT, and DNAT |
| icmp | Match based on ICMP type | --icmp-type [!] *type*<br>Type can be any ICMP type or its corresponding number |
| iprange | Match if message's source or destination falls within the given range | [!]--src-range *range*<br>[!]--dst-range *range*<br>Range denoted as *address1-address2* |
| length | Match if message's length is equal to, or within the range, provided | --length [!] *length*[:*length*]<br>Examples:<br>--length 500:1000<br>--length ! 0 |
| limit | Limit the number of received messages | --limit-burst *number* |
| time | Match if specified time is met | --timestart *value*<br>--timestop *value*<br>--days *days*<br>--datestop *days*<br>*value* is a time given in hh:mm format and *days* is Monday, Tuesday, Wednesday, and so on |

Aside from -A rules, you can also specify -P, -F, and -L rules. The -P rules are default rules. These will be applied if the current chain being followed exhausts its list of rules. For instance

```
-P INPUT REJECT
```

states that if no INPUT rule matches, then reject the message. The -F rules are used to flush all existing rules so that we start with a new set of -A rules. You might use -F if you are rewriting rules from the command line, or as the first rule in iptables. The -L rules are used to list all the rules in the firewall. You would typically only use -L from the command line and you would also more commonly use -F from the command line.

As noted above, aside from editing the iptables file itself, you can issue changes to the firewall rules from the command line through the `iptables` command. This would look like any rule in iptables except that you enter it at the command line, preceded by the word `iptables`. For instance, to add the rule INPUT -i eth0 -p udp -j ACCEPT, you would enter

```
iptables -A INPUT -i eth0 -p udp -j ACCEPT
```

With iptables, you could then flush the entire set of rules in the firewall with `iptables -F`.

## 12.6.3 Examples of Firewall Rules

Here, we will examine a basic set of rules that we might find in an iptables file and explore what the rules mean. Each rule is explained in comments following the rule.

```
-P INPUT REJECT      # default for incoming packets
-P OUTPUT ACCEPT     # allow all outgoing messages
-P FORWARD DROP      # do not perform forwarding
```

Forwarding is commonly used for routers, not workstations.

```
-A INPUT -i lo -j ACCEPT        # accept anything over "lo"
-A INPUT -p tcp —dport 22 -j ACCEPT
                                # accept incoming ssh
-A INPUT -m state —state ESTABLISHED,RELATED -j ACCEPT
                                # accept messages that are
                                # continuations of previously
                                # established connections
-A INPUT -p icmp -j ACCEPT      # accept ICMP messages
-A INPUT -i eth0 -s 10.11.12.0/24 -j ACCEPT
                                # accept messages from
                                # subnet 10.11.12.0
-A INPUT -p tcp -s facebook.com —sports 80, 443 -j DROP
                                # drop webpage responses from
                                # facebook.com
```

The idea behind the above rules is to accept any incoming message that fits the given criteria with the exception of the last entry, which drops the message. However, our very first rule establishes the default, which is to reject any incoming message that does not pass any of the accept rules. Why therefore do we need a separate rule to drop Facebook messages? The answer is based on the different action that takes place. By default, we REJECT messages, but the Facebook message is DROPped. You might recall earlier that a REJECT action will notify the sender that the message was received. With a DROP action, we do not respond to the sender.

Let us consider another situation. We are not explicitly accepting messages over port 80, which is HTTP. How then is the user able to view web pages? The answer comes from the third rule that deals with an ESTABLISHED state. The above –A rules are all INPUT rules. We saw that all OUTPUT messages are automatically ACCEPTed via the –P default rule. Therefore, the user is free to send out messages without the firewall stopping them.

For an HTTP interaction, it is the user who initiates the process (either by entering a URL in a web browser's address box or clicking on a link in the web browser's current web page). This action results in an outgoing HTTP message sent to a web server. In doing so, this establishes a connection with the server so that the state becomes ESTABLISHed. Any response from the web server is of an already ESTABLISHed (or possibly RELATEd) message. So, the state of the incoming message from the web server has a state acceptable by our –m state rule. In fact, until the connection is ended (at our end) or times out, the state remains ESTABLISHED.

On the other hand, we are not allowed to receive HTTP requests as the only protocols that we accept are tcp over port 22 and icmp. This is fine if we are not running a web server as any incoming HTTP requests would be nonsense. If we do decide to run our own web server, we would need to add rules to permit messages from port 80 and quite possibly other related ports like 8080 and 443. The following might be added to permit HTTP and HTTPS requests:

```
-A INPUT -p tcp -m multiport —dports 80,8080,443
        -m state —state NEW,ESTABLISHED -j ACCEPT
```

This rule states that incoming TCP packets over any of the ports 80, 8080, or 443 that have either a NEW or ESTABLISHED state are accepted.

We might want to further refine this or other rules if we are worried that HTTP requests could cause us damage. One example is to add unwanted IP addresses to a list of REJECT or DROP actions. For instance, let us assume that we have received several denial of service attacks from the subnet 1.80.0.0/13. We might then add the following rule that should be placed earlier than the ACCEPT rules to ensure that this rule triggers; otherwise, an ACCEPT rule might trigger and the INPUT chain would be followed no further so that the firewall would not reach this rule.

```
-A INPUT -s 1.80.0.0/13 -j DROP
```

Another way to prevent possible denial of service attacks is to restrict the number of incoming messages permissible from any one source. We could add the following rule so that there is a limit to the number of accepted messages per minute or per established connection.

```
-A INPUT -p tcp -m multiport —dports 80,443 -m limit
       —limit 25/minute —limit-burst 100 -j ACCEPT
```

The remaining ACCEPT rules in our earlier example will permit access into our computer over the lo interface, port 22 (using ssh), messages using the icmp protocol (both ping and traceroute use this), and anything from the 10.11.12.0 subnet that arrives over the Ethernet card.

By permitting icmp messages, we open ourselves up to possible reconnaissance attacks. Through ping, a potential attacker could identify IP addresses in a network. While we do not necessarily want to stop all incoming icmp traffic, we might want to discourage ping and traceroute queries. To accomplish this, we might add the following two rules. The first rule states that incoming ping/traceroute requests will be dropped while the second rule states that outgoing ping/traceroute responses will be dropped.

```
-A INPUT -p icmp —icmp-type echo-request -j DROP
-A OUTPUT -p icmp —icmp-type echo-reply -j DROP
```

Finally, we might set up explicit logging rules. We could define our own chain, rather than the preestablished INPUT, OUTPUT, and FORWARD chains. For instance, –N LOGGING will define a new set of rules under the chain LOGGING. We then add (-A) rules to this chain, for instance:

-A LOGGING --log-level 7 -j LOG

-A LOG -j DROP

These rules state that anything to be dropped should also be logged at log level 7 (the number 7 is used by syslog).

As stated earlier, we could modify our firewall by issuing these commands from the command line. If we do this, then the new rules are available now, but unfortunately, not the next time we reboot the computer. To save the changes entered at the command line, you can issue the instruction /sbin/iptables-save. Alternatively, you can directly edit the iptables file (/etc/sysconfig/iptables) in vi and save the results. Once done, restart the iptables service.

## 12.7 WRITING YOUR OWN NETWORK SCRIPTS

As with any aspect of Linux, there are gaps in the operating system that you can fill by writing your own scripts. This section is meant to convey some possible scripts. It would be up to you to determine what you might need and implement them. We will examine several ideas for scripts. Each of these scripts will be written using the Bash scripting language.

### 12.7.1 A Script to Test Network Resource Response

Our first script will test to see if a given network resource is responding. We will test the resource with ping. The command `ping  ipaddress` will result in a number of messages being displayed that specify the amount of time and packet sizes returned from the resource, *ipaddress*. As ping concludes (either when the user ends ping by typing `control+c` or because the user issued `-c  num` to limit the number of attempts), ping responds with a summary that includes the number of packets transmitted, received, the amount of loss, and the time in milliseconds. What we would like is to either obtain the number of received packets or the percentage loss. We will focus only on the received packets.

We can avoid receiving the remainder of the output using the –q option in ping to work in quiet mode. This still leaves a summary. The awk command can help us here by searching for only the line with the word "received." Our resulting command for some *ipaddress* stored in the variable `ip` is

```
ping -c $num $ip | awk '/received/ {print $0}'
```

This instruction, using print $0, will output the entire row, giving us feedback such as

```
3 packets transmitted, 2 received, 33% packet loss,
      time 0 ms
```

Let us prune this down to just the number received by using {print $4} rather than {print $0} in awk. This would return 2 for the above example. We store this in the variable `x`. Now, we compare $x to 0. If we received 0 packets, either the device at the given *ipaddress* is not responding at all, or we did not give ping enough attempts. Assuming that we want to accumulate a list of all unresponsive devices, we will output this device's IP address to a file, along with the current time/date and the number of attempts tried. We will put this into a function as follows:

```
try_ip() {
    ip=$1
    num=$2
    x=`ping -c $num -q $ip | awk '/received/{print $4}'`
    if [ $x -eq 0 ];    then
        echo "$ip not responsive on `date` with $num tries"
            >> /root/net_stats/non_responding_devices.txt
        fi
}
```

The function expects to receive two parameters: the IP address and the number of attempts that ping should make. We could also pass in the destination file's name to record unresponsive devices, but instead, we have redirected this to the file `/root/net_stats/non_responding_devices.txt`.

To use this function, we might define an array of IP addresses that we wish to test and then, using a for loop, call `try_ip` with each address. The value 10 below in the try_ip function call is the number of attempts per device for the ping command.

```
list=(10.11.12.13 10.11.12.14 10.11.12.15
      10.11.21.22 10.11.38.83 10.11.0.1)
for ip in $list; do
      try_ip $ip 10
done
```

### 12.7.2 A Script to Test Internet Access

We might wish to implement a script to test our own computer's Internet access. Reasons for a lack of Internet access include the network service being stopped, problems with our interface device(s), and problems with our local area network such as the router being unavailable. Rather than attempting to determine where our problem might exist, we could simply issue an instruction that requires external network response and see if we obtain such a response.

There are several ways to test our network, for instance, by using ping as shown in the previous script. Instead though, we are going to use a simpler approach, by attempting to wget files and seeing if we were successful. We will need to identify several sites that we know we can obtain files from. For simplicity, we will use only search engine sites (e.g., Google, Yahoo, and Bing) and query them for their home pages (index.*).

The wget program is a noninteractive way to download a file from a web server without going through a web client. wget responds with information about the retrieval process. We are not interested in this, only in receiving the file. We will redirect wget's output to /dev/null. We will use the --tries option so that wget makes several attempts in case the first attempt is unsuccessful because of a spurious error. We will also issue wget commands to several sites until we have success. We can stop as soon as we have our first successful retrieval because this will confirm Internet access. As with the previous script, we will break our script into two parts, the retrieval from one site that will be written as a function, and a while loop to continue to issue wget commands until we achieve success.

```
try_wget() {
      filename=$2.index
      /usr/bin/wget -q --tries=$1 $2 -O /root/$filename
      if [ -e $filename ]
            then
                  rm /root/$filename
                  return 0
            else
                  return 1
      fi
}
```

```
urls=(www.google.com www.yahoo.com www.bing.com)
numAttempts=0
contact=0
while [[ $contact -eq 0 && $numAttempts -lt ${#urls[@]} ]]
        do
                u=${urls[numAttempts]}
                try_wget 5 $u

                if [ $? -eq 0 ]; then
                        contact=1
                fi
                numAttempts=$((numAttempts+1))
done
if [ $contact -eq 1 ]; then
        echo Warning, Internet connection appears to be down
fi
```

In the above script, we store a series of web server addresses in the array urls. We are using search engines here assuming that at least one will always be available and because these pages are dynamic, they should not be cached locally. We iterate through all the URLs in the array urls. For each url, u, we pass this URL along with the value 5 to try_wget. The function attempts to perform wget on the URL's home page, using the number 5 for the number of attempts. If an attempt is successful, the file (the index page for the webserver) is downloaded to /root/$u.index where $u is the URL's name, for example, the file could be /root/www.google.com.index.

After the wget attempts have been made, the try_wget function tests to see if the file exists. If so, the file is deleted and the function returns 0 for an error code. If not, then obviously, wget failed and the function returns the error code 1. In the while loop, after try_wget is called, we examine its return value to see if it was successful (0) or not (1). If successful, we set the variable contact to store 1 that will let us out of the while loop because we have found that our Internet connection is in fact working. We increment numAttempts so that, if no attempts work, we can still exit the while loop once we have tried all the URLs. Upon exiting the while loop, if $contact is 1, we have Internet access; otherwise, with the URLs we tried, we apparently do not have Internet access.

The automated nature of the above script allows us to schedule such a test at intervals so that we can log Internet access results. Assuming this script is /root/try, we might issue a crontab job such as

```
0 * * * *   /root/try >> /root/internet-access.log
```

so that the try script executes every hour on the hour with results being stored to a log file. We should modify the try script so that the output error message includes the current time/date.

### 12.7.3 Scripts to Compile User Login Information

Another interesting script is one that compiles a list of all users remotely logged in. We can obtain this information from the who command. This will give you all users logged in, including yourself. You can remove your own listing by piping the result to an egrep command.

```
who | egrep -v $USER
```

Let us go a little further and actually compile the list of user names who are logged into the system. The response from who gives us extra information such as the terminal window, date and time of the last login, and the location that they are logging in from. We will prune this list down to just the user name. We can prune all the extra information away by using awk '{print $1}'.

In the following script, we use who to generate all logged-in users, discard those users whose name matches $USER, and then obtain just their login names. Next, we test to see if that user ($user) has already been added to the variable list or not. If not, concatenate $user onto the list. At the end of the script, the value of $list, along with the time/date, is output to the log file /root/logged_in_users.txt.

```
list=
count=0
for user in `who | egrep -v $USER | awk '{print $1}'`; do
        if [ -z `echo $list | grep $user |
                awk '{print $1}'` ]; then
                        list="$list $user"
        fi
done
echo "Users at `date` are $list"
        >> /root/logged_in_users.txt
```

As you will see in Chapter 14, log files retain authentication and log-in events; so, in fact, we are already logging information about who has logged in. The above script has the advantage that it is only storing who is currently logged in so that you would not have to parse through a much longer file that includes many other types of events. As with the script earlier where we tested Internet availability every hour, we could do the same here by scheduling the above script with crontab.

A related script to the previous one is to collect the IP addresses of users remotely logged in. Using who once again, we obtain a list of all logged-in users. Those with IP addresses will be the ones remotely logged in. The IP address will have the form #.#.#.# where # is an octet (number between 0 and 255). We can use egrep along with a regular expression for IP addresses to isolate just those items from who that are of these remotely logged-in users. For instance, we might use the following instruction:

```
who | egrep '[0-9]{1,3}\.[0-9]{1,3}\.[0-9]{1,3}\.[0-9]{1,3}'
```

To obtain just the IP address, we can pipe the result of the above instruction to awk '{print $5}' that will give us all of the user's IP addresses, although they will be placed in parens like the following:

> (10.11.12.13)
> (10.11.14.15)
> (10.11.15.16)

Now, let us imagine that we want to perform a reverse nslookup to obtain the IP aliases of the clients logged in. We first have to remove the parens from around the IP addresses. We can accomplish this using sed or the `expr substr` operation (see Chapter 7). Here is the code to accomplish this using substring.

```
x=`who | egrep '[0-9]{1,3}\.[0-9]{1,3}\.[0-9]{1,3}\.[0-9]{1,3}'
     | awk '{print $5}'`
for ip in $x; do
     len=`expr length $ip`
     len=$((len-2))
     y=`expr substr $x 2 $len`
done
```

Now, we perform a reverse lookup on $y using host –i. Our full script is given below.

```
#!/bin/bash
x=`who |'[0-9]{1,3}\.[0-9]{1,3}\.[0-9]{1,3}\.[0-9]{1,3}'|
     awk '{print $5}'`
for ip in $x; do
     len=`expr length $ip`
     len=$((len-2))
     y=`expr substr $ip 2 $len`
done
echo `host -i $y` >>/root/remote_users_by_ip_alias.txt`
```

Obviously, these are just a few of the possible scripts you could write to monitor your network's activity.

## 12.8 CHAPTER REVIEW

Concepts and terms introduced in this chapter:

- Address resolution—the process of converting an IP alias into an IP address using some resolver such as a DNS server, a DNS cache, or the entries stored in /etc/hosts.

- Computer network—a collection of computers and computing resources connected together to facilitate communication between resources.

- DHCP server—a device (usually a router or gateway but possibly a computer) set up to issue IP addresses dynamically upon request to devices on its subnet.

- Domain name system—the collection of servers and resolution information that permits the use of IP aliases on the Internet rather than IP addresses. The DNS includes DNS servers, caches, and local resolving programs.

- Dynamic IP address—an IP address issued to your computer temporarily (for instance, for a few days).

- Ethernet—a technology for local area networks.

- Firewall—software that helps enforce security. In some cases, a firewall is both hardware and software if an organization dedicates a computer to the server solely as a firewall.

- Gateway—a broadcast device responsible for connecting local area networks of different types together.

- Hub—a broadcast device operating on a subnetwork that, when it receives a message, broadcasts that message to all devices on that subnet.

- IP address—a unique address (number) assigned to a computing resource on the Internet. There are two types of IP addresses, version 4 (IPv4) and version 6 (IPv6).

- IPv4 address—32-bit address usually written as four octets of numbers between 0 and 255, separated by periods, as in 10.11.12.13.

- IPv6 address—a 128-bit address offering far greater range of addresses. Usually written as 32 hexadecimal digits. IPv6 is a protocol created to replace IPv4 because IPv4 is outmoded and because we have run out of most available IPv4 addresses. IPv6 includes features such as security and autoconfiguration that are not directly available in IPv4.

- IP alias—a name given to a computer to use in lieu of an IP address. The IP alias is a collection of usually short words separated by periods much like the IP address that is a series of numbers separated by periods. IP aliases are much easier to remember but since routers and gateways cannot use IP aliases, the use of IP aliases requires address resolution.

- Loopback device—an interface in Linux machines that allows software to communicate to the computer as if the messages were coming over the network. The loopback device does not send messages onto a network.

- MAC address—the media access control address given to devices such as Ethernet cards. This address is used at the lowest level of the TCP/IP protocol and is used by switches.

- Name server—a computer with the responsibility of performing address resolution. Typically, a name server is an authority for the domain of which it is a part of and not an authority for any other domain.

- Netmask—a binary number used to AND with an IP address to obtain the network address for the device.

- Octet—an 8-bit number, typically written as an integer between 0 and 255. Four octets are used to make up an IPv4 address.

- Point to point—a type of network connection in which two devices are directly connected together, rather than an Ethernet-style network. A point-to-point connection might exist between your computer and a printer if you have connected the printer directly to the computer.

- Port—an address assigned to a type of communication protocol. This address is used to identify the proper protocol and application for a given message. It can also be used for security purposes to determine whether a message should be permitted through a firewall.

- Protocol—the formal definition is a set of rules used to describe how entities should interact and/or communicate. In networks, a protocol describes the activities that a device must take to prepare the message for transmission and how the recipient is to interpret the message. TCP/IP is a protocol stack in that it consists of several protocols.

- Router—a broadcast device that examines a message's destination IP address and routes the message onto the proper network or subnetwork as the next link in the chain of the communication.

- Static IP address—an IP address assigned to a computing resource permanently or at least for a long period of time. The static address is not expected to change. Changing it will require modifying DNS tables.

- Subnet—a subset of a local area network where all computers on the subnet share the same broadcast device (e.g., switch or router) and share the same netmask and therefore the same network address.

- Switch—a broadcast device operating on a subnetwork; when it receives a message, it broadcasts that message to a single device on the subnetwork using an MAC address for addressing.

- TCP/IP—a commonly used network protocol that lets computers access the Internet. TCP/IP is known as a protocol stack, comprising several lesser protocols.

- Tunnel—a temporary dedicated network communication link between two resources that is persistent, longer than typical network communications.

- Zero-configuration service—a network service that can locate network resources such as a DHCP server.

Linux commands covered in this chapter:

- ifconfig—older network command to configure or obtain network information such as IP address and router address.

- ip—newer network command that encapsulates the operations available in lesser programs such as ifconfig, route, and iptunnel.

- netstat—older network command to output statistics about network usage. Has been superseded with ss.

- ping—program to constantly send messages to another network-based resource to test for its availability.

- route—displays local router tables. Command replaced by ip.

- ss—socket investigation program.

- traceroute—like ping, used to determine availability of network-based resource. Differs because traceroute outputs the network addresses of routers and other devices that the request message(s) encounters on the way.

- xinetd (or inetd)—a superserver capable of invoking appropriate network services based on the ports of incoming messages.

Linux services and files of note covered in this chapter:

- certmonger—manages Internet digital certificates, replacing those that become outdated.

- dnsmasq—lightweight DNS server, primarily used as a DNS cache.

- netfs—used to permit remote access to local file systems.

- network—used to provide any form of network access. Among its duties are to bring up network interfaces (e.g., eth0) and establish the /etc/resolv.conf file.

- nfs—used to permit access to remote file systems for mounting.

- portrelease—with portreserve, manages port addresses that need to be reserved for usage by an application. Portrelease releases a reserved port.

- portreserve—reserves a port address for an application until portrelease releases the reserved port.

- sshd—service that permits ssh access into your computer.

- /etc/hosts—stores IP alias to IP address mapping information for resources that your computer will often communicate with.

- /etc/resolv.conf—stores IP addresses of your local name server(s).

- /etc/sysconfig/iptables—stores the Linux IPv4 firewall rules.

- /etc/sysconfig/iptables-config—stores the Linux IPv4 firewall configuration directives.

- /etc/sysconfig/ip6tables—stores the Linux IPv6 firewall rules.

- /etc/sysconfig/ip6tables-config—stores the Linux IPv6 firewall configuration directives.

- /etc/sysconfig/network-scripts/network-functions—stores various functions that are used by network scripts.

- /etc/sysconfig/network-scripts/ifcfg-eth0—data file for the Ethernet device (in this case, eth0), including an IP address if a static IP address is being used.

- /etc/sysconfig/network-scripts/ifcfg-lo—data file for the loopback device.

- /etc/sysconfig/network-scripts/ifdown—script to bring down interface devices.

- /etc/sysconfig/network-scripts/ifdown-eth—script to bring down your Ethernet interface device.

- /etc/sysconfig/network-scripts/ifup— script to bring up interface devices.

- /etc/sysconfig/network-scripts/ifup-eth—script to bring up your Ethernet interface device.

- /etc/xinetd.conf—the configuration file for the xinetd service.

## REVIEW QUESTIONS

1. You have a local area network containing four subnetworks. When a message reaches a subnet, you want to broadcast that message to only the appropriate destination computer. Which type of broadcast device would you use?

2. Following up from #1, what type of broadcast device would you use to connect the four subnets together assuming that the four subnets share the same network protocol?

3. Following up from #2, you want to connect the device from #2 to the Internet. What type of broadcast device might you use?

4. What is the difference between eth0 and lo?

5. What takes place in TCP/IP's application layer?

6. In what layer of the TCP/IP protocol stack are MAC addresses used?

7. In what layer of TCP/IP do routers get involved?

8. To maintain a connection between two computers on the Internet, which layer of the TCP/IP protocol stack is involved?

9. What is a checksum?

10. What is the difference between UDP and TCP?

11. Let us assume there are 7 billion people on the planet. Explain why IPv4 addressing does not provide a sufficient number of IP addresses.

12. Again, assume there are 7 billion people on the planet. How many different IPv6 addresses could each person be awarded given the 128-bit size for IPv6?

13. Why are there port numbers that are not reserved?

14. What Linux service would you run so that an application could claim a nonreserved port number and use it temporarily? What service would you run once that application completed so that the port number could be made available again?

15. On the Internet, what is a domain? List some top-level domains.

16. Which Linux service needs to be running for you to have access to the Internet?

17. What does the script ifdown do?

18. What is the difference between the script ifdown and the script ifdown-eth0?

19. In the interface device configuration file (e.g., ifcfg-eth0), you might find the entry BOOTPROTO, what does this value indicate?

20. Of the following, which is(are) necessary for the ifcfg-eth0 file? ONBOOT, HWADDR, BROADCAST, IPADDR, or BOOTPROTO?

21. Of the following services, which one(s) are useful to support network security? avahi-daemon, certmonger, dnsmasq, iptables, ip6tables, postfix, netfs, or nfs?

For questions 22 through 25, assume the default information provided in Section 12.3 for xinetd and the following entry:

```
service ftp
{
        socket_type             = stream
        instances               = 10
        wait                    = no
        user                    = root
        server                  = /usr/sbin/in.ftpd
        server_args             = -l -a
        no_access               = 1.0.0.0 2.0.0.0 3.0.0.0
        access_time             = 04:00-22:59
        log_on_success          += USERID
        log_on_failure          += USERID
        disable                 = yes
}
```

22. What are the items logged on an ftp success? On an ftp failure?

23. How many instances of ftp can xinetd handle at once? Does this differ from other services?

24. What does wait = no mean?

25. What limitations are there in accessing ftp ?

26. Explain the role of the /etc/hosts table.

27. Your /etc/resolv.conf file is empty. Does this mean that you cannot access the Internet at all? If not, what restriction(s) might this place on your Internet usage?

28. Why might you want to issue a static IP address to a computer?

29. What are the advantages and disadvantages of using dynamic IP addresses?

For questions 30 through 35, assume your computer's IP address is 10.145.201.12. Compute your network address given each of the following netmasks:

30. 255.255.255.0

31. 255.255.192.0

32. 255.255.128.0

33. 255.255.0.0

34. 255.240.0.0

35. 255.224.0.0

36. Why might you want a DHCP server to lease IP addresses rather than assign them?

37. What command would you issue to obtain the IP address of all of your interfaces?

38. What command would you issue to obtain the IP address of your eth0 interface?

39. What command would you issue to change the IP address of your eth0 interface?

40. What command would you issue to obtain the IP addresses of any routers in your router tables?

41. What command would you issue to flush your local router table?

42. Why might you be discouraged from using the command ifconfig?

43. Under what circumstance(s) might you want to implement a firewall to prevent some outgoing messages from being sent?

44. Are the orders that rules are listed in the iptables file important? Why or why not?

45. What does -P indicate in an iptables rule?

46. What is the difference between a target of REJECT and a target of DROP in the Linux iptables firewall?

47. The example rules from Section 12.6 do not explicitly accept incoming messages from a web server (which is usually over port 80). How then could the user of this firewall perform web browsing?

48. Assume your computer has a point-to-point connection with another device over interface ppp0. You want to permit all messages to come in from that device. What iptables rule would you specify to permit this?

49. Define an iptables rule to accept messages from IP address 10.11.53.1 over any of ports 20, 21, 22, 23, 53, or 80.

50. Define an iptables rule to reject messages over your Ethernet interface whose length is 0.

51. Define an iptables rule to accept messages over your Ethernet interface that use the UDP protocol and come from IP addresses 10.11.12.1, 10.11.12.2, or 10.11.12.3 between 9 a.m. and 5 p.m.

52. Define an iptables rule to reject any outgoing messages intended for the IP address 31.13.69.130.

53. Does iptables permit rules that can prevent a denial of service attack? If so, how?

54. Write a script similar to the first script from Section 12.7 that computes the total number of packets lost given a single IP address.

55. In the function `try_wget` developed in Section 12.7, explain why we need to actually save the downloaded file retrieved from wget.

# CHAPTER 13

# Software Installation and Maintenance

THIS CHAPTER'S LEARNING OBJECTIVES are

- To know the questions you should consider when selecting and installing software

- To know the various methods of software installation in Linux

- To understand the concept of packages, package managers, and package dependencies

- To understand the installation process for open source software

- To understand the compilation process and the use of gcc

- To understand the role of the system administrator in maintaining installed software

- To understand the history and significance of the open source movement

## 13.1  INTRODUCTION

Software installation in any operating system has become a simple task for the user. In the Windows operating system, installation wizards are almost completely automated with the user only having to answer a few questions. In Linux, software installation can range from simple if the software is bundled in a package, to complex when dealing with open source software.

In this chapter, we begin by considering questions that you should answer before installing software. We look at several different approaches to installing software in Linux. We examine the open source initiative that supports most of the Linux software. We also take a look at gcc, the GNUs C Compiler. We finish the chapter with a brief look at software maintenance, including updating and removing software and forms of software support.

## 13.2 SOFTWARE INSTALLATION QUESTIONS

Before you install any software, there are questions that you should consider.

- Do you need the software? Software will take up disk space, and depending on the source of the software, there could be risks in installing it. For instance, some software could potentially interfere with already-installed software. Microsoft software has interfered in the past with some non-Microsoft titles. Additionally, if the software is from an untrusted source (an unknown third party), you take the risk that the software may not be as it appears. Such software could include a Trojan horse, spyware, or other malware.

- Are there other software titles worth examining? This is particularly relevant in Linux where you might find open source versions of equivalent software or alternatively proprietary software that comes with support. Your choice of software title might be based on whether you want something for free or you want support and are willing to pay for it.

- What resources does the software require? All software requires some level of resources, including hard disk storage space, main memory capacity, and specific platform (e.g., Windows or desktop Linux). Software also requires a minimum processing capability to function smoothly. Other resources might include access to the network, input from pen tablet or microphone, some specific type of graphics card, and so forth. Make sure you have the resources necessary before purchasing or acquiring the software.

- What does installation entail? You might find that to install a piece of software, you have to first install a number of supporting files. If the installation process becomes unwieldy, you might prefer to obtain a different piece of software.

- Will this software title run on a stand-alone computer or a server, or be used on multiple computers? This question does not so much suggest *whether* you should install the software but *how* and *where* it should be installed. Installing software on multiple computers can require a good deal of time. Many organizations will install the software on a shared file server. Another alternative is to "push" the software onto all computers over the network.

- Finally, do you have enough disk space in the destination partition(s)? In Linux, most software will be installed under /usr (in /usr/bin most likely) or possibly /opt, but the software may also create files in /etc and /var.

Over the next three sections, we look at different mechanisms by which we can install software. First, we will look at installation through the Add/Remove Software GUI application available in CentOS. Second, we will look at three package managers: `rpm` and `yum` for Red Hat Linux and `apt` for Debian Linux. Finally, we will look at installing software from source code (open source software), requiring the use of the `make` command.

## 13.3 INSTALLING SOFTWARE FROM A GUI

### 13.3.1 Add/Remove Software GUI in CentOS

Recall during CentOS installation (Chapter 8) you were given a choice of installation types (minimal, desktop, server, etc.). You were also given the opportunity to select additional software packages. It is not necessary to select software at that juncture because you can always add software packages at a later point in time using the same mechanism.

From the System menu, select Administration and then Add/Remove Software. The Add/Remove Software GUI appears. You have three possible ways to proceed. First, if you know the software's package name, enter it in the search box. Second, click on one of the four selections in the upper left-hand pane to select a group of packages, or third, select one of the triangles in the lower left-hand pane to open up a particular software type. Figure 13.1 illustrates the selection of all packages. In the right-hand pane, you see that there are a few packages visible and far many for you to scroll through. Of these packages, "Automated bug detection and reporting tool" (abrt-2.0.4-14.el6.centos) is the only one currently installed.

The areas in the lower left-hand pane are groupings of types of packages. Applications contains selections for Emacs, Graphics Creation Tools, Internet Applications, Internet Browser, Office Suite and Productivity, Tex support (for the TeX and LaTeX text formatting program), and Technical Writing. Selecting any of these brings up all related packages. For instance, Graphics Creation Tools includes the GIMP program and supporting packages, X Windows tools, and drawing programs. If packages have already been installed, they will appear with check marks next to them.

The Base System packages include programs and tools that help support the Linux user and system administrator. These include the Bacula backup utility, authentication programs, including Kerberos, PAM, and LDAP; network programs such as FTP, Samba, nscd

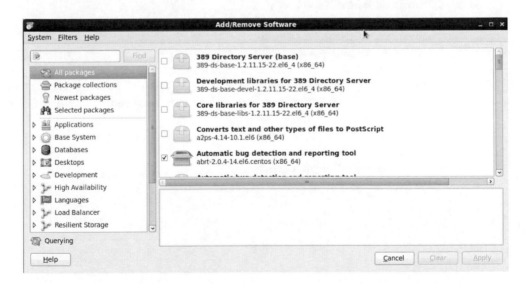

FIGURE 13.1   Add/Remove Software GUI.

(name service caching daemon), and mail programs; the Java run-time environment (JDK); support for CUPS (common unix printing service); and many other types of packages.

The Desktops entry contains a number of different windowing systems for Linux. You have most likely already-installed Gnome. Available here are also KDE, Nautilus, Motif, and X Windows support utilities. There are packages for font libraries and remote desktop clients. There is also support for different languages (human languages).

The Development entry contains packages to support the software developer through different programming language environments, including compiler support for C, C++, Java, FORTRAN, and Ada 95 to name a few. Additionally, this entry contains C library files.

There are separate entries for Databases and Servers. Under Databases, you will find MySQL and PostgreSQL client and server software. Server types include backup servers, directory servers, Email servers, FTP servers, NFS servers, NIS servers, print servers, and administration tools. Separately, under Web Servers, you will find the Apache web server, the Squid proxy server, and supporting files and modules.

Installation from any one of these packages is simple. Select the item(s) to be installed and click on the Apply button. Any package might have *dependencies*, that is, require additional files. In the case that those dependencies are not already installed, the installer will find the appropriate package(s) to install and ask you if the additional package(s) should also be installed. If you do not install the dependent packages, your installation will fail.

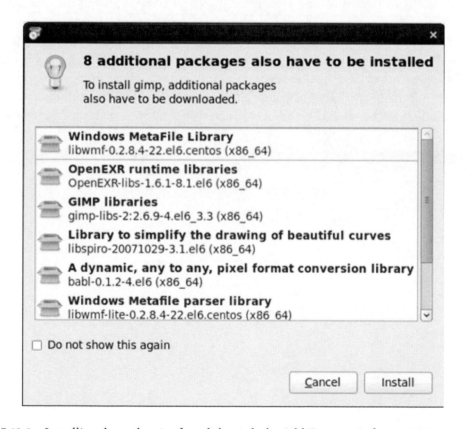

FIGURE 13.2   Installing dependencies found through the Add/Remove Software GUI.

FIGURE 13.3    Ubuntu Software Center.

Figure 13.2 shows the dependencies required to complete the installation for GIMP (the GNU Image Manipulation Program). In this figure, we see that eight additional packages need to be installed. Fortunately, installation is not any more challenging than selecting the Install button. The only downside to installing these additional packages is the time and disk space that they will take up. We will return to the Add/Remove Software GUI in Section 13.7 as we can use it to remove software packages.

### 13.3.2  Ubuntu Software Center

Ubuntu has taken a different approach to software installation, through the Software Center as shown in Figure 13.3. From this GUI, you can explore installed software or all software titles that are currently available. All software includes titles provided by Ubuntu, those provided by "canonical partners," proprietary software and software provided by independent programmers. You can view software based on type: Accessories, Books & Magazines, Developer Tools, Education, Fonts, Games, Graphics, Internet, Office, Science & Engineering, Sound & Video, System, Themes & Tweaks, and Universal Access. Within these categories, software may be divided into subcategories. For instance, Science & Engineering is divided into categories like Astronomy, Biology, Chemistry, Computer Science & Robotics, and Mathematics. Developer Tools is subdivided into categories like Debugging, Graphical Interface Design, IDEs, Java, Lisp, Perl, and Web Development.

Once you have selected a software title, you are asked to authenticate. Recall in Ubuntu, you are not provided the root password. However, the initial user is made an administrator.

So this user is able, through sudo, to perform software installation. This user should be the one who authenticates. When done, installation takes place and the new software title is installed. Installation will most likely take place under /usr.

## 13.4 INSTALLATION FROM PACKAGE MANAGER

A package manager is a program used by users or system administrators to install, update, upgrade, or remove software. In Red Hat Linux, the primary forms of package managers are `rpm` and `yum`. Rpm stands for Red Hat Package Manager and yum stands for Yellowdog Updater Modifier. Rpm is the more primitive approach requiring more effort by the user/administrator. Yum on the other hand calls upon rpm to accomplish its tasks. As such, yum is a much simpler way to install software. In Debian Linux, the primary package manager is `apt` (Advanced Packaging Tool). Rpm, yum, and apt are now available in other Linux distributions, so it is useful to learn all three tools.

In general, a package manager, or a package management system, provides several different useful functions for both installation and maintenance of software. The package manager can be used to verify the correctness of the contents of a package. Correctness means that the files themselves have proper checksum values (for error detection) and proper signatures. A digital signature is not required for software installation. But when one is provided with a software package, you can use this to ensure that the software is legitimate. The package manager operates as an archival tool in that it can unpack and uncompress the files in the package. Since the packages typically consist of executable files, library files, and supporting files (e.g., documentation and data files), there is no need to compile the files. However, the files must be placed correctly. The package manager performs installation by testing files and moving them to their destination directories. Testing of files involves checking dependencies. For instance, if one file requires a particular library file to function, then that library file must be available in your Linux system or the dependency is not met.

Libraries consist primarily of .so files. The so extension stands for "shared object" where shared means that the item is to be shared among multiple programs and object means that it is an object (already compiled) file.

The .so files play a similar role in Linux as the .dll files (dynamic linked library) in Windows. However, in Linux, so files are versioned, meaning that they are stored based on the version number of the Linux system or software that will use them. Thus, the same named library may exist under slightly different names. In Windows, a newer version of the same dll file will replace an older version so that software requiring the given library is forced to use the most recent version. This may or may not complicate matters in running the older software. This leads to a situation that many Windows programmers have dubbed "DLL Hell."

The majority of Linux library files are found under either /lib or /lib64. These directories are divided into subdirectories for classes of library files. For instance, you might find security library files under `/lib/security` and/or `/lib64/security` while most library files are found directly under /lib64. Most of these files have the name lib*software-version*.so or lib*software*.so.*version* such as `libgmodule_2.0.so`, `libselinux.so.1`, or `libgcrypt.so.11.5.3`. We will briefly discuss the idea of linking and libraries when we examine gcc later in this chapter.

Returning to the package manager, it can be used to update installed software by installing newer versions, patches, or additional library files. The package manager can also be used to remove software packages if they are no longer of use. This is the antithesis of installation in that the files placed in various directories must be identified, and those that are not involved with dependencies of other packages are deleted.

### 13.4.1 RPM

For our examination in this section, we start with the more primitive tool, rpm. To use rpm, you must first have an rpm file. The rpm file is an archive compressed using gzip. You can obtain rpm files from an *RPM repository*. There are many RPM repositories. For CentOS, the primary website is `centos.karan.org`. For non-CentOS Red Hat, you can find RPMs at `rpmfind.net/linux/RPM`. RPM Fusion (`rpmfusion.org`) advertises RPMs that the Fedora Project and Red Hat Enterprise Linux do not want to ship. You can also find numerous RPM files for a variety of distributions at `rpm.pbone.net`. At `packages.debian.org`, you can find Debian apt repositories that have been ported to the RPM format.

The RPM files are given very expressive names divided into four sections. The first three sections are separated by hyphens to express the software's name (title), version number, and release number. You might, for instance, see a package name like *title*-2.3.1-3. This would be the *title*'s version 2.3.1, release number 3. Following these three are a period followed by the intended architecture type. As the RPM contains compiled software in the form of executable files, the RPM is intended for a particular platform. For instance, `i386` means Intel 386 (or later), `i686` means Intel 686 (Pentium II or later), `ppc` for PowerPC processors, and `sparc` is used for Sun Sparc workstations. The entry `noarch` for an architecture means that the RPM should run on any architecture.

Once you have selected and downloaded the rpm file(s), you can use the rpm instruction. The basic form of the command is

```
rpm [option] file(s)
```

where the file(s) is(are) the rpm file(s). The options are

- -i  (or --install) to install the package

- -U (or --upgrade) to upgrade the package

- -F  (or --freshen) to freshen the package

- -e  (or --erase) to remove the package

There is little difference between upgrade and freshen. Both will update the given package with a newer version. However, with freshen, a previous version must have already been installed whereas upgrade can either update the already-installed package or install from scratch. If you are unsure if a package already exists, upgrade is the better choice to either install or freshen. In fact, since install allows you to install multiple instances of the same package rather than replace preexisting versions (or skip installation), using -U/ --upgrade is preferred in all cases.

Installation using rpm should be straightforward. However, you might find that any given package has *dependencies*. A package dependency means that there is some capability that the RPM file needs in order to install and run properly. A dependency is usually a *library* that is missing (libraries are explained in the next paragraph). It is now your responsibility to figure out what package(s) contains the dependent item(s) and install it(them).

The reason that dependencies exist is that programmers will rely on other, already-implemented software components to simplify their own programming process. Without this reliance, programmers would each have to implement numerous additional tasks such as program code that creates windows, deals with virtual memory and handles security. Therefore, programmers utilize library components that are collections of functions (or classes). The problem is that to install their software, you may not have already installed those dependent components.

Unfortunately, it is not necessarily as simple as downloading and installing another RPM file to fulfill a dependency. You might find that to fulfill a dependency, you need to download a package that itself has dependencies that require that you install a package, which itself has dependencies that require... This is a situation that some call *dependency hell* because it can be a challenge to track down any missing dependencies.

You can test an RPM for dependencies by adding `--test` to your install command. For instance, `rpm -i --test somepackage.version.rpm` will test to see if *somepackage* can be installed. Below is the output of such a report. In this case, the software title is replaced with — — — — — — — — — -.

```
error: Failed dependencies:
     libSDL-1.2.so.0 is needed by — — — — — — — — — -
     libSDL_mixer-1.2.so.0 is needed by — — — — — — — — — -
     libc.so.6 is needed by — — — — — — — — -
     libc.so.6(GLIBC_2.0) is needed by — — — — — — — — — -
     libc.so.6(GLIBC_2.1) is needed by — — — — — — — — — -
     libc.so.6(GLIBC_2.1.3) is needed by — — — — — — — — — -
     libm.so.6 is needed by — — — — — — — — — -
     libpthread.so.0 is needed by — — — — — — — — — -
     libpthread.so.0(GLIBC_2.0) is needed by — — — — — — — — — -
```

This message informs us that this package has numerous unmet dependencies. Unfortunately, the listed capabilities, `libSDL-1.2.so.0`, `libSDL_mixer-1.2.so.0`, `libc.so.6`, `libm.so.6`, and `libpthread.so.0` are library names, not package names. We must find one or more packages that contain these libraries and install that/those package(s). But which package(s)?

Fortunately, to simplify this task, there is an RPM resource finder website, `rpmfind. net`. By entering the needed library, say `libc.so.6`, into the search box from this website, we are provided a page of links to obtain the necessary RPM file. See Figure 13.4, which provides a partial view of this page. From here, we select a link that matches our platform,

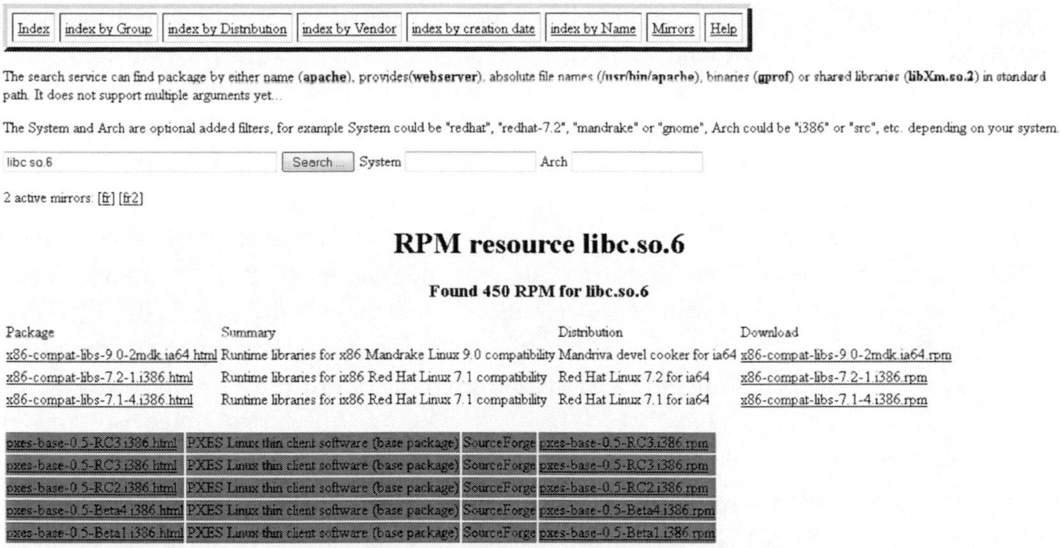

FIGURE 13.4    RPM finder webpage providing links to missing libc.so.6 library.

as in ix86 Red Hat Linux 7.1 for ia64. As the rpm file contains executables, we have to match the platform. Alternatively, we could download the source code from SourceForge and build the missing libraries ourselves.

So we download the rpm file that contains the missing library(ies). Now we install that or those rpm file(s). With the dependency(ies) resolved, we can return to installing our original RPM file. We issue the rpm -i *packagename*. Figure 13.5 provides a verbose output of this command (note that the specific rpm filename has again been replaced

```
D:   read h#     1094 Header sanity check: OK
D:   Requires:  libc.so.6                                 YES (db provides)
D:   Requires:  libc.so.6(GLIBC_2.0)                      YES (db provides)
D:   Requires:  libc.so.6(GLIBC_2.1)                      YES (db provides)
D:   Requires:  libc.so.6(GLIBC_2.3)                      NO
D:   package ------------------ has unsatisfied Requires: libc.so.6(GLIBC_2.3)
D:   Requires:  libm.so.6                                 YES (db provides)
D:   Requires:  libm.so.6(GLIBC_2.0)                      YES (db provides)
D:   Requires:  libncurses.so.5                           YES (db provides)
D:   Requires:  rpmlib(CompressedFileNames) <= 3.0.4-1    YES (rpmlib provides)
D:   Requires:  rpmlib(PayloadFilesHavePrefix) <= 4.0-1   YES (rpmlib provides)
D:   opening db index       /var/lib/rpm/Conflictname rdonly mode = 0x0
error: Failed dependencies:
        libc.so.6(GLIBC_2.3) is needed by -----------------------
D:   closed  db index       /var/lib/rpm/Conflictname
D:   closed  db index       /var/lib/rpm/Providename
D:   closed  db index       /var/lib/rpm/Basenames
D:   closed  db index       /var/lib/rpm/Name
D:   closed  db index       /var/lib/rpm/Packages
D:   closed  db environment /var/lib/rpm
```

FIGURE 13.5    Output of RPM installation.

with — — — — — — -). Notice that not all dependencies have been fulfilled in the output. Specifically, GLIBC_2.3 is needed.

Many RPM files contain a digital signature. This signature is used to ensure the authenticity of the file. This allows you to feel safe in that what you are installing is a legitimate piece of software. Should the signature be absent or incorrect, you will receive either an error or warning. These messages can be cryptic if you do not know why they arise. Errors can arise if the signature is not recognized or does not match what is expected and warnings are issued if there is no signature. Error messages include the word "BAD" to indicate that the signature was deemed bad while the warning messages include the word "NOKEY." The warning will not prevent the package from being installed/upgraded while the error will.

The rpm program has a number of installation/upgrade options. A few of the more useful ones are listed in Table 13.1.

Other options are available if you are querying or verifying a package. Querying a yet-to-be-installed package allows you to find out information about that package, including, for instance, the executable files, document files, configuration files, or script files enclosed in the package, package information (name, version, description), dependencies, and the state of the files in the package (e.g., installed, not installed, replaced). Verifying a package compares the information as stated about the files to the actual files. This includes testing the checksum value, comparing the listed and actual file types, permissions, owners, and modification times. You can also check the signature explicitly using rpm --checksig.

Obviously, there is a lot to using rpm. However, as a system administrator, it can be easy to install software from the RPM should the dependencies check out ok. Otherwise, you may find yourself searching for RPM files that satisfy dependencies, and this can be annoying and time consuming.

## 13.4.2 YUM

Using yum to install software is far simpler than using rpm. The main reason for this is that yum will seek out dependencies of a package and find packages to install that resolve those dependencies. In this way, installing a single package may actually require installing numerous packages. But once you issue the yum command, you can be blissfully ignorant of the amount of work involved!

TABLE 13.1    RPM Installation and Upgrade Options

| Option | Meaning |
| --- | --- |
| --excludedocs | Install without any documentation (if any of the files are explicitly listed as documentation); this includes omitting man pages. |
| --force | Force installation even if some of the files have already been installed. |
| --ignoresize | Do not check for sufficient disk space in given partition before installing. |
| --ignorearch | Perform installation even if the binary files do not match the host computer's architecture. |
| --includedocs | Install documentation files; this is selected by default. |
| --nosignature | Do not verify signature so that errors/warnings do not arise from bad or missing signatures. |
| --nodeps | Install without performing dependency check; the software may install but may not run correctly. |

Yum is also handy for upgrading software without your involvement. Whereas in RPM, to upgrade a package you would have to initiate the request, you can schedule yum to regularly search for upgrades and install them without your interaction. We examine this use of yum in Section 13.5.

The yum program does not require a file already downloaded to perform its task unlike rpm (although rpm could download the file first, if a URL is provided in place of a file name). With yum, instead, it searches prespecified repositories and mirror sites for the package named in the instruction. The yum instruction essentially has three parts:

- The command: install, update, remove, list
- The package name: this is not a file name, but a less specific software title such as gcc (the C compiler), emacs, httpd (the Apache web server)
- Options

For example, you might issue the command

```
yum install emacs
```

or

```
yum update httpd
```

The `list` command does not require a package name at all. If one is provided, you will be given all packages that match that name. Wildcard characters of * and [] can be included with the package name so that globbing takes place. Two examples follow:

```
yum list *gnome*
yum list gimp[0-9]
```

The first of these will list any packages that contains "gnome" in the title. The second will list any packages starting with the characters "gimp" followed by a digit.

Aside from install, update, remove, and list, there are dozens of other possible commands. Table 13.2 highlights a few of the more common commands. The additional parameters are typically either package names or filenames. As with install and list, these can include wildcards.

yum has a number of options although only a few are worth mentioning. First, the option -y (or --assumeyes) will respond 'y' to any "yes/no" question that yum asks. This allows yum to run without any user interaction. The options -d and -e set debug and error levels to control output. Other options control aspects of the installation such as --installroot=*dir* to alter the installation location to *dir* as opposed to the default and --enablerepo=*repos* and --disablerepor=*repos* to enable and disable specific repositories to be or from being accessed for installation.

TABLE 13.2   Yum Commands

| Command | Meaning | Additional Parameters |
|---|---|---|
| check-update | Check to see if there are any upgrades available for installed software. This allows you to test to see if you need to perform a `yum install` at this time. | None |
| clean | Clean up the yum cache directory. | One of: `packages`, `metadata`, `expire-cache`, `rpmdb`, `plugins`, `all` |
| downgrade | Attempt to downgrade a package from the current version to the previous version (if available). | Package(s) |
| groupinstall, grouplist, groupremove, groupupdate | Same as individual commands (install, list, remove, update) but on a list of packages. We discuss update and remove in Section 13.7. | Group(s) |
| info | Provide description and summary of the available package(s). | Package(s) |
| provides | List any package(s) that provides a given feature or file. | Feature/file(s) |
| reinstall | Uninstall and then install anew the package(s) specified. | Package(s) |
| repolist | List available repositories. | One of: `all`, `enabled`, `disabled` |
| resolvedep | List packages that fulfill a given dependency. | Dependency(ies) |
| search | Search all packages' summaries for the given string for matches. Useful if you know something about the package but not its name (somewhat similar to Linux `apropos`). | String |
| update-to | Update an installed package to the version specified. | New version |

Let us take a look at the output from yum. Here, we issue the command

```
yum -y install emacs
```

This command will find the emacs package in one of the repositories and begin installation without waiting for the user because of the –y option. The output provided is shown below with comments interspersed to explain what is going on.

```
Loaded plugins: fastestmirror, refresh-packagekit, security
Loading mirror speeds from cached hostfile
* base: ftp.linux.ncsu.edu
* extras: mirror.serversurgeon.com
* updates: mirror.linux.duke.edu
```

Up to this point, yum has contacted the repositories to locate the necessary packages.

```
Setting up Install Process
Resolving Dependencies
--> Running transaction check
---> Package emacs.x86_64 1:23.1-21.el6_2.3 will be installed
--> Finished Dependency Resolution
Dependencies Resolved
```

Here, we see that yum has located all dependencies of the package to be installed.

```
================================================================================
 Package         Arch           Version              Repository          Size
================================================================================
Installing:
 emacs           x86_64         1:23.1-21.el6_2.3    base                2.2 M

Transaction Summary
================================================================================
Install 1 Package(s)

Total download size: 2.2 M
Installed size: 11 M
```

At this point, yum would usually pause and seek confirmation on the installation. The reason for pausing is to let the system administrator see the size of the package(s) and version(s) to be installed. Using –y causes yum to skip the interaction and proceed with the installation.

```
Downloading Packages:
emacs-23.1-21.el6_2.3.x86_64.rpm             | 2.2 MB          00:00
Running rpm_check_debug
Running Transaction Test
Transaction Test Succeeded
Running Transaction
  Installing : 1:emacs-23.1-21.el6_2.3.x86_64                    1/1
  Verifying : 1:emacs-23.1-21.el6_2.3.x86_64                     1/1
```

A progress bar is shown during these steps, which is erased upon completion of each step. We see at this point that yum is installing an RPM file. That is, emacs is being installed through rpm. The file passes the various tests (debugging, verification, dependencies).

```
Installed:
      emacs.x86_64 1:23.1-21.el6_2.3

Complete!
```

And now yum has completed the installation. The entire process takes a few moments, perhaps as much as a couple of minutes. Emacs is a rather small package; larger packages could take several minutes or longer. If the repository (or its mirrors) is unavailable, yum will fail.

You might wonder how you could identify the package's name. Most package names are the name of the software, such as emacs, gimp, gcc, httpd, and squid. You can query the repository for available packages using yum list *name* where *name* can contain wildcard characters.

### 13.4.3 APT

For Debian Linux, we have apt (for Advanced Packaging Tool). Like yum, apt is a tool that calls upon lesser programs. There are currently projects to implement apt for Red Hat, but we will only report on apt in Debian. apt operates on deb packages and utilizes the Debian-based dpkg program. dpkg is like rpm while apt is like yum.

There are also variants of the dpkg: ipkg, opkg, and wpkg. The ipkg and opkg are actually the same format although opkg is no longer used and ipkg stands for "Itsy" Package Management System. This format is used for storage-constrained Linux environments such as embedded devices. The wpkg is a format used to install Debian packages for the Windows operating systems.

We concentrate only on apt here, leaving the readers to explore dpkg and the other formats on their own. apt works like yum in that it contacts one or more repositories that store packages. In this case, the packages are .deb packages. The file /etc/apt/sources. list stores the URLs of repositories to try. You might, for instance, find dpkg repositories at us.archive.ubuntu.com or archive.canonical.com. Another file, /etc/apt/preferences, is used to control preferred locations from the sources.list file for different software versions. Retrieved packages and status information are stored in caches under /var/cache/apt/archives and /var/lib/apt/lists, respectively. One of the most complex components in apt is a program that performs *topological sorting* to work out interpackage dependencies and determine the order that packages should be installed. Topological sorting is a graph algorithm often used to organize a sequence of tasks such that any task that has dependencies is executed only after those dependent tasks are executed.

apt is not really a program itself but is instead the name given to a package that contains several software maintenance tools. The primary program for software installation is called apt-get. The apt-get program uses the following syntax:

```
apt-get [options] command package(s)
```

where *command* is one of install, remove, update, upgrade, or dist-upgrade. These are all self-explanatory except for dist-upgrade, which handles changing dependencies between versions of packages. There are also commands available to check for dependencies and clean the local repositories.

Aside from apt-get, there are three other apt programs of note. First, apt-cache is used to query the APT package cache, which consists of downloaded packages and portions of packages. While this command will not alter the stored data, it can be used to retrieve and summarize information on what has been downloaded. Next is apt-file, which can inspect a package to find out what specific files are included. Finally, apt-secure can be used to ensure the integrity and authenticity of a package through digital signature checking.

While apt is a suite of programs, these programs all operate via the command line. Other, more recent management tools are available, which use apt just as apt draws upon dpkg. The popular front-end tool aptitude provides a GUI-like interface (it is still text-based with menus and mouse interaction).

Figure 13.6 provides an example of this interface. In aptitude, a number of choices are presented to the user to select from: Security Updates, Upgradable Packages, Installed Packages, Not Installed Packages, Virtual Packages, and Tasks. Within any one of these categories are subcategories and subsubcategories. Additionally, the menus offer the ability to search for packages, resolve packages, and so forth.

Other GUI front-end programs are the Ubuntu Software Center (see Figure 13.3), the Synaptic Package Manager, and the Adept Package Manager. The Synaptic Package is shown in Figure 13.7. Much like the CentOS Add/Remove Software GUI, this GUI provides types of software under Sections. Clicking on a specific package provides its description in the lower right window.

Commands available include determining a package's status (installed, not installed, upgradable), location (origin), filters (e.g., broken, community maintained, missing recommends), and available architectures. Selecting a package then allows you to install, reinstall, upgrade, or remove the package.

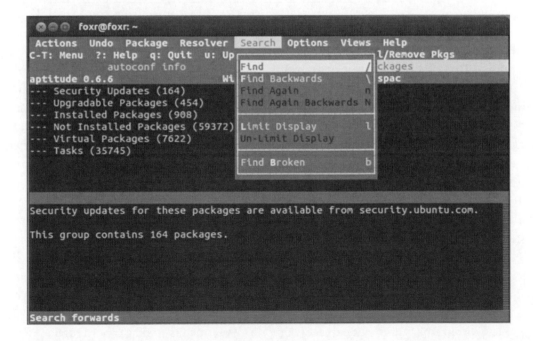

FIGURE 13.6   Debian-based aptitude interface.

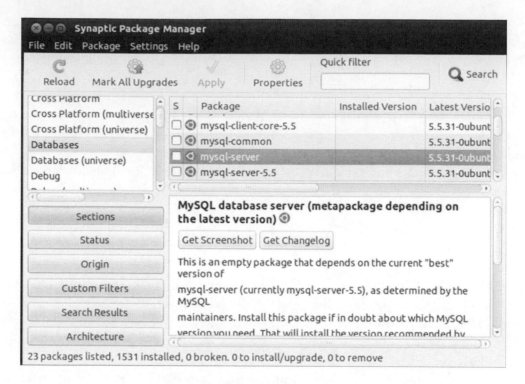

FIGURE 13.7   Debian-based Synaptic Package Manager.

## 13.5 INSTALLATION OF SOURCE CODE

One of the greatest strengths of Linux is its openness. This comes from the availability of both the operating system and much of the application software in source code format. With the source code, individuals and organizations can make modifications to software. These modifications may be improvements, new features, or simply alterations of features available. Of course any modification to source code requires the expertise in programming to first understand the source code to figure out how to make the desired changes.

Because of the availability of source code, many administrators will download software packages not in executable form but in the source code form. Such packages are often stored in tar files and then zipped (using gzip or bzip2). This allows for easy storage and transport (over network) of the packages. However, because most of the software we are dealing with consists of many files, the compilation and installation process can be very challenging. To simplify the operation, Linux has a suite of instructions that permit the system administrator to perform the installation with ease. Here, we look at this process.

Note that most open source software is written in C or C++. You must have installed the GNU's C compiler (gcc) to use the approach described here. You can see if gcc is already installed by typing which gcc. It should be installed in /usr/bin. If it is not available, you can install it using yum by typing yum -y install gcc. Note that gcc used to solely compile C programs but now it can also compile C++ code (as well as other languages; we discuss gcc in the next section).

### 13.5.1 Obtaining Installation Packages

The first step in installing an open source software package is to obtain the package. There are numerous repositories often found on websites managed by software developers of the software titles. For instance, the Apache webserver is available for download at `httpd.apahce.org` while the Squid proxy server is available at `www.squid-cache.org`.

In addition to websites dedicated to the software product, most open source developers make versions of their software available via the `SourceForge` site at `sourceforge.net`. This site offers thousands of open source software titles. The browse page contains the list of products available, sorted by popularity (number of downloads), name, last modification, or rating. Alternatively, you can search for applications based on categories of Audio & Video, Business & Enterprise, Communications, Development, Home & Education, Games, Graphics, Science & Engineering, Security & Utilities, or System Administration.

Upon reaching some particular software's page, you will find a variety of information on the software. First, you should find a download link. This link downloads an installation program while also taking you to a form whereby you can be placed on a mailing list for one of SourceForge's sponsors. The software's main page will also contain statistics of the number of total downloads to date and the rating of the software (on a scale of 5). You will find several tabs to select from that include a summary of the software, reviews, support, and latest news. There may be other tabs, including a wiki dedicated to the software, a list of files available, add-ons, or plugins, and ways to donate to the project.

To download the source code, select the Files tab, and from there, you are presented with various versions. Once you have selected the version desired (you should always look for a stable version, not necessarily the newest as the newest may be a *beta-test release*, meaning that the software is not fully developed or debugged), you are presented with a list of files to select from.

There may also be a subdivision in the software listed based on the intended architecture and platform. This becomes necessary if the code uses operating system functions. More commonly, there will be a division of operating system platform but not of processor type. You may find both executable programs and source code. The executable programs are already compiled, meaning that they are easily installed but not configurable. The source code consists of many files and subdirectories packaged using the tar program. Most of the time, such a bundle will also be compressed using the gzip format or the bz2 format so that these files will end with the extension .tar.gz or .tar.bz2. You might also find encrypted versions using one of PGP, MD5, or SHA1 (among others).

### 13.5.2 Extracting from the Archive

Once the package has been downloaded, your next step is to extract the files from its archive. This is typically taken care of by issuing a single tar command. If, for instance, you are installing the package *somepackage* version 3.1.8, you will have probably downloaded a file like *somepackage*-`3.1.8.tar.gz` (the numbers represent the version and release numbers). Now you issue the command

```
tar -xzf somepackage-3.1.8.tar.gz
```

The x option performs file extraction from the archive, the z option uncompresses the file using gunzip, and the f indicates that the accompanying filename should be used.

As the package itself most likely contains a directory of files (and subdirectories), extracting from the package results in an entire file structure being created and populated with the directories and subdirectories from the tar file. Assuming our file was called somepackage-3.1.8.tar.gz, the above command should create a directory called somepackage-3.1.8. With the directory created, we cd into that directory.

Inside the software's installation directory, we will most likely see a great number of files and subdirectories (depending upon the size of the software package). Among the files will be a README file. This text file contains the instructions for installation, alerting the system administrator to any specific requirements and options for installation. There may be other text files such as LICENSE, CHANGES, NOTIFICATION (or NOTICE), ABOUT, and VERSION. There will also be at least one script file present, makefile (alternatively Makefile or Makefile.in). The makefile is used to compile and install the software. There may also be a configure file. The configure file is a script containing Bash instructions to create or modify the makefile file(s).

Among the subdirectories you will most likely find build (containing C program code), docs (man page documentation or other forms of documentation), include (C library files), and modules (containing add-on code that the system administrator might wish to add to increase the functionality of the software).

Figure 13.8 comes from untarring the open source package apg (automated password generator). Here, we see that most of the source code (the .c and .h files) is located in the top-level directory. The subdirectories of bfconvert, cast, and sha contain additional source code for extra functionality. The perl subdirectory contains perl code to be used if you want apg to run on a server. The files whose names are all in capital letters are instructions, acknowledgments, and a to-do list. The two scripts, install-sh and mkinstalldirs, are controlling scripts like those found in /etc/init.d. That is, they are used to start and stop the apg program as if it were a service. Finally, Makefile contains the compilation and installation operations. The apg software was packaged without a configure script, so any changes that you want to make to the compilation and installation process would have to be done by hand by altering Makefile yourself.

Assuming the package contains all necessary code, and assuming your system has the proper software already installed to compile and install the given package, the next steps should be straightforward. If not, it usually means that your system is not set up as the installation process expects (i.e., the original programmer(s) expects certain files in certain

| apgbfm.c | convert.c | getopt.h | perl | README.CYGWIN | THANKS |
| apg.c | convert.h | INSTALL | php | restrict.c | TODO |
| bfconvert | COPYING | INSTALL.CYGWIN | pronpass.c | restrict.h | |
| bloom.c | doc | install-sh | pronpass.h | rnd.c | |
| bloom.h | errors.c | Makefile | randpass.c | rnd.h | |
| cast | errs.h | mkinstalldirs | randpass.h | sha | |
| CHANGES | getopt.c | owntypes.h | README | smbl.h | |

FIGURE 13.8   Contents of untarred open source archive.

directories and if that is not the case, the installation may fail). If the installation is not as expected, you as the system administrator may have to make changes to the configure file, makefile, or other files. It is therefore critical that you review the README and other installation documentation files first.

### 13.5.3  Running the configure Script

The first step may be optional, which is to run the configure script. If there is already a makefile present and you are not making modifications, you may be able to skip the configure step. If however there is no file named makefile (or Makefile), then you must run configure. You might find, for instance, Makefile.in. This is not a make-file. It is instead a partial makefile that will be used as a starting point by configure. configure will output a new file called makefile to be used in the next step. Additionally, if you wish to alter the default installation, you would specify these changes through the configure script as a series of options.

If you are dealing with a large piece of software, chances are either the README file or configure itself will include instructions on how to use configure. To execute configure, since it is a script, you would specify ./configure. To obtain help, you might try ./con-figure  -h (or --help). Some of the options available in configure might be to specify destination directories and modules to compile. You might find, for instance, the following options available:

- --bindir=*DIR* – specify location of the binary directory as *DIR*
- --libdir=*DIR* – specify location for object code (library files) as *DIR*
- --infodir=*DIR* – specify location for documentation as *DIR*
- --prefix=*DIR* – specify location for all directories (this will ensure all directories are placed in one location rather than distributing them) as *DIR*
- --enable-all – enable all available modules
- --enable-module=*module1 module2 module3 ...* – enable specified modules
- --disable-module=*module1 module2 module3 ...* – disable specified modules
- --with-*feature*=*value* – establish that variable *feature* should have *value*

The configure step may take seconds or minutes depending on its complexity. The result will not only be a great deal of output sent to your terminal window, but also the creation of a makefile.

### 13.5.4  The make Step

Once the makefile is successfully generated, your next step is quite easy. Just type make. The make command in Linux causes the makefile/Makefile script to execute. There are numerous options available for make, but perhaps the only significant ones are –i to ignore errors that may occur during compilation, -k to continue to work as much as possible even if errors arise, and –I *dir* to specify an include directory. If there are include statements, this directory is searched in place of the current directory. As with the configure step, the make step may take seconds or minutes and will result in numerous messages being displayed.

Most makefiles are written in parts. These parts may include the compilation section, an installation section, a clean-up section, and a tar section. To run the compilation section, just issue the make command by itself. To perform the installation separately (which is usually required), use the command make install. If you have already attempted to compile and/or install the software and it failed, before trying again, issue make clean. Finally, if you want to take the files and wrap them up (or back up) in a tar file, use make tar. The command make all should perform a make clean, make, and make install all in one. Not all makefiles will have all of these sections and you can find out which sections are available in the README file.

Here, we examine the components of a typical makefile, albeit a very simple one. Interspersed between the lines of the makefile file are explanations of the script's code.

```
CC=gcc
FLAGS=-Wall
LIBS=-lcrypt
LIBM=-lm
```

The above lines define several compilation variables. gcc is the GNUs C/C ++ compiler. This will be the command that the makefile eventually issues to compile the source code. –Wall is a popular option for the compiler, which enables all warnings. LIBS and LIBM define library options.

```
INSTALL_PREFIX=/usr/local
BIN_DIR=/bin
MAN_DIR=/man/man1
BIN_DIR_D=/sbin
```

These variables define installation directories. The BIN_DIR_D is used to store the daemon version of the program assuming that you want to create both a normal executable and a daemon.

```
PROGRAM_NAME = . . .
SOURCES = . . .
HEADERS = . . .
OBJECTS = . . .
```

These four variables define the names of the files to be used by gcc. Specifically, PROGRAM_NAME is the English description of the program and will be the name of the compiled file. SOURCES are the C/C++ source code files (.c or.cpp). HEADERS lists one or more header files (.h). Finally, OBJECTS are files that store already-compiled C/C++ code. These are known as object files (.o). The object files are often used as libraries, containing the compiled code of commonly referenced functions.

In this case, let us assume that we have the following definitions for SOURCES and OBJECTS, respectively:

```
SOURCES = file1.cpp file2.cpp file3.cpp
OBJECTS = file4.o file5.o file6.o
```

Continuing with our makefile, we reach the portion that actually performs the operation(s) desired.

```
all: name
name:
        ${CC} ${FLAGS} -o ${PROGRAM_NAME} ${SOURCES}
            ${LIBS} ${LIBM}
install:
        ...
clean:
        rm -f ${PROGRAM_NAME} ${OBJECTS} *core*
```

Here, we see the make options defined. With this `makefile`, you can perform any of make, `make all`, `make install`, or `make clean`. The `make all` command merely calls `make name` where *name* is the program's name (i.e., instead of *name*, you would fill in the program's name although in fact you could use any string as long as it is consistent in both occurrences of *name*).

The command listed under *name* uses the variables from earlier, creating the full compilation command as shown below.

```
gcc -Wall -o name file1.cpp file2.cpp file3.cpp
        file4.o file5.o file6.o -lcrypt -lm
```

In this command, `file1.cpp`, `file2.cpp`, and `file3.cpp` are the C files and `file4.o`, `file5.o`, and `file6.o` are object files.

### 13.5.5 The make install Step

With make complete, we can now run `make install`. These instructions are missing from the above `makefile`. The install section would primarily be composed of Linux operations to move the files to their destination. The destination locations would be based on the previously defined directory variables. For instance, the executable program might be placed in /bin as /bin/*name*. If the program is for use only by the system administrator, or there is a corresponding service that goes with the program, this executable might be placed in /sbin/*name* (or perhaps /sbin/*name*d). Documentation would be placed in /man/man1.

Instead of using mv or cp instructions, the make  install section might use the Linux install command. This command has the ability to not only move files but also change the permissions of those files. In this way, the files created by the make command are not limited to the permissions of the gcc compiler, nor will the author of the makefile have to include chmod commands.

For instance, we might see an instruction like

```
install -m 0755 name $BIN_DIR/$PROGRAM_NAME
```

The filename *name* is provided in the above gcc instruction. Without the −o *name* portion, gcc will automatically name the compiled file a.out.

Let us assume that our make command resulted in the creation of several files: an executable to be placed in $BIN_DIR_D, a script file to control the executable, to be placed in a location called $SCRIPT (which presumably will be /etc/init.d), a man page to be placed in MAN_DIR, a conf file to be placed in $INSTALL_PREFIX/conf/, and finally a group of icons (.png files) to be placed in $INSTALL_PREFIX/icons.

The install section of the makefile might then look like the following. The test commands are used to determine if a given directory exists and if not, create it. This is necessary before we attempt to move files into those directories via the install commands.

```
test -d $INSTALL_PREFIX || mkdir $INSTALL_PREFIX
test -d $INSTALL_PREFIX/conf || mkdir $INSTALL_PREFIX/conf
test -d $INSTALL_PREFIX/icons || mkdir $INSTALL_PREFIX/icons
install -m 0700 a.out $BIN_DIR_D/$PROGRAM_NAME
install -m 0700 $PROGRAM_NAME.sh $SCRIPT/$PROGRAM_NAME
install -m 664 $PROGRAM_NAME.man $MAN_DIR
install -m 644 $PROGRAM_NAME.conf $INSTALL_PREFIX/conf
install -m 644 *.png $INSTALL_PREFIX/icons
```

Your last step in the installation process, if desired, is to issue make  clean. You would do this either if an error arose during compilation or installation, or to finalize the process so that any temporary files created during this process can be cleaned up. If the makefile does not have a specific clean section, then you might have to perform the removal operation yourself.

Alternatively, most of the temporary files are produced in the directory created when you untarred the original source code package. If you believe your compilation and installation process is complete, you can cd up one level and type rm  −rf  somepackage-3.1.8. This will remove the entire directory and all of its contents. You might then place the original tar.gz file in some location to archive it so that, if you ever need to reinstall the package, you can bypass the original download.

Of course, by the time you might need to reinstall, there could be newer versions available for you to download. Alternatively, you might issue make  tar (if this section exists in the makefile) to automatically package up the important files produced during the compilation process. Naturally, you would do this before removing the directory.

## 13.6 THE GCC COMPILER

This optional section is provided for readers who are interested in learning more about the gcc compiler. gcc is the GNU Project C and C++ compiler. It is an essential tool in Linux because most open source software is written in C or C++. Although you can install a good deal of Linux software from executable code, to modify software you will need to obtain the source code, modify it, and compile it. Thus, you will need gcc. gcc may not be part of your current installation depending on which installation you selected.

To obtain gcc, you can either use the Add/Remove Software GUI and install it from the Development selection (Development tools subselection), or you can install it by issuing the instruction `yum -y install gcc`.

To use gcc, you will need to have some C or C++ source code. The gcc program then allows you to perform preprocessing, compilation, assembly, and linking of the gcc files.

Compilation is the process of converting source code of a program into executable code. This is a challenging problem, requiring multiple distinct phases. Compilers handle these in order and usually if one phase cannot be completed because of an error, the compiler abandons the attempt and provides the programmer with syntax error messages.

### 13.6.1 Preprocessing

In C/C++, preprocessing occurs before compilation. It executes pound (#) directives written in your program. Table 13.3 describes some of the more common # directives. Preprocessing is performed by the compiler, but it does so prior to any actual compilation of the program.

### 13.6.2 Lexical Analysis and Syntactic Parsing

Next, the program code is broken into distinct elements. These are sometimes referred to as lexemes. Each lexeme is then categorized using a token. In English, we might use tokens such as "noun," "verb," and "adjective."

In C/C++, we might identify these lexemes as "identifier," "literal," "for-operator," "semicolon," and "arithmetic-operator." The "for-operator" would literally be the word "for," used to denote the beginning of a for loop while an "arithmetic-operator" would be any legal arithmetic operator in C/C++ such as +, −, *, /, % (mod), << (shift), or & (bitwise AND).

Lexical analysis may result in errors if any particular lexeme does not have a corresponding category. For instance, a misspelled reserved word such as For (capitalization) or whil (misspelled) would not be properly identified. Another source of lexical error is of an identifier that was not properly declared. For instance, if we declare the variable x to be an int (integer) and later reference X, we have a lexical error in that x was declared by X is not x.

With lexemes identified and labeled, syntactic parsing organizes these elements together into instructions. In doing so, it identifies groupings of lexemes into large categories and then those groupings can be grouped together.

TABLE 13.3   Common C/C++ Compiler Directives

| Directive | Meaning | Example |
|---|---|---|
| #define | Macro substitution definition to replace a given string with another string | `#define MAX 1000`<br>`#define getmax(a,b)`<br>`((a>b)?(a):(b))` |
| #include | Include header file into this file before compilation | `#include <stdio.h>`<br>`#include "myheader.h"` |
| #ifdef... #endif | If parameter has been defined, perform the following task | `#ifdef MAX`<br>`int a[MAX];`<br>`#endif` |
| #ifndef...#endif | If parameter is not defined, perform the following task | `#ifndef MAX`<br>`#define MAX 1000`<br>`#endif` |
| #if...#else...#endif | Test given parameter, perform associated action (#elif clauses are also permitted) | `#if MAX > 500`<br>`int a[500];`<br>`#else`<br>`int a[MAX];`<br>`#endif` |

For instance, the if statement below would first be described as an if-operator, a condition, and a block. Inside the block, we have two assignment statements. The first assignment statement is a group of an identifier, an assignment operator, and an arithmetic expression, which itself is an identifier, an arithmetic operator, and a literal.

```
if(x>y - z) {
      x = x + 1.5;
      z + +;
}
```

The compiler creates what is known as a parse tree to represent the syntactic structure of the code. Figure 13.9 illustrates the parse tree for the previous C code.

Syntactic parsing may discover errors and so parsing would be abandoned. Slight variations of the above code could result in errors. For instance, omitting the first semicolon would lead the compiler to view the if clause to be $x = x + 1.5$ z++, which is lacking an arithmetic operator between 1.5 and z++. Alternatively, omitting one of the + in z++ would similarly be flagged as an error. Another error would arise if either the open or close parenthesis was omitted.

## 13.6.3 Semantic Analysis, Compilation, and Optimization

Semantic analysis is next performed by ensuring that the instructions are being used appropriately. Does the comparison in the if statement make sense? Are we comparing compatible types? Can the types being compared be directly compared? If the types are structs, then the comparison is invalid. Do the assignment statements in the if clause make

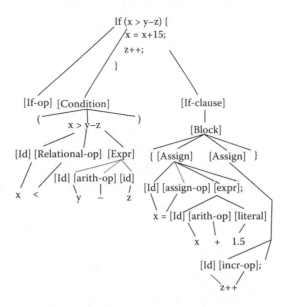

FIGURE 13.9    Parse tree for C instruction.

sense? For instance, if z is an array, you cannot apply ++ to it whereas if x is an int, you cannot add 1.5 to it.

As with the previous two phases, semantic errors can be caught during this phase. While lexical and syntactic errors are misuses of the programming language itself, semantic errors are misapplications of the programming language.

If the semantic analysis completes without error, the code can be converted. Typically, as an intermediate stage, the code is first converted to assembly language; thus we have an assembly component. Next, the assembly code is converted by assembler to machine language code (executable code). Even here, gcc is not finished. Optionally, you can ask that gcc optimize the code. This is particularly useful for today's processors that offer a number of parallel processing hardware elements that can execute program code more effectively if that code is compiled specifically to match the architecture. Optimization itself may require several additional passes through the code. The result is an executable program (or multiple executable programs).

## 13.6.4 Linking

Earlier in this chapter, we examined the role of the library file and the need to track down dependencies. The C/C++ programming languages use library files extensively. This allows you to utilize pieces of precompiled code without having to write your own versions. Libraries in both C and C++ are extensive and include such necessary functions as input and output operations and string and character operations but also include random number generation, dynamic memory allocation functions, math functions, type conversion operations, and diagnostic and testing functions to name just a few.

In order to utilize a function written in an external file, you must include the function's prototype. This is a definition of the function, similar to a function header. For instance, the function prototype for the printf output function is

```
int printf(const char *format, ...);
```

meaning that this function will return an int value and expects to receive at least one parameter, a char pointer (string). The ... indicates that optional parameters may also be passed to the function.

In order to simplify the use of library functions, header files collect together information that the compiler will need in order to use those functions. These will include at a minimum the function prototypes but may also include constants and data types defined for use in those functions as well as global variables. These are collectively placed into a text file known as a header file, labeled with a .h extension. You include header files by using `#include <headername.h>` or `#include "headername.h"` depending on whether the header file is part of the C/C++ library or one of your own header files, respectively.

If you are utilizing library file functions, then another step in the compilation process must take place. This is known as linking. Linking can take place either during compilation, in which case it is known as static linking, or at run-time, in which case it is known as dynamic linking. Linking takes already-compiled functions and loads them into your program.

If linking is done statically, then your program will be completed during the compilation process. This means that the program, when executed, is not slowed by the necessity of run-time linking or loading. It also means that your program can be completely optimized at compile time. However, if your compiler performs static linking, it means that the executable program will contain *all* functions called in your program.

Consider the following C statement:

```
if (x < 0) {
     foo1();
     foo2();
}
```

If at run-time x is not less than 0, then these two functions are not called. If in fact these functions are only called from this code and nowhere else in the program and if x is not expected to be less than 0, then static linking will increase your program's size somewhat unnecessarily.

Instead, with dynamic linking, library functions are only linked to and loaded on demand (as needed). While this may slow down the processing of the program as it has to wait to load additional content at run-time, it makes up for the problem above where functions may not be necessary. On the other hand, reliance on run-time linking and loading can also lead to run-time errors if the needed libraries are not available on the system. This problem may only be discovered at run-time rather than compile time if you are using dynamic linking.

In C/C++, functions default to being dynamically linked but this can be overridden by using the reserved word `static`. When compiling your program, to specify statically linked files, use the .a extension while dynamically linked files use the .o extension. As noted here, linking is a separate compilation step. In the next subsection, we examine gcc specifically and see that we can include linking with compilation or perform linking separately.

### 13.6.5 Using gcc

The gcc command can range from simple to convoluted depending upon the operations that the user wishes to perform. At its simplest, you specify gcc [compilation-option] *filename*.

The compilation option will be one of –c, -S, or –E, where –c performs compilation or assembly but no linking, -S performs compilation but not assembly, or –E stops after the preprocessing stage. If no compilation option is given, then compilation performs preprocessing, compilation, linking, and assembly combined.

By using –c, the output produced by the compiler is an object file (with a .o extension). You would use –c to create a library file that other programs might link to. By itself, the .o file is not executable but can only be used by the compiler to produce other compiled programs.

If not specified in the gcc command, the resulting executable file will be stored as a.out in the current directory. To alter the output file name, add the option –o *filename*. Thus, gcc foo.c will produce a.out while gcc foo.c –o foo will produce the executable file foo.

There are a number of different options available to specify the type of language that gcc is to compile. These include -ansi for ansi standard C, -std=*standard* to specify the standardized version desired, and –x *language*, where *language* specifies which language is included with the C code. For instance, it is possible to include Ada or FORTRAN code with your C source code.

To go along with this theme, there are a number of naming conventions used for file extensions. The .c and .cpp files stand for C and C++ source codes, respectively. A .i or .ii file contains C and C++ source codes that should not be preprocessed. The .m extension is used with objective-C source code. The .h and .o files are C/C++/Objective-C header and object files, respectively. And then there are extensions for FORTRAN source code, including .f, .for, and .ftn while .f90 and .f95 represent FORTRAN 90 and FORTRAN 95, respectively. Ada code is placed in .ads or .adb files. Assembly code is placed in .s or .S files.

As gcc is expected to compile very large files, compiler warnings can be crucial in discovering potential problems or flaws with the source code. There are a great number of different warning options available. Each warning option is preceded by –W. These options include the aforementioned –Wall which provides all warning options. Other more specific options include –Wcast-align, -Wfatal-errors, -Wno-overflow, -Wnonnull, -Wpointer-arith, -Wsign-conversion, -Wtype-limits, -Wunused-value, and –Wunused-variable to mention but a few.

Similarly, there are a number of debugging options available. These are best explored by programmers who are looking for specific debugging assistance. Finally, gcc provides for a large number of optimization choices. These are specified using the –O options (-O, -O2,

-O3, -Os, -O0). Aside from the general optimizations, you can also specify a machine-dependent option to indicate the intended destination architecture. These include mainframe computers produced by DEC, Intel-based architectures like i386 and x86-64, and ARM processors often found in handheld devices.

What follows are some example gcc operations and explanations.

- `gcc -Wall hello.c -o helloworld`—compile the program hello.c into the executable helloworld using the –Wall warning option.

- `gcc -Wall -O -mpower hello.c -o helloworld`—same as the first one but optimize for the Power PC processor.

- `gcc -Wall hello1.c hello2.c hello3.c -o helloworld`—same as the first one but compiles the program stored in three different source code files.

- `gcc -Wall -g hello.c -o helloworld`—same as the first one but include debugging help.

- `gcc -Wall -I ../lib hello.c -o helloworld`—same as the first one except that hello.c uses #include statements whose libraries are stored in the directory lib located one level above the current directory.

- `gcc -c hello.c -o hello.o`—same as the first but the output is an object file instead of an executable file. The object file can then be used as a library file for other C programs.

One last comment about gcc: There are alternate programs available that also compile C/C++ programs. First is g++, a variant of gcc, which will treat any C file as a C++ file instead. Although it is a subtle difference, many C++ programmers prefer to use g++ as it is specifically tuned to compile C++ rather than C. This is especially true if your C++ program links to the standard libraries in C++ (which gcc is not capable of handling). GNAT is a version of gcc which specifically will compile Ada programs. While gcc can handle multiple languages, GNAT is specific to Ada. There is also gdb, the GNUs Debugger, which is not a replacement for gcc but an addition to gcc.

## 13.7 SOFTWARE MAINTENANCE

Once software is installed, the system administrator's task is over, right? No, not really. Although the system administrator may not use the software, others will and because software is often being revised over time, upgrades, updates, and patches need to be installed. Linux, like Windows, allows much of this software to be upgraded with little to no intervention. In other cases, the system administrator might explicitly perform the upgrade.

### 13.7.1 Updating Software through rpm and yum

Updating software can be performed in much the same way as installation, through rpm or yum. If you wish to upgrade through rpm, you must obtain the RPM file. If you had deleted the RPM file after installation, you will have to download it anew from the repository. With the RPM file available, the command to upgrade the package is `rpm –U packagename` or `rpm –F packagename`. The –U is for upgrade, the –F is for freshen. As stated in Section 13.3, the only difference between upgrade and freshen occurs when you are installing a new package. Freshen will not work if a package has not yet been installed whereas upgrade can be used to either install the first time or upgrade the installed package.

In addition, if you are upgrading/freshening an installed package, you can add further options: --replacepkgs, --replacefiles, --oldpackage, and --force. These options force the installation of a package during the upgrade even if that particular package already exists, force the installation of files that already exist, downgrade from the currently installed package to a previous one, and perform all three of them combined, respectively.

It is obviously easier to issue a yum command than an rpm command because you do not need to first have the RPM file available, nor do you have to worry about file dependencies. The format to upgrade software with yum is

```
yum upgrade packagename [options]
```

Yum has a number of different update options.

- check-update—checks the repository(ies) for any new updates to the package. If you do not specify any package name with check-update, you receive a list of all packages updated since your last update.

- update—upgrade the given software package (if an upgrade is available). If no package name is specified, this will attempt to update every currently installed package. Unless otherwise specified, if any packages are obsolete, they will be skipped. If the obsoletes flag is set to true (or the user specifies --obsoletes), then these packages will also be included.

- update-to—in this case, you specify a version number as well as the package name. It will update up to the given version number but not beyond it.

- upgrade—same as update except that the obsoletes flag is set to true.

- upgrade-to—same as update-to except that the obsoletes flag is set to true.

- groupupdate—performs updates on a group of packages based on a type. Recall from the Add/Remove Software approach to installing software, packages are grouped together. *Note*: there is also a groupinstall available.

- reinstall—reinstalls from scratch a package that is already installed.

- downgroup—will downgrade to the most recent upgrade available prior to the currently installed package. This may not work if an older version is not available.

### 13.7.2 System Updating from the GUI

In CentOS, you can perform a software update through the GUI. Along the top of the Gnome desktop menu bar, if software updates are available, you will see this symbol: ❋. Clicking on it will bring up the software update window as shown in Figure 13.10.

You can select which packages to update, update all of them or update none of them. In this case, there are 90 updates available totaling 255.9 MB. These software updates not only are of applications software but also operating system updates. As the window reminds us, software updates "correct errors, eliminate security vulnerabilities, and provide new features." There is no reason not to do the updates unless you are running very low on disk space. Even if you do not have time for the updates, you can launch them the next time you have to leave your workstation.

Other forms of software maintenance will include troubleshooting the software, producing user documentation for the software, monitoring the system's performance while executing software, and removing software. We will explore some of your options for troubleshooting software in the next chapter.

We have already examined various monitoring tools available in Linux, such as the System Monitor, top, vmstat, iostat, and ps. Another handy command is pmap, which reports on process memory usage. You can also explore many performance characteristics of a running process in the /proc directory. Recall that every running process has its

FIGURE 13.10 CentOS update window.

own subdirectory, named after the PID. Within that subdirectory, you can find information on the process' CPU utilization (cpuinfo), memory usage (meminfo), file system usage (mounts), and numerous other useful pieces of information.

In some cases, software will come with its own monitoring program(s). For instance, the Apache web server can store performance information in log files or create reports that can be made available from the command line or even via a web browser.

### 13.7.3 Software Documentation

Producing documentation for software may or may not be part of a system administrator's responsibility. To write useful documentation, learning how to perform technical report writing could be advantageous. The role of software documentation goes beyond merely "how to" use the software. Information placed in such documentation might include any or all of the following:

- How to install the software—although it is the system administrator's role to install the software, this information could potentially help users who will install their own versions or future system administrators who might take over for the current administrators.

- How to configure the software—the system administrator may be responsible for configuring the software although this task may also fall on individual users. Or, the individual users may want to reconfigure the software to better fit their needs or style.

- How to maintain the software—upgrading the software, reporting bugs, creating or updating accounts and so forth are all tasks that may fall on the administrator or individual users.

- Training manuals—here, the administrator may write a step-by-step description of how to use the features of the software, including some of the more basic features for naïve users and more advanced features for those who will use the software extensively.

- Errata—known problems in the software and solutions either derived locally or from the software vendor.

And of course, if the system administrator learns the software to produce documentation, this makes the administrator a go-to person for questions from the users of the software.

### 13.7.4 Software Removal

The system administrator may at some point have to remove software. This occasion arises when the software is obsolete (has been replaced by something better) or the intended users of the software are no longer with the organization. It may also be necessary if free disk space is running low and a particular piece of software which is taking up too much space is deemed not worth keeping.

562 ■ Linux with Operating System Concepts

Removing software can be a very tricky proposition when that software consists of files distributed across the file system. This is why the system administrator should always use the proper removal tool rather than just deleting files. All of the package manager programs that we have discussed have facilities to remove software.

Using the Add/Remove Software GUI, by default, Install is selected. To remove software, if the package has already been installed, once you select it you choose Remove. This is located under the Selection menu. As expected, this will remove all of the files that correspond to that particular package with the exception of files that are dependencies from other packages. You have an option of whether to retain or delete those files (if other files depend on these files, they should be retained).

In rpm, the –e or --erase option will erase the specified package. The command must include the package name, although in this case, it is merely the software title's name, not the name of the RPM file (i.e., there is no version or release number and you do not end the name with .rpm). For yum, the option is remove as in `yum remove packagename`. As an alternative, you can also use the option `erase`. There is also a `groupremove`, which removes all packages of the same group.

Whichever approach is taken to remove the software package, make sure that your users know of its removal. They might have a shock if their favorite piece of software suddenly is missing! Additionally, if there are any symbolic links pointing at the software, these should all be removed.

## 13.8 THE OPEN SOURCE MOVEMENT

The simplicity of yum (or apt-get) seems to make rpm and make/make install obsolete. In fact, the former is generally true but the latter is not the case. Why would you want to download source code only to have to compile it prior to installation? This added step can be time consuming and also involve effort on your part if the makefile requires modification.

On the other hand, one of the greatest strengths of Linux is that so much of its software is available in source code. In this section, we explore the open source movement and the open source community. Why does it exist? Why do people contribute? What does it mean to you as a system administrator?

### 13.8.1 Evolution of Open Source

The open source movement can be traced back to perhaps two groups of programmers. There was Richard Stallman at MIT's artificial intelligence laboratory who created the GNUs project. GNU stands for GNU Not Unix, a recursive definition (recursion is a programming tool commonly used in artificial intelligence). What Stallman had in mind was the creation of a Unix-like operating system and its application software. GNU would be offered entirely as source code, free of charge but also free to be modified by whomever desired to work on it. Programmers could modify, enhance, or alter the code as they saw fit. Whatever they produced would then be made available in source code format.

The second group of programmers was centered at the University of California at Berkeley, who were working on their own dialect of Unix, fronted by one of the original Unix creators. This version of Unix became known as the Berkeley Standard Distribution, or BSD Unix. While funded by the government to create this version of Unix, they also called up other Unix developers to assist for free. They also had the intention of freely distributing BSD Unix.

Both groups worked on their respective operating systems starting in the late 1970s, working through the 1980s. While GNU Unix was never completed in the sense that a formal release has not been distributed, BSD Unix became very popular. At this time, no one was referring to this ideology as open source. But in the mid-1980s, Richard Stallman created the free software movement. His specific definition of free software was that it was "free as in free speech," not "free as in beer." He opposed proprietary software and saw the software field as one where all programmers could or should contribute to the development of software. Software, he felt, were based on ideas and ideas could not be owned. Therefore, proprietary software was immoral.

Stallman created the GNU General Public License (GNU GPL). The GPL requires that software published under the GPL must be free for anyone to use for any purpose, free to be studied, free to be changed, free to be redistributed, and free to be improved. The proviso is that anything created by GPL software would also be published under the GPL so that further distribution of such software would also be available as source code allowing others the same freedoms.

In 1991, Linux Torvalds created Linux and freely distributed the source code to other programmers who might be willing and able to contribute. Again, the notion is that the community as a whole can contribute to and improve the operating system.

Open source was further enhanced when the creators of the Netscape web browser followed this ideology. In fact, it was the Netscape group that coined the term open source and the open source initiative.

With the inclusion of Linux and Netscape, the free software movement fell into dispute. Torvalds did not require that modified versions of software be made freely available. Instead, he felt that if a person modified a piece of open source code, the modified code could continue to be freely distributed as source code, or it could be freely distributed but as executable code, or it could even be sold for profit. This created a rift between various open source contributors.

On the one hand, contributors felt that any use of open source software should result in open source products. That is, someone who is modifying or otherwise contributing to open source software would be required to contribute their products back to the open source community. Others felt that some contributions could be marketed commercially or otherwise have restrictive licenses applied to the products.

The rift caused the community to splinter into two groups: the Open Source Initiative (OSI) and the Free Software Foundation (FSF). The difference being that the OSI is willing to accept copyrights on some software that restrict the freedoms over the GPL while the FSF generally feels that anything created from a GPL produced product must also be made available in open source under the GPL.

Today, we find roughly 50% of Red Hat Linux code (both the operating system and applications software) is published under the GPL. This means that half of Red Hat is freely available for use and modification and half is at least somewhat protected by a more restrictive copyright. Some of this latter group of software is available for free but with distribution restrictions or available for free but not in source code format, and some is not free. Table 13.4 provides a list of notable software published under the GPL.

It should be noted that Stallman has referred to the GPL as a *copyleft*, not a copyright. The copyleft is not the antithesis of a copyright as the copyleft is still a legal mechanism that requires that software enhanced or modified from GPL software be published under the GPL copyleft (for details on the GPL, visit the GPL website at http://www.gnu.org/copyleft/gpl.html). The opposite of a copyright presumably would be public domain, which means there are no laws at all governing its usage.

Today, we find many outside of the Linux community embracing the open source initiative and using the GPL or a similar mechanism. There are many software developers contributing to open source. These include, for instance, employees at Microsoft who are now producing some software for free in source code format. Thus, Microsoft has become a partner in the development of some open source software. In fact, there are many ongoing open source projects at Microsoft, including those that impact Apache, the dot net (.net) languages and platform, Silverlight, and SQL database tools to name a few. A summary of Microsoft's involvement can be found at http://www.microsoft.com/opensource/directory.aspx. Today, we find open source contributions impacting all of the operating system platforms: Windows, Mac OS, Linux, and Unix.

This shift in perspective not only has impacted software developers' perspective of open source but companies as well. Early on, open source software was primarily used

TABLE 13.4 Popular Open Source Software Titles

| Software Title | Software Type | Platforms |
|---|---|---|
| 7-Zip | File archiver and compression/decompression utility | Linux (command line only), Windows |
| Audacity | Sound editor | Linux, Mac OS X, Windows |
| Blender | 3D image tool (animation, rendering) | BSD Unix, Linux, Mac OS X, Windows |
| FileZilla | FTP clients and servers | Linux, Mac OS X, Windows |
| (Mozilla) Firefox | Web browser | All platforms |
| FullSync | File synchronization and backup tool | Windows |
| GIMP | Image editor | Linux, Windows |
| Inkspace | Vector graphics editor | Linux (Ubuntu), Mac OS X, Windows |
| LibreOffice | Productivity software office suite (originated as OpenOffice but this software has since been developed by others) | Linux, Unix, Mac OS X, Windows |
| Notepad + + | Text editor for CSS, C/C++, C#, Java, JavaScript, SQL, HTML, XML, PHP | Windows |
| (Apache) OpenOffice | Productivity software office suite | Linux, Windows |
| TrueCrypt | Disk encryption software | Linux, Mac OS X, Windows |
| Ubuntu | Linux operating system | N/A |
| VLC Media Player | Multimedia player for streaming video/audio files and DVD/CD player | All platforms |
| Xvid | Video compression/decompression | Linux, Windows |

by hobbyist and people in the open source community. Open source software was largely shunned by professional organizations. There was a fear that any open source product would have errors that could potentially harm the organization because the software would not work when it was needed leading to possible delays, inaccuracies, and other problems. Additionally, most proprietary software comes with support and a guarantee that errors would be fixed in a timely manner.

Open source software competes regularly with proprietary software in part because the open source community has been found to be as or more responsive than many of the companies producing proprietary software. In fact, a comparison between open source software and proprietary software often illustrates that open source software has better security, is of higher or equal quality, has a greater degree of interoperability (available on multiple platforms), and provides a reasonable amount of support. The support might be available for free, or for a fee but the fee would in most cases still be far below the expense of the proprietary software and its support.

### 13.8.2 Why Do People Participate in the Open Source Community?

Why are people willing to contribute their time to the open source community? It is especially puzzling because contributions are often made by programmers who develop software for a livelihood. If they are willing to freely contribute, then they are in essence producing for free something that they could be paid for instead. In fact, it is possible that the software they are helping to produce might compete against software that they are paid to produce or maintain. And yet being involved in the open source community could help their careers. How?

Here are several reasons to contribute to open source projects.

- Learn about new technologies.

- Learn about different software development approaches.

- Participate in large group projects (with potentially hundreds of developers).

- Learn to work in groups that are geographically separated.

- Produce tools that might be useful to the individual at work or even at home.

- Desire to see a given project be successful.

- Make contact with other developers.

- Demonstrate to current (or future) employers a wide range of skills, some of which may not be displayed through their current job.

Participation in open source development also gives the programmer an opportunity to give back to the open source community, or even a larger audience, all computer users. Mozilla's Firefox, for instance, is such a success that a contribution to its development is truly a contribution to much of humanity.

Contributions are not only provided by software developers. There are a lot of open source programmers who program for a hobby and not as part of their job. Their motivations may be similar to those who want to further their career: learning, working with other developers, producing something useful. But in their case, the impact may be more of a personal nature over improving their career.

Contributions are not limited to programming. For those who are not programmers, software testing is also an avenue of contribution. Most open source software consists of very complex programs. Testing is a crucial aspect of software development and it is typical that testers are not the same group as the developers. Testing can be performed at different levels. Software testers commonly run the software on a number of test-cases, that is, groups of input. They then compare the output to the expectations and write up reports where the output diverges from expectation. In other cases though, testers are just users who compile lists of problems that they detect when using the software. For instance, if the software crashes when opening a file or if a particular feature operates too slowly, these cases are noted and sent back to the developers. Aside from testing, there is also a need for documentation writing.

The open source ideology has extended beyond software development to other community-involved pursuits. Wikipedia is one of the best examples of this. The online encyclopedia receives contributions from the common person who wants to share knowledge. Unlike open source software that requires some expertise as a programmer, Wikipedia only requires knowledge of a particular topic and the willingness to contribute. With Wikimedia, knowledge-based contributions extend to other media, namely, images (whether drawings or photographic). We also see artists sharing their products through the Internet using mechanisms such as Myspace, Facebook, and other open forums and social media.

## 13.9 CHAPTER REVIEW

Concepts and terms introduced in this chapter:

- Digital signature—a binary file included in software packages to ensure the integrity of the software product. If this signature is not available in the installation package, software like rpm might report this as a warning or error.

- Free software foundation—a movement by programmers headed by Richard Stallman who felt that software should be made freely available both in cost and in open source format.

- GPL—GNU's General Public License, created by Richard Stallman for the open source community so that software produced by the community could remain free and open.

- Open source initiative—the idea that people can freely contribute to produce and support software that is both reliable and highly usable. People from all over the world contribute time and effort to the open source community.

- Open source software—software produced by the open source community and made available both for free and in source code format (also available for free in executable format).

- Package dependency—software libraries required to complete the installation of a given piece of software. If the dependencies are not met by already-installed software, then the new piece of software will not be installed correctly (or at all). Some software will locate these dependencies and automatically install the missing libraries for you (e.g., yum) while others require that you perform these tasks (e.g., rpm).

- Package manager—a program that helps facilitate the installation, upgrading and removal of software with little effort on the user's part. These include command-line programs like yum and apt-get, text-based menu-driven programs like aptitude, and GUI programs like the Add/Remove Software tool in Red Hat or the Ubuntu Software Center.

- Package repository—a website storing software packages that can be installed either by hand through a package manager like rpm or automatically through a package manager like yum.

- Red Hat Add/Remove Software—GUI-based software program that allows you to find and install, update, or remove software in Red Hat Linux.

- Software package—a bundle of the files necessary to perform the installation of some software package onto a computer. The package will either be in source code format, coupled with the scripts needed to compile and install the software, or will be already-compiled executable programs, coupled with an installation program to move files to their appropriate locations.

- Software update/upgrade—upgrading already-installed software by downloading new versions of some of (or all of) the files that the software uses to run. An update usually entails adding patched software to fix known problems like bugs and security holes. An upgrade installs a new version of the software that may have new features.

- Source code—the program code in the originally written language such as C, C++, Java, or Ada. In such a form, the program cannot be executed but it can be modified. Source code must be compiled into an executable form before it can be used on a computer. Software developers work on the source code while most users typically only interact with the executable code.

- Sourceforge—a website storing thousands of software packages in both source code and executable forms.

- Synaptic Package Manager—a GUI-based software program to easily install, upgrade, and remove software packages for Debian Linux.

- Ubuntu Software Center—a GUI-based software program to easily install, upgrade, and remove software packages for Ubuntu Linux.

- Version—a specific release of some software package, identified by a naming and numbering scheme. Many software titles have major and minor releases that might be denoted with two numbers, separated by a period as in *sometitle*.5.3, meaning the 5th major version and the third minor release of the 5th major version.

Linux commands and files covered in this chapter:

- apt—an umbrella term for the apt command-line package manager tools for Debian Linux.

- apt-get—the portion of apt used to download and install dkpg packages.

- aptitude—a text-based but menu-driven version of apt.

- configure—a script written using Linux instructions to generate a makefile as part of the process of installing open source software.

- dpkg—the Debian package manager program that operates on dpkg files. This is a more primitive tool that does not perform dependency handling unlike apt.

- gcc—the GNU's C/C++ compiler that is a requisite program for installing most open source software.

- install—similar to mv but allows you to change a program's permissions before moving the program.

- make—a command to execute the makefile script that is in charge of compiling large pieces of open source software.

- makefile—the script called upon by the make command to compile and install open source software. Many makefiles will come with sections responding to commands such as `make`, `make all`, `make clean`, `make install`, and `make tar`. Some pieces of open source software do not come with a makefile, requiring that you run the configure script first.

- make all—typically, this command handles all of the parts of the make operation (make clean, make, make install, and possibly make tar).

- make clean—when a make command fails, it is best to clean up any temporary or partially created files before trying again. The make clean command will clean up after a failed or successful make attempt.

- make install—this portion of the make process is used to move the produced executable and supporting files to their destination folders such as /bin, /sbin, /usr/bin, and /usr/sbin.

- make tar—some makefiles contain a portion that allows you to package the result of the compilation process into a single file.

- rpm—the Red Hat package manager command-line program that can install software from an RPM file.

- tar—the tape archive program used to package together and unpackage files and directories. While it is used for backup purposes, it is commonly used by the open source community to create software packages for easy transport over the Internet.

- test—used to test if a directory or file already exists.

- yum—the Yellowdog Updater, Manager is a front-end tool for rpm so that the user has a minimal amount to issue when trying to install RPM files. yum, unlike rpm, will track down and install dependent packages.

## REVIEW QUESTIONS

1. Before installing software, what questions should you ask about your organization? What questions should you ask about your system?

2. What is a package dependence?

3. Using the Add/Remove Software tool, what happens if a dependence is found?

4. Who is allowed to install software through the Ubuntu Software Center tool?

5. What is the relationship between yum and rpm?

6. What is the RPM Fusion project?

7. What types of packages can you find at packages.debian.org?

8. An RPM file has the name mysoftware-3.1.5-2.i386.rpm. What is the version number of this software package? What is the release number?

9. What does noarch mean when found in an RPM title?

10. What is the difference between upgrading and freshening an RPM package?

11. What happens if, when using rpm to install a package, there are dependencies that are unmet?

12. What happens if, when using rpm to install a package, there is no digital signature?

13. What happens if, when using rpm to install a package, the digital signature is not recognized?

14. How can you test for an RPM file's digital signature before trying to install it?

15. What does the instruction yum list * do?

16. What does yum's downgrade option do?

17. What is the difference between yum install *somesoftware* and yum -y install *somesoftware*?

18. What is the Debian equivalent to yum?

19. Why might you want to install open source software from source code rather than an already-created executable or installation program?

20. What program will you probably need to have in your Linux system to successfully install open source software from its source code?

21. Why might you want to run ./configure?

22. What types of options might you specify when running the configure script?

23. What is the difference between make and makefile?

24. When might you run make clean?

25. What is the difference between make and make install?

26. Draw a parse tree for the following C instruction:

    x = y*z++;

27. Draw a parse tree for the following C instruction:

    while(x >=0)
        x = x-y;

28. What does optimization mean with respect to compilation?

29. In gcc, what does -Wall mean?

30. Why might you use the –c option in gcc so that compilation does not include linking?

31. What is the difference between using the –o option in gcc and not using it?

32. If you were asked to produce documentation on how to use a piece of installed software, what information might you include?

33. Which of these operating systems was produced and released entirely by the open source community? BSD Unix, GNUs, Linux.

34. Richard Stallman has said that software should be "free as in free speech" and not "free as in beer." What did he mean by this?

35. What dispute exists between the Free Software Foundation and the Open Source Initiative?

36. Examine the software titles in Table 13.3. Come up with a list of five other titles that you or someone you know uses that are open source titles.

37. Why might a programmer contribute to the open source community?

38. Even if you are not a programmer, how could you contribute to the open source community?

# CHAPTER 14

# Maintaining and Troubleshooting Linux

THIS CHAPTER'S LEARNING OBJECTIVES are

- To understand the importance of backups and how to implement backup strategies

- To be able to use the Linux performance monitoring tools

- To utilize the scheduling programs of at, batch, and crontab

- To understand the role of log files and how to apply this information

- To understand how to perform disaster recovery planning

- To understand how to perform operating system troubleshooting

## 14.1 INTRODUCTION

So, you have successfully installed Linux, created accounts, mounted file systems, automated the initialization process, configured your network, and installed software. There is nothing more to do and you can coast for the rest of your career, right? Wrong. These steps have established Linux and given users the ability to use Linux. But this is only half of your job.

Now, you have to maintain the system. This requires keeping tabs on system resources to ensure that you have adequate resources for the user load. You have to monitor for external threats over the network. You have to ensure the integrity of the data files in the system through backups. You will also want to automate as many tasks as possible. To perform maintenance, you will have to learn

- How to perform backups of the various file systems

- How to monitor system resources

- How to schedule tasks

- How to keep track of events through log files

Whenever something goes wrong, you will be the first to hear about it! Now, you need to learn to take the next step in Linux: how to troubleshoot a system that is not running efficiently or correctly, or is not running at all.

In this chapter, we examine four activities that are all related in that they are tasks revolving around maintenance and troubleshooting:

- Performing backups

- Scheduling tasks

- Monitoring the system

- Reviewing log files

We also discuss disaster planning and recovery. This chapter wraps up with some troubleshooting scenarios: problems, ways to discover the cause(s), and solutions.

## 14.2  BACKUPS AND FILE SYSTEM INTEGRITY

In Chapter 10, we introduced the tools to perform backups. Here, we examine when, why, and how you might perform backups. Before you undertake this task, you must ask yourself which partitions really need to be backed up.

- For /home, it depends on the number of users and how frequently they use/store files

- For /var, it depends on the software you are running; which programs store data files in /var and how frequently is that software being used?

- For /usr, it generally depends on how often you install and remove software

- For most of the root partition (/), changes will only occur when you upgrade the system

### 14.2.1  Examining File System Usage

You should determine the frequency that the particular partition should be backed up: daily, weekly, or monthly (or even less frequently). The /home directory will be the partition that requires the most frequent backups. If you have dozens (or more) users who regularly use the system, then this file space will change daily. In such a case, daily backups would not be a bad thing. Alternatively, as you may have to bring the partition down (unmount it) to back it up, you might perform backups a few days during the week such as every third day at midnight or every Sunday night and every Wednesday night.

If a workstation is primarily used by an individual user, weekly backups might be more practical. An alternative approach is to use a network file server to mount all user directories. Local disk storage would store the operating system and application software along with software-based data files. This allows you to perform nightly backups of a single, centralized location. This solution has its own detractors. First, you might need to purchase several file servers to support the number of users. Second, file servers are more expansive than workstations, requiring a greater investment to support this solution. Third, you would unmount the network file system before backing it up, thus making user home directories inaccessible during the backup period. By having several file servers and distributing users across the file servers, you can then alternate which server is backed up each night.

The other file systems do not need to be backed up nearly as often. The/var directory would be the next most common file system in need of backing up. As with /home, when /var is backed up, it is best to unmount it. With log rotation, you can ensure that there is adequate disk space. Once per week, you save the current log files and then start the new week with new log files. After a few weeks, you back up the log files to storage, enabling you to delete those in /var. The mail directory might tend to grow rapidly. However, you can solve this problem by instituting a disk quota on mail log files, forcing users to regularly delete or save to their home directory any important emails. Alternatively, you could implement the mail server elsewhere and not on this Linux system at all.

## 14.2.2 RAID for File System Integrity

The next question to consider is why are you performing a backup? This, of course, has an obvious answer, to protect the data in case of hard-disk problems. Additionally, you may wish to archive old files (e.g., save log files so that they can be examined some time later). However, there are other potential answers here that lead to differing backup strategies.

If you are attempting to secure your data, should you perform daily/weekly backups or use some form of redundant storage? RAID storage provides you the ability to restore data should a disk surface become unreadable. This is possible because other disks are storing enough information to restore what has been lost. In this way, every time you save something, extra data in the form of redundancy are being saved.

There are several different forms of RAID. These are known as *RAID levels*. The levels differ from each other such that the levels do not progressively improve upon the prior level. There are seven commonly cited RAID levels (although not all are used) along with two "combined" levels and a hybrid level. Table 14.1 provides a comparison between all the RAID levels. We will concentrate on two different RAID approaches that are found in RAID 1 and RAIDs 3–6.

RAID level 1 provides a complete mirror of your storage space. Let us imagine that you are using file servers for disk storage. Every user's home directory is stored on a file server. If this file server uses RAID 1, then every time the user stores some data, they are stored twice, once to one set of disks and once to a second set. Should one set of disks ever fail, the data are still available and can be restored easily. The mirror might be implemented inside one cabinet, or alternatively, the mirror may be a separate file server located elsewhere.

A nice side effect of the mirror is to support multiple disk accesses simultaneously. Specifically, two read accesses could take place at the same time, one from one set of disks

TABLE 14.1    RAID Levels

| Level | Description | Advantages/Disadvantages | Usage |
|---|---|---|---|
| 0 | Striping at block level, no redundancy | A: Improved disk performance over standard disk drive<br>D: No redundancy | For superior disk performance without redundancy, where increased cost is not a concern |
| 1 | Complete mirror | A: Provides 100% redundancy and improves disk access for parallel reads<br>D: Most costly form of RAID | Safest form of RAID if cost is not a factor |
| 2 | Striping at bit level, redundancy through Hamming codes | A: Fast access for single-disk operation<br>D: Hamming codes are time consuming to compute | Not used in practice because of Hamming codes |
| 3 | Striping at byte level, parity bit redundancy | A: Fast access for single-disk operation, compromise between expense and redundancy<br>D: All drives active for any single access; so, cannot accommodate parallel accesses | Useful for single user systems |
| 4 | Striping at block level, single-parity disk | A: Larger stripes accommodate parallel accesses (like RAID 0) but improve over RAID 0 because of redundancy<br>D: Single-parity disk is a bottleneck defeating advantage gained by striping | Not used in practice because the parity disk is a bottleneck |
| 5 | Striping at block level, parity distributed across disks | A: Same as 4<br>D: None | Useful for multiuser systems (e.g., file servers) |
| 6 | Striping at block level, parity distributed across disks and duplicated | A: Same as 4 and 5 except that with double the parity information, it provides a greater degree of redundancy<br>D: More expensive than RAIDs 3–5 | Same as 5 |
| 7 | RAID 3 (or 4) with real-time operating system controller | A: Faster single-disk access over RAID 3<br>D: More expensive | Same as 3 |
| 10 | Striping at block level, complete mirror (RAID 0 and RAID 1 combined) | A: Same as 0 and 1 combined<br>D: Requires twice as much disk drive as RAID 1 | For file servers that require both parallel disk access and redundancy |
| 53 | Extra disks to support RAID 3 and RAID 5 striping | A: Best overall access as one can access the RAID 3 or the RAID 5 set of disks<br>D: Requires more disks and so is more expensive than 3 or 5 | Useful for multiuser systems (e.g., file servers) |
| S | Proprietary form of RAID 5 with high-speed disk cache | A: Improves over RAID 5 access<br>D: More expensive than RAID 5 | Useful for multiuser systems (e.g., file servers) |

and one from the mirror. However, if one access involves a write, then a second access would not be permitted simultaneously because the write must involve both sets.

RAID levels 3–6 use parity information to record redundancy. This requires a brief explanation. Let us imagine that we have the following four bytes of information to store, each of which is stored on a separate disk.

```
00000000
11110000
10101010
00101111
```

To compute the parity information, we will use the XOR Boolean operation on the 4 bits of each column (i.e., we will XOR the first bit of each of the 4 bytes, we will XOR the second bit of each of the 4 bytes, etc.). XOR can be applied to implement an even parity computation. Even parity means that the number of 1 bit in the sequence will always be an even number. We will perform the following operation given four bits, b0, b1, b2, and b3:

(b0 XOR b1) XOR (b2 XOR b3)

For the first column, we would compute

(0 XOR 1) XOR (1 XOR 0) = 1 XOR 1 = 0

So, our first parity bit will be 0. For the second column, we have

(0 XOR 1) XOR (0 XOR 0) = 1 XOR 0 = 1

For the third column, we have

(0 XOR 1) XOR (1 XOR 1) = 1 XOR 0 = 1

The entire parity byte for our four bytes above is computed as shown below.

```
00000000
11110000
10101010
00101111
- - - - - - - -
01110101
```

We store the parity information on a separate disk. So, in this case, we would have five disks, the first four storing data, and the fifth storing parity. Should a disk block fail on any of the five drives, we have enough information to restore it by using XOR on the surviving data. In fact, with this approach, we could restore data if we lost an entire disk. We could even restore data lost when multiple disks fail as long as the failures are of different bytes (RAID 3) or blocks (RAID 4–6).

In Figure 14.1, we see an illustration of a RAID drive. There are five disk drives, each with four surfaces. Let us assume that a disk block's data are stored on four of the surfaces with the parity information on a fifth surface where each block is stored on the same location of different drives. For instance, we might find all five blocks of a file on track 32, sector 7, and surface 1 on the five drives. In the figure, we see that this RAID drive contains three failed blocks (denoted as black rectangles). Since the bad blocks are not of the *same* track and sector, the bad blocks are of different file blocks. Therefore, for any bad block, four other blocks are not bad, allowing us to restore the bad block from the other four. It does not matter whether the bad block is of original data or parity data.

Unfortunately, what RAID will not save us from is a complete failure of many disks. Even RAID 1, with its complete mirror, would not help if we lost both sides of the mirror. So, while RAID is a solution to promote redundancy and provide some protection from data loss, it is not necessarily the best solution. We should still perform backups. However, with RAID available, we may be able to get away with less frequent backups. Additionally, the combination of RAID and backups provide us with a more flexible means of ensuring data integrity and availability. This is discussed in Section 14.6.

### 14.2.3 Backup Strategies

Now that we have decided to perform backups, we should also explore how long a backup should be saved. Imagine that you back up some partition every day. You use magnetic tape for each backup. You save each day on a separate tape and each week, you recycle the tapes. You take each day's backup and move it to a separate location (e.g., a safe, or some off-site location). This seems more than satisfactory because you are backing up daily and you have the data held securely.

But consider this situation. A user deletes a file. Two weeks later, the user realizes that he wants the file back. He cannot reclaim it from the backups because the backups only cover the previous week. Should backups be retained for a longer period of time? We will attempt to answer this with a strategy in the following paragraphs. Notice that there is a difference between saving data for *security* and saving data for *archival* purposes.

The next question is whether backups should be *whole* or *incremental*. We discussed briefly how to perform incremental backups in Chapter 10. It seems like incremental backups are preferable because they will take less time to perform than full backups. But there is another advantage to the incremental backup. An incremental backup will take less space (say on tape). This would allow you to retain backup information for a lengthier period of time than say 1 week.

Consider the following strategy. Once per week, we perform a whole backup of /home. This is stored on tape 1. Then, for the next week, we perform incremental daily backups

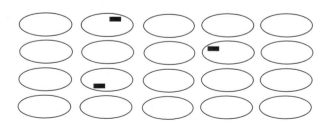

FIGURE 14.1   RAID 5 with some failed blocks.

TABLE 14.2    Daily Backup Strategy

| Tape Number | Week | Usage |
|---|---|---|
| 1 | 1 | Full backup |
| 2 | 1 | Daily incremental backups for 3 days |
| 3 | 1 | Daily incremental backups for 3 days |
| 4 | 2 | Full backup |
| 5 | 2 | Daily incremental backups for 3 days |
| 6 | 2 | Daily incremental backups for 3 days |
| 7 | 3 | Full backup |
| 2 | 3 | Daily incremental backups for 3 days |
| 3 | 3 | Daily incremental backups for 3 days |
| 5 | 4 | Full backup |
| 6 | 4 | Daily incremental backups for 3 days |
| 1 | 4 | Daily incremental backups for 3 days |

stored on tapes 2 and 3. We are making the assumption that incremental backups will not require nearly as much storage as a full backup. With this pattern of saving onto tapes, we would continue to use an incremental tape for several days before we had to reuse a full backup tape. See Table 14.2 that shows in 1 week that we are using only three tapes. We must reuse our first tape only in week 3 and tape 1 is not reused until week 4.

What happens if you need to restore from an incremental backup? Here is the drawback to incremental backups, and the solution depends on what needs to be restored. Are you restoring individual files or the full file system? In the former case, your solution is to scan the incremental backups for the most recent version of the file(s) in question. If there are multiple files to recover, you may find that you have to restore from more than one tape and in fact, you may have to search all the tapes to locate the file(s) to restore. You would search these tapes backward from the most recent incremental backup back to the full backup. The reason for this is that once you have found the file(s), you can stop.

If you are restoring the entire file system, you will proceed in the opposite order. First, you restore the file system from the latest full backup. Then, you will restore individual files found from each incremental backup going forward in time. This approach may result in your restoring the same file several times, but each time, you will be restoring a newer version. Restoration of the full file system may take a good deal more time than performing either the full backup or any individual incremental backup.

The final question that we posed originally was how you perform the backup. The answer to this question will impact other answers. An easy solution to backing up all of /home is to mount a tape drive, writable CD drive or external file system, and perform a tar operation on /home resulting in a tar file stored to the mounted drive.

Tar has several benefits over other approaches. For instance, you could potentially run tar without unmounting the file system being backed up, leaving it accessible during the backup. Tar is also fairly easy to use and can compress its contents during archiving. Tar is also capable of incremental backups although it is not as easily controlled as through dump. One way to handle an incremental backup is to specify in a separate file, *filename*, all the files (and directories) that have been changed (modified, added to, and deleted) since the last backup and apply the option −g *filename* or --listed-incremental=*filename*.

The more recent archiving programs are cpio and dump. Both cpio and tar have an advantage over dump in that you can select specific files or directories or both to archive with either cpio or tar. Additionally, tar allows you to specify criteria for file selection while cpio can use find to obtain the files to backup. For instance, we might want to back up only text files or files created by a particular user.

Dump operates only on the level of file systems and so is unable to create an archive in a piecemeal manner. There is also a --newer=*date* option available in tar that will backup files newer than the supplied *date* (or alternatively, --newer=*file* where tar stores files newer than *file*).

All three programs (dump, cpio, and tar) can perform incremental backups. Both cpio and tar have more primitive techniques for handling incremental backups though; so, we might prefer to use dump. For tar, we mentioned above using a file to compare against the stored files to determine recency. With cpio, you could use a find command as stated in the previous paragraph, but in this case, use −anewer or −cnewer resulting in much the same behavior.

Dump provides different levels of backup. A full backup is denoted as level 0. After making a full backup, you can next specify a different level that then performs a backup of all changes to the file system up to the most recent incremental backup whose level is less than the current level number. For instance, a follow-up backup of level 1 would find anything new since the full backup. The next backup at level 2 would find all changes since the level 1 backup.

In addition, dump has a feature superior to tar and cpio in terms of indexing the contents of an archive. While both tar and dump allow you to view the contents of an archive, dump also lets you interactively find *and* restore specific files.

## 14.3 TASK SCHEDULING

As a system administrator, you will most likely want to schedule processes in an attempt to further automate tasks. Scheduling may be for one-time execution or recurring in some pattern such as daily, weekly, or monthly. By automating tasks through scripts and scheduling them, you free your time up for other, more important tasks such as system monitoring, troubleshooting emergencies, and continuing education. There are many tools available to schedule tasks. We have already explored or mentioned some of these. Here, we will review them.

The earliest a task can be automated is when the system is initialized. Recall from Chapter 11 that during initialization, the scripts in /etc/rc.d are executed. The last script to execute during this process is rc.local. In this script, the system administrator is able to specify his or her own initialization tasks that were not performed by other start-up scripts.

For instance, imagine that your Linux machine is to run a Squid proxy server. You may wish to create the Squid caches and then start the proxy server. These start-up commands can be placed in the rc.d script.

Another use of this script is to automatically mount remote file systems. This could be a critical step in the initialization process if you are using remote file systems as part of or all of /usr or /home.

There may be other tasks that you wish to run after initialization. These could include

- Log rotation

- Execution of scripts that obtain statistical information

- Data-mining analysis on log files

- Examination of user disk space for poorly protected files, users who are using too much space, files that have become corrupted, or files that might contain a virus

The next tool for scheduling activities is the anacron service. Anacron will examine crontab scheduled processes to see if any were supposed to execute during the downtime from which you have just booted. If any are found, then they are executed. Anacron and crontab are available to schedule recurring tasks. Additionally, at and batch are available to schedule one-time tasks.

For the remainder of this section, we will concentrate on how to use crontab and at for scheduling. We start with at, which is the easier of the two to understand.

## 14.3.1 The at Program

The at program is used for one-time scheduling. The command requires that you specify the time that the command should execute. The time includes both the date and the time on that date. Time and date are indicated by either absolute values or relative to when the command is being issued.

For an absolute specifier, the time format is HH:MM, optionally followed by AM or PM. If AM or PM is omitted, then the time is interpreted as being in *military notation*. Thus, 12:00 PM and 12:00 are treated the same but 1:00 PM would be equal to 13:00 and 11:59 PM would be indicated as 23:59. For 12:00 AM, the military time is indicated as 0:00. If you use AM or PM, you can use either uppercase or lowercase letters. You can also specify the time as noon, midnight, and teatime (4 PM).

The date is specified using any one of the following four formats. Each of these is followed by an example for November 12, 2013.

- MMDDYY as in 111213

- MM/DD/YY as in 11/12/13

- DD.MM.YY as in 12.11.13

- YYYY-MM-DD as in 2013-11-12

If you do not specify a date, the event is scheduled to take place at the next occurrence of the specified time. If it is 3:35 p.m. and you schedule for 14:30 (2:30 PM), then the event is scheduled to take place at 2:30 PM of the next day. You can also use the words `today` or `tomorrow` for the date, as in `2:30 PM tomorrow` or `23:59 today`.

The alternative for the time/date is to specify a time/date relative to the time that the at command is submitted. This is specified using the notation: `now + count unit`, where *count* is an integer and *unit* is a temporal unit such as minutes, hours, or days. Here are some examples:

- now + 5 minutes

- now + 2 hours

- now + 7 days

- now + 4 weeks

You can combine relative and absolute methods but it may not make much sense. For instance, `now + 12:00 PM` makes no sense at all but `now + 3 days` would schedule the event at the current time in 3 days. Errors in the time/date specification will yield the message `Garbled time`. Note that at schedules events to the minute, not to the second. You would never specify the seconds unit for either an absolute or a relative time specification.

The event to be scheduled can be either specified in a separate file or through the command line. As an example, you have written the activities to be scheduled in the shell script `my_event.sh`. You would then schedule this using

```
at time/date specifier -f ./my_event.sh
```

where *time/date specifier* is as explained above. The –f option indicates that the event(s) to take place is(are) stored in the accompanying file. The file does not have to be a script. It could comprise Linux commands or invoke a script.

The alternate approach to specifying the event(s) in a file is to omit the –f option. In this case, upon hitting < enter >, you are dropped into an `at>` prompt. You now enter your commands one line at a time, pressing the < enter > key to end each line. To exit the `at>` prompt, press `control+d`. For instance, you might specify

```
at now + 1 minute
```

Since you did not include –f filename, you now see

```
at>
```

You might then enter

```
at> mount 10.11.12.13:/home/backup/mnt/backup
at> tar -xzf /home/* /mnt/backup/home-backup.tar
at> umount /mnt/backup
at> control+d
```

Thus, in 1 minute time, you will use tar to create a backup on a mounted file server and then unmount the file server when done.

Once events are scheduled, you can examine them using atq. This lists each waiting event scheduled through at or batch. You can remove any scheduled item using atrm. If there is only one item in the event queue, atrm will delete it. Otherwise, when you use atq, each event is numbered and you will have to specify the number of the event with atrm as in atrm 2 to remove the second item in the scheduled list.

The batch command is identical to the at command except that you do not specify a time. If you specify batch by itself, you are dropped into the at> prompt; otherwise, you would specify batch -f filename.

The atd daemon is responsible for determining when a scheduled event should execute. For a command scheduled with at, it executes at the time specified. For a command scheduled with batch, it executes when the system load drops below 80%. If you are working on a stand-alone workstation, chances are the load will usually be below 80%; so, batch could potentially run the scheduled event within the next minute.

The at command makes a note of the current shell. This is the value stored in the environment variable SHELL of the user submitting the command. This shell's interpreter is utilized to execute the line(s) in the scheduled event. For instance, if you switch from bash to csh, issue an at command and then exit back to bash, the scheduled command when executed is executed using the csh interpreter rather than the bash interpreter.

The at command may or may not be set up to be used by normal users. You may find in your system that it can only be executed by root. Alternatively, the system administrator has access to two files to control who can use at: /etc/at.allow and /etc/at.deny. If you are defaulting to only allowing root to execute at commands, you can add individual users to the /etc/at.allow file to permit specific users to access at. Alternatively, if at can be used by every user, you can exclude specific users from using at by listing them in /etc/at.deny.

## 14.3.2 The crontab Program

While at and batch are easy to use and very useful, they only permit one-time scheduling of events. The cron program is more useful because it schedules recurring events. But to use cron, the approach is significantly different from that of at/batch. The program that executes scheduled events is called crond. The service is stored in /usr/sbin while the script to control the service is in /etc/init.d. To issue a scheduling command, you use the command crontab.

Crontab requires that you specify a file that contains a list of events to schedule. Each listed event is in a row of the file by itself. The row indicates the times and dates that the event is to be executed and the event itself. The file can contain multiple times/events, one per row. The actual command will look like crontab filename where *filename* contains the scheduling information.

Since each line must be self-contained, the event(s) scheduled will need to be limited in size to fit in one line. To avoid this restriction, you might want to place the individual commands in a script and then place the script name as the event to execute. In this way, you are free to express a much more detailed series of instructions that are not limited to one line in the text file.

For instance, if you wish to unmount a partition, backup that partition and then remount the partition, you would specify at least three commands. This would most likely not be appropriate for one line in the crontab file. You would move the operations to a shell script and then specify the script from the crontab file.

The crontab file must specify the recurrence for scheduling the event. This is where crontab becomes complicated. There are five specifiers for the time and date. These are indicated through the first five columns of the row. The format looks like this:

```
1       2       3       4       5       command
```

The values 1–5 specify, in order:

1. Minute of the hour (0–59)

2. Hour of the day (0–23, military time)

3. Day of the month (1–31 but make sure that the day matches the month if you use 29–31)

4. Month of the year (1–12)

5. Day of the week (0–7 where 0 means Sunday, 1–6 means Monday through Saturday, and 7 means Sunday again)

For any one of these five entries, you can use the wildcard * to indicate "every time."
For instance,

```
30      12      15      *       *
```

would define 12:30 p.m. on the 15th with month and day being "every." This indicates 12:30 p.m. on the 15th of every month.

You would usually not specify a date and month if you specify a particular day of the week. The entry

```
30      12      *       *       0
```

represents every Sunday at 12:30 pm. If you instead use

```
30      12      15      *       0
```

you would be specifying every Sunday, the 15th, at 12:30 p.m. This latter pattern may only match a couple of days out of the year.

You can specify multiple occurrences within any of these five values by separating the values with commas. The list 0,30 for the minute schedules the event for both on the hour and half past the hour. If you have specified 0 for the hour, the event would take place at both 12:00 a.m. and 12:30 a.m. Alternatively, if the event is to recur more often and you do not want to enumerate all of the times, you can use the notation /x to indicate the amount of time to elapse before the next recurrence.

Consider the following:

```
*/10          0,12          *          *          *
```

This indicates that the event would recur at 12:00 a.m., 12:10 a.m., 12:20 a.m., 12:30 a.m., 12:40 a.m. and 12:50 a.m. and again at 12:00 p.m., 12:10 p.m., 12:20 p.m., 12:30 p.m., 12:40 p.m., and 12:50 p.m.

Each user is able to submit a single crontab job. Any new submission will replace any earlier scheduled job for the user. To submit multiple events, you must therefore add them to the crontab file. Each line will be a separate event with its own scheduled time and recurrence.

What follows is an example of a crontab file that root might submit.

```
0      0      *      *      *      ./backup /home
*/5    *      *      *      *      ./intruder_alert
30     0      15     *      *      ./usage_report >> disk_data.dat
15     3      1      1      *      ./end_of_year_statistics
```

Here, we see four different scheduled events, each event calling a fictional script. First, every night at midnight, the script *backup* is executed, using /home as a parameter. Second, every 5 min of every day, the intruder_alert script is executed. Third, every 15th of the month at 12:30 a.m., the usage_report script executes, appending the output to the file disk_data.dat. Finally, at 3:15 a.m. on every January 1st, the script end_of_year_statistics is run.

If you would like to add more scheduled events, you could modify the submitted file and resubmit the file using crontab. Crontab would replace the old scheduled events with the ones in the new version of the file. To view the scheduled crontab event(s), the command is crontab -l. This lists whatever you, as a user have scheduled.

To view what other users have scheduled, add -u *username*. For instance, as root, you might query crontab using

```
crontab -u foxr -l
```

to find out what foxr has scheduled. Only root is allowed to use the -u option so that one user could not inspect another user's scheduled events.

The command crontab -r will delete whatever is currently scheduled for the current user. An alternative to resubmitting your crontab file when you want to change, add, or delete scheduled events is to directly edit the events. By specifying crontab -e, you are placed into the vi editor with the currently scheduled crontab job. This allows you to edit and save what is scheduled. Upon exiting, crontab will enact the new version. You could also use `crontab -e` to enter all events that you want scheduled and avoid having to write the file of events.

Similar to at, when crontab runs a task, it must do so with an interpreter. When a crontab job is submitted, saved with it is the user's current environment. The environment variables saved are HOME, LOGNAME, PATH, and SHELL. This allows the user to specify an event that uses items in his or her home directory, items in the directories of the user's PATH variable, and the log-in shell of that user.

There are `/etc/cron.allow` and `/etc/cron.deny` files that serve the same role as the allow and deny files for at. If all users can access crontab, then root would modify cron. deny to deny certain users access. Alternatively, if no users are allowed to access crontab, then adding users to cron.allow will provide permission for those users.

Another scheduling approach is to combine crontab with the program `run-parts`. The run-parts program is given a directory and it executes all the scripts in that directory. Now, with crontab, we might create an entry that looks like this:

```
0     0     1     *     *      run-parts /somedirectory
```

This will execute the scripts found in */somedirectory* at midnight on the first of every month. That is, the scripts execute monthly.

One such file that implements this approach is found in /etc/cron.d called 0hourly. It specifies that run-parts should execute the contents of /etc/cron.hourly at 01 * * * *. This will cause crond to execute the contents (scripts) found in /etc/cron.hourly at 1 minutes past every hour. If you want to run a script every hour, place that script into the directory `/etc/cron.hourly`. Similarly, there are directories of `/etc/cron.daily`, `/etc/cron.weekly`, and `/etc/cron.monthly` that are set up to execute any scripts in those directories every day, week, and month, respectively.

## 14.4 SYSTEM MONITORING

The next step for the system administrator is to ensure that the system is running smoothly. Specifically, user processes should make progress toward completion in any situation. Although the system administrator does not have to monitor all running processes all the time, the system administrator will be called upon if the system is not running as users expect and at that time, the system administrator should examine the processes.

### 14.4.1 Liveness and Starvation

The idea that a process is making progress toward completion is known as *liveness*. To promote liveness, resources that the process needs to execute must be made available such that the process is not forced to wait an indefinite period of time for access to that/ those resource(s). The opposite situation is referred to as *starvation*, which means that the resources that the process needs are continually being withheld from the process.

Resources include access to the operating system, a sufficient amount of main memory, access to the file system and specific files within the file system, network access, and most especially, execution cycles from the CPU. The reason that starvation may arise is subtle but has to do with concurrent processing of operating systems (refer back to Chapter 4). Let us examine why.

Imagine that we have two running processes, P0 and P1, both of which need to operate on a shared file, F0, which stores the single value 0. P0 is going to read the file and update the datum by adding 3 to it. P1 is going to read the file and update the datum by subtracting 2 from it. Both processes will write the datum back to the file. No matter which order P0 and P1 execute, the result should be the value 1 written to the file (0 + 3 − 2 = 1).

However, if the operating system is multitasking, consider the following situation:

1. P0 begins executing, reads the datum from the file, and stores the datum in a local variable, X.

2. P0 adds 3 to X. X is now 3 (the file is still storing 0).

3. The CPU is interrupted and the operating system performs a context switch to P1.

4. P1 begins executing, reads the datum from the file, and stores the datum in a local variable, Y.

5. P1 subtracts 2 from Y. Y is now -2 (the file is still storing 0).

6. P1 writes Y back to the file (the file now stores -2).

7. The CPU is interrupted and the operating system performs a context switch to P0.

8. P0 writes X (3) back to the file (the file now stores 3).

If step 8 occurs before the interruption in step 3, the file will be ok. Alternatively, if step 7 occurs before step 6, then step 6 (P1 writing Y to the file) may occur after step 8, resulting in the file storing -2. So, we see three possible outcomes from this situation. The file stores 1 (the correct answer), the file stores -2, or the file stores 3.

Since multitasking processes could corrupt data when multiple running processes are sharing that data, we need to implement a mechanism to ensure that this cannot happen. We do so by making access to a shared datum or file *mutually exclusive*. This means that if a process has access to a resource, any other process that wants to also access that resource is forced to wait.

In the above example, this will be handled by forcing P1 to wait before accessing file F0. Therefore, before step 4, P1 is forced to wait. With P1 waiting, the operating system will move on to another process (perhaps P0 again). The operating system is in charge of enforcing mutually exclusive access to resources by granting access to a process upon request but only if no other process is currently holding onto the resource.

This leads to another situation that we need to resolve. Let us go back to our prior example with P0 and P1 but slightly modify it. There are two files that both processes are going to access, F0 and F1. We have the following pattern:

1. P0 begins executing

2. P0 requests access to F0

3. As no other process is using F0, the operating system grants P0 access

4. A context switch forces the CPU to switch to P1

5. P1 begins executing

6. P1 requests access to F1

7. As no other process is using F1, the operating system grants P1 access

8. A context switch forces the CPU to switch to P0

9. P0 resumes executing

10. While holding onto F0, P0 requests access to F1

11. The operating system denies access to P0 as P1 is holding onto F1. P0 enters a waiting state and the CPU performs a context switch to P1

12. P1 resumes executing

13. While holding onto F1, P1 requests access to F0

14. The operating system denies access to P1 as P0 is holding onto F0. P1 enters a waiting state

In our example, we have P0 waiting for F1 to become available and P1 waiting for F0 to become available. As P0 is holding onto F0, F0 will not become available for P1 until P0 resumes. P0 itself cannot resume until F1 becomes available but F1 is being held by P1 and P1 is also waiting. P0 and P1 are in a state of *deadlock*. Neither process can continue but because each process needs the resource it is holding onto, neither resource can be used. Neither P0 nor P1 are making progress toward completion; so, they are both starving. Even worse though is that any other process that might need F0 or F1 also becomes deadlocked.

Operating systems handle deadlock in differing ways. Some aggressively prevent deadlock from arising by keeping processes that might cause a deadlock waiting before they can be executed. Others prevent deadlock by not allocating a resource to a process if the action might result in a deadlock. Others ignore the possibility of causing a deadlock but look for deadlocks on occasion (such as every few minutes). If a deadlock is detected, the deadlock is resolved by arbitrarily killing off some of the processes. Yet other operating systems ignore deadlock entirely, leaving it up to the user (or administrator) to detect and handle.

Linux handles deadlock in two ways. First, for kernel processes, deadlocks are avoided because kernel processes are granted access to resources in a specific order. A particular resource will only be granted to a kernel process when it is that process' turn to access any or all resources. Second, for user processes, Linux implements the Ostrich algorithm. This is a sarcastic response meaning that Linux buries its head in the sand to ignore deadlocks. So, deadlocks are free to arise for any nonkernel set of processes!

## 14.4.2 Fairness

Related to liveness and starvation is another concept known as *fairness*. All processes need to be treated fairly so that no process is allowed to monopolize a resource. Without fairness, a process could potentially starve because the needed resources are always being granted to other processes.

While the operating system is set up to ensure these criteria (fairness, availability of resources), it is up to the system administrator to monitor the processes and resources to make sure that the operating system is doing what we expect. Toward that end, Linux provides us with a number of programs to monitor the processes and system resources. Although we have examined several such tools to this point of the chapter, we will reexamine them and look at several other programs.

Let us consider an example of a poorly performing system. When an operating system is spending much of its time performing page swapping, this is known as *thrashing*. Recall that paging accesses the hard disk that incurs a large penalty in processor performance because the hard disk is so much slower than the processor. Figure 14.2 illustrates a simple although unrealistic example of thrashing. In this case, there are so many processes running that this particular process is given only two frames of memory, 315 and 612.

The process contains a for loop that consists of two sets of code: the loop mechanism (initializing i to 0, comparing i to n, and incrementing i at the end of the loop) and the assignment statement to modify array element a[i]. In assembly code, the assignment statement would also require computing the memory location of a[i]. Coincidentally, both sets of the for loop code are stored on two different but consecutive pages and thus when moving from one page to the next, if a frame is not available in memory, the current page must be discarded in favor of the new page (because the process has been given only two frames in memory). The data used by this code are stored in two separate but consecutive pages as well.

Every iteration through the loop causes four page faults. First, the for loop mechanism is needed to initialize i and compare it to n requiring the page with the for loop. To execute this instruction also requires that the first page of data (with i and n) must be loaded into memory. If i is less than n, then the assignment statement is executed. Thus, the page with this statement, which is not currently in memory, is required. This causes a page fault.

Now, to compute a[i] = a[i] + b, we need to access array a. This causes another page fault. After completing a[i] = a[i] + b, we resume with the loop increment and loop comparison, again, not in memory because it was discarded. This requires access to both i and n, also

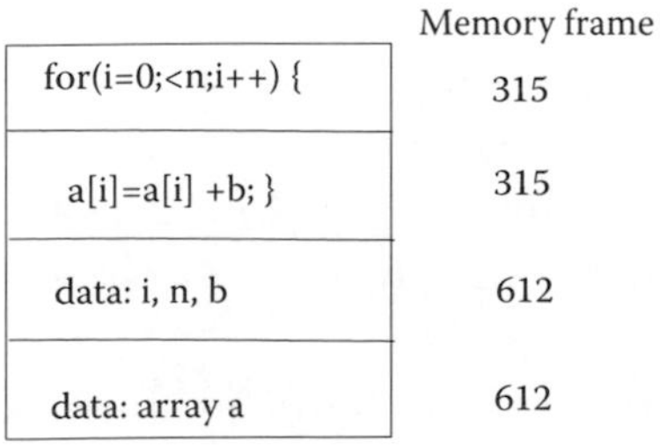

| | Memory frame |
|---|---|
| for(i=0;<n;i++) { | 315 |
| a[i]=a[i] +b; } | 315 |
| data: i, n, b | 612 |
| data: array a | 612 |

FIGURE 14.2   Thrashing example.

not currently in memory. Two more page faults occur within a single iteration of the loop. If there are four page faults per iteration, then this set of code could result in a total of 4*n total page faults!

It should be obvious that this example is artificial and should never arise. For one reason, a process will be given far more than two frames. Another is that it is unlikely that the division within a program would fall exactly as shown in this example (although this is possible). However, as unlikely as the example is, it illustrates the potential problem of thrashing.

### 14.4.3 Process System-Monitoring Tools

The system monitor program provides us with both an overview and detail of what is going on with many system resources. Specifically, we can see the amount of CPU usage, memory usage, network usage, virtual memory usage, and file system usage. For CPU, memory, swap space, and network utilization, the tool shows us the changes taking place over the past minute.

As a system administrator, a quick look at this tool can indicate that your system is running smoothly. You would hope to see CPU load is not approaching or at 100% for any length of time and that swap space is not approaching 100% utilization. Further, if memory use has not come close to 100%, you should expect swap space usage to remain low. If you find swap space usage to be high and remain high, it means that your system is doing a lot of very inefficient paging.

Through the system monitor, you can inspect the running processes (the Processes tab) and see the status, CPU, and memory usage of the processes. If you see many processes with high CPU utilization, you might inspect these processes. Which processes are owned by root that can be postponed? Which processes are owned by users that you might deem of low priority that could be moved either to the background or have their priority lowered? You could even select some user process(es) to run later via scheduling.

If memory utilization is approaching capacity causing a lot of virtual memory usage, you might again select some of the processes to move into the background or schedule them later. Moving processes to the background permits the operating system to temporarily swap their content out of memory and thus free up space for other processes. Lowering the priority of running processes will not impact their memory usage as they would remain in memory. If you find that memory usage remains very high, it might indicate that either you have too many users for your Linux system or that you need more memory in your computer.

Top also provides information about processes and system resource usage. While top and the system monitor will tell you much of the same information, top displays process and resource utilization on one screen, rather than having to switch between tabs. Additionally, top can be tailored to display different types of information. See Chapter 4 for more details on top.

Other tools for querying the system are command line programs. These include ps (see Chapter 4) as well as `mpstat`, `sar`, `pidstat`, `uptime`, `vmstat`, `free`, `stat`, `df`, `du`, `netstat`, `nmap`, `ip`, `ss`, `lnstat`, `iostat`, and `lpstat`.

Another command that might provide useful information is called `strace`, which traces the system calls made by a particular Linux command. We have already looked at some of these programs earlier in the chapter (stat, df, du, ip, and ss). We will examine the remainder here.

The first set of commands explores the processor's performance and the active processes. The mpstat command reports processor-related statistics for each of the system's processors. It is implied that mpstat is used on a multiprocessor system, but it can also be used for a single-processor computer.

The mpstat command provides for each CPU the percentage of time it spends on user activities, system activities, hardware interrupts, software interrupts, guest activities, ideal time, cycle stealing, and periods of input or output when it is forced to wait. The user activity is broken into two different values: ordinary user activity and user activity with niced processes (those given lower priority). Guest activities are actions that the CPU takes to run a virtual processor. Cycle stealing occurs when the CPU is given a lower priority to accessing resources than the disk drive (this is described in more detail later).

Similar to mpstat, sar reports on CPU utilization. The primary difference is that mpstat reports an average over some duration of time. Sar on the other hand, outputs a series of snapshots of processor utilization over a period of time. The output shows, row by row, the processor's utilization at each timed interval.

By default, the interval is every 10 minutes. This information is stored in log files under the directory `/var/log/sa`. You might find numerous entries in this directory with names such as `sa#` or `sar#` where # is a number.

For each interval, sar provides you with the CPU number (or statistics on all CPUs), percentage of time the CPU was used in user and nice modes, system usage, wait time, cycle-stealing time, and ideal. You are also given averages over the entire recorded time period. You can limit the intervals by specifying a different start and/or end time for the report, or alter the interval. You can also ask sar to specify different types of statistics than CPU usage. These include I/O transfer rate statistics, memory usage statistics, paging statistics network statistics, and block device usage.

Another command that reports on the processor's performance is pidstat. This program provides CPU statistics for each running process. These statistics are broken into user, system, guest, and CPU usage where CPU usage is the sum total of user, system, and guest. You are also told which processor is running the particular process (for a multiprocessor system). So, while mpstat reports on each processor, pidstat shows similar information broken down at the process level. pidstat has a number of options that allow you to specify which type of information is output.

The –d option shows I/O usage, including the amount of reading and writing to hard disk. The –r option provides virtual memory information, showing the number of minor and major page faults that occur per second (a minor fault does not require a page to be loaded from disk while a major fault does), the virtual memory size, and resident size.

Finally, the –w option displays task-switching activity. In this case, you are shown for each process the number of voluntary context switches made per second and the number of nonvoluntary context switches. Voluntary context switches arise when the process willingly surrenders the processor to the next process because of I/O requirements while a nonvoluntary context switch occurs when the timer, which is timing the amount of CPU time each process receives, reaches 0.

The uptime program reports on the duration that the system has been running (since the last boot/reboot). The output is presented in a single line that states the current time, the number of days, hours, and minutes of uptime, the number of current users (logged in), and the average load over the past 1, 5, and 15 minutes.

The strace program may or may not be useful as a troubleshooting tool, but it is an interesting program. As stated in Chapter 8, applications software and portions of the operating system invoke the kernel through system calls. System calls are the same as function calls in a C program except that the calls are made to the kernel rather than to a function within the running software. The strace program will map the number and nature of system calls of a given Linux command.

As an example, using strace on pwd results in a listing of the system calls made by pwd. The output includes the parameters passed to those functions. Pwd calls, in order, brk, mmap, access, open, fstat, mmap, close, open, read, fstat, mmap, mprotect, mmap (4 times), arch_prctl, mprotect (twice), munmap, brk (twice), open, fstat, mmap, close, getcwd, fstat, mmap, write, close, munmap, close, and exit_group.

What strace is doing is invoking the Linux command and tracing what the command does while it executes. It intercepts and records the system calls to list them as the result of strace. Strace is primarily used as a debugging tool for those who are writing Linux commands but can be used by system administrators to troubleshoot problems within the operating system itself. Chances are, if you have installed a stable version of Linux and you are not modifying Linux, you would not need to resort to strace. However, it can help you identify what a command is doing.

### 14.4.4 Memory System-Monitoring Tools

The program vmstat reports on memory and virtual memory usage. The default report is broken into six sections:

- Process information

- Memory usage

- Swap space usage

- I/O usage

- System usage

- CPU usage

These values are all averages, as computed since the last time the system was booted. What follows is an example of vmstat's output (edited to fit within the space below).

```
procs- - - -memory- - - - -swap- - - -io- - - system- - - cpu- - -
r b swpd free    buff   cache  si so bi bo in   cs us sy id wa st
1 2 64    112976 65360 413114 0  0  71 30 203 60  2 1  85 0  0
```

Table 14.3 describes the meaning of the abbreviations in the header. Explanations follow for some of the terms following the table.

As Linux is a multitasking operating system, it is possible and likely that more than one process will be in some state of execution at any time. As covered in Chapter 4, this means that there are several active processes although the CPU is only executing one process at any moment. It switches off between them. The number of waiting processes (r) are those that are active (loaded in memory) but not being executed at that moment by the CPU while those in uninterruptible sleep (b) are uninterruptible processes that are waiting on hardware. Interrupting such a process when it is executing could damage data. Portions of the kernel run in uninterruptible sleep.

The cs value indicates how many context switches occur per second. Recall that there are several reasons for a context switch. In multitasking, the system timer will interrupt the CPU to force a context switch between a running process and a waiting process. A switch can also occur when one process terminates or requires time-consuming I/O and so voluntarily surrenders the CPU. Additionally, altering process priority, while not forcing a context switch, will alter the amount of time a process executes before it undergoes a switch.

The four values us, sy, id, and wa provide the amounts of time that the processor is processing user processes, is processing system processes, is idle (no process executing), and is waiting on I/O. Finally, *cycle stealing* is a situation where the CPU is forced to wait for some I/O process to complete. This often happens with DMA (direct memory access) I/O when

TABLE 14.3    vmstat Output Information

| Header | Meaning |
| --- | --- |
| r | Number of processes waiting for run time |
| b | Number of processes in uninterruptible sleep |
| swpd | Amount of virtual memory used (in KB) |
| free | Amount of free RAM (in KB) |
| buff | Amount of RAM used as buffers (in KB) |
| cache | Amount of RAM used to cache hard-disk data (in KB) |
| si | Average KB/second of data swapped into memory from disk |
| so | Average KB/second of data swapped out of memory to disk |
| bi | Average blocks/second of data swapped into memory from disk |
| bo | Average blocks/second of data swapped out from memory to disk |
| in | Number of interrupts/second |
| cs | Number of context switches/second |
| us | User time (nonkernel) |
| sy | System (kernel) time |
| id | Idle time |
| wa | Wait time |
| st | Cycle-stealing time |

there is data movement between a disk drive and memory. If a disk transfer is taking place, the CPU may be forced to wait as the transfer will be slower than CPU access. Thus, cycles are stolen away from the CPU. Cycle stealing used to be more predominant but modern computers utilize multiple buses and multiple memory modules these days to reduce any such contention. As can be seen in this example, the value for cycle stealing (st) is 0.

The snapshot of the vmstat command shown previously indicates a system with a light load. There are few processes running, and a more than sufficient amount of main memory so that disk swapping is not needed at the moment. In this system, a majority of the CPU time is spent idling rather than either running the user processes or system processes.

The program free displays the free and used memory. In some ways, this is like the report provided by df (see Chapter 10) but the information provided is about memory rather than the file system. And like df, the information output by free is the current utilization, unlike vmstat that reports on an average over a period of time.

In free's output (see below), it specifies Mem (memory) and Swap as well as −/+ buffers/cached. Mem and Swap refer to main memory and swap space, respectively. The line that contains −/+ buffers/cached describes the amount of main memory that is allocated for a buffer in support of some application(s) or as disk cache.

The report specifies the total amount of memory, how much is free, and how much is used. You are also given, for memory, the amount that is being shared, the amount that is being buffered, and the amount that is cached. In the example output shown below, notice that swap space is only one-half a GB. This is unusual in that swap space (virtual memory) is usually at least the size of main memory. This particular Linux system was placed into a virtual machine of limited size so that swap space was similarly limited in size.

```
                 total       used       free     shared     buffers     cached
Mem:           1020648     907160     113488     0           65436      471488
-/+ buffers/cached:        370236     650412
Swap:  524280      8          524272
```

### 14.4.5 I/O System-Monitoring Tools

There are a number of "stat"-type programs used to report on I/O performance. These are iostat to report on the file system, lpstat to report on the printer, and network programs such as ip, ss, netstat, nstat, rtacct, and nmap. We have explored the stat, df, and du (file system) commands in Chapter 10.

Both nstat and rtacct provide network interface statistics. These are useful if you want to examine the types of messages that have been received, for instance, IcmpInErrors, IpOutRequests, TcpActiveOpens, UdpOutDatagrams, and Ip6InDelivers. The lnstat program provides network-routing cache statistics and additional network statistics from the Linux kernel, as stored in the directory /proc/net/stat. You are able to obtain information from the ARP cache, the RT cache, and the ndisk cache.

The nmap program is a network exploration tool. Its main function is to search for network hosts and scan the ports that are available. It does this by sending out IP packets and examining the responses. By determining port access, it can report on the services that the given computer offers (e.g., ssh, http). Additionally, it can report on firewall activity, operating system type, and numerous other features of a given computer. With nmap, you can investigate a network's security (or the individual computer's network access security) as well as monitor network components. You can also use nmap to accumulate network statistics.

The only argument required for nmap is the address to be investigated. This can be an IP address or an IP alias of a single device, or a range of network addresses. For instance, nmap 10.11.12.5-205 would investigate all devices in the range from 10.11.12.5 through 10.11.12.205. We can also specify a full subnet using nmap 10.11.12. In response, for each device contacted, you will receive a report. The report will provide details of accessible ports indicating their status (open, closed), the service implemented on that port, and the version (if available). For instance, the command nmap 10.11.12.13 might receive a report like the following:

```
Nmap scan report for 10.11.12.13
Host is up (0.000120s latency).
Not shown: 993 closed ports
PORT        STATE      SERVICE
22/tcp      open       ssh
25/tcp      closed     smtp
80/tcp      open       http
111/tcp     open       rpcbind
113/tcp     closed     auth
631/tcp     open       ipp
```

The nmap program has a number of options available. By using –A, you can obtain more detail on version types for the various services and the operating system. For instance, from the above report, we might instead see the following entries for ssh and http:

```
22/tcp        open           ssh    OpenSSH 5.3 (protocol 2.0)
| ssh-hostkey: 1024... (DSA)
| 2048... (RSA)
80/tcp        open           http   Apache httpd 2.2.4 ((CentOS))
```

The ... in the above entries are specifications of keys, omitted for space.

Other options allow you to specify the scan type in place of the default scan using Ping. There are a wide variety of IP-based scans that can be used including TCP SYN, ACK, Maimon scans, UDP scans, and FTP bounce scans. Another group of options is available for obtaining timing and performance information. And yet other options are available for analyzing firewall protection and can help you analyze the security of the network. There is not enough space to cover nmap in detail here; so, you are encouraged to explore it on your own should you find yourself serving as a network administrator.

The iostat program gives you the flexibility to obtain three types of reports: CPU utilization, device utilization, and network filesystem utilization. The CPU utilization provides the percentage of CPU time used for ordinary and niced user applications, system operations, cycle-stealing time, I/O wait time, and idle time. The device report lists each type of connected block device. This report displays the average number of transfers per second, blocks read per second, blocks written per second, and total number of blocks read and written.

If there are any mounted file systems (including USB- or CD-ROM-mounted devices), iostat also provides a report for each of these indicating the filesystem name, number of blocks read and written per second, and total number of blocks read and written. You can obtain more detailed information on any one device by using -x *devicename*, as in iostat -x /dev/sda5. This report includes the number of sectors read/written per second, the average size of requests, and the average wait time.

One last program worth noting is who. Although this merely lists the logged users of the system, if you use this program often, you can see how many users are commonly using the system and where they are logged in from. You can also obtain this information by viewing log files (see the next section) that store log-in attempts.

With all these programs, you are able to obtain several views of the system. First, you can see current statistics to see if there are immediate problems such as too heavy a demand because of too many processes or too many users. Second, you can view the statistics over an interval of time to see if heavy loads are occurring frequently or infrequently. Third, you can view averages over the interval. Most intervals are from the last boot.

You might save some of the statistics into files before rebooting so that you can compare performance over a greater period of time. By viewing statistics accumulated over weeks or months, you can obtain evidence that your system might need an upgrade.

See Table 14.4 for a summary of the various tools discussed in this section.

## 14.5 LOG FILES

We explored the syslogd (rsyslogd) service in Chapter 11. Through this daemon, we are able to log events that are generated from systems and applications software. Additionally, klogd logs kernel events. In most cases, kernel events are logged to the same collection of files. When you review one of these log files, you will find each entry preceded by one of syslogd (rsyslogd) or kernel to indicate what initiated the event.

Log files can be used to help the system administrator interpret the current and previous status of the system. Searching log files is a critical part of ensuring that your system is running smoothly and of troubleshooting problems. You may or may not inspect log files often but it is a good place to look when issues arise.

There are, in fact, a great many pieces of information that can be gleaned from examining log files. These include

- Who is and has tried to log in.
- What services were successfully started or stopped at system initialization time.
- Events related to hardware or software errors.
- What yum updates and installations occurred (successfully or unsuccessfully).
- What jobs successfully ran via crontab.

TABLE 14.4  Various System Administration Monitoring Programs

| Name | Processor Info | Process Info | Memory Info | VM Info | File System Info | I/O Info | Network Info | Comments |
|---|---|---|---|---|---|---|---|---|
| df | | | | | * | | | |
| du | | | | | * | | | |
| free | | | * | * | | | | |
| iostat | * | | | * | * | * | | |
| lpstat | | | | | | * | | |
| mpstat | * | | | | | | | |
| netstat | | | | | | | * | Obsolete, replaced by ss |
| nmap | | | | | | | * | |
| nstat | | | | | | | * | |
| pidstat | * | * | | * | | | | |
| ps | * | * | * | * | | * | | |
| rtacct | | | | | | | * | |
| System monitor | * | * | * | * | * | * | * | Graphical, persistent |
| sar | * | * | * | * | * | * | * | |
| ss | | | | | | | * | |
| stat | | | | | * | | | |
| strace | * | * | | | | | | |
| top | * | * | * | * | | | | Persistent |
| uptime | * | | | | | | | System uptime |
| vmstat | * | * | * | * | | | | |
| who | | | | | | | * | Lists logged-in users |

- Who has accessed your Apache web server and which pages have been requested (if you are running Apache).

Here, we look at some specific log files: what software wrote to them and what information you might find there.

## 14.5.1  syslogd Created Log Files

We return to the /etc/syslog.conf (or /etc/rsyslog.conf) file covered in Chapter 11. In this file, you will find logging events and logging locations. You might find the following entries:

```
*.info                  /var/log/messages
mail.none               /var/log/messages
authpriv.none           /var/log/messages
cron.none               /var/log/messages
authpriv.*              /var/log/secure
mail.*                  /var/log/maillog
cron.*                  /var/log/cron
uucp,news.crit          /var/log/spooler
local7.*                /var/log/boot.log
```

You might also find critical messages are being sent to *, meaning every open console. Let us step through these files to explore what you can expect to find in each (we omit maillog and spooler).

Log files tend to store entries that have the following information:

- Date and time of the event

- Host name of the computer on which the event arose

- Name of the program that generated the log message

- A (short) description of the event

The message will generally be short (less than one line). If the process that generated the message is not one of syslogd (rsyslogd) or the kernel, the PID is also included.

As the messages file contains entries generated from both syslogd (rsyslogd) and the kernel via klogd, you might find the entries here overlap the information obtained by dmesg. Recall from Chapter 11 that dmesg responds with the kernel ring, a complete listing of the events during the last kernel initialization. Any syslogd (rsyslogd) generated events should be listed in the messages file after the kernel initialization messages. The syslogd generated events are provided by various email (mail), authentication (authpriv), and scheduling (cron) programs whose events are denoted with a priority of none. You might notice that all other events generated by these three types of programs are sent to other files (/var/log/maillog, /var/log/secure, and /var/log/cron). Also sent to messages are all events with a priority of info.

The secure log stores information based on authorization and authentication events. These events occur when users attempt to log in to the system or must otherwise provide a password (e.g., when using su or sudo). The secure file will contain entries that look like the following:

- Nov 23 10:29:16 mycomputer sshd[1781]: Server listening on 0.0.0.0 port 22.

- Nov 23 10:29:41 mycomputer pam: gdm-password[2041]: pam _ unix(gdm-password:session): session opened for user foxr by (uid = 0)

- Nov 23 10:32:18 mycomputer su: pam _ unix(su:session): session opened for user root by foxr(uid = 500)

These three messages have taken place within a few minutes of each other. The first message indicates that someone is attempting to log in using ssh over port 22. The next message indicates that a log-in attempt was made through the PAM authentication module by user foxr. The entry uid = 0 means that root handled the log-in attempt. The next line indicates that a user attempted to use su. In this case, the user was foxr (uid = 500) attempting to log in as root. Note that mycomputer is the hostname of the computer that generated these messages.

The following entries show messages in the `secure` log file that arose because of a failed log in to root via su.

- `Nov 23 11:37:20 mycomputer su: pam_unix(su:session): session opened for user root by foxr(uid = 500)`

- `Nov 23 11:37:27 mycomputer unix_chkpwd[4993]: password check failed for user (root)`

- `Nov 23 11:37:27 mycomputer su: pam_unix(su:auth): authentication failure; logname = foxr uid = 500 euid = 0 tty = ptrs/1 ruser = foxr rhost = user = root`

Here, we see a logged event of an su attempt followed by two messages indicating failed authentication, one from `unix_chkpwd` and one from `su`. Logged with the events are the user's username who attempted to but failed to log in and the username that this user attempted to su to.

The log file `/var/log/cron` contains entries generated from crond, anacron, and run-parts. These log entries, like the previous examples contain the date/time, host, program, and event. In the case that the event was executed by crond, then the user who issued the crontab job is also recorded. For instance, you might find entries like the following:

- `Nov 20 11:01:01 mycomputer anacron[5013]: Anacron started on 2012-11-20`

- `Nov 20 15:10:01 mycomputer CROND[5042]: (foxr) CMD (./my_ scheduled_script >> output.txt)`

- `Nov 20 16:43:01 mycomputer CROND[5311]: (foxr) CMD (echo "did this work?")`

- `Nov 20 16:44:01 mycomputer CROND[5314]: (foxr) CMD (echo "did this work?")`

The first entry informs us when anacron is starting. The other three entries are of crontab jobs submitted by user foxr. The first submission executes a script while the other two occur one minute apart and merely perform an echo statement.

Next, we see two logged events from run-parts that are running programs found in /etc/cron.daily. For any events run by run-parts, there should be two entries, one that indicates when the event started and one of when it finished. The two entries are of the same event, logrotate.

- `Nov 18 03:49:01 mycomputer`

  `run-parts(/etc/cron.daily)[14361]: starting logrotate`
- `Nov 18 03:49:01 mycomputer`

  `run-parts(/etc/cron.daily)[14364]: finishing logrotate`

The boot.log file contains entries of logged events during the previous boot attempt, specifically initialization steps. These entries, unlike the previous log file entries, will be segmented into groups based on the phase of the initialization step. Rather than logging these by time, they are listings of the events and the results (success or failure). Below are some excerpts from this file, with comments interspersed.

```
Starting udev:
Setting hostname mycomputer:                    [ OK ]
```

Message indicates hostname has been established.

```
Checking filesystems
/dev/sda1: clean, 93551/256000 files, 784149/1024000 blocks
/dev/sda5: recovering journal
/dev/sda5: clean, 472/40320 files, 26591/16180 blocks
                                                [ OK ]
Remounting root file system in read-write mode:  [ OK ]
Mounting local filesystems:                     [ OK ]
Enabling local file system quotas:              [ OK ]
Enabling /etc/fstab swaps:                       [ OK ]
```

The root file system along with the /home file system is checked by fsck. We see both file systems are clean (no corruption found). This is followed by mounting the file systems. First, the root file system is remounted in read–write mode as earlier it was read-only mode. All file systems mounted as per the /etc/fstab file. File system disk quotas enabled (if any).

```
Iptables: Applying firewall rules:              [ OK ]
Bringing up loopback interface:                 [ OK ]
Bringing up interface eth0:                      [ OK ]
```

Network service started bringing up both available interface devices. Firewall started with no errors (errors could be caused by having syntactically improper rules).

```
Starting auditd:                                [ OK ]
Starting portreserve:                           [ OK ]
Starting system logger:                         [ OK ]
Starting irqbalance:                            [ OK ]
Starting crond:                                 [ OK ]
Starting atd:                                   [ OK ]
Starting certmonger:                            [ OK ]
```

Other services started successfully.

## 14.5.2 Audit Logs

A log subdirectory of note is /var/log/audit. This subdirectory stores the audit daemon log files. The audit daemon (auditd) is responsible for logging general activities of the users. Such activities will include logins, login attempts, failed logins, opening of

TABLE 14.5    Aureport Options

| Option | Meaning |
| --- | --- |
| -au | Authentication attempts |
| -c | Configuration changes |
| -e | Events |
| -f | File operations |
| -i | Convert numeric (UID, GID, etc.) entries into text |
| -l | Login attempts |
| -m | Modification of user accounts |
| -n | Anomalous events |
| -p | Process-initiated events |
| -s | System calls |
| -u | User-initiated events |
| -x | Processes executed |

TABLE 14.6    Ausearch Options

| Option | Meaning |
| --- | --- |
| -a *EID* | All entries for event # *EID* |
| -gi *GID* | All entries of processes owned by group *GID* |
| -i | Convert numeric (UID, GID, etc.) entries into text |
| -k *string* | All entries that contain *string* |
| -m *type* | All entries whose message type is listed in *type* |
| -p *PID* | All entries generated by process *PID* |
| -pp *PID* | All entries generated by process whose parent is *PID* |
| -sc *name* | All entries generated by the system call *name* (name may either be a string or number) |
| -ui *UID* | All entries generated by user *UID* |
| -x *name* | All entries generated by the executable program *name* |

terminal windows, launching of executable programs, system calls, failed system calls, cryptographic events, and cryptography keys generated among numerous other activities.

The audit log files are quite large; so, we will view their content using two handy programs. First, there is `aureport` that summarizes the log entries, providing totals for the various event categories. Second is `ausearch` that the system administrator will use to query the audit log files for specific types of event information.

For ausearch, you specify a type of event you are interested and in some cases, a specific value to match against. You might, for instance, request any log entry of a user given the UID or any log entry of a given process given the PID. Tables 14.5 and 14.6 provide options for aureport and ausearch, respectively. The ausearch program is more complex but provides specific audit entries rather than summaries from aureport.

Running aureport with no options gives the following output:

```
Summary Report
======================
Range of time in logs: 03/19/2013 10:11:02.774 -
    04/16/2013 10:21:15.081
```

```
Selected time for report: 03/19/2013 10:11:02 -
     04/16/2013 10:21:15.081
Number of changes in configuration: 18
Number of changes to accounts, groups, or roles: 47
Number of logins: 20
Number of failed logins: 1
Number of authentications: 164
Number of failed authentications: 5
Number of users: 3
Number of terminals: 16
Number of host names: 5
Number of executables: 21
Number of files: 0
Number of AVC's: 0
Number of MAC events: 20
Number of failed syscalls: 0
Number of anomaly events: 3
Number of responses to anomaly events: 0
Number of crypto events: 68
Number of keys: 0
Number of process IDs: 4787
Number of events: 29145
```

Running aureport with an option will provide a listing of all events corresponding to the option(s) supplied. For instance, `aureport` –au generates all authentication events. Three such events are listed below. The columns represent the authentication event number, the date and time, the user account responsible for the event, the host and terminal window from which the event was generated, the executable program responsible for the event, whether it was successful or not, and the overall event number.

```
2. 03/19/2013 10:11:16 foxr ? :0 /usr/libexec/gdm-session-
     worker yes 35602
3. 03/19/2013 10:15:59 foxr ? ? /usr/sbin/userhelper yes
     35609
4. 03/19/2013 10:22:35 root ? pts/0 /bin/su yes 35617
```

Below is a brief excerpt generated by the command `aureport` –e. The –e option provides all events. In this case, we see events from the morning of March 19, 2013. The first column, as with the list above, indicates the number within this listing while the number in the fourth column is the overall event number. The fifth column is the type of event followed by the user's UID and whether it was successful or not. A UID of -1 would indicate "unset."

```
10. 03/19/2013 10:11:16 35607 USER_START 500 yes
11. 03/19/2013 10:11:16 35608 USER_LOGIN 500 yes
12. 03/19/2013 10:15:59 35609 USER_AUTH 500 yes
13. 03/19/2013 10:15:59 35610 USER_ACCT 500 yes
```

Here, we see the same output with the option −i added to −e so that numeric values are translated into names.

```
10.  03/19/2013 10:11:16 35607 USER_START foxr yes
11.  03/19/2013 10:11:16 35608 USER_LOGIN foxr yes
12.  03/19/2013 10:15:59 35609 USER_AUTH foxr yes
13.  03/19/2013 10:15:59 35610 USER_ACCT foxr yes
```

If you refer back to the previous output, you will see the last event has an event number of 35610. Let us examine that event in more detail using ausearch. We issue the command ausearch -a 35610 and we are given the following output:

```
time-> Tue Mar 19 10:11:16 2013
type = USER_START msg = audit(1363702276.674:35607):
user pid = 1948 uid = 0 auid = 500 ses = 2
subj = system_u:system_r:xdm_t:s0-s0:c0.c1023
msg = 'op = PAM:session_open acct = "foxr"
exe = "/usr/libexec/gdm-session-worker" hostname = ? addr = ?
terminal = :0 res = success'
```

We see the same information that aureport provided us, but we have more information. We see the date, type of event, message specifier, PID and UID of the event, AUID (which is the effective user, or the user on whose behalf the PID executed), a session number, a subject, and the full message that includes more detail such as the specific executable program and terminal window that the event originated from, and the result (success in this case). The above example is a successful login by foxr. Below is another event, an unsuccessful attempt to su to root.

```
time-> Tue Apr 16 11:15:48 2013
type = USER_AUTH msg = audit(1366125348.228:13171):
user pid = 23909 uid = 500 auid = 500 ses = 1
subj = unconfined_u:unconfined_r:unconfined_t:s0-s0:c0.c1023
msg = 'op = PAM:authentication acct = "root" exe = "/bin/su"
hostname = ? addr = ? terminal = pts/0 res = failed'
```

### 14.5.3 Other Log Files of Note

There are numerous log files with the name anaconda such as anaconda.log, anaconda.ifcfg.log, anaconda.syslog, and anaconda.yum.log. These log files are generated upon system installation. They all store events related to Linux installation. For instance, anaconda.log is the log file of the Linux installation itself while anaconda.ifcfg.log stores information about the network as it is initially configured and anaconda.yum.log contains information on upgrades to the initial installation.

The following are additional log files of note:

- Xorg—log files generated by X Windows, including modules loaded when X Windows is started and failures of X Windows components. There are also messages from the audit client for window opening and closing events.

- maillog—messages generated by the mail system. Even if your computer is not running email, email messages are automatically generated and sent to root or users of the current system when errors or warnings arise.

- yum.log—entries here indicate yum operations (installations, updates, and deletions).

- lastlog—last log-in attempts for users, this information is not stored in the text; so, you cannot view it directly.

- httpd/access_log—the default log file generated by Apache for every received http request. This log file can be highly useful website administrators who want to view user habits when visiting their site.

- httpd/error_log—also generated by Apache, this file contains events that result in errors such as bad or lost connections, bad links and server side script errors.

- cups—messages logged from printer requests and printer errors.

- utmp, wtmp, and btmp—these log files show who is currently logged into the system and failed log-in attempts.

## 14.5.4 Log File Rotation

You will no doubt find multiple files whose names are similar to the above but with numeric extensions or a hyphenated date added to the name. These extensions/dates indicate log files that have been rotated. The `logrotate` program is set up to *rotate* log files. This means that a new, empty, log file is created and the older log file(s) is(are) renamed.

The renaming strategy differs by log file. Most log files are renamed by adding the date of the rotation to the end of the file name, such as changing `messages` to be `messages-20130829`. The older log rotation approach was to number each file such as `messages` (the new or newest file), `messages.1`, `messages.2,` and so forth. This would require that logrotate not only change one file's name but all file names by rotating the numbers.

Many rotations are scheduled via anacron to be performed daily, weekly, or monthly. Log files are retained for some time. The frequency of log rotations and the number of log files retained is usually established by the file `/etc/logrotate.conf`. If set to `weekly` and `rotate 4` (the defaults), logrotate will rotate log files once per week, retaining not more than four backup logs at a time (thus keeping five total log files, the current file, and the four older files). Upon the next rotation, the oldest log file is discarded. As a system administrator, you might want to retain log files indefinitely by backing up /var/log at least once per month. Alternatively, you can change logrotate.conf to retain more than four log files at a time.

The system administrator should know about log rotation to understand when log files should be archived so that no logged data are lost. The system administrator should come up with a log rotation plan. This log rotation plan should be consistent with the logrotate configuration file, /etc/logrotate.conf. In this file, default configurations are provided that include the rotate rate, the number of log files to retain, whether to create a new (empty) file upon log rotation or not, and whether to compress log files. Then the file lists directives for specific log files. You might, for instance, find default values of `weekly`, `rotate 4`, `create`, and `#compress` (i.e., the compress command is commented out). This may be followed by a specific logrotate entry such as the following for the log file wtmp:

```
/var/log/wtmp {
      monthly
      create 0664 root utmp
      rotate 1
}
```

As a simple example, the system administrator may choose to back up /var/log every 4 weeks to correspond with the number of default log files that are retained at a time. However, this may itself be insufficient. As we saw above, some log files' names are altered to indicate the date of rotation. In such a case, saving the log file to a backup location does not present a problem. But in the case of log files whose names are altered by extending the name with a digit as in `somelog.1` being rotated to `somelog.2`, then just saving these to a location where there are already some log entries would cause older log files to be deleted. The system administrator should come up with a naming convention.

Speaking of log names, it should be noted that the naming convention for the Xorg log files is different from the others described above. Rather than numeric values or dates to indicate the log files' ages, the log files' names for Xorg are generated based on the terminal window that the file corresponds to. For instance, `Xorg.0.log` is the log file for console 0 whereas `Xorg.1.log` is the log file for console 1.

Now that we have logging taken care of, what about reviewing log files? How often should a system administrator examine the contents of these files and which files? Although there is no set or recommended policy, a plan should be devised by the system administrator(s) based on the size and nature of the organization. The simplest plan is to examine the log files whenever the system is running poorly or you suspect a problem has arisen.

We want to work to prevent problems; so, we should be proactive in examining log files. Searching log files daily though will not typically be required, with the possible exceptions of secure and messages. Assume we have a Linux system of dozens of users. We might use the following schedule for examining log files, unless of course, a situation arises requiring more immediate attention:

- secure—examine daily to see pattern of log ins and see if there are incidents of reported failures that might hint at a hacking attempt.

- messages—examine daily to see if there are services or software reporting failures.

- boot—examine after system boot and initialization to ensure no errors doing the boot process.

- cron—examine weekly to ensure that all scheduled cron jobs are executing as expected.

- yum—examine weekly to see if updates and installation are being handled without errors.

- maillog, spooler—examine only as needed if there are email or printer problems.

In addition, run aureport weekly to examine incidents of failures and anomalies. If there are an unusual number of entries, use ausearch to find out more.

## 14.6 DISASTER PLANNING AND RECOVERY

*Risk assessment* is a necessary undertaking for any organization that has assets to protect. In risk assessment, the organization identifies its assets and the vulnerabilities of those assets. They also identify their organizational goals. From the vulnerabilities, a *threat analysis* can be performed. Next, the threats are organized by seriousness based on their likelihood of occurring, the potential damage that the threat can cause, and the impact on the organization based on the prioritized goals. The organization can judge the risk that its assets are under from the identified threats and formulate strategies to reduce these risks.

One risk that any organization will face is that of a disaster, whether natural (flood, fire, and earthquake) or man made (explosion caused by armed conflict or terrorism, sabotage, and theft). Of course, not all threats are disasters but it is important to plan for disasters. Thus, we have *disaster planning*. Disaster planning is the process of ensuring that the organization can still function amid disaster. Disaster recovery consists of the steps necessary to bring the organization back up to full (or even partial) strength after a disaster.

While disaster planning and recovery can encompass assets of the organization that include personnel, infrastructure, reputation, and so forth, here, we will solely concentrate on information technology (IT). We can divide the IT assets into roughly three components:

- Hardware
- Software
- Data

Although hardware and software are the assets that the company will purchase, it is data that are the most critical. The hardware and software are replaceable whereas the data may not be. So, the most important part of disaster planning is to ensure that the data are safeguarded.

In this chapter, we have already explored backups and RAID technologies. However, even painstaking backups can be defeated without some common sense applied to the process. Let us consider. You as a system administrator backup all of /home and /var weekly and then perform incremental backups every night. You retain backups for weeks and you perform a full archive once every 6 months, retaining the archives for 10 years. Your system uses RAID 1 technology so that information is fully duplicated in case of a problem before you have a chance for the increment evening backup. Yet, this plan, as cautious as it sounds, may not be enough.

Your RAID file server is located on the ground floor of the building. You back all of the file system to magnetic tapes that you leave in your office behind a locked door and you place the archival backups in a safe, also in your office. Your office is also on the first floor of the building. The river next to your building crests at 10 feet above flood stage that enters your building, flooding 3 feet of the first floor. The RAID server is on the floor while the magnetic tapes and safe are also low to the ground. This disaster leaves you with no data except perhaps any data saved on local workstations in higher floors of the building. Oops! Now, imagine instead that your office is located on the fifth floor. This will most likely save you from flooding, but a fire could destroy the entire building.

A simple solution to this scenario is to make sure that your backups are never on the same site as your original data. Where might you keep the data? If your organization has offices in other locations, move the backup media there, whether by physically transporting them or performing the backups over a network. If your organization does not include other locations, then take your backup media to a bank and lock it up for a week at a time (unless a disaster requires that you retrieve it).

Another strategy is to extend the idea behind RAID 1. Use two file servers, one being a mirror. Locate the second file server off-site. If you do not have a second site, rent file server space from some organization that offers storage area network support.

Making sure data are backed up and available after a disaster is only part of a solution. For most organizations, extended downtime can be extremely costly. If the company has a web portal and the webserver is brought down during a disaster, the company is unable to do any business in that time. If a company offers credit information on clients, it needs to be able to respond to telephone or electronic requests at any time. Downtime will impact this. Any form of downtime can not only damage the company's ability to business but in the long run, its reputation.

One solution is to distribute the data and processing to multiple sites. It is unlikely that more than one site would be impacted by a disaster. Although this is an expensive solution, if the company is large enough, it would be located in several sites in any event. The cost then is one of ensuring that the data and processing centers can all function together or separately. One additional advantage to distributing the processing across centers is that load balancing can be implemented so that servers are kept equally busy with incoming requests.

Another threat to data is keeping them secure. If there is sensitive information being maintained such as the customer's credit card numbers, it is essential that this information must not be accessible by any unauthorized personnel (whether within the organization or hacker). Authentication of course is the common solution to ensure that access is only granted to authorized users. Encryption is another common solution. Encryption technologies are readily available and can be applied to individual files, entire file systems, or messages broadcast over the network. Chapter 5 discussed openssl and briefly mentioned other encryption tools available in Linux.

Another concern in maintaining the security of data comes from authorized users of the data in the form of a *disloyal* employee. Such a person may copy, alter, move, or delete data in an attempt to gain from the action. For instance, a rival company may pay to have data stolen. A disgruntled employee may attempt sabotage or extortion. There is no IT-related solution to this problem; however, educating the employees can limit such a problem. Additionally, monitoring access to sensitive files can help track employee movements. As stated earlier in the chapter, you can obtain such information through log files, particularly the audit logs.

Finally, there are physical threats to IT equipment. You might have theft or vandalism. This is more common in settings where computers are available to the public (e.g., a library) or in schools where computers may not be monitored. Solutions are to monitor computer usage through lab monitors (personnel) or cameras and to lock equipment to the desks. Physical threats also arise because of fire, flood, smoke, and heat damage.

One solution to a fire is to use sprinklers. However, water can damage the equipment as quickly as the fire! Fire-retardant chemicals are sometimes used in place of sprinklers as they may cause less damage to hardware although they tend to be far more expensive. Smoke and heat will have less of an impact on computer hardware but a sufficient amount of smoke or a hot enough environment can still lead to damage.

In addition, electrical power surges can destroy computer chips. A simple solution is to make sure that all electronic equipment is plugged in through surge protectors.

To recover from disaster, a plan must specify how to proceed from the point that the disaster ends. Such a plan must include how to restore data from backups and how to replace damaged or destroyed equipment. The plan must clearly map out how to proceed so that the organization's IT can reach full capability. The plan will no doubt have different steps based on the type of and degree of disaster. As there are any number of possible disasters that you might face, you will have different plan steps per disaster. Each disaster plan will have its own team with a selected team leader. The personnel that make up the team might overlap (for instance, we might expect IT personnel to be involved in all plans and perhaps the same person).

Let us consider a partial plan. First, to implement the plan, we need to make sure that we have information available to enact the plan steps. We should keep electronic and hard-copy versions of variety of information at multiple sites. The information should include at a minimum the following:

- Contact information for our personnel including home addresses and phone numbers, especially for our emergency response team(s).

- A full inventory of our IT infrastructure (servers, workstations, network components, software including versions installed, licenses, and description of our data).

- A copy of our plan.

Now, if you have a disaster, you follow your plan.

Let us consider as an example a fire that results in damage to the building housing some of the organization's employees, data-processing equipment, and storage. This is only one site though and other sites have overlapping employees, equipment, and storage.

The first step is to make sure the building has been evacuated and that the fire department is responding. Next, determine if the disaster is real. If not, return to normal business and possibly cancel the call to the fire department (this is optional as there may be a good reason to have the fire department investigate the cause of an alarm). Otherwise, initiate the fire disaster-recovery plan by contacting the fire disaster team's leaders and members. Now, the team takes over.

Their plan should include contacting other sites of the organization to let them know their status. These other sites should expect a heavier load and to handle calls and emails intended for the site affected. Additionally, if this site has data that are not securely backed up elsewhere, the team needs to work out how to recover that data or what to do in the event of lost data.

The fire itself would presumably be put out within an hour or two. The next step is to assess the damage. The team would not be permitted on the premises until the fire inspector gives an approval. The team, in conjunction with the fire inspector, could determine, for instance, if the building will be safe to work in or will require extensive repair. This would provide management with an estimate for how long the organization would have to go without access to the building. If the team is allowed to move through the building, they might collect items that would help recover from the disaster such as any storage media that might have been held in a secure location (e.g., safe) or was not damaged by the fire.

Now, it is about recovery. The team will work to ensure that the off-site centers are handling the load and that there was no loss of data. A press release would be placed with the media and/or directly to clientele to indicate any ongoing problems such as a limitation to processing or a reduction in website availability. If the building requires more than a few days' worth of repairs, then an alternate site should be established during this period.

The team will also explore whether new equipment must be purchased to replace anything damaged. This new equipment may have to be housed separately during building repair. Once the building is available, any new equipment must be moved. Finally, the disaster team, along with management should explore their plan and update it based on any failures that may have been identified during the disaster.

Once disaster planning is complete, the organization should review its plans, perhaps annually. New threats may arise as the organization changes its assets. Over time, threats may change as the organization implements security schemes to defeat old threats.

You might wonder in this section why this topic is included in a text on Linux. As a system administrator you may not be responsible for disaster planning and recovery, but you should certainly be knowledgeable about the process as management might (and should) seek your input. Without IT staff's input, any disaster-recovery plan that covers IT will no doubt be lacking.

## 14.7 TROUBLESHOOTING

You now have the tools available to troubleshoot your system. Which tools do you apply and when? This section will examine a number of technical problems and discuss system administration efforts to resolve these problems.

For each problem, we look at steps to further identify the cause of the problem followed by easy or short-term solutions and then more involved or long-term solutions. These solutions described in these scenarios are not intended to be complete but should illustrate some of the types of efforts that you, the system administrator, should take when faced with similar problems.

## PROBLEM 1: SYSTEM IS RUNNING INEFFECTIVELY

Description: Simple tasks are taking too long to execute. Log in is taking more time than expected. There is a delay between issuing a command and seeing its result.

Steps to determine the problem:

1. Use top, ps, the system monitor, and/or mpstat, sar, and pidstat to view the running processes.

    a. Is CPU load heavy, approaching 100%?

    b. Are there processes that are taking most of the CPU time or are there many processes taking little time but combined that cause a heavy load?

2. Use vmstat and free to examine main memory and swap space utilization.

    a. Is main memory full?

    b. Is the system spending a lot of time swapping?

    c. Are there too many processes in memory?

3. Use uptime to see how long the system has been running without a reboot. While Linux seldom needs to be rebooted, a reboot may resolve the problem.

Short-term solutions: Identify processes that can be halted and scheduled for later. Identify processes whose priorities can be lowered through renice, or those processes that could be moved to the background. Alternatively, can you contact the users and ask some of them to discontinue their processes and/or log off? Also, reboot the computer if the above steps do not solve the problem.

Long-term solutions: Purchase more main memory, increase the size of the swap partition (possibly add a second hard disk to contain more of the swap partition), and purchase a more powerful processor (or additional processors).

## PROBLEM 2: SERVICE NOT RESPONDING

Description: Some service fails to respond such that users are receiving error messages when they attempt to use it. For instance, http requests to a web server result in server not found.

Steps to determine the problem:

1. Identify the service in question (from the users).

2. See if the service is running (e.g., `/sbin/service` *servicename* `status`).

3. See if the service's supporting programs are running.

4. View the configuration file for the service to see if it is configured correctly.

5. View the log file(s) generated as a response to the service.

Short-term solutions: Restart the service. Reconfigure the service. Restart supporting services. If needed, reboot the system.

Long-term solution: If the problem is related to a server (not service) that is not functioning correctly, reinstall the server, or view the server's web sites to see if there are solutions to the given problem.

## PROBLEM 3: INADEQUATE HARD-DISK SPACE

Description: One or more of the file systems is filling up or has become full. Users cannot save files. Or, swap space is commonly low on available space.

Steps to determine the problem:

1. Use df to view how full each file system is.

2. Use find to search for inordinately large user files and core dumps (if /home is low on space), or log and spool files (if /var is low).

3. If swap space is low, examine swap history using vmstat, sar, and pidstat.

Short-term solutions: Back up the file system that is running out of space. Delete overly large files and warn the user/owner of the files (e.g., "I have removed several core dumps found in your directory"). Ask users to clean up their file space.

Long-term solutions: Back up all file systems. Purchase additional hard disks and either segment users onto different partitions (e.g., /home/1 and /home/2) or repartition the file system so that the partition can be moved either to the new hard disk or to be split across the hard disks. Implement disk quotas if necessary to prevent user spaces from filling up in the future. Initiate mail quotas. Move large log files to an archive.

## PROBLEM 4: SUSPICIOUS SYSTEM BEHAVIOR

Description: Services or programs are not working as they should. System might be too slow. Files might have disappeared.

Steps to determine the problem:

1. Examine your log files, particularly secure, lastlog, and btmp, to look for unusual patterns of logins.

2. Look for running processes with peculiar ownership.

3. Use ausearch to look at authentication events, particularly failed ones.

4. Look for evidence of computer virus or Trojan horse.

Short-term solutions: Kill any suspicious processes (with apologies to any users who own these processes). Run antiviral software. Reboot the computer if needed. Examine your firewall to make sure it is running.

Long-term solutions: Implement a more secure authentication system and a more secure firewall. Implement an intrusion detection system. Discuss account protection with your users. Require all users to change passwords at the next log in. Delete any suspicious user accounts.

## PROBLEM 5: SCHEDULED EVENT DID NOT TAKE PLACE

Description: An event scheduled through crontab, at, batch, or anacron did not take place as expected.

Steps to determine the problem:

1. Examine the cron log file to see if the event took place.

2. See if system downtime occurred when the event was scheduled.

3. Check to see if the event took place at the next boot (via anacron).

4. Check to see if the event failed because the scheduler did not have adequate permission.

5. See if the event took place but generated an error, through the audit or cron log.

Short-term solution: Reexecute the scheduled event.

Long-term solution: Move the event from cron to anacron. Change the scheduled event date and time. If a user needs permission to run cron or at jobs, add the user to /etc/at.allow and/or /etc/cron.allow.

## PROBLEM 6: NETWORK NOT RESPONDING

Description: You are unable to reach other computers via your web browser or other network tools.

Steps to determine the problem:

1. See if the network service is running.

2. Use ip to check the status of your interface device(s), do you have MAC addresses? IP addresses? Do you have a router or gateway connection?

3. Check the physical connection to the network to see if there is something wrong with the cable or port.

4. Use ping and/or traceroute to see if you can reach your gateway. If successful, use ping/traceroute to reach a computer on your local area network and then a computer on the Internet.

5. Test to see if you can reach computers using IP addresses but not IP aliases (check your resolv.conf file).

6. See if other users are also unable to communicate via the network.

Short-term solution: Restart the network service. Check your network configuration files (e.g., ifcfg-eth0). If you are using DHCP, make sure your DHCP server is responding. Make sure your name servers are responding. If they are unavailable, you may still be able to reach the network using IP addresses (instead of aliases), or place the mapping information in your /etc/hosts file. If none of this works, reboot your computer. Reboot the DHCP server.

Long-term solutions: Reconfigure the network itself by replacing the DHCP server and/ or network gateway. Test your network cables. Try an alternate network interface device (e.g., replace your Ethernet card with a new one).

### PROBLEM 7: SOFTWARE NOT RUNNING CORRECTLY

Description: Some software that was previously running is no longer running correctly.
Steps to determine the problem:

1. View the messages and audit logs to see if the software is causing errors.

2. Check your yum log or your updates to see if the software is in need of an update.

3. Examine the software's configuration and/or rules file(s) (if any) and see if they have been inappropriately modified.

4. View the software's website to see if known errors have been found recently.

Short-term solution: Using yum, upgrade the software. If this fails, attempt to remove the software and install it anew, possibly moving to a newer version.

Long-term solution: If the software comes with support, contact the support team and detail the problem. Alternatively, if there is a website that permits problem uploads, upload the problem there. Explain what you tried to do when it failed and any modifications made to the software since the installation.

### PROBLEM 8: DEVICE FAILURE

Description: Hardware device is not functioning.
Steps to determine the problem:

1. Check the device's physical connection(s) to the computer.

2. Make sure the device is turned on, or if the device has its own power source, make sure it is plugged in or that the battery is adequate.

3. Make sure the device's driver software is installed.

4. Check the boot log and messages log file to see if an error message was generated at the last reboot (or if this is a plug-and-play device, when the device was connected).

5. If the device uses a service, see if the service is running.

Short-term solution: Restart any necessary service. Disconnect and reconnect the device and check the power. Reboot the computer.

Long-term solution: Connect the device to another computer to see if it works from there. If so, check the port/connection on the computer where it was not working. Replace the device driver on that computer. If the device still does not work on different computers, then replace the device.

## PROBLEM 9: SYSTEM DOES NOT INITIALIZE CORRECTLY

Description: Upon boot/reboot, the operating system does not come up in a usable mode.
Steps to determine the problem:

1. Check dmesg for errors during system boot.

2. If the system initializes to Linux, see what runlevel /etc/inittab is set to.

3. Does /sbin/init exist?

4. Is the root file system being mounted?

5. Is vmlinuz available?

6. Is GRUB configured correctly?

Short-term solution: If errors arose during boot (from dmesg), try to diagnose the cause of these errors (bad device, bad kernel image) and reboot. If the system came up in the wrong runlevel, run telinit to change the runlevel and alter the inittab file's default statement to modify the default runlevel. If there are errors arising during the init process, you might need to repartition one or more of your file systems. If the system does not come up at all, check the GRUB command line (shortly after booting, press 'c' to interrupt the process and drop to a command line prompt).
Long-term solution: Reinstall the OS.

## PROBLEM 10: CANNOT ACCESS USER FILE SYSTEM

Description: Root file system available and other file system(s) unavailable.
Steps to determine the problem:

1. Use df and examine /etc/mtab to see if the partition is mounted.

2. If the file system is to be remotely mounted, make sure netfs and network are running and if not, start them and manually remount.

3. Run fsck on the file system to make sure it is not damaged.

Short-term solution: Make sure netfs and network start automatically at the default run-level and add the mount information to /etc/fstab (or alternatively, as a separate mount command in /etc/rc.local).

Long-term solution: Repartition to create a new version of the file system (if it is damaged), and add hard-disk space. If you are remotely mounting the file system, contact the system administrator of the remote system and make sure the file system is available.

Although the above 10 troubleshooting scenarios cover serious problems, they are only a few possible situations that you may encounter. Hopefully, in reading this chapter, you know enough now to explore problems and find solutions. Remember that there are tens of thousands of Linux users and administrators who regularly contribute to the community. Do not be afraid to ask your questions on a Linux website.

## 14.8  CHAPTER REVIEW

Concepts and terms introduced in this chapter:

- Backup—storing files onto a media separate from your file system for data security and archival purposes.

- Backup strategy—a plan that dictates when to perform full and incremental backups, how to rotate and reuse storage media, and where to store your backups once saved.

- Cycle stealing—a situation where the CPU is forced to wait for some I/O to take place.

- Deadlock—a situation where two or more processes are holding onto resources that the other processes need whereby no process is able to continue without access to one of the held resources but none of the processes are willing to free up the resources they are holding.

- Disaster-recovery plan—the strategy employed by an organization to recover from a disaster that prevents the organization from doing business. In the context of this chapter, we focus on disaster-recovery planning for the organization's IT infrastructure (hardware, software, and data).

- Full backup—backing up an entire file system at one time.

- Incremental backup—backing up only those files in the file system that are new or have been modified since the last full or incremental backup.

- Liveness—the quality of a process in which is it making progress toward eventual completion and termination.

- Log file—a collection of messages that describe events that may be of use to the system administrator. Log files are generated automatically by logging software. In Linux, this includes klogd, syslogd, and auditd.

- Log rotation—the process of starting a new log file so that no single log file grows too large. Log rotation will rename the current log file (and possibly older log files) before creating the new file. Log rotation may be planned to occur daily or monthly but the default is typically weekly.

- Mirror—using two sets of disk drives so that any file is stored on both sets. This provides 100% redundancy so that if damage occurs to one of the sets of disk drives, any file is still available from the other. In addition to offering redundancy, the mirror also permits multiple disk accesses in parallel.

- Mutually exclusive access—a resource that must not be accessed by two or more processes in an overlapped manner or else it will result in data corruption. A process, once granted access to the resource, holds it until it no longer needs the resource. Data corruption can arise because in multitasking, a process could be interrupted while accessing data of the resource.

- Parity—the use of extra bits to record redundancy information. This is used in many of the forms of RAID technology.

- RAID—redundant array of independent (or inexpensive) disks is a form of technology often used today so that file system storage can be made more secure. Through redundancy, bad sectors will not necessarily cause a loss of data as redundant data allow the lost data to be restored. There are several different forms of RAID technology known as levels.

- Risk assessment—a strategy by an organization to determine its own assets and the vulnerabilities and threats to those assets so that the organization can protect itself better. One step in a risk assessment is to create a disaster-recovery plan.

- Scheduling—the ability to control when processes will run in the future. Linux has several different programs for scheduling, including one-time scheduling using at and batch and recurring scheduling using crontab and anacron.

- Starvation—a situation where a process is not able to progress toward completion because resources that it needs are continually being held by other processes. If a process is starving, its liveness is uncertain.

- Thrashing—a situation in which the operating system is spending too much time paging between main memory and virtual memory and so little-to-no processing is accomplished.

Linux commands and files covered in this chapter:

- anacron—a service that runs scheduled tasks upon system initialization.

- at—a program to schedule tasks one time at the time and date specified.

- atd—the daemon used by at and batch.

- auditd—the daemon that logs all software events.

- aureport—a program that queries audit log files for summaries of types of events.

- ausearch—a program that queries audit log files for details of types of events.

- batch—a program to schedule tasks one time to take place when system load drops below 80%.

- cpio—a backup utility that can store individual files, directories, or partitions. It has a crude incremental backup facility.

- crond—the crontab daemon.

- crontab—a program to schedule recurring tasks. Each user is able to submit a single crontab job that can include many different schedules and tasks.

- dump—a backup utility that only operates on file systems but can perform full and incremental backups.

- free—a program to report on the amount of used and available memory.

- klogd—the kernel-logging daemon.

- mpstat—a program that reports on processor utilization, primarily intended for multiprocessor systems.

- pidstat—a program that reports on processor utilization broken down by process.

- sar—a program that reports on statistics compiled into the sa log files. Logged events are snapshots of the system at specified intervals (usually every 10 minutes).

- strace—a program to trace through the system calls made by an executing Linux program, useful for debugging the kernel.

- syslogd—a logging daemon for nonkernel system events.

- tar—the tape archive program for creating archives and backups.

- uptime—reports on the time since the last boot.

- vmstat—reports on the amount of memory and virtual memory utilization during the current uptime.

- who—outputs who is currently logged into the system.

- /etc/logrotate.conf—configuration file to control logrotate for rotating log files.

- /etc/syslog.conf—configuration file to control the syslogd daemon. It lists directives that express what types of events from which software should be logged to given log files.

- /var/log/audit/—directory storing the audit log files.

- /var/log/boot.log—log file storing information related to the last boot.

- /var/log/cron—log file storing all scheduled events executed by anacron and crontab.

- /var/log/lastlog—log file storing users' last login information.

- /var/log/maillog—log file storing messages related to the mail program.

- /var/log/messages—log file storing system events as logged by syslog and klogd.

- /var/log/secure—log file storing authentication events as generated by any authentication program, including when users log in and when they use su and sudo.

- /var/log/Xorg—log file storing X Windows information. Different log files pertain to different console windows.

- /var/log/yum.log—log file storing yum events for software install, upgrade, and removal.

## REVIEW QUESTIONS

1. What is the advantage of using incremental backups rather than full backups?

2. What is the disadvantage of using incremental backups?

3. For a stand-alone Linux workstation, how often should you back up /home? /var?

4. For a file server storing dozens of users' home directories, how often should you back up /home?

5. How does RAID 1 differ from RAID 0?

6. How does RAID 3 differ from RAID 5?

7. How does RAID 6 differ from RAID 5?

8. If you were to purchase RAID for your home computer simply to have faster disk access but you were worried about the cost, which RAID level would you select and why?

9. If you were to purchase RAID for your home computer to have some degree of redundancy available but you were worried about the cost, which RAID level would you select and why?

10. Why is RAID 4 not used?

11. Compute the parity byte for the four bytes 10111000, 11100101, 00111101, and 01111000.

12. Compute the parity byte for the four bytes 00001111, 11110000, 11111111, and 10101010.

13. What are the advantages of using tar over dump for backups?

14. What are the advantages of using dump over tar for backups?

15. You have three tapes to store backups on. You plan to put a full backup on one tape and then incremental backups on the other two tapes doing the incremental backups every other day. You estimate that you can place 4 days' worth of incremental backups on each tape. How long will it be before you run out of tapes and must overwrite the full backup on the first tape?

For questions 16 through 19, assume you have a process called myscript in your home directory. Provide the proper at command to execute myscript as specified.

16. Tonight at 11:59 p.m.

17. Tomorrow morning at 3:15 a.m.

18. On July 10, 2014 at 6 p.m.

19. At 4 p.m. on September 30, 2014.

For questions 20 through 24, interpret the given recurrence information for crontab by stating the time(s)/dates/days when the task will execute. The specific tasks are omitted from the problem.

| | | | | | |
|-----|-----|------|-----|-----|-----|
| 20. | *   | *    | 1   | *   | 5   |
| 21. | 0   | 1,13 | *   | *   | *   |
| 22. | 30  | 19   | *   | 6   | *   |
| 23. | */5 | 23   | *   | *   | *   |
| 24. | 59  | 11   | 15  | 3   | *   |

25. What is the difference between crontab  -e and crontab  *filename*  where *filename* is a file storing crontab scheduling information?

26. What does run-parts do and which service does it call upon?

27. Why might you use /etc/at.allow?

28. What is the role of the directory /etc/cron.daily?

29. How does free's output differ from vmstat's?

30. Which of the programs from Section 14.4 might you use to determine how much virtual memory your system is or has been using?

31. Which of the programs from Section 14.4 might you use to view how your system's performance has been changing over a few minutes' time?

32. Which of the programs from Section 14.4 might you use to view how your system's processor is performing?

33. Which of the programs from Section 14.4 might you use to determine if any single process is using an inordinate amount of system resources?

34. Which of the programs from Section 14.4 might you use to compare your processor's time spent on user processes, system processes, and idle time?

35. How does sar differ in terms of the data it uses from ps? From top?

36. The who program does not in itself inform you about system performance. Why as a system administrator might you use it if you are investigating your system's performance?

37. You suspect a possible intruder in your system. What log file(s) might you examine?

38. You suspect a scheduled process did not run correctly. What log file(s) might you examine?

39. Why might you examine the boot.log files?

40. Provide a command to display all audit log messages from the user whose UID is 501.

41. Assume you want to explore all the events logged by running processes that are children of /sbin/init. Provide the proper ausearch command for this (hint: all these processes have something in common, init is their parent).

42. Given the following audit log entry, explain each part that is underlined.

```
time-> Wed Jul 17 12:20:10 2014
type = CRED_ACQ msg = audit(1374078010.023:59603): user
pid = 3102 uid = 501 auid = 501 ses = 1
subj = unconfined_u:unconfined_r:unconfined_t:s0-s0:c0.c1023
msg = 'op = PAM:setcred acct = "root" exe = "/bin/su" hostname = ?
addr = ? terminal = pts/2 res = success'
```

43. The `aureport` −m option displays user modification events. What is a user modification event and what programs can you think of that would generate such an event?

44. What would ausearch −sc open do?

45. Why would a system administrator be involved in disaster-recovery planning?

46. Why would it be unwise to store file system backup media in the same building as the computer or file server that is being backed up?

47. Why would distributing file systems and processing across multiple locations be a wise precaution for an organization that could afford it?

48. What types of preventative measures could an organization take to prevent damage or destruction of hardware from a natural disaster?

49. How might a disaster plan for a flooding situation differ from that of a fire?

50. How might a disaster plan for a lockdown because of a gunman on the premises differ from natural disasters such as fire, flood, and earthquake?

51. What role does encryption play in risk assessment?

For questions 52 through 56, similar to Section 14.7, describe the types of tests that you should take to troubleshoot the given problem.

52. A user is not able to log into his or her account even though other users can.

53. Installation using yum of an available software package has failed.

54. Remotely mounted file system no longer accessible.

55. Network access showing heavy usage.

56. Too much disk swapping seems to be taking place.

# Bibliography

Abhari, A., Dandamudi, S., and Majumdar, S. Web Object-Based Storage Management in Proxy Caches, *Future Generation Computer Systems*, Vol. 22, Issue 1–2, pp. 16–31, Amsterdam: Elsevier, 2006.

Abrams, M., LaPadula, L., Eggers, K., and Olson, I. A Generalized Framework for Access Control: An Informal Description, *Proceedings of the 13th National Computer Security Conference*, pp. 135–143, 1990.

Accetta, M., Baron, R., Bolosky, W., Golub, D., Rashid, R., Tevanian, A., and Young, M. Mach: A New Kernel Foundation for Unix Development, *Proceedings of the Summer 1986 USENIX Conference*, pp. 93–112, 1986.

Adelstein, T., and Lubanovic, B. *Linux System Administration*, CA: O'Reilly, 2007.

Aho, A., Kernighan, B., and Weinberger, P. *The AWK Programming Language*, MA: Addison-Wesley, 1988.

Al-Hafeedh, A., Crochemore, M., Ilie, L., Kopylova, E., Smyth, W., Tishcler, G., and Yusufu, M. A Comparison of Index-Based Lempel-Ziv LZ77 Factorization Algorithms, *ACM Computing Surveys*, Vol. 45, Issue 1, 5:9–5:17, Article 5, 2012.

Allen, N. *Network Maintenance and Troubleshooting Guide: Field Tested Solutions for Everyday Problems*, MA: Addison-Wesley, 2009.

Almesberger, W. Booting Linux: The History and the Future, *Proceedings of the Ottawa Linux Symposium*, 2000.

Almgren, M., Debar, H., and Dacier, M. A Lightweight Tool for Detecting Web Server Attacks, *Networks and Distributed System Security Symposium*, pp. 157–170, 2000.

Anderson, K. Convergence: A Holistic Approach to Risk Management, *Network Security*, Vol. 2007, Issue 5, pp. 4–7, Amsterdam: Elsevier Science Publishers, 2007.

Aulds, C. *Linux Apache Web Server Administration*, CA: SYBEX, 2000.

Bach, M. *The Design of the Unix Operating System*, NJ: Prentice-Hall, 1996.

Balasubramanian, K., and Johnson, D. Linux Memory Management Overview, *The Linux Kernel Hacker's Guide*, http://www.redhat.com:8080/HyperNews/get/memory/memory.html, 1993.

Bar, M. *Linux File Systems*, NY: McGraw-Hill, 2001.

Barrett, D., Silverman, R., and Byrnes, R. *Linux Security Cookbook*, MA: O'Reilly, 2003.

Bembenek, J., and Klus, A. *Grep Pocket Reference*, MA: O'Reilly, 2009.

Bennett, H. CD-E: Call It Erasable, Call It Rewritable, but Will It Fly? *CD-ROM Professional*, 1996.

Benvenuti, C. *Understanding Linux Network Internals*, MA: O'Reilly, 2006.

Berry, D. *Copy, Rip, Burn: The Politics of Copyleft and Open Source*, London: Pluto Press, 2008.

Black, D., Golub, D., Julin, D., Rashid, R., Draves, R., Dean, R., Forin, A., Barrera, J., Tokadu, H., Malan, G., and Bohman, D. Microkernel Operating System Architecture and Mach, *Proceedings of the USENIX Workshop on Micro-Kernels and Other Kernel Architectures*, pp. 11–30, 1992.

Blum, R., and Bresnahan, C. *Linux Command Line and Shell Scripting Bible*, NJ: Wiley & Sons, 2011.

Bonaccorsi, A., and Rossi, C. Why Open Source Software Can Succeed, *Research Policy*, Vol. 32, Issue 7, pp. 1149–1292, Elsevier, 2003.

Bourne, S.R. The UNIX Shell, *The Bell System Technical Journal*, NJ: American Telephone and Telegraph Company, Vol. 57, Issue 6, pp. 1971–1990, 1978.

Bovet, D. *Understanding the Linux Kernel*, MA: O'Reilly, 2005.

Bradley, D. The Divergent Anarcho-Utopian Discourses of the Open Source Software Movement, *Canadian Journal of Communication*, Vol. 30, pp. 585–611, 2005.

Bridger, R. *Introduction to Ergonomics*, FL: CSC Press, 2008.

Brinch, P. *Classic Operating Systems: From Batch Processing to Distributed Systems*, NY: Springer-Verlag, 2001.

Brookshear, J. *Computer Science: An Overview*, NJ: Prentice-Hall, 2011.

Brown, F. *Boolean Reasoning: The Logic of Boolean Equations*, NY: Dover, 2012.

Brown, G. *zOS JCL (Job Control Language)*, NJ: John Wiley & Sons, 2002.

Burtch, K. *Linux Shell Scripting with Bash*, MA: Pearson, 2004.

Callaghan, B. *NFS Illustrated*, MA: Addison-Wesley, 2000.

Ceruzzi, P. *A History of Modern Computing*, MA: The MIT Press, 2003.

Chau, P. Selection of Packaged Software in Small Business, *European Journal of Information Systems*, Vol. 3, Issue 4, pp. 292–302, 1994.

Chen, P., Lee, E., Gibson, G., Katz, R., and Patterson, D. RAID: High-Performance, Reliable Secondary Storage, *ACM Computing Surveys*, Vol. 26, Issue 2, pp. 145–185, 1994.

Chervenak, A., Vellanki, V., and Kurmas, Z. Protecting File Systems: A Survey of Backup Techniques, *Proceedings of the Joint NASA and IEEE Mass Storage Conference*, 1998.

Cheung, W., and Loong, A. Exploring Issues of Operating Systems Structuring: From Microkernel to Extensible Systems, *Operating Systems Review*, Vol. 29, pp. 4–16, 1995.

Ciampa, M. *Security+ Guide to Network Security Fundamentals*, MA: Thomson Course Technologies, 2011.

Clements, A. *The Principles of Computer Hardware*, NY: Oxford, 2000.

Cole, E. *Network Security Bible*, NJ: Wiley and Sons, 2009.

Comer, D. *Internetworking with TCP/IP*, NJ: Prentice-Hall, 1996.

Comer, D. *Internetworking with TCP/IP*, Vol. I, MA: Addison-Wesley, 2013.

Corner, D. *Computer Networks and Internets*, NJ: Prentice-Hall, 2008.

Cox, R., Muthitacharoen, A., and Morris, R. Serving DNS Using a Peer-to-Peer Lookup Service, *Lecture Notes in Computer Science*, Vol. 2429, pp. 155–165, NY: Springer, 2002.

Cramer, R., and Shoup, V. Design and Analysis of Practical Public-Key Encryption Schemes Secure against Adaptive Chosen Ciphertext Attack, *SIAM Journal on Computing*, Vol. 33, pp. 167–226, 2001.

Crawley, D. *The Accidental Administrator: Linux Server Step-by-Step Configuration Guide*, WA: CreateSpace, 2010.

Dallheimer, M., and Welsh, M. *Running Linux*, MA: O'Reilly, 2005.

Davis, D., and Swick, R. Network Security via Private-Key Certificates, *ACM SIGOPS Operating Systems Review*, Vol. 24, Issue 4, pp. 64–67, 1990.

Dean, T. *Network+Guide to Networks*, MA: Thomson Course Technology, 2009.

De Goyeneche, J. Loadable Kernel Modules, *IEEE Software*, Vol. 16, Issue 1, pp. 65–71, 1999.

Denning, P. Thrashing: Its Causes and Prevention, *Proceedings of the December 9–11 1968 AFIPS Fall Join Computer Conference*, pp. 915–922, NY: ACM, 1968.

Denning, P. Virtual Memory, *ACM Computing Surveys*, Vol. 2, Issue 3, pp. 153–189, 1970.

Denning, P. A Short Theory of Multiprogramming, *Proceedings of the 3rd International Workshop on Modeling, Analysis and Simulation of Computer and Telecommunication Systems*, pp. 2–7, 1995.

Dijkstra, E. The Structure of the THE Multiprogramming System, *Communications of the ACM*, Vol. 11, Issue 5, pp. 341–346, NY: ACM, 1968.

Doeppner, T. *Operating Systems in Depth: Design and Programming*, NJ: Wiley and Sons, 2010.

Dougherty, D., and Robbins, *A. sed & awk*, MA: O'Reilly, 1997.

Droms, R. Automated Configuration of TCP/IP with DHCP, *IEEE Internet Computing*, Vol. 3, Issue 4, pp. 45–53, 1999.

Easttom, W. *Computer Security Fundamentals*, IN: Que, 2011.

Economides, N., and Katsamakas, E. Linux vs. Windows: A Comparison of Application and Platform Innovation Incentives for Open Source and Proprietary Software Platforms, *The Economics of Open Software Development*, J. Bitzer and P. Schroder (eds), Amsterdam: Elsevier, 2006.

Elbroth, D. *The Linux Book*, NJ: Prentice-Hall, 2001.

Elmasri, R., Carrick, A., and Levine, D. *Operating Systems: A Spiral Approach*, NY: McGraw-Hill, 2009.

Fenwick, P. The Burrows-Wheeler Transform for Block Sorting Text Compression: Principles and Improvements, *Computer Journal*, Vol. 39, Issue 9, pp. 731–740, NY: Oxford University Press, 1996.

Fitzgerald, B. The Transformation of Open Source Software, *MIS Quarterly*, Vol. 30, Issue 30, pp. 587–598, 2006.

Forouzan, B. *TCP/IP Protocol Suite*, NY: McGraw-Hill, 2002.

Forouzan, B. *Data Communications and Networking*, NY: McGraw-Hill, 2006.

Fox, R. *Information Technology: An Introduction for Today's Digital World*, FL: CRC Press, 2013.

Fox, T. *Red Hat Enterprise Linux 5 Administration Unleashed*, IN: Sams, 2007.

Friedl, J. *Mastering Regular Expressions*, MA: O'Reilly, 2006.

Frisch, E. *Essential System Administration*, MA: O'Reilly, 2002.

Gallegos, F., Senft, S., Manson, D., and Gonzales, C. *Information Technology Control and Audit*, FL: Auerbach Publications, 2004.

Gajewska, H., Manasse, M., and McCormack, J. Why X Is Not Our Ideal Window System, *Software—Practice & Experience*, Vol. 20, Issue S2, 1990.

Gancarz, M. *Linux and the Unix Philosophy*, FL: Digital Press, 2003.

Garfinkel, S., Spafford, G., and Schwartz, A. *Practical Unix and Internet Security*, MA: O'Reilly, 2003.

Garrels, M. *Bash Guide for Beginners*, CA: Fultus Corporation, 2004.

Garrido, J., and Schlesinger, R. *Principles of Modern Operating Systems*, MA: Jones and Bartlett, 2007.

Gay, J., Stallman, R., and Lessig, L. *Free Software, Free Society: Selected Essays of Richard M. Stallman*, WA: CreateSpace, 2009.

Gibson, D. *Management Risk in Information Systems*, MA: Jones and Bartlett, 2010.

Gillay, C. *Linux User's Guide: Using the Command Line and Gnome Red Hat Linux*, OR: Franklin, Beedle & Associates, 2003.

General Public License, http://www.gnu.org/copyleft/gpl.html.

Goldman, D., and Bonzini, P. *Definitive Guide to sed: Tutorial and Reference*, WA: EHDP Press, 2013.

Gomberg, M., Evard, R., and Stacey, C. A Comparison of Large-Scale Software Installation Methods on NT and Unix, *Proceedings of USENIX Technical Program on Large Installation System Administration of Windows NT Conference*, pp. 37–48, 1998.

Goodheart, B., and Cox, J. *The Magic Garden Explained: The Internals of UNIX System V Release 4 An Open Systems Design*, NJ: Prentice-Hall, 1994.

Goossens, M., Mittelbach, F., and Samarin, A. *The LaTeX Companion*, MA: Addison-Wesley, 1994.

Goyal, V., Horman, N., Ohmichi, K., Soni, M., and Garg, A. Kdump: Smarter, Easier, Trustier, *Ottawa Linux Symposium*, 2007.

Goyvaerts, J., and Levithan, S. *Regular Expressions Cookbook*, MA: O'Reilly, 2009.

Grampp, F., and Morris, R. UNIX Operating-System Security, *AT&T Bell Laboratories Technical Journal*, Vol. 63, pp. 1649–1672, 1984.

Green, R., Baird, A., and Davies, J. Designing a Fast, On-line Backup System for a Log-Structured File System, *Digital Technical Journal of Digital Equipment Corporation*, Vol. 8, Issue 2, pp. 32–45, 1986.

Gregg, J. *Ones and Zeros: Understanding Boolean Algebra, Digital Circuits, and the Logic of Sets*, NJ: Wiley-IEEE Press, 1998.

Groom, F. The Structure and Software of the Internet, *Annual Review of Communications*, Vol. 50, pp. 695–707, 1997.

Guttman, E. Service Location Protocol: Automatic Discovery of IP Network Services, *IEEE Internet Computing*, Vol. 3, Issue 4, pp. 71–80, 1999.

Hagen, S. *IPv6 Essentials*, MA: O'Reilly, 2006.

Halsall, F. *Data Communications, Computer Networks, and Open Systems*, MA: Addison-Wesley, 1996.

Hamacher, C., Vranesci, Z., Zaky, S., and Manjikian, N. *Computer Organization and Embedded Systems*, NY: McGraw–Hill, 2012.

Hansen, P. (editor). *Classic Operating Systems: From Batch Processing to Distributed Systems*, NY: Springer, 2010.

Harker, J., Brede, D., Pattison, R., Santana, G., and Taft, L. A Quarter Century of Disk File Innovation, *IBM Journal of Research and Development*, Vol. 25, Issue 5, pp. 677–689, 1981.

Harkins, M. *Managing Risk and Information Security: Protect to Enable*, NJ: Apress, 2012.

Hartig, H., Hohmuth, M., Liedtke, J., Schonberg, S., and Wolter, J. The Performance of Micro-Kernel-Based Systems, *Proceedings of the 16th ACM Symposium on Operating Systems Principles*, 1997.

Hecker, F. Setting up Shop: The Business of Open-Source Software, *IEEE Software*, Vol. 16, Issue 1, pp. 45–51, 1999.

Helmke, M. *Ubuntu Unleashed*, IN: Sams, 2012.

Hicks, B., Rueda, S., St. Clair, L., Jaeger, T., and McDaniel, P. A Logical Specification and Analysis for SELinux MLS Policy, *ACM Transactions on Information and System Security*, Vol. 13, Issue 3, pp. 26–31, Article 26, 2010.

Hill, B., Burger, C., Jesse, J., and Bacon, J. *The Official Ubuntu Book*, NJ: Prentice-Hall, 2011.

Holcombe, C., and Holcombe, J. *Survey of Operating Systems*, NY: McGraw-Hill Osborne Media, 2002.

Huitema, C. *IPv6: The New Internet Protocol*, NJ: Prentice-Hall, 1998.

Hunt, C. *TCP/IP Network Administration*, MA: O'Reilly, 2002.

Jacko, J. *Human–Computer Interaction Handbook: Fundamentals, Evolving Technologies, and Emerging Applications*, FL: CRC Press, 2012.

Jacob, B., and Mudge, T. Virtual Memory in Contemporary Multiprocessors, *IEEE Micro Magazine*, Vol. 18, pp. 60–75, 1998.

Jacob, B., and Wang, D. *Memory Systems: Cache, DRAM, Disk*, CA: Morgan Kaufmann, 2007.

Jang, M. *Security Strategies in Linux Platforms and Applications*, MA: Jones and Bartlett, 2010.

Kamel, M., Keast, J., and Pal, C. Concrete Architecture of the Linux Kernel, technical report, University of Waterloo, Waterloo, Ontario, N2L 3G1. Department of Electrical and Computer Engineering, Department of Computer Science, 1998.

Katsicas, S. Chapter 35, *Computer and Information Security Handbook*, CA: Morgan Kaufmann, 2009.

Katz, J., and Lindell, Y. *Introduction to Modern Cryptography*, FL: CRC Press, 2007.

Kernighan, B., and Pike, R. *The Unix Programming Environment*, NJ: Prentice-Hall, 1984.

Kernighan, B., and Ritchie, D. *The C Programming Language*, NJ: Prentice-Hall, 1988.

Kehtarnavaz, N., and Gamadia, M. *Real Time Image and Video Processing: From Research to Reality*, CA: Morgan & Claypool Publishers, 2006.

Kiddle, O., Stephenson, P., and Peek, J. *From Bash to Z Shell: Conquering the Command Line*, NJ: Apress Media LLC, 2004.

Kirtch, O. *Linux Network Administrator's Guide*, MA: O'Reilly, 1995.

Knuth, D. Dynamic Huffman Coding, *Journal of Algorithms*, Vol. 6, Issue 2, pp. 163–180, Amsterdam: Elsevier, 1985.

Krishnamurthy, B., and Rexford, J. *Web Protocols and Practice: HTTP/1.1, Networking Protocols, Caching and Traffic Measure*, MA: Addison-Wesley, 2001.

Kumar A. Migration from Microsoft to Linux on Servers and Desktops, *Proceedings of the 20th Annual Conference of the National Advisory Committee on Computing Qualifications*, S. Mann and N. Bridgeman (eds), pp. 117–123, 2007.

Kurose, J., and Ross, K. *Computer Networking: A Top-Down Approach*, MA: Addison-Wesley, 2009.

Langfeldt, N. *The Concise Guide to DNS and Bind*, IN: Que Corp, 2000.

Laurent, A. *Understanding Open Source and Free Software Licensing*, MA: O'Reilly, 2004.

Leach, R. *Advanced Topics in UNIX: Processes, Files and Systems*, NJ: Wiley and Sons, 1994.

LeBlanc, D., and Yates, I. *Linux Install and Configuration Little Black Book: The Must-Have Troubleshooting Guide to Installing and Configuring Linux*, AZ: Coriolis Open Press, 1999.

Lempel, A., and Ziv, J. On the Complexity of Finite Sequences, *IEEE Transactions on Information Theory*, Vol. 22, Issue 1, pp. 75–81, 1976.

Lerner J., Pathak P. A., and Tirole, J. The Dynamics of Open Source Contributors, *American Economic Review*, Vol. 96, Issue 2, pp. 114–118, 2006.

Lewine, D. *POSIX Programmer's Guide: Writing Portable Unix Programs*, MA: O'Reilly, 1992.

Li, Y., Li, W., and Jiang, C. A Survey of Virtual Machine Systems: Current Technology and Future Trends, *Proceedings of the Third International Symposium on Electronic Commerce and Security*, 2010.

Liedtke, J. On Micro-Kernel Construction, *Proceedings of the Fifteenth ACM Symposium on Operating Systems Principles*, 1995.

Liedtke, J. Toward Real Microkernels, *Communications of the ACM*, Vol. 39, Issue 9, pp. 70–79, 1996.

Limoncelli, T., Hogan, C., and Chalup, S. *The Practice of System and Network Administration*, MA: Addison-Wesley, 2007.

Lin, Y., Hwang, R., and Baker, F. *Computer Networks: An Open Source Approach*, NY: McGraw-Hill, 2011.

Linux Cross Reference Guide, lxr.linux.no.

Liu, C., and Albitz, P. *DNS and BIND*, MA: O'Reilly, 2003.

Loscocco, P., and Smalley, S. Integrating Flexible Support for Security Policies into the Linux Operating System, *Proceedings of the FREENIX Track: 2001 USENIX Annual Technical Conference, The USENIX Association*, pp. 29–42, CA: USENIX Association, 2001.

Love, R. *Linux Kernel Development*, NJ: Addison-Wesley, 2010.

Luotonen, A., and Altis, K. World-Wide Web Proxies, *Computer Networks and ISDN Systems*, Vol. 27, Issue 2, pp. 147–154, Amsterdam: Elsevier, 1994.

Mann, S., and Mitchell, E. *Linux System Security: The Administrator's Guide to Open Source Security Tools*, NJ: Prentice-Hall, 2000.

Mansfield, K., and Antonakos, J. *Computer Networking for LANs to WANs: Hardware, Software and Security*, MA: Thomson Course Technology, 2009.

Markatos, E., Katevenis, M., Pnevmatikatos, D., and Flouris, M. Secondary Storage Management for Web Proxies, *Proceedings of USITS 99: The 2nd Conference on USENIX Symposium on Internet Technologies and Systems*, Vol. 2, pp. 93–114, CA: USENIX, 1999.

Marpe, D., and Wiegand, T. The H.264/MPEG4 Advanced Video Coding Standard and Its Applications, *IEEE Communications Magazine*, Vol. 44, Issue 8, pp. 134–144, 2006.

Matotek, D. Startup and Services, *Pro Linux System Administration*, pp. 145–173, NJ: Apress, 2009.

Matthews, J. *Computer Networking: Internet Protocols in Action*, NJ: Wiley and Sons, 2005.

Mauerer, W. *Professional Linux Kernel Architecture*, NJ: Wrox, 2008.

McCarty, B. *SELinux: NSA's Open Source Security Enhanced Linux*, CA: O'Reilly, 2004.

Meeker, H. *The Open Source Alternative: Understanding Risks and Leveraging Opportunities*, NJ: Wiley & Sons, 2008.

Melve, I., Slettjord, L., Bekker, H., and Verschuren, T. Building a Web Caching System—Architectural Considerations, *Proceedings of the 1997 NLANR Web Cache Workshop*, CA: National Laboratory for Applied Network Research, 1997.

Menezes, A., Oorschot, P., and Vanstone, S. *Handbook of Applied Cryptography (Discrete Mathematics and Its Applications)*, FL: CRC Press, 1996.

Miles, S. Linux Closing in on Microsoft Market Share, Study Says, CENT News.com, http://news.com.om/2100-1001-243527.html?tag=mainstry, 2000.

Miller, F., Vandome, A., and McBrewster, J. *Comparison of Windows and Linux: Linux, Microsoft Windows, Operating System, Proprietary Software, Free Software, Binary Blob, Embedded System, Supercomputer, Desktop Computer, Free Software Community*, London: Alpha Press, 2009.

Mockapetris, P., and Dunlap, K. Development of the Domain Name System, *SIGCOMM 88 Symposium Proceedings on Communications Architectures and Protocols*, pp. 123–133, NY: ACM, 1988.

Moody, G. *Rebel Code: Linux and the Open Source Revolution*, NY: Basic Books, 2002.

Morris, R., and Thompson, K., Password Security: A Case History, *Communications of the ACM*, Vol. 22, Issue 11, pp. 594–597, NY: ACM, 1979.

Moshe, B. *Linux File Systems*, NY: McGraw-Hill, 2001.

Mustonen, M. Copyleft—The Economics of Linux and Other Open Source Software, *Information Economics and Policy*, pp. 99–121, Vol. 15, Issue 1, Amsterdam: Elsevier, 2003.

Negus, C., and Boronczyk, T. *CentOS Bible*, NJ: Wiley and Sons, 2009.

Negus, C., and Bresnahan, C. *Linux Bible*, NJ: Wiley and Sons, 2012.

Nemeth, E., Snyder, G., Hein, T., and Whaley, B. *Unix and Linux System Administration Handbook*, NJ: Prentice-Hall, 2010.

Newham, C. *Learning the bash Shell: Unix Shell Programming*, MA: O'Reilly, 2005.

Null, L., and Lobur, J. *The Essentials of Computer Organization and Architecture*, MA: Jones and Bartlett, 2012.

Odom, W. *Computer Networking First-Step*, IN: Cisco Press, 2004.

Paar, C., and Pelzl, J. *Understanding Cryptography*, NY: Springer. 2010.

Pate, S. *UNIX Filesystems: Evolution, Design and Implementation*, NJ: Wiley and Sons, 1988.

Patterson, D., Gibson, G., and Katz, R. A Case for Redundant Arrays of Inexpensive Disks (RAID), *Proceedings of SIGMOD '88*, pp. 109–116, NY: ACM, 1988.

Patterson, D., and Hennessy, J. *Computer Organization and Design: The Hardware/Software Interface*, CA: Morgan Kaufmann, 1998.

Peek, J., Powers, S., O'Reilly, T., and Loukides, M. *Unix Power Tools*, MA: O'Reilly, 2007.

Petersen, R. *Linux: The Complete Reference*, NY: McGraw-Hill, 2007.

Peterson, L., and Davie, B. *Computer Networks: A Systems Approach*, CA: Morgan Kaufmann, 2011.

Pfleeger, C., and Pfleeger, S. *Security in Computing*, NJ: Prentice-Hall, 2006.

Plank, J. A Tutorial on Reed-Solomon Coding for Fault-Tolerance in RAID-like Systems, *Software Practice and Experience*, Vol. 27, Issue 9, pp. 995–1012, NJ: Wiley and Sons, 1997.

Portnoy, M. *Virtualization Essentials*, NJ: Sybex, 2012.

Prabhakaran, V., Arpaci-Dusseau, A., and Arpaci-Dusseau, R. Analysis and Evolution of Journaling File Systems, *USENIX '05 Online Proceedings*, pp. 105–120, 2005.

Puryear, D. *Best Practices for Managing Linux and Unix Servers*, NY: Penton, 2006.

Quarterman, J., and Hoskins, H. Notable Computer Networks, *Communications of the ACM*, Vol. 29, Issue 10, pp. 932–971, NY: ACM, 1986.

Quigley, E. *UNIX Shells by Example with CDROM*, NJ: Prentice-Hall, 1999.

Ramey, C., and Fox, B. Bash Reference Manual, http://www.gnu.org/software/bash/manual, Network Theory LTD, 2012.

Rash, M. *Linux Firewalls: Attack Detection and Response with iptables, psad and fwsnort*, CA: No Starch Press, 2007.

Raymond, E. *The Cathedral and the Bazaar: Musings on Linux and Open Source by an Accidental Revolutionary*, MA: O'Reilly, 2001.

Red Hat Linux 8.0: The Official Red Hat Linux Reference Guide, Raleigh, NC: Red Hat, Inc, 2002.

Rescorla, E. An Introduction to openssl Programming, *Linux Journal*, Vol. 2001, p. 3, Issue 89, Article 3, 2001.

Reynolds, G. *Ethics in Information Technology*, KY: Cengage, 2011.

Riehle, D. The Economic Motivation of Open Source: Stakeholder Perspectives, *IEEE Computer*, Vol. 40, Issue 4, pp. 25–32, April 2007.

Ritchie, D. The Evolution of the UNIX Time-Sharing System, *Language Design and Programming Methodology, Lecture Notes on Computer Science*, Vol. 79, Berlin: Springer-Verlag, 1979.

Ritchie, D., and Thompson, K. The UNIX Time-Sharing System, *Communications of the ACM*, Vol. 17, Issue 7, pp. 365–375, 1974.

Robbins, A. *Bash Pocket Reference*, MA: O'Reilly, 2010.

Robbins, A., and Beebe, N. *Classic Shell Scripting*, MA: O'Reilly, 2005.

Robbins, A., and Dougherty, D. *Sed and Awk*, MA: O'Reilly, 1997.

Rose, R. Survey of System Virtualization Techniques, *Technical Report, Oregon State University*, 2004.

Royon, Y., and Frenot, S. A Survey of Unix Init Schemes, Technical report arXiv:0706.2748v2, Cornell University Library, 2007.

Rusen, C. *Networking Your Computers & Devices Step by Step*, WA: Microsoft Press, 2011.

Rusling, D. The Linux Kernel, http://sunsite.unc.edu/Linux/LDP/tlk/tlk.html, 2001.

Ryan, P. Linux Market Share Set to Surpass Win 98, OS X Still Ahead of Vista, http://arstechnica.com/apple/2007/09/linux-marketshare-set-to-surpass-windows-98, 2007.

Saini, K. *Squid Proxy Server 3.1: Beginner's Guide*, Birmingham: Packt Publishing, 2011.

Salas, P. *The Daemon, the GNU and the Penguin*, Groklaw, 2005.

Salus, P (ed). *A Quarter Century of Unix*, MA: Addison-Wesley, 1994.

Samar V. Unified Login with Pluggable Authentication Modules, *Proceedings of the 3rd ACM Conference on Computer and Communications Security*, pp. 1–10, ACM Press, 1996.

Sandberg, R., Goldberg, D., Kleiman, S., Walsh, D., and Lyon, B. Design and Implementation of the Sun Network File System, *Proceedings of the 1985 USENIX Summer Conference*, pp. 119–130, 1985.

Sandhu, R., and Samarati, P. Access Control: Principles and Practice, *IEEE Communications*, Vol. 32, pp. 40–48, 1994.

Sarwar, S., and Koretsky, R. *Unix: The Textbook*, MA: Addison-Wesley, 2004.

Sawicki, E. *Guide to Apache*, MA: Thomson, 2008.

Schach, S., Jin, B., Wright, D., Heller, G., and Offutt, A. Maintainability of the Linux Kernel, *IEE Proceedings—Software*, Vol. 149, Issue 1, pp. 18–23, London: Institute of Engineering and Technology, 2002.

Scheifler, R., and Gettys, J. *X Window System: Core and Extension Protocols: X Version 11, Releases 6 and 6.1*, FL: Digital Press, 1996.

Schwartz, M. Linux Job Scheduling, *Linux Journal*, Vol. 2000, Issue 77, Article 8, 2000.

Sebesta, R. *Concepts of Programming Languages*, MA: Addison-Wesley, 2012.

Shoch, J., Dalal, Y., Redell, D., and Crane, R. Evolution of the Ethernet Local Computer Network, *Computer*, Vol. 15, Issue 8, pp.10–27, 1982.

Shotts Jr., W. *The Linux Command Line: A Complete Introduction*, CA: No Starch Press, 2012.

Siever, E., Weber, A., Figgins, S., Love, R., and Robbins, A. *Linux in a Nutshell: A Desktop Quick Reference*, MA: Riley, 2005.

Silberschatz, A., Galvin, P., and Gagne, G. *Operating System Concepts*, NJ: Wiley & Sons, 2012.

Silva, S. *Web Server Administration*, MA: Thomson Course Technology, 2008.

Sloan, J. *Network Troubleshooting Tools*, MA: O'Reilly, 2001.

Smalley, S., and Fraser, T. A Security Policy Configuration for the Security-Enhanced Linux, *NAI Labs Technical Report*, 2001.

Smith, J., and Nair, R. *Virtual Machines: Versatile Platforms for Systems and Processes*, CA: Morgan Kaufmann, 2005.

Sobell, M. *A Practical Guide to Linux Commands, Editors, and Shell Programming*, NJ: Prentice-Hall, 2009.

Sobell, M. *A Practical Guide to Ubuntu Linux*, NJ: Prentice-Hall, 2010.

Spafford, E. The Internet Worm: Crisis and Aftermath, *Communications of the ACM*, Vol. 32, Issue 6, pp. 678–687, 1989.

St. Laurent, A. *Understanding Open Source and Free Software Licensing*, MA: O'Reilly, 2004.

Stallings, W. *Computer Organization and Architecture: Designing for Performance*, NJ: Prentice-Hall, 2003.

Stallings, W. *Cryptography and Network Security: Principles and Practices*, NJ: Prentice-Hall, 2010.

Stallings, W. *Data and Computer Communications*, NJ: Prentice-Hall, 2010.

Stallings, W. *Operating Systems: Internals and Design Principles*, NJ: Prentice-Hall, 2011.

Stallings, W., and Brown, L. *Computer Security: Principles and Practices*, NJ: Prentice-Hall, 2011.

Stallman, R. The GNU Operating System and the Free Software Movement, *Open Sources: Voices from the Open Source Revolution*, C. DiBona, S. Ockman and M. Stone (eds), CA: O'Reilly, 1999.

Stallman, R. Why 'Free Software' Is Better than 'Open Source', http://www.gnu.org/philosophy/free-software-for-freedom.html, 2007.

Stallman, R., and Gay, J. *Free Software, Free Society: Selected Essays of Richard M. Stallman*, NY: SoHo Books, 2002.

Stankovic, J. Software Communication Mechanisms: Procedure Calls versus Messages, *Computer*, Vol. 15, Issue 4, pp. 19–25, 1982.

Stevens, W. *UNIX Network Programming: The Sockets Networking API*, NJ: Prentice-Hall, 1998.

Stewart, J. *Network Security, Firewalls, and VPNs*, MA: Jones and Bartlett, 2010.

Tanenbaum, A. *Structured Computer Organization*, NJ: Prentice-Hall, 1999.

Tanenbaum, A. *Modern Operating Systems*, NJ: Prentice-Hall, 2007.

Tanenbaum, A. *Computer Networks*, NJ: Prentice-Hall, 2010.

Tanenbaum, A., Herder, J., and Bos, H. Can We Make Operating Systems Reliable and Secure? *Computer*, Vol. 39, Issue 5, pp. 44–51, 2006.

Tevanian, A., Rashid, R., Golub, D., Black, Dl, Cooper, E., and Young, M. Mach Threads and the UNIX Kernel: The Battle for Control, *Proceedings of the Summer 1987 USENIX Conference*, 1987.

The Var Guy, Gartner: Microsoft Windows 15% Mobile, Desktop market Share? http://www.linuxtoday.com/upload/gartner-microsoft-windows-15-mobile-desktop-market-share-130625183006.html, 2013.

Toigo, J. *Disaster Recovery Planning: Preparing for the Unthinkable*, NJ: Prentice-Hall, 2002.

Tominaga, A., Nakamura, O., Teraoka, F., and Murai, J. Problems and Solutions of DHCP, *Proceedings of INET 95*, 1995.

Toxen, B. *Real World Linux Security: Intrusion Prevention, Detection, and Recovery*, NJ: Prentice-Hall, 2003.

Ts'o, T., and Tweedie, S. Planned Extensions to the Linux Ext2/Ext3 Filesystem, *Proceedings of the FREENIX Track: 2002 USENIX Annual Technical Conference*, pp. 235–243, 2002.

Uti, N. Real Time Mobile Video Streaming Using Wavelet Transformation and Run-Length Coding, *Proceedings of the 2009 International Conference on Wireless Networks*, 2 (ICWN '09), pp. 368–373, 2009.

Vacca, R. *Computer and Information Security Handbook*, CA: Morgan Kaufmann, 2009.

Vahalia, U. *UNIX Internals: The New Frontiers*, NJ: Prentice-Hall, 1996.

Warford, J. *Computer Systems*, MA: Jones and Bartlett, 2009.

Watt, A. *Beginning Regular Expressions (Programmer to Programmer)*, NJ: Wrox, 2005.

Welch, Terry A., A Technique for High Performance Data Compression, *IEEE Computer*, Vol. 17, Issue 6, pp. 8–19, 1984.

Wells, N. *The Complete Guide to Linux System Administration*, MA: Thomson Course Technology, 2005.

Wessels, D. *Squid: The Definitive Guide*, MA: O'Reilly, 2005.

Whitesitt, J. *Boolean Algebra and Its Applications*, NY: Dover, 2010.

Williams, S. *Free as in Freedom: Richard Stallman's Crusade for Free Software*, WA: CreateSpace, 2009.

Wirzenius, L. (ed). Linux System Administrator's Guide, Linux Documentation Project, http://www.tldp.org/LDP/sag/html/index.html.

Wright, C. Linux Security Module Framework, *Proceedings of the Linux Symposium*, pp. 604–610, 2002.

Wrightston, K., and Merino, J., *Introduction to Unix*, CA: Richard D. Irwin, 2003.

Wu, C., Gerlach, J., and Young, C. An Empirical Analysis of Open Source Software Developers' Motivations and Continuance Intentions, *Information and Management*, Vol. 44, pp. 253–262, Amsterdam: Elsevier, 2007.

Ziv, J., and Lempel, A. Compression of Individual Sequences via Variable-Rate Coding, *IEEE Transactions on Information Theory*, Vol. 24, Issue 5, pp. 530–536, CA: IEEE, 1978.

# Appendix: Binary and Boolean Logic

THIS APPENDIX'S LEARNING OBJECTIVES are

- To understand the binary numbering system

- To be able to convert numbers between binary and decimal

- To understand the difference between binary, decimal, octal, and hexadecimal

- To understand storage capacities

- To be able to apply Boolean operators and the netmask

## A.1  BINARY NUMBERING SYSTEM

Our computers process and store information by means of electrical current flowing through digital circuits. Although this detail is not something a typical Linux user needs to know, it does influence a number of aspects of our computer usage. Most significantly, current is in one of two states, high current or low (no) current. Our circuits store and process information that must be representable in one of two states. For convenience, we refer to these states as 1 (on) and 0 (off). Thus, what a computer does is store and process binary information. Binary is the base 2 numbering system.

A numbering system is a means of representing numbers. There are an infinite number of numbering systems available to us, but most of our experience is with the decimal (base 10) numbering system. A numbering system of base $k$ (where $k$ is some integer greater than 0) consists of a range of digits from 0 to $k - 1$ where each digit represents a power of $k$. For instance, if $k$ is 10, then our digits are 0–9 and the digits in a number represent powers of 10 (1, 10, 100, 1000, 10000, and so forth from right to left). The number 362 is really 3 in the "one hundred's column," 6 in the "ten's column," and 2 in the "one's column." For the most part, people only use decimal unless they are dealing with computer programming, computer networks, computer or electrical engineering, mathematics, or perhaps philosophy of logic. In most of these cases, people may use binary, octal (base 8), or hexadecimal (base 16). The reason for octal and hexadecimal usage is that both of these bases allow us to represent more readable binary values.

Using our definition for base k above, binary (base 2) will consist solely of digits 0 and 1. We refer to a single *binary digit* as a bit. The digits represent powers of 2. The powers of 2 are 1, 2, 4, 8, 16, 32, 64, 128, 256, 512, 1024, and so forth. These values are computed by using $2^i$, where i is 0, 1, 2, 3, 4, 5, 6, 7, 8, 9, 10, respectively, for the values in the list. If you do not see the relationship between these numbers, we are multiplying each value by 2 to get the next one in the sequence, or we are doubling the numbers. The above list represents the first 11 powers of 2 starting with $2^0$ and running through $2^{10}$. For the most part, we can limit our understanding of binary to the first 11 powers of 2 for an interesting reason that we will consider later. Table A.1 shows the first 11 powers of 2 and the equivalent binary value as described in this paragraph.

You might notice in the third column of Table A.1 that each successive power of 2 is represented by adding a 0 at the end. That is, 2 is 10 and 4 is 100, which we got by adding a 0 after 10. We are not really "adding a 0." Instead, we are *shifting* the number one bit to the left. A left shift moves all of the digits one bit (location) to the left and the vacated spot is filled in with a 0. So, for instance, 10 is 2, left shifting it gives us 10_ and the gap is filled in with a 0. The left shift then multiplies a number by 2. Similarly, a right shift, which is shifting each bit one position to the right, divides by 2. So, 10000, which is 16, becomes 1000 when right shifted, or 8.

Using digital circuitry, we construct a "cell" to store a single bit. We need to connect several cells together to store a meaningful piece of information as a single 0 or 1 does not tell us much.

The typical size of data storage is measured in either *bytes* or *words*. A byte is 8 bits. Usually, a computer program is unable to inspect the individual bits in a byte so processes will look at entire bytes at a time. The byte stores a range of values from 00000000 to 11111111. These represent the decimal numbers 0 to 255, so there are 256 total combinations of 0s and 1s that can be stored in a byte (we confirm this because $2^8 = 256$).

TABLE A.1    Powers of 2

| i | $2^i$ | Written in Binary |
|---|---|---|
| 0 | 1 | 1 |
| 1 | 2 | 10 |
| 2 | 4 | 100 |
| 3 | 8 | 1000 |
| 4 | 16 | 10000 |
| 5 | 32 | 100000 |
| 6 | 64 | 1000000 |
| 7 | 128 | 10000000 |
| 8 | 256 | 100000000 |
| 9 | 512 | 1000000000 |
| 10 | 1024 | 10000000000 |

Many computers use *word-level addressing* instead of *byte-level addressing*. This means that the smallest unit that the computer operates on is a word. The word size varies computer-by-computer. Older computers had 8-bit or 16-bit word sizes. More recent computers used a 32-bit word size but many computers today have 64-bit word sizes. In Section A.3, we will discuss storage capacities.

How do we represent meaningful information using binary numbers? For instance, what does 00000000 mean if stored in the computer? We want to store the following types of information in our computer.

- Positive (unsigned) integer numbers

- Positive and negative (signed) integer numbers

- Numbers with fractional parts (decimal points)

- Strings of characters

- Program instructions

- Images

- Animated images (video)

- Sounds (audio)

Each of these types of information will require a form of binary representation so that we can store them in the computer.

In Section A.2, we will examine how to store unsigned integer numbers in binary. Although it is interesting to see how signed integer numbers and fractions can be stored in binary, we will omit those details in this text because it is not particularly relevant with respect to Linux. Strings of characters are usually stored as individual characters, each stored in one byte or word. A string would then be stored in consecutive memory locations until we reach an "end of string" marker such as \0, which is used in C/C ++ to denote the end of a string. Each character is stored in binary using a code. The most common codes are ASCII and UNICODE. In a similar way, we use codes to store program instructions. Images are stored using a bitmap or a compressed bitmap. Sounds use their own formatted file types. Again, while these types of storage are interesting, they are beyond the scope of this text. So, in the next section, we will concentrate on unsigned integer numbers.

## A.2 CONVERTING UNSIGNED NUMBERS

Let us assume we are going to store a number in a single byte. As stated in the previous section, a byte can store any combination of eight 1s and 0s such as 00000000, 00001110, 01010101, 11110000, 11111100, and 11111111. So, how do these 1s and 0s represent numbers that are meaningful to us? Let us reexamine Table A.1. The value $2^0$ is 1, the value $2^1$ is 2, and the value $2^2$ is 4. In binary, we indicate these values as 1s in the 0th, 1st, or 2nd column of the binary number, counting from right to left. The number 100 (in binary) is $2^2$ or 4.

Rather than writing "in binary," we will note binary numbers by ending them with a 2 subscript as in $100_2$ (as opposed to 100, which would be one hundred). If $100_2$ is 4 and $10_2$ is 2, then combining these two gives us $110_2$, which is $4 + 2 = 6$. This leads us to an easy way of converting numbers from binary to decimal.

We will use a simple formula for the conversion. In this case, we are assuming that our binary number is 1 byte long (consists of only 8 bits). The formula is

$$\text{Value in decimal} = \sum_{i=0}^{7} x_i * 2^i$$

This formula says to sum the products of $x_i*2^i$ for each i from 0 to 7. What is $x_i$ and what is $2^i$? The value $x_i$ is the ith bit of the number, where $x_7$ is the leftmost bit and $x_0$ is the rightmost bit. For instance, the number $11001010_2$ will have $x_7 = 1$, $x_6 = 1$, $x_5 = 0$, $x_4 = 0$, $x_3 = 1$, $x_2 = 0$, $x_1 = 1$, $x_0 = 0$. The value $2^i$ is the value 2 raised to the power i, or the ith value in the sequence of the powers of 2. If you are not familiar with powers of 2, you can get the values from Table A.1.

Now that we have explained the formula, let us apply it and convert $11001010_2$ to decimal.

$$11001010_2 = 1*2^7 + 1*2^6 + 0*2^5 + 0*2^4 + 1*2^3 + 0*2^2 + 1*2^1 + 0*2^0$$
$$= 1*128 + 1*64 + 0*32 + 0*16 + 1*8 + 0*4 + 1*2 + 0*1$$
$$= 128 + 64 + 8 + 2 = 202$$

We can simplify this formula by noting that 0*a number = 0. Thus, we can eliminate any of the 0 digits. Redoing our conversion, we have $11001010_2 = 1*2^7 + 1*2^6 + 1*2^3 + 1*2^1 = 202$. We could even eliminate the 1 bits because 1*a number is that number. In this case, we see that $1*2^7$ is $2^7$, $1*2^6$ is $2^6$, and so on. So, we perform our conversion by adding powers of 2 for the columns in the binary number that contain a 1. This gives us $11001010_2 = 128 + 64 + 8 + 2 = 202$.

Let us try a few more:

- $00011101_2 = 0*2^7 + 0*2^6 + 0*2^5 + 1*2^4 + 1*2^3 + 1*2^2 + 0*2^1 + 1*2^0 = 16 + 8 + 4 + 1 = 29$

- $00001111_2 = 8 + 4 + 2 + 1 = 15$

- $10101010_2 = 128 + 32 + 8 + 2 = 170$

- $11111111_2 = 128 + 64 + 32 + 16 + 8 + 4 + 2 + 1 = 255$

- $00000011_2 = 2 + 1 = 3$

If we are going to convert values larger than 8 bits, we have to extend our formula by increasing the upper limit for i. If the value is a 16-bit number, we sum up $x_i*2^i$ for i between 0 and 15. Unfortunately, Table A.1 does not extend beyond i = 10 but you should have little difficulty extending the table yourself. For instance, for i = 11, $2^i = 2*1024 = 2048$. Can you

TABLE A.2    Large Powers of 2

| i | $2^i$ | Abbreviation and Meaning |
|---|---|---|
| 10 | 1024 | 1K (kilo), approximately 1 thousand |
| 20 | 1048576 | 1M (meg or mega), approximately 1 million |
| 30 | 1073741824 | 1G (gig or giga), approximately 1 billion |
| 40 | 1099511627776 | 1T (tera), approximately 1 trillion |
| 50 | 1125899906842624 | 1P (peta), approximately 1 quadrillion |
| 60 | 1152921504606846976 | 1E (exa), approximately 1 quintillion |

figure out $2^i$ for i = 12, 13, 14, and 15? This question is given as a review problem at the end of the appendix.

Another way to cope with computing a power of 2 where i is greater than 10 is to use some algebra. In algebra, we learn that $2^{x+y} = 2^x*2^y$. This looks horribly abstract, how is it useful? Let us look at an example. We want to determine what $2^{15}$ is. Since 15 = 5 + 10, $2^{15} = 2^5*2^{10}$. How does this help us? Well, $2^5$ and $2^{10}$ are in Table A.1. Looking these up, we see that the two values are 32 and 1024, respectively. So, $2^{15} = 32*1024$.

Although 32*1024 looks like a challenge to compute, it really is not once we realize that 1024 is equal to 1K. So $2^{15} = 32*1K = 32K$. Now that we know this handy little algebraic formula, we can put it to use with one more piece of useful information. It turns out that all of the powers of 2 where i is a multiple of 10 (10, 20, 30, 40, etc.) have handy abbreviations. We see this in Table A.2.

Given Tables A.1 and A.2, we should be able to determine any power of two up through $2^{69}$. Let us see a few examples.

- $2^{19} = 2^9*2^{10} = 512*1K = 512K$ (roughly 512 thousand)

- $2^{22} = 2^2*2^{20} = 4*1M = 4M$ (roughly 4 million)

- $2^{36} = 2^6*2^{30} = 64*1G = 64G$ (roughly 64 billion)

- $2^{58} = 2^8*2^{50} = 256*1P = 256P$ (over 256 quadrillion)

Let us put this together and convert a 16-bit binary number into decimal. We will use the number $0101010101010101_2$. Ignoring the 0s in the binary number, we have $2^{14} + 2^{12} + 2^{10} + 2^8 + 2^6 + 2^4 + 2^2 + 2^0 = 2^4*2^{10} + 2^2*2^{10} + 2^{10} + 2^8 + 2^6 + 2^4 + 2^2 + 2^0 = 16*1024 + 4*1024 + 1024 + 256 + 64 + 16 + 4 + 1 = 16384 + 4096 + 1024 + 256 + 64 + 16 + 4 + 1 = 21845$.

How about converting from decimal into binary? There are two different algorithms used for this conversion. One uses division by 2 and one uses subtraction. People will favor one or the other based on whichever is easier for them to perform. The division approach is the more common version.

The first algorithm requires dividing the original decimal number by 2 and recording both the quotient and remainder. You repeat this on the quotient from the most recent division until the quotient is 0. The binary number is then the remainder in the opposite order of what was recorded.

TABLE A.3    Example Converting 81 to Binary

| Number | Quotient | Remainder | Explanation |
|---|---|---|---|
| 81 | 40 | 1 | 81/2 = 40 and 1/2 |
| 40 | 20 | 0 | 40/2 = 20 and 0/2 |
| 20 | 10 | 0 | 20/2 = 10 and 0/2 |
| 10 | 5 | 0 | 10/2 = 5 and 0/2 |
| 5 | 2 | 1 | 5/2 = 2 and 1/2 |
| 2 | 1 | 0 | 2/2 = 1 and 0/2 |
| 1 | 0 | 1 | 1/2 = 0 and 1/2 |

Let us try an example. We will convert 81 from decimal to binary. The steps are shown in Table A.3. Here, we see 81 is divided by 2 yielding a quotient of 40 and a remainder of 1 (81/2 = 40 and 1/2, the remainder then is 1). We record both values. Now we divide the quotient, 40, by 2, which gives us a quotient of 20 and a remainder of 0 (40/2 = 20 and 0/2, so there is no remainder). 20/2 = 10 and 0/2 (quotient of 10, remainder of 0). 10/2 = 5 and 0/2 (quotient of 5, remainder of 0). 5/2 = 2 and 1/2 (quotient of 2, remainder of 1). 2/2 = 1 and 0/2 (quotient of 1, remainder of 0). Finally, 1/2 = 0 and 1/2 (quotient of 0, remainder of 1). Our binary equivalent to 81 then is the collection of remainders from the end of the process to the beginning, or $1010001_2$.

Note in our example from Table A.3 our result is a 7-bit number, $1010001_2$. If we wanted to store this in 8 bits, we would add a leading 0 giving us $01010001_2$. Given our binary result, we can confirm that it is correct by converting it back to decimal: $01010001_2 = 64 + 16 + 1 = 81$.

The other approach to convert decimal to binary is in essence the process of converting binary to decimal in reverse. Rather than adding up the powers of 2 indicated by the 1 bits in the binary number, we must determine which powers of 2 make up the decimal number and record those bits as 1s.

For instance, 81 is the result of $64 + 16 + 1 = 0*2^7 + 1*2^6 + 0*2^5 + 1*2^4 + 0*2^3 + 0*2^2 + 0*2^1 + 1*2^0 = 01010001_2$. To solve the problem through subtraction, we determine the largest power of 2 that we can subtract from the original number. With 8 bit numbers, the largest power of 2 will be 128. Let us practice a few, starting with 219.

- Largest power of 2 less than or equal to 219 is 128

  - 219 − 128 = 91

- Largest power of 2 less than or equal to 91 is 64

  - 91 − 64 = 27

- Largest power of 2 less than or equal to 27 is 16

  - 27 − 16 = 11

- Largest power of 2 less than or equal to 11 is 8

  - 11 − 8 = 3

- Largest power of 2 less than or equal to 3 is 2

    - $3 - 2 = 1$

- Largest power of 2 less than or equal to 1 is 1

    - $1 - 1 = 0$

- Done

So we see that 219 is made up of 128, 64, 16, 8, 2, 1, or $219 = 128 + 64 + 16 + 8 + 2 + 1 = 2^7 + 2^6 + 2^4 + 2^3 + 2^1 + 2^0$, so our binary equivalent has 1s in the 7th, 6th, 4th, 3rd, 1st, and 0th columns, or $11011011_2$.

- Convert 104 to binary: $104 = 64 + 32 + 8$, or $104 = 2^6 + 2^5 + 2^3 = 01101000_2$.

- Convert 60 to binary: $60 = 32 + 16 + 8 + 4$, or $60 = 2^5 + 2^4 + 2^3 + 2^2 = 00111100_2$.

- Convert 255 to binary: $255 = 128 + 64 + 32 + 16 + 8 + 4 + 2 + 1 = 2^7 + 2^6 + 2^5 + 2^4 + 2^3 + 2^2 + 2^1 + 2^0 = 11111111_2$.

Notice that 255 is the largest value that we can store in 8 bits. This should make sense given that $2^8 = 256$. This means that we can store 256 different values in 8 bits. The smallest value is $00000000_2$ or 0, and the largest value is $11111111_2$ or 255. The range 0 to 255 consists of 256 numbers.

What if we want to store a number larger than 255? If we only have 8 bits, we cannot do so. This is known as an *overflow*. Alternatively, we can use more memory by placing the value into multiple bytes. The word size of the computer is the common size for data storage. Today, most computers have word sizes of 32 or 64 bits. This would allow us to store $2^{32}$ or $2^{64}$ different values in a word. $2^{32}$ is $2^{2}*2^{30} = 4*1G = 4G$ (roughly 4 billion). Can you figure out what $2^{64}$ is?

Let us assume we can store a number in 16 bits. Try to convert each of the following from binary to decimal or decimal to binary. The answers are given; see if you can properly compute them.

- $0000111100001111_2$ to decimal = 3855

- 56313 to binary = 1101101111111001

- $1010101010101010_2$ to decimal = 43690

- 9999 to binary = 0010011100001111

- $0101111101010000_2$ to decimal = 24400

- 44044 to binary = 1010110000001100

As a Linux user, you may never need to understand binary. However, network addresses consist of binary numbers. Your IP address (version 4) is a collection of 4 bytes, each byte known as an *octet*. The octets are usually presented to you in decimal. But you may be required to know how to convert between binary and decimal.

In addition, netmasks are applied on the binary versions of the IP addresses numbers. We explore netmasks in Section A.5. It should be noted that IPv6 addresses are 128 bits long and usually displayed using hexadecimal notation. We briefly examine hexadecimal and a related representation, octal, in the next section.

Before we conclude this section, notice that we only considered how to store positive integers in binary. Recall from earlier that a computer is required to store all kinds of data in binary, including strings of characters, sounds, pictures, program instructions, and nonpositive integer numbers (negative numbers, fractions, and real numbers). We use a representation called two's complement to store signed numbers (numbers that can be either positive or negative) and a variation of scientific notation to store fractional numbers as floating point numbers. We will briefly examine character storage (e.g., letters, punctuation marks) and image storage in Section A.4 of this appendix.

## A.3 HEXADECIMAL AND OCTAL REPRESENTATIONS

Looking at lengthy strings of binary digits is challenging. A person having to inspect data in this form will probably lose their place and have to concentrate carefully on every bit. For convenience, we can group bits together and represent them using more readable characters. It is easy to convert binary into either the octal (base 8) or hexadecimal (base 16) numbering system. In octal, we have 8 different digits available to us, numbered 0 through 7. You might notice that the numbers 0 through 7 can all be represented using three binary bits ($0_8 = 000_2$, $7_8 = 111_2$). Table A.4 provides the octal and binary values for 0–7.

To convert from binary to octal, all we need to do is group our binary number into groups of three bits and convert each set of bits into its equivalent octal number. We start at the right end of the number and group three bits at a time and convert them. If the leftmost portion of the number consists of fewer than three bits, pad them with leading zeroes.

For instance, the binary number 10111110 would be grouped as 10 111 110 and we would add a leading 0 to give us 010 111 110. Now we substitute the octal equivalent of the three-bit binary number: 010 is 2, 111 is 7, 110 is 6, so $10111110_2 = 276_8$. To convert from octal to binary, just replace each octal digit with the three bit binary equivalent. Depending upon the size of storage available, add leading 0s as needed. For instance, if we want

TABLE A.4  Octal and Binary for the First Seven Values

| Octal | Binary |
| --- | --- |
| 0 | 000 |
| 1 | 001 |
| 2 | 010 |
| 3 | 011 |
| 4 | 100 |
| 5 | 101 |
| 6 | 110 |
| 7 | 111 |

to store $45102_8$ in 16 bits, we would do the following: $4 = 100$, $5 = 101$, $1 = 001$, $0 = 000$, $2 = 010$, so $45102_8 = 100\ 101\ 001\ 000\ 010 = 100101001000010_2$. Since the five octal digits is stored in 15 binary bits, we had to add a leading 0 to give us a total of 16 bits, giving us $0100101001000010_2$.

Let us try a few more for practice.

- Convert $00011100_2$ to octal: 00 011 100, add a leading 0:000 011 100 = 0 3 4 = $034_8$.

- Convert $1111110010110111011_2$ to octal: 1 111 110 010 110 111 011, add two leading 0s: 001 111 110 010 110 111 011 = 1 7 6 3 6 7 3 = $1763673_8$.

- Convert $4154_8$ to 16 bit binary: 100 001 101 100, add 4 leading 0s: $0000100001101100_2$.

- Convert $260_8$ to 8 bit binary: 010 110 000 = $010110000_2$. Notice that our answer is really 9 bits. We drop the leading 0 to give us $10110000_2$. We have to be careful here, had our octal number been $460_8$, we could not store it in 8 bits, this would cause an overflow.

While octal is a useful representation to reduce the quantity of digits when representing data in binary, we could reduce the digits even further using hexadecimal, or base 16. Recall from Section A.1 that a base $k$ numbering system uses digits 0 through $k − 1$. In binary, $k$ is 2 and we use digits 0 and 1. In octal, $k$ is 8 and we use 0 through 7. In hexadecimal, $k$ is 16. This implies that we use 0 through 15.

Our numbering system requires that each digit be expressed in a single column of the number. That is, each digit needs to be specified using a single character. What should we do about 10, 11, 12, 13, 14, and 15 for hexadecimal? We need to represent these multidigit values using single digits. Instead of trying to use a digit, we replace each with a single character. In this case, we will use the first six letters, A, B, C, D, E, F. For instance, the number 11 would be represented as B in hexadecimal and 14 would be represented as E in hexadecimal.

Notice that 16 is twice 8, so there are twice as many possible combinations needed to store a single hexadecimal over that of a single octal digit. By doubling the number of combinations, we need an additional bit of storage. Why? Consider in 3 bits we store values of 000 to 111 (0 to 7, or eight total values). By adding a bit, we now store _000 to _111 where the blank can be a 0 or a 1. So this gives us 0000 to 0111 and 1000 to 1111. Thus, increasing the storage capacity by 1 bit doubles the number of sequences.

The 4-bit binary values are shown in Table A.5 along with their equivalent decimal values and hexadecimal values. In looking at the 4-bit numbers, you might think of them as 0 followed by the 3-bit binary numbers and 1 followed by the 3-bit binary numbers of Table A.4. In this table, note that in place of single digits for the values 10 through 15, we use the letters A through F to represent the last 6 hexadecimal values.

To convert from binary to hexadecimal, use the same approach as converting from binary to octal but group the binary bits into sets of four bits instead of three. Add leading zeroes to the leftmost group as needed. To convert from hexadecimal to binary, convert

TABLE A.5   Decimal, Binary, and Hexadecimal for First 16 Numbers

| Decimal | Binary | Hexadecimal |
|---------|--------|-------------|
| 0 | 0000 | 0 |
| 1 | 0001 | 1 |
| 2 | 0010 | 2 |
| 3 | 0011 | 3 |
| 4 | 0100 | 4 |
| 5 | 0101 | 5 |
| 6 | 0110 | 6 |
| 7 | 0111 | 7 |
| 8 | 1000 | 8 |
| 9 | 1001 | 9 |
| 10 | 1010 | A |
| 11 | 1011 | B |
| 12 | 1100 | C |
| 13 | 1101 | D |
| 14 | 1110 | E |
| 15 | 1111 | F |

each hexadecimal digit to its equivalent 4-bit binary value. Add leading 0s as needed to fill up the storage size. Below we have several examples.

- $10111011111101_2$: 10 1110 1111 1101, add 2 leading 0s = 0010 1110 1111 1101 = 2 E F D = $2EFD_{16}$

- $1011000000010_2$: 1 0110 0000 0010, add 3 leading 0s = 0001 0110 0000 0010 = 16 0 2 = $1602_{16}$

- $A3B4_{16}$ = 1010 0011 1011 0100 = $1010001110110100_2$

- $123F_{16}$ = 0001 0010 0011 1111 = $01001000111111_2$

Four hexadecimal digits convert to exactly 16 binary bits. Converting between hexadecimal and binary is somewhat simpler than between octal and binary because we often will not need to add leading bits.

## A.4  STORAGE CAPACITIES

While we are exploring binary, we should consider the amount of information we can store using a binary representation. The term *storage capacity* conveys the size of the information that can be placed into/onto a storage device. When we hear about gigabytes of main memory or terabytes for disk storage, it sounds impressive. But what do these values really mean?

We will start off at the most primitive unit of storage, the *bit*. One bit stores a single binary digit and as we saw in Section A.1, a binary digit stores either a 0 or a 1. Thus, a bit stores one of two possible values. We can confirm this because in 1 bit, we have $2^1$

combinations, $2^1 = 2$. This is the smallest unit of storage in a computer in that we build a storage device called a memory *cell* capable of storing exactly 1 bit.

As 1 bit cannot store almost anything useful, we group memory cells together to create larger storage locations. As we already discussed, 8 bits make up a *byte*. 8 bits give us $2^8 = 256$ different combinations for storage. In 1 byte, we can store a number from 0 to 255, or a number within the range $-128$ to $+127$, or alternatively one of 256 characters using a character code, or one of 256 different programming instructions.

While the smallest unit of storage is a bit, the smallest *addressable unit* is either a byte or a word. That is, when the CPU requests something from memory, it is at least the size of a byte. We therefore use the term byte to describe storage sizes. Over time, computers were able to move and process more than 1 byte at a time. They began working on 2 or 4 bytes at a time. The size of the standard datum is known as the *word size* of the computer. Early computers had 8-bit and 16-bit word sizes. Modern computers have at least 32-bit word sizes and most today are now 64-bit word sizes (in some cases, there are even 128-bit word sizes). Many computers today are known as *word addressable* in that they no longer operate on individual bytes but instead on words.

Let us consider the amount that can be stored in a word of 32 bits. The number of possible combinations of sequences of 1s and 0s in 32 bits is $2^{32}$. This is 4,294,967,296 telling us that a 32-bit word can store any of over 4 billion values! As an integer number, we would be able to store an unsigned (positive only) number in the range of 0 to $2^{32} - 1 = 4,294,967,295$ (or if signed number in the range $-2,147,483,648$ to $+2,147,483,647$).

What about storing characters? Recall from the end of Section A.1 that we can store characters using ASCII or Unicode. The ASCII character set uses 7 bits out of a byte. With 7 bits, we can store any of $2^7 = 128$ characters. This is more than enough different combinations of bits to represent the English alphabet (differentiating between uppercase and lowercase letters) to include the 10 digits, all of the punctuation marks on the keyboard as well as white space characters (space, tab, enter). There are many leftover combinations available to also represent special characters like control characters. If ASCII uses 7 bits of a byte, what happens to the 8th bit? It is either ignored so that we store the value 0, or we use it for *parity*. We will discuss parity in Section A.5.

With the worldwide popularity of computers and the Internet, we need to also be able to represent the characters found in other languages. There is not enough room for this in ASCII, so Unicode was developed. Unicode stores a character in 16 bits instead of 7. With 16 bits, there are $2^{16} = 65,336$ combinations of bits, or the ability to store well over 65,000 characters. The first 128 characters of Unicode are the same as the 128 characters of ASCII so that our textfiles that use ASCII do not have to be altered to be presentable to software that is using Unicode.

Let us consider a sequence of text such as the following sentence:

The small, brown mouse ran on 4-ever in his cage to catch up with the cheese!

How much storage space would the sentence require? Counting the number of characters informs us that there are 58 letters, 1 digit, and 3 punctuation marks. There are also

TABLE A.6   Equivalent Text for Storage Sizes in Bytes

| Storage in Bytes | Abbreviation | Equivalent Text |
| --- | --- | --- |
| 1 | 1 byte | 1 character |
| 1024 ($2^{10}$) | 1 kilobyte | 1000 characters, approximately 200 words in English which is approximately equal to 1 page of text |
| 1 million ($2^{20}$) | 1 megabyte | 1 million characters or approximately 1000 pages of text, or between 1 and 5 books of text |
| 1 billion ($2^{30}$) | 1 gigabyte | 1 billion characters would represent a few thousand books, a library |
| 1 trillion ($2^{40}$) | 1 terabyte | A thousand libraries! |

15 spaces. Thus, to store the sentence would take $58 + 1 + 3 + 15 = 77$ total characters. This would take 77 bytes in ASCII and 154 bytes in Unicode (because each character is stored in 2 bytes). If we include the end-of-line characters (often denoted as \n), we have 78 characters requiring 78 bytes (ASCII) or 156 bytes (Unicode).

What about storing thousands or millions of characters? It is useful to extend our vocabulary of sizes to express these larger capacities. Table A.6 illustrates for several different orders of magnitude of storage capacity the amount of text that we can store.

A terabyte of disk storage is readily available today. So we could easily, on our own home computers, store text equivalent to millions of books. What if the books have pictures? Image storage is not nearly as concise as text storage. To store an image, we usually use a representation called a *bitmap*.

The bitmap must encapsulate every point of the image by denoting its color, or if a black-and-white image, its shade of gray. Alternatively, we can describe its hue and intensity. We will call each point a *pixel* (short for picture element). If we assume a single bit for each pixel, then we can store whether that pixel is on (white) or off (black) to store a simple black-and-white image. If the image consists of 100 rows and 100 columns, then we need 10,000 bits of storage. In bytes, we divide the bits by 8, or 10,000 bits/8 bits per byte = 1250 bytes, which is slightly more than 1 kilobyte.

Most images will be larger than $100 \times 100$, for instance, $1000 \times 1000$ instead. We will also desire color images over black and white. To store a color image, we will represent each pixel by the amount of red, green, and blue (RGB) that make up that point. RGB storage is a very common way to represent images. We will use 1 byte to store the amount of each of the red, green, and blue of the pixel, or 3 bytes per pixel. If we assume a $1000 \times 1000$ image using RGB, we would need 3 bytes per pixel*1000*1000 pixels or 3 million bytes, roughly 3 megabytes.

There are compression techniques to reduce the storage size and so rather than storing raw bitmaps, we use compression to store the file in one of several other formats such as gif, jpg, or png. These compression techniques can reduce the image file's size to as little as about 1/3 its original size. Instead of 3 megabytes, a common image size is around 1 megabyte. Notice that 1 megabyte could store 1 million characters, or about 200,000 English words. Is a picture really worth only a thousand words?

If our average book of 250 pages can be stored as text in 250 kilobytes, what happens when we insert pictures into the text? Let us assume that each picture is stored in 1 megabyte and that we have a picture on every other page. This gives us 125 images or

125 megabytes of storage. Now instead of 250 kilobytes, we need 500 times the storage space! That is the cost of images.

What about video? To create video, we take a series of still images (pictures) and display them rapidly where each individual image will be nearly identical to the image that came before it. The number of images captured per second is known as the *scan rate*. The higher the scan rate, the higher the quality of the overall video. The human eye is able to detect the flicker in video if the frame rate is below a certain threshold (about 12 frames or images per second), so we typically want a scan rate higher than this. Frame rates of at least 24 per second are desired and high-quality film has as many as 72 per second.

Let us assume we use a frame rate of 24 images per second and each image is $1000 \times 1000$ in color using a bitmap. This means 3 megabytes per frame, or $3*24 = 72$ megabytes per second. If the video lasts for 1 minute, we require $60*72$ megabytes = 4320 megabytes or over 4 gigabytes. Now we are talking about some serious storage sizes.

If you have a computer that stores 8 gigabytes of memory, we could only store up to 2 minutes of video in memory. Alternatively, we could fill a 1 terabyte hard disk with just about 4 hours' worth of video. We need to use our disk storage space more wisely, particularly when we want to move these video files over the Internet.

A number of compression techniques are applied to video storage. For instance, by storing the first video frame wholly and then for successive frames, only encode the differences. For instance, if frame 1 and frame 2 are identical except for three pixels, then frame 1 is stored in its entirety but frame 2 is denoted by the three changed pixels. In this way, while we need a lot of storage for frame 1, we need almost nothing for frame 2.

There are key frames in the video that take place when we are changing the image entirely (e.g., a different camera angle or a different scene) and so these key frames must also be stored wholly. With proper compression, the typical 45-min video can be stored in approximately 350 megabytes.

Without compression, 350 megabytes would only store about 5 seconds' worth of video! Let us now compare the storage sizes from Table A.6 with compressed images or compressed video. See Table A.7.

We wrap up this section by considering the typical amounts of storage capacity for memory and storage devices. We start with the memory hierarchy (see Chapter 1, Section 1.8).

- On-chip cache: 16–32 kilobytes for most computers, less or none for handheld devices.

- Off-chip cache: 1–2 megabytes for most desktop and laptop computers, more for larger computers and less or none for handheld devices.

TABLE A.7    Sizes for Image and Video

| Storage in Bytes | Abbreviation | Equivalent Compressed Image/Video |
| --- | --- | --- |
| 1 | 1 byte | Nothing |
| 1024 ($2^{10}$) | 1 kilobyte | A $32 \times 32$ color image |
| 1 million ($2^{20}$) | 1 megabyte | A $1000 \times 1000$ color image |
| 1 billion ($2^{30}$) | 1 gigabyte | 2½ hours of compressed video, or ¼ of a movie on DVD |
| 1 trillion ($2^{40}$) | 1 terabyte | 250–1000 movies (depending on the quality) |

- DRAM (main memory): usually 8 gigabytes but as much as 32 gigabytes for desktop and laptop computers, 1 terabyte or more for mainframe and supercomputers, under 4 gigabytes for handheld devices.

The storage devices consist of hard disk drives, flash drives, optical disks, and magnetic tape. In order of smallest to largest capacity, we have the following:

- Flash memory: 1–256 gigabytes, usually in the range of 2–8 gigabytes.

- Optical disk: 750 megabytes up to 8 gigabytes for DVD and up to 128 gigabytes for Blu Ray.

- Hard disk: internal hard disk drives typically are between 500 gigabytes and 1 terabyte, external hard disk drives can have up to 3 terabytes.

- Magnetic tape: up to 5 terabytes.

## A.5 BOOLEAN LOGIC*

The computer operates by applying Boolean operations on bits. The Boolean operators are AND, OR, NOT, and XOR. We do not typically think at such a low level and instead view data using more abstract operations such as addition, multiplication, and comparison. Yet it is still worth learning the Boolean operators as you will need to understand logic when writing your own code (see, for instance, the use of &&, ||, and ! from Chapter 7). You also need to understand the use of AND to apply a netmask (covered in Chapter 12). So here we look at these four Boolean operators.

We apply AND, OR, and XOR on two bits. The NOT operator is applied to a single bit. Before we go through some examples, let us formally define the four operators.

- AND—outputs 1 if both of the two inputs are 1, otherwise outputs 0.

- OR—outputs 1 if either of the two inputs are 1, otherwise outputs 0.

- XOR—outputs 1 if the two inputs differ, otherwise outputs 0.

- NOT—outputs 1 if the input is 0, otherwise outputs 0. We could also define NOT as flipping the bit (0 becomes 1, 1 becomes 0).

Table A.8 provides the truth tables for these four operators. The truth table shows all possible combinations of inputs and their outputs. With two bits, for instance, we might have 1 OR 0 = 1 or 1 AND 0 = 0. NOT only applies to one bit, for instance, NOT 1 = 0.

As we discussed in Section A.1, having single bit storage is not very useful, so we group storage cells together. Similarly, performing Boolean operations on single bits is also not

---

* In this section, we omit the subscript 2 after any binary number. Assume all numbers in this section are written in binary.

TABLE A.8    Truth Table for Boolean Operators

| X | Y | X AND Y | X OR Y | X XOR Y | X | NOT X |
|---|---|---------|--------|---------|---|-------|
| 0 | 0 | 0 | 0 | 0 | 0 | 1 |
| 0 | 1 | 0 | 1 | 1 | 1 | 0 |
| 1 | 0 | 0 | 1 | 1 | | |
| 1 | 1 | 1 | 1 | 0 | | |

particularly useful. Our Boolean operators are applied pair-wise so that given two binary values we apply the operator to each corresponding pair of bits of the two numbers. Thus, we apply AND on the leftmost two bits of the numbers, then the two bits of the next column and so forth. This is shown in Figure A.1 where we apply AND on the two bytes 00111111 and 01010101.

What follows are the pair-wise application of OR and XOR on the same two binary numbers.

```
        0 0 1 1 1 1 1 1              0 0 1 1 1 1 1 1
OR      0 1 0 1 0 1 0 1      XOR     0 1 0 1 0 1 0 1
        0 1 1 1 1 1 1 1              0 1 1 0 1 0 1 0
```

The NOT operator only applies on a single binary number. This operation can be thought of as flipping each bit (0 becomes 1, 1 becomes 0). For instance, NOT 00111111 = 11000000 and NOT 01011010 = 10100101.

Let us put this into practice. In Chapter 12, we learn about the *netmask*. Given a computer's IP address and its netmask, we can determine the network address for the computer. The IP address encapsulates two pieces of information, the computer's network address and the computer's address within that network. Assume our computer's IP address is 10.11.12.13. If we have a netmask of 255.255.240.0, we can apply AND to obtain the network's computer address.

First, we must convert each of these octets from decimal to binary. Then, we apply AND. The resulting binary number can be converted back to decimal to obtain the network address.

$$255.255.240.0 = 11111111.11111111.11110000.00000000$$
$$10.11.12.13 = 00001010.00001011.00001100.00001101$$

ANDing the two numbers = 00001010.00001011.00000000.00000000 = 10.11.0.0. This tells us our network address is 10.11.0.0.

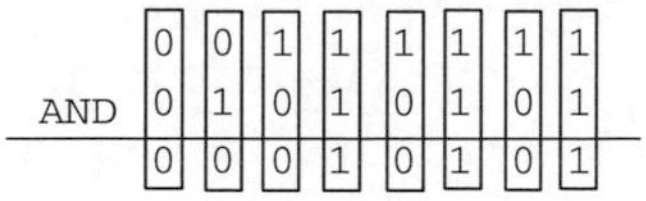

```
        0 0 1 1 1 1 1 1
AND     0 1 0 1 0 1 0 1
        0 0 0 1 0 1 0 1
```

FIGURE A.1    Demonstration of pair-wise AND operation.

If you think that the network address then is merely the first two octets, you may not be correct. Using the same netmask on this IP address 128.58.221.39 gives us a very different network address:

$$255.255.240.0 = 11111111.11111111.11110000.00000000$$
$$128.58.221.39 = 10000000.00111010.11011101.00100111$$

ANDing these two numbers = 10000000.00111010.11010000.00000000 = 128.58.208.0.

To obtain the computer's address, we can NOT the netmask and AND the result to the IP address. NOT 255.255.240.0 = NOT 11111111.11111111.11110000.00000000 = 00000000 .00000000.00001111.11111111.

$$00000000.00000000.00001111.11111111$$
$$\underline{AND \quad 00001010.00001011.00001100.00001101}$$
$$00000000.00000000.00001100.00001101 = 0.0.12.13$$

$$00000000.00000000.00001111.11111111$$
$$\underline{AND \quad 10000000.00111010.11011101.00100111}$$
$$00000000.00000000.00001101.00100111 = 0.0.13.39$$

If you are not dealing with either programming or network addressing, you may never need to apply Boolean logic.

We mentioned the use of parity earlier with respect to ASCII. Parity is a simple means of performing error checking. As bits of data move around the computer, through storage, and across computer networks, it is possible that some data become corrupt. For instance, imagine that we are storing 11110110 and send it from component A to component B in the computer. It arrives as 11010110. This could lead to an incorrectly executed program, a run-time error or worse. We can add a simple error checking mechanism to discover if such a byte contains an error. We do this by adding a parity bit to each byte. The parity bit encodes whether the byte has even or odd parity. In even parity, the total number of 1 bits in a value is even. In odd parity, the total number of 1 bits in a value is odd.

Let us assume we want to use even parity. Since the value 11110110 has 6 1-bits and 2 0-bits, it has even parity. To add a parity bit to this byte, we need to maintain even parity. We do this by adding a parity bit of 0. Now, the parity bit plus byte is 0 11110110, which still has even parity. Upon receiving this information, if the component receives 0 11010110, there is obviously some error because this parity bit plus byte has odd parity (5 1-bits).

In ASCII, we can add the parity bit to the byte itself to make use of the 8th bit. Alternatively, we can compute and tack on the parity bit whenever we want to move a byte of information around the computer. If our computer implements even parity, then an error is known to have arisen whenever we have an odd number of 1 bits when we combine the parity bit and the byte. Some computers use odd parity instead of even parity, in which case the number of 1 bits should always be an odd number.

Note that with the parity bit, we can detect an error but we cannot resolve what bit is wrong. It might be the case that one of the 8 bits in the byte is wrong, but it could also be

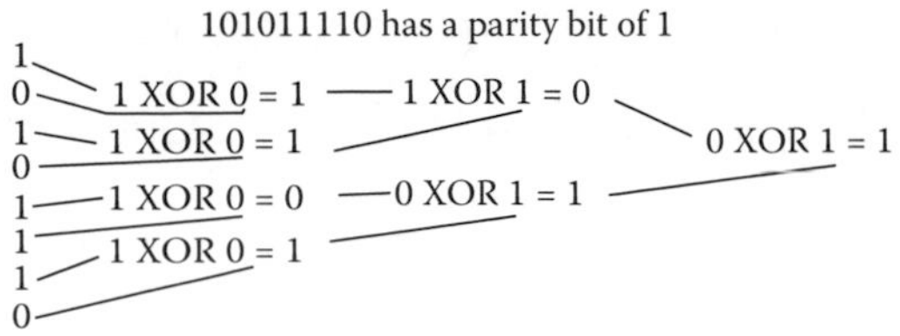

101011110 has a parity bit of 1

FIGURE A.2    Computing the parity bit using XOR operations.

the parity bit that is wrong. For instance, if we have the byte 10101110, we would add a parity bit of 1. If some component receives 10101110 and a parity bit of 0, it is actually the parity bit that is erroneous, but we cannot determine this from the 9 bits. The solution is to just ask the sending component to resend the 9 bits. What if two errors arise in the same 9 bits? The single parity bit cannot resolve this problem. But it is very unlikely to have a single error let alone two so we do not concern ourselves with this possibility.

Parity is computed using XOR. Given a byte, XOR the 8 bits together. You might remember that XOR only operates on two bits. Thus, we have to cascade the XOR operations together. Returning to our previous example of 10101110, we XOR each pair of bits as shown in Figure A.2.

We see in the figure that each pair of bits are XORed together. The first two are 1 and 0, or 1 XOR 0 = 1. This is the same for the next two bits (and the last two bits). The fifth and sixth bits are 1 and 1 giving us 1 XOR 1 = 0. Now, we take the four resulting bits and XOR them in pairs, or 1 XOR 1 = 0 and 0 XOR 1 = 1. Finally, we XOR these resulting bits giving us 0 XOR 1 = 1. Thus, the parity bit for 10101110 is 1. We confirm this by counting the total number of 1 bits in the byte plus parity bit, 6, an even number. To compute odd parity, use the exact same process but NOT the result.

Parity is also used in RAID technology to compute redundancy information. We briefly discuss RAID technology in Chapter 10 but go over it in more detail in Chapter 14 where we return to this idea of parity computations.

## A.6  CHAPTER REVIEW

Concepts introduced in the appendix:

- ASCII—common character representation using 7 bits. Superseded by Unicode.

- Binary—numbering representation using two digits, 0 and 1. Nearly all computers store information in binary.

- Bit—a single binary digit storing a 0 or a 1.

- Bitmap—a common representation for images where each pixel is individually represented.

- Boolean operator—one of four operations applied to binary bits, AND, OR, NOT, XOR. Computer hardware operates using these operators.

- Byte–8 bits combined together for a larger unit of storage.

- Decimal—the base 10 numbering system, used by people.

- Frame rate—the number of images displayed per second in video. A sufficiently high rate must be used to trick the human eye into thinking the images are moving.

- G—abbreviation for a giga, $2^{30}$, or about 1 billion.

- Hexadecimal—the base 16 numbering system, primarily used to combine binary bits for easier examination. IP version 6 addresses are stored in hexadecimal.

- K—abbreviation for a kilo, $2^{10}$, or about 1 thousand (1024 precisely).

- M—abbreviation for a mega (or meg), $2^{20}$, or about 1 million.

- Netmask—a 32-bit number (4 octets) that, when ANDed with an IP address, provides the computer's network address.

- Octal—the base 8 numbering system, sometimes used to simplify binary numbers.

- Parity—a means of determining an error in the transmission of binary data by determining whether the number of 1 bits in a number is even or odd.

- Pixel—a single point or dot of an image stored by computer.

- Powers of 2—the values of 2 raised to an integer value such as $2^0 = 1$, $2^1 = 2$, $2^2 = 4$, $2^3 = 8$, and $2^4 = 16$.

- T—abbreviation for a tera, $2^{40}$, or about 1 trillion.

- Unicode—expansion of the ASCII character representation to 16 bits so that we can encode characters of many languages and other useful symbols.

- Word—the typical data storage size for a computer. Modern computers have a 32- or 64-bit word size.

## REVIEW QUESTIONS

1. Compute $2^{25}$.

2. Compute $2^{37}$.

3. Compute $2^{51}$.

4. What range of numbers, starting at 0, can be stored in 8 bits? In 10 bits? In 14 bits? In 24 bits?

5. Extend Table A.1 by computing $2^{12}$ through $2^{16}$.

For questions 6–17, convert the given binary numbers to decimal. The $_2$ subscripts are omitted.

6. 00000111

7. 01001001

8. 11110000

9. 10101010

10. 00110011

11. 10111101

12. 0000000000110011

13. 0000000100000000

14. 0011001100110011

15. 1001001010000101

16. 1111111100000000

17. 1111111111111111

For questions 18–24, convert the given decimal values to 8-bit binary values. Use the division approach.

18. 45

19. 61

20. 99

21. 115

22. 180

23. 200

24. 241

For questions 25–31, convert the given decimal values to 8-bit binary values. Use the subtraction approach.

25. 40

26. 51

27. 89

28. 121

29. 194

30. 221

31. 250

For questions 32–36, convert the given decimal values to 16-bit binary values. Use the division approach.

32. 4985

33. 6613

34. 12345

35. 23456

36. 34567

For questions 37–41, convert the given decimal values to 16-bit binary values. Use the subtraction approach.

37. 2000

38. 7811

39. 13343

40. 33333

41. 60000

For questions 42–48, convert the given binary number to octal and to hexadecimal.

42. 11110000

43. 10101010

44. 11000101

45. 1110011110

46. 10000101001011

47. 1001110101011110

48. 1111101111010100

49. Assume a book comprises 500 pages of text where each page contains 1000 characters. If we were to store this book using Unicode, approximately how large would the file's size be?

50. We encode the color of an RGB pixel using 3 bytes, one for each of the amount of red, the amount of green, and the amount of blue. This gives us the ability to store 16,777,216 colors for each pixel. Explain where the number 16,777,216 came from.

We often encode the RGB values using hexadecimal notation. For instance, full Red, no Green, and full Blue would be FF00FF where FF is the hexadecimal equivalent to the decimal value 255. This would be the color purple (all red and all blue and no green). Given this description, explain what color each of the following hexadecimal values represents for questions 51–56.

51. 00FF00

52. 000000

53. 00FFFF

54. 0000FF

55. 8888FF

56. FFFFFF

57. If our IP address is 153.47.138.201 and our netmask is 255.255.240.0, what is our network address?

58. If our IP address is 153.47.138.201 and our netmask is 255.224.0.0, what is our network address?

59. If our IP address is 192.14.243.1 and our netmask is 255.255.240.0, what is our computer's address on the network?

60. If our IP address is 192.14.243.1 and our netmask is 255.255.192.0, what is our network address?

For questions 61–69, compute each of given Boolean operations.

61. 11110000 AND 11001100

62. 11110000 OR 11001100

63. 11110000 XOR 11001100

64. NOT 00110000

65. 11000011 AND 10101010

66. 11000011 OR 10101010

67. 11000011 XOR 10101010

68. (NOT 11110000) AND (NOT 11001100)

69. (NOT 11000011) XOR 10101111

For questions 70–73, compute the parity bit assuming even parity.

70. 10101010

71. 11110001

72. 00000000

73. 01011011

For questions 74–77, state whether there is an error in the parity bit + byte assuming even parity.

74. 0, 00001111

75. 1, 00111101

76. 1, 01010110

77. 0, 00000000

# Index*

---

* Items in Courier New font are command, directory, file, script or program names. Items ending in / are directory
  names.